AF576051

Perspectives on Bacterial Flagellar Motor

Perspectives on Bacterial Flagellar Motor

Editors

Tohru Minamino
Keiichi Namba

MDPI • Basel • Beijing • Wuhan • Barcelona • Belgrade • Manchester • Tokyo • Cluj • Tianjin

Editors
Tohru Minamino
Osaka University
Japan

Keiichi Namba
Osaka University
Japan

Editorial Office
MDPI
St. Alban-Anlage 66
4052 Basel, Switzerland

This is a reprint of articles from the Special Issue published online in the open access journal *Biomolecules* (ISSN 2218-273X) (available at: https://www.mdpi.com/journal/biomolecules/special_issues/Flagellar_Motor).

For citation purposes, cite each article independently as indicated on the article page online and as indicated below:

LastName, A.A.; LastName, B.B.; LastName, C.C. Article Title. *Journal Name* **Year**, *Volume Number*, Page Range.

ISBN 978-3-0365-1338-6 (Hbk)
ISBN 978-3-0365-1337-9 (PDF)

Cover image courtesy of Fumiaki Makino and Tomoko Yagamuchi

Contents

About the Editors

Tohru Minamino received his Ph.D. degree in Molecular Microbiology from Hiroshima University in 1997. He then worked on the flagellar type III protein export system as a postdoctoral associate for three years in the laboratory of the late Robert M. Macnab at Yale University. He then joined ERATO Protonic NanoMachine Project funded by the Japan Science and Technology Agency (JST) as Research Staff (1999–2002) and became Group Leader of ICORP Dynamic NanoMachine Project also funded by JST (2002–2008), which were both directed by Keiichi Namba. He has been working as Associate Professor in the Graduate School of Frontier Biosciences at Osaka University. His main research interests include self-assembly, protein export and energy conversion of biological nanomachine.

Keiichi Namba received his Ph.D in Biophysical Engineering from Osaka University in 1980. He spent five years in the laboratories of Donald L. D. Caspar at Brandeis University and Gerald Stubbs at Vanderbilt University, and then became Group Leader of ERATO Molecular Dynamic Assembly Project funded by JRDC, Research Director of Panasonic IIAR (1992–1999) and Research Director of Panasonic Advanced Research Laboratories (1999-2002). He directed the ERATO Protonic NanoMachine Project (1997–2002) and ICORP Dynamic NanoMachine Project (2002–2008) as Project Director. He was Professor of the Graduate School of Frontier Biosciences at Osaka University (2002–2017) and has been Specially Appointed Professor of Osaka University since 2017. He has also been Deputy Director of the RIKEN SPring-8 Center and Team Leader of the RIKEN Center for Biosystems Dynamics Research since 2018. His primary research interests include structures and functions of biological macromolecular assemblies as a nanomachine.

Editorial

Recent Advances in the Bacterial Flagellar Motor Study

Tohru Minamino [1,*] and Keiichi Namba [1,2,3,*]

1 Graduate School of Frontier Biosciences, Osaka University, 1-3 Yamadaoka, Suita, Osaka 565-0871, Japan
2 RIKEN SPring-8 Center and Center for Biosystems Dynamics Research, 1-3 Yamadaoka, Suita, Osaka 565-0871, Japan
3 JEOL YOKOGUSHI Research Alliance Laboratories, Osaka University, 1-3 Yamadaoka, Suita, Osaka 565-0871, Japan
* Correspondence: tohru@fbs.osaka-u.ac.jp (T.M.); keiichi@fbs.osaka-u.ac.jp (K.N.)

Citation: Minamino, T.; Namba, K. Recent Advances in the Bacterial Flagellar Motor Study. *Biomolecules* **2021**, *11*, 741. https://doi.org/10.3390/biom11050741

Received: 13 May 2021
Accepted: 14 May 2021
Published: 17 May 2021

The bacterial flagellum is a supramolecular motility machine that allows bacterial cells to swim in liquid environments. The flagellum is composed of the basal body, which acts as a rotary motor, the filament, which functions as a helical propeller, and the hook, which connects the basal body and filament and works as a universal joint to smoothly transmit torque produced by the motor to the filament. The flagellar motor is composed of a rotor ring complex and multiple transmembrane stator units, each of which acts as an ion channel to couple the ion flow through the channel to torque generation. The flagellar motor is placed under the control of sensory signal transduction networks, thereby allowing bacterial cells to migrate towards more desirable environments for their survival. The entire structure of the flagellum and flagellar component proteins are highly conserved among bacterial species. However, novel and divergent structures associated with the flagellar motor are clearly observed by in situ structural analyses of flagellar motors derived from different bacterial species [1–3].

The scope of this Special Issue is to cover recent advances in our understanding of the structures and functions of the bacterial flagellar motor derived from different bacterial species. This Special Issue includes ten review articles [4–13] and eleven original research papers [14–24] from well-known experts in the field.

All review articles provide both expert and non-expert readers with advances in understanding the structures and functions of the bacterial flagellum. They highlight the most recent observations and illustrate perspectives for future research [4–13].

The amino acid sequence of the distal rod protein FlgG is very similar to that of the hook protein FlgE. The FlgG rod structure is straight and rigid, whereas the hook adopts a curved form with high bending flexibility. Saijo-Hamano et al. solved a crystal structure of the FlgG fragment missing both N- and C-terminal disordered regions and fitted the atomic model of the FlgG fragment into a density map of the FlgG rod by electron cryomicroscopy (cryoEM). They found that an N-terminal short segment called L-stretch stabilizes intermolecular packing interactions, making the rod straight and rigid. As a result, the rod functions as a drive shaft of the flagellar motor [14]. Horváth and Kato et al. carried out cryoEM image analysis of the straight polyhook structure and provided structural evidence that domain Dc of FlgE with a long β-hairpin structure connecting domains D0 and D1 not only contributes to the structural stability of the hook but also allows the bending flexibility of the hook so that the hook can function as a universal joint [15].

Salmonella enterica has two distinct flagellin genes, namely *fliC* and *fljB*, on the genome and autonomously switches their expression at a frequency of 10^{-3}–10^{-4} per cell per generation. Yamaguchi et al. carried out functional and structural analyses of the filaments formed by either FliC or FljB and provided evidence that domain D3 of flagellin molecules plays an important role not only in changing the antigenicity of the filament but also in optimizing the motility function of the filament as a propeller under different environmental conditions [16].

To construct the flagellum on the bacterial cell surface, the flagellar type III secretion system (fT3SS) transports flagellar building blocks from the cytoplasm to the distal end of the growing flagellar structure. Terashima et al. developed in vitro protein transport assays using inverted membrane vesicles and provided direct evidence that coordinated flagellar protein export and assembly can occur at the post-translational level [17].

A non-flagellated bacterium *Lysobacter enzymogenes* OH11 moves on solid surfaces using type IV pili. Interestingly, this bacterium encodes highly homologous fT3SS genes on its genome. Fulano et al. constructed fT3SS-knockout mutant strains and provided evidence that some fT3SS components are required for the twitching motility of *L. enzymogenes*. Thus, the homologous components of the fT3SS seem to have acquired a divergent function that controls the twitching motility [18].

MotA and MotB form a transmembrane proton channel complex to couple the proton flow through the channel with torque generation. The MotAB stator complex autonomously controls its proton channel activity in response to changes in the environment. Morimoto et al. provided experimental evidence that the N-terminal cytoplasmic tail of MotB regulates the gating of the MotAB proton channel [19]. Furthermore, Naganawa and Ito provided an interesting clue of how the stator unit selects the coupling ion to drive flagellar motor rotation [20].

Onoe et al. showed that the *Paenibacillus* MotAB complex, which was originally thought to conduct divalent cations such as Ca^{2+} and Mg^{2+} to drive flagellar motor rotation, can work as a stator unit in the *E. coli* flagellar motor and that this stator unit directly converts the energy released from the proton influx to motor rotation in *E. coli* [21].

The chemotaxis signaling protein, namely CheY-P, binds to a rotor of the flagellar motor to switch its rotational direction from counterclockwise to clockwise in a highly cooperative manner. The cytoplasmic level of CheY-P largely fluctuates so that *E. coli* cells respond to changes in the environment rapidly and efficiently to migrate toward more desirable conditions. Che et al. analyzed the coordination of directional switching between flagellar motors on the same cell and provided evidence suggesting that the fluctuation of the cytoplasmic CheY-P level coordinates rotation among flagellar motors and regulates steady-state run-and-tumble swimming of cells to facilitate efficient responses to environmental changes [22].

A motile *Methylobacterium* ME121 strain is more motile when they grow together with a non-motile *Kaistia* 32K strain. Usui et al. purified a swimming acceleration factor from the culture supernatant and found that extracellular polysaccharides, which they named the K factor, facilitate the flagellar motor function of the ME121 strain [23].

Lysophosphatidic acid acyltransferase (LPAAT) introduces fatty acyl groups into the sn-2 position of membrane phospholipids. *E. coli* has another LPAAT homolog named YihG in addition to PlsC, which is essential for the growth of *E. coli*. Toyotake et al. constructed a *yihG* null mutant (Δ*yihG*) and provided evidence suggesting that YihG has specific functions related to flagellar assembly through the modulation of the fatty acyl composition of membrane phospholipids [24].

Thus, the studies included in this Special Issue illustrate various examples of the recent progress in the studies on the conserved structure and function of the flagellar motor as well as its structural and functional diversities among different bacterial species.

Finally, we would like to thank all authors for their great contributions to this Special Issue and Fumiaki Makino and Tomoko Yamaguchi for creating the cover image.

Conflicts of Interest: The authors declare no conflict of interest.

References

1. Minamino, T.; Namba, K. Self-assembly and type III protein export of the bacterial flagellum. *J. Mol. Microbiol. Biotechnol.* **2004**, *7*, 5–17. [CrossRef]
2. Minamino, T.; Imada, K. The bacterial flagellar motor and its structural diversity. *Trends Microbiol.* **2015**, *23*, 267–274. [CrossRef]
3. Terashima, H.; Kawamoto, A.; Morimoto, Y.V.; Imada, K.; Minamino, T. Structural differences in the bacterial flagellar motor among bacterial species. *Biophys. Physicobiol.* **2017**, *14*, 191–198. [CrossRef] [PubMed]

4. Nakamura, S.; Minamino, T. Flagella-driven motility of bacteria. *Biomolecules* **2019**, *9*, 279. [CrossRef] [PubMed]
5. Kojima, S.; Terashima, H.; Homma, M. Regulation of the single polar flagellar biogenesis. *Biomolecules* **2020**, *10*, 533. [CrossRef]
6. Halte, M.; Erhardt, M. Protein export via the type III secretion system of the bacterial flagellum. *Biomolecules* **2021**, *11*, 186. [CrossRef]
7. Zhuang, X.Y.; Lo, C.J. Construction and loss of bacterial flagellar filements. *Biomolecules* **2020**, *10*, 1528. [CrossRef] [PubMed]
8. Carroll, B.; Liu, J. Structural conservation and adaptation of the bacterial flagellar motor. *Biomolecules* **2020**, *10*, 1492. [CrossRef]
9. Khan, S. The architectural dynamics of the bacterial flagellar motor switch. *Biomolecules* **2020**, *10*, 833. [CrossRef]
10. Chu, J.; Liu, J.; Hoover, T.R. Phylogenetic distribution, ultrastructure, and function of bacterial flagellar sheaths. *Biomolecules* **2020**, *10*, 363. [CrossRef]
11. Zhang, W.J.; Wu, L.F. Flagella and swimming behavior of marine magnetotactic bacteria. *Biomolecules* **2020**, *10*, 460. [CrossRef] [PubMed]
12. Nakamura, S. Spirochete flagella and motility. *Biomolecules* **2020**, *10*, 550. [CrossRef]
13. Camarena, L.; Dreyfus, G. Living in a foster home: The single subpolar flagellum Fla1 of *Rhodobacter sphaeroides*. *Biomolecules* **2020**, *10*, 774. [CrossRef] [PubMed]
14. Saijo-Hamano, Y.; Matsunami, H.; Namba, K.; Imada, K. Architecture of the bacterial flagellar rod and hook of *Salmonella*. *Biomolecules* **2019**, *9*, 260. [CrossRef]
15. Horváth, P.; Kato, T.; Miyata, T.; Namba, K. Structue of Salmonella flagellar hook reveals intermolecular domain interactions for the universal joint function. *Biomolecules* **2019**, *9*, 462. [CrossRef] [PubMed]
16. Yamaguchi, T.; Toma, S.; Terahara, N.; Miyata, T.; Ashihara, M.; Minamino, T.; Namba, K.; Kato, T. Structural and functional comparison of Salmonella flagellar filaments composed of FljB and FliC. *Biomolecules* **2020**, *10*, 246. [CrossRef]
17. Terashima, H.; Tatsumi, C.; Kawamoto, A.; Namba, K.; Minamino, T.; Imada, K. In vitro autonomous construction of the flagellar axial structure in inverted membrane vesicles. *Biomolecules* **2020**, *10*, 126. [CrossRef]
18. Fulano, A.M.; Shen, D.; Kinoshita, M.; Chou, S.H.; Qian, G. The homologous componenets of flagellar type III protein apparatus have acuired a novel function to control twitching motility in a non-flagellated biocontrol bacterium. *Biomolecules* **2020**, *10*, 733. [CrossRef]
19. Morimoto, Y.V.; Namba, K.; Minamino, T. GFP fusion to the N-terminus of MotB affects the proton channel activity of the bacterial flagellar motor in *Salmonella*. *Biomolecules* **2020**, *10*, 1255. [CrossRef] [PubMed]
20. Naganawa, S.; Ito, M. MotP subunit is critical for ion selectivity and evolution of a K^+-coupled flagellar motor. *Biomolecules* **2020**, *10*, 691. [CrossRef]
21. Onoe, S.; Yoshida, M.; Terahara, N.; Sowa, Y. Coupling ion specificity of the flagellar stator proteins MotA1/MotB1 of *Paenibacillus* sp. TCA20. *Biomolecules* **2020**, *10*, 1078. [CrossRef] [PubMed]
22. Che, Y.S.; Sagawa, T.; Inoue, Y.; Takahashi, H.; Hamamoto, T.; Ishijima, A.; Fukuoka, H. Fluctuations in intracellular CheY-P concentration coordinate reversals of flagellar motors in *E. coli*. *Biomolecules* **2020**, *10*, 1544. [CrossRef]
23. Usui, Y.; Wakabayashi, Y.; Shimizu, T.; Tahara, Y.O.; Miyata, M.; Nakamura, A.; Ito, M. A factor produced by *Kaistia* sp. 32K accelerated the motility of *Methylobacterium* sp. ME121. *Biomolecules* **2020**, *10*, 618. [CrossRef]
24. Toyotake, Y.; Nishiyama, M.; Yokoyama, F.; Ogawa, T.; Kawamoto, J.; Kurihara, T. A novel lysophosphatidic acid acyltransferase of *Escherichia coli* produces membrane phospholipids with a *cis*-vaccenoyl group and is related to flagellar formation. *Biomolecules* **2020**, *10*, 745. [CrossRef] [PubMed]

Review

Flagella-Driven Motility of Bacteria

Shuichi Nakamura [1] and Tohru Minamino [2,*]

[1] Department of Applied Physics, Graduate School of Engineering, Tohoku University, 6-6-05 Aoba, Aoba-ku, Sendai 980-8579, Japan

[2] Graduate School of Frontier Biosciences, Osaka University, 1-3 Yamadaoka, Suita, Osaka 565-0871, Japan

* Correspondence: tohru@fbs.osaka-u.ac.jp; Tel.: +81-6-6879-4625

Received: 27 June 2019; Accepted: 12 July 2019; Published: 14 July 2019

Abstract: The bacterial flagellum is a helical filamentous organelle responsible for motility. In bacterial species possessing flagella at the cell exterior, the long helical flagellar filament acts as a molecular screw to generate thrust. Meanwhile, the flagella of spirochetes reside within the periplasmic space and not only act as a cytoskeleton to determine the helicity of the cell body, but also rotate or undulate the helical cell body for propulsion. Despite structural diversity of the flagella among bacterial species, flagellated bacteria share a common rotary nanomachine, namely the flagellar motor, which is located at the base of the filament. The flagellar motor is composed of a rotor ring complex and multiple transmembrane stator units and converts the ion flux through an ion channel of each stator unit into the mechanical work required for motor rotation. Intracellular chemotactic signaling pathways regulate the direction of flagella-driven motility in response to changes in the environments, allowing bacteria to migrate towards more desirable environments for their survival. Recent experimental and theoretical studies have been deepening our understanding of the molecular mechanisms of the flagellar motor. In this review article, we describe the current understanding of the structure and dynamics of the bacterial flagellum.

Keywords: bacterial flagellum; chemotaxis; ion motive force; ion channel; mechanochemical coupling; molecular motor; motility; torque generation

---

1. Introduction

Bacterial motility is an extremely intriguing topic from various scientific aspects. For example, motility can be a crucial virulence attribute for pathogenic bacteria, such as *Salmonella enterica* (hereafter referred to *Salmonella*) and *Helicobacter pylori* [1,2]. Bacterial motility also plays a significant role in mutualistic symbioses [3,4]. Furthermore, motile bacteria are also a representative example for understanding the underlying physical principles that form the basis of energy conversion, force generation and mechanochemical coupling mechanisms [5]. Active motilities of bacteria are represented by movement in liquid (e.g., swimming motility in *Escherichia coli* and *Salmonella*) and on solid surfaces (e.g., flagella-driven swarming motility in *Proteus mirabilis* and *Vibrio parahaemolyticus*, gliding motility in *Mycoplasma mobile*, and twitching motility in *Pseudomonas aeruginosa*), and passive motility is typically actin-based locomotion (e.g., *Listeria monocytogenes* and *Shigella* spp.) [6]. Since bacterial motility varies among bacterial species, bacteria utilize their own motility system optimized for their habitats.

E. coli and *Salmonella* use flagella viewable from the cell exterior as a thin, long, helical filament (Figure 1a). On the other hand, the flagella of spirochetes reside within the periplasmic space, and so they are called periplasmic flagella [7]. Whether the bacterial flagella are exposed to the cell exterior or are hidden within the cell body, the flagellum is divided into three structural parts: the basal body as a rotary motor, the hook as a universal joint and the filament as a molecular screw in common (Figure 1b), and flagellar formation and function involves more than 60 genes [8–10].

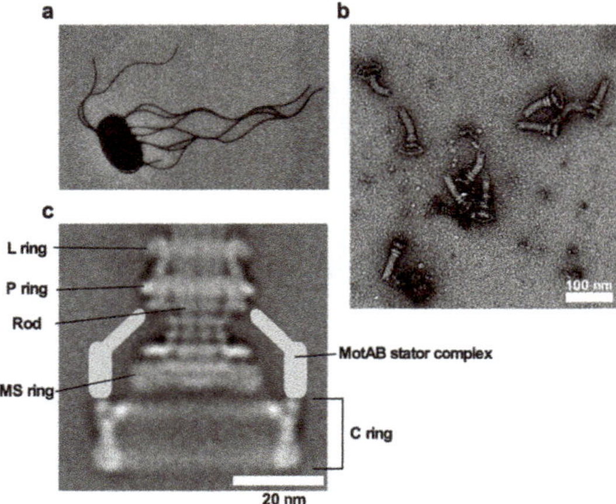

Figure 1. *Salmonella* flagellum. (**a**) Electron micrograph of *Salmonella* cell. The micrograph was taken at a magnification of ×1200. (**b**) Electron micrograph of hook-basal bodies isolated from *Salmonella* cells. (**c**) CryoEM image of purified basal body. Purified basal body consists of the L, P, MS and C rings and the rod. A dozen MotAB complex are associated with the basal body to act as a stator unit in the motor but is gone during purification.

The bacterial flagellar motor is powered by the transmembrane electrochemical gradient of ions, namely ion motive force (IMF) and rotates the flagellar filament to generate thrust to propel the cell body. The maximum motor speed reaches 300 revolutions per second in *E. coli* and *Salmonella* [11] and 1700 revolutions per second in a marine bacterium *Vibrio alginolyticus* [12]. Thus, the rotational speed of the flagellar motor is much faster than that of a manufactured car engine such as formula one car. The flagellar motor is composed of a rotor and multiple stator units. Each stator unit acts as a transmembrane ion channel to conduct cations such as protons (H^+) or sodium ions (Na^+) and applies force on the rotor [13,14].

The flagellar motors of *E. coli* and *Salmonella* rotate in both counterclockwise (CCW) and clockwise (CW) without changing the direction of ion flow. *E. coli* and *Salmonella* cells can swim in a straight line by bundling left-handed helical filaments behind the cell body (run) when all of them rotate in CCW direction. When one or multiple motors switch the direction of rotation from CCW to CW, the flagellar bundle is disrupted, enabling the cell to tumble and change the swimming direction. *E. coli* and *Salmonella* cells sense temporal changes in nutrients, environmental stimuli, and signaling molecules to coordinate the switching frequency of the motor. Transmembrane chemoreceptors, energy-related taxis sensors and intracellular phosphotransferase systems detect environmental signals and then convert them into intracellular signals. Then, an intracellular signal transduction system transmits the signals to the flagellar motor to switch the direction of motor rotation from CCW to CW. The cells repeat a run–tumble pattern to explore more favorable environments for their survival [15]. This review article covers our current understating of flagella-driven motility mechanism in *E. coli* and *Salmonella*. We also describe the structural and functional diversities of the bacterial flagella.

2. Axial Structure

The axial structure of the bacterial flagellum is commonly a helical assembly composed of 11 protofilaments and is divided into at least three structural parts: the rod, the hook and the filament from the proximal to the distal end. The rod is straight and rigid against bending and twisting and acts as a drive shaft. The hook is supercoiled and flexible against bending and acts as a universal joint to smoothly transmit torque produced by the motor to the filament. The filament is also supercoiled but stiff against bending. The filament is normally a left-handed supercoil to act as a helical screw to produce thrust for swimming motility. The filament undergoes polymorphic transformation from the left-handed supercoil to right-handed ones when bacterial cells tumble and change swimming direction [16].

2.1. Flagella Filament

The flagellar filament of *E. coli* is formed by ~30,000 copies of flagellin, FliC. *Salmonella* has the *fljB* gene encoding another flagellin subunit in addition to the *fliC* gene. Because flagellin is a major target of host immune system (H-antigen), such an additional flagellin subunit enables *Salmonella* cells to escape from adaptive immune response of the host more efficiently compared to *E. coli* cells [17]. The FliC-type filament structure derived from *Salmonella* has been solved at the atomic level [18–20]. *Salmonella* FliC is composed of four domains D0, D1, D2 and D3, arranged from the inner to the outer part of the filament structure. Domains D0 and D1 are well conserved among bacterial species whereas domains D2 and D3 are variable even among *Salmonella* spp., because these two domains are the major targets of antibodies [21].The supercoiled forms of the filament structure are generated by combinations of two distinct left-handed (L-type) and right-handed (R-type) helical conformations of flagellin molecule and packing interactions of the L- and R-type protofilaments, and so the helical properties of each supercoil are determined by a ratio of L-type protofilaments to R-type ones in the filament structure [22,23]. The intermolecular distance along the L-type straight filament consisting of all L-type protofilaments is 0.8 Å longer than that of the R-type one composed of all R-type protofilaments [24]. Since a conformational change of a β-hairpin in domain D1 generates the 0.8 Å difference in repeat distance, this β-hairpin is thought to be responsible for the supercoiling switching [17]. Therefore, it seems likely that an abrupt reversal of motor rotation applies mechanical stress on each protofilament to induce the sliding motion between flagellin subunits along the protofilament, thereby changing the filament structure from the L-type supercoil to R-type one to disrupt the flagellar bundle for tumbling of the cell body [18]. Recent high-resolution electron cryomicroscopy (cryoEM) imaging analyses of L- and R-type straight filaments derived from *Bacillus subtilis* and *P. aeruginosa* have shown that the switching of the supercoiled forms of these flagellar filaments occurs in a way similar to the *Salmonella* filament [25].

Although the flagellar filaments of *E. coli* and *Salmonella* are formed by a single flagellin subunit, many bacterial species have multiple flagellins for the synthesis of flagellar filaments. The single polar flagellum of *Caulobacter crescentus* is composed of six flagellins, FljJ, FljK, FljL, FljM, FljN, and FljO [26]. Although the function of each flagellin subunit and their organization are not yet characterized, they are not essential for filament formation because some flagellin defects are compensated by others [26]. The flagellar filament of *Sinorhizobium meliloti* consists of four flagellins, FlaA, FlaB, FlaC, and FlaD, and that of *Rhizobium lupini* contains just three of them FlaA, FlaB, and FlaD. For the flagella of these soil bacteria, FlaA is the principal component, and others are secondary ones [27]. The flagellar filament of *Rhizobium leguminosarum* comprises three major proteins, FlaA, FlaB, and FlaC, and four minor proteins, FlaD, FaE, FlaH, and FlaG [28]. *Agrobacterium tumefaciens* also possesses four flagellins, FlaA, FlaB, FlaC, and FlaD; FlaA and FlaB are abundant in the filament in comparison with FlaC and FlaD, and the swimming ability of *A. tumefaciens* is considerably decreased by a loss of FlaA but not by that of FlaB [29]. *Bradyrhizobium diazoefficiens* has two flagella systems: One is subpolar flagella, of which filament is composed of four flagellins, FliC1, FliC2, FliC3, and FliC4, whereas the other is lateral flagella, of which filament is made up of two flagellins, LafA1 and LafA2 [30]. The bi-polar flagellar filaments of *Campylobacter jejuni* comprise two distinct FlaA and FlaB subunits, both of which share 92.3% sequence identity. The FlaB filament grows first and then FlaA filament grows on the FlaB filament [31]. Consistently, two different flagellins, FlaA and FlaB (86% sequence identity) form the single polar flagellar filament in *Shewanella putrefaciens*, and FlaA forms a proximal part of the filament whereas FlaB makes the remaining portion [32]. The spatial assembly by these two distinct flagellin subunits benefits motility under a various range of environmental conditions [32]. Because the assembly of the flagellar filament by multiple flagellins affects its mechanistic properties for flagellar function in different environments [26,32–35], the composition of the flagellar filament structure would be optimized for environmental conditions, in which the bacteria live and survive.

2.2. Hook and Rod

The *Salmonella* hook is formed by about 120 copies of the hook protein FlgE. *Salmonella* FlgE consists of three domains D0, D1, and D2, arranged from the inner to outer parts of the hook structure and the Dc region connecting domains D0 and D1 [36,37]. The hook forms several supercoils, and axial interactions between a triangular loop of domain D1 and domain D2 are responsible for hook supercoiling [36,38,39]. However, a truncation of neither the triangular loop nor the D2 domain affects the bending flexibility of the hook structure [38]. Since there are gaps not only between D1 domains but also between D0 domains, these gaps make the hook flexible for bending. The amino-acid sequence of FlgE of *C. jejuni* (864 a.a for strain NCTC 11168) is much longer than that of *Salmonella* (402 a.a), and so FlgE of *C. jejuni* has two additional outer domains, D3 and D4, and these two domains are involved in the interaction within and between protofilaments, conferring stiffness and robustness on the *C. jejuni* hook structure to act as a universal joint under highly viscous condition [40].

The bending flexibility of the hook structure is required for the formation of a bundle structure behind the cell body of *E. coli* and *Salmonella* [41,42]. The hook length is also important for maximum stability of the flagellar bundle. Shorter hooks are too stiff to function as a universal joint whereas longer hooks buckle and create instability in the flagellar bundle [43]. The hook length is controlled by the molecular ruler protein FliK, which is secreted via a type III protein export apparatus during hook assembly [44].

The elasticity of the hook is also important for changing swimming direction in *V. alginolyticus*, which is a monotrichous bacterium. When *V. alginolyticus* cell changes swimming from forward to backward by the switching of direction of flagellar motor rotation from CCW to CW, the hook undergoes compression and buckles, resulting in an axis mismatch between the flagellar filament and the cell body to induce a flicking motion of the cell body. As a result, the swimming direction changes by ~90° [45].

The rod is composed of three proximal rod proteins, FlgB, FlgC, FlgF, and the distal rod protein FlgG [46,47]. FliE is postulated to connect the MS ring and the most proximal part of the rod formed by FlgB [48]. These four rod proteins and FliE are well conserved among bacterial species [9,10]. Domains D0 and D1 of *Salmonella* FlgG show high sequence and structural similarities to those of FlgE, thereby allowing direct connection of the rigid rod with the flexible hook [49]. However, one major structural difference between the rod and hook is the orientation of their D1 domains relative to the tubular axis, and so axial packing interactions between domains D1 of FlgG are tight whereas those of FlgE are loose. As a result, such a structural difference is likely to be responsible for the bending rigidity of the rod and flexibility of the hook [49]. The Dc region of FlgG has a FlgG specific sequence (GSS; YQTIRQPGAQSSEQTTLP). Since the GSS insertion into the Dc region of FlgE makes the hook straight and rigid, the GSS contributes to the rigidity on the rod structure [42]. However, since FlgE of *B. subtilis* and *C. jejuni* has the GSS-like sequence in their Dc region [32,34], it remains unknown how the hook of *B. subtilis* and *C. jejuni* can form a curved structure with bending flexibility.

3. Type III Protein Export Apparatus

The assembly of the axial structure begins with the rod, followed by the hook and finally the filament. A type III protein export apparatus transports axial component proteins from the cytoplasm to the distal end of the growing flagellar structure to construct the axial structure beyond the cellular membranes [50]. The type III protein export apparatus consists of an export gate complex made of five transmembrane proteins, FlhA, FlhB, FliP, FliQ and FliR, and a cytoplasmic ATPase ring complex consisting of FliH, FliI and FliJ [51–53]. The transmembrane export gate complex is located within the basal body MS ring and acts as a H^+–protein antiporter to couple an inward-directed H^+ translocation through the export gate with an outward-directed protein export [54,55]. FliP, FliQ, and FliR form a right-handed helical assembly with a 5 FliP to 4 FliQ to 1 FliR stoichiometry inside the MS ring, and FliO is required for efficient assembly of the FliPQR complex [52,56,57]. The FliPQR complex has a central channel with a diameter of 1.5 nm [57]. Since FliP and FliR are likely to interact with FliE [51,57,58], the central channel of the FliPQR complex is postulated to be a protein translocation pathway. FlhA and FlhB associate with the FliPQR complex [52]. FlhA forms a nonameric ring structure through its C-terminal cytoplasmic domain [59–61] and forms an ion channel to conduct both H^+ and Na^+ [62]. FliH, FliI and FliJ form the cytoplasmic ATPase ring complex with a 12 FliH to 6 FliI to 1 FliJ stoichiometry [63–65]. The ATPase ring complex is associated with the basal body through interactions of FliH with FlhA and a C ring protein FliN [66–69]. The $FliI_6$ ring hydrolyzes ATP to activate the transmembrane export gate complex, thereby driving H^+-coupled flagellar protein export by the export gate [55,70].

4. Basal Body Rings

The basal body has multiple ring structures, namely L ring, P ring, MS ring, and C ring [71] (Figure 1c). The L and P rings, which are formed by the lipoprotein FlgH and the periplasmic protein FlgI, respectively, are embedded in the outer membrane and the peptidoglycan (PG) layer, respectively, and they together act as a bearing for the rod. The LP ring complex is missing in the basal body of gram-positive bacteria such as *B. subtilis* [9]. In contrast, the MS and C rings are well conserved among bacterial species [9,10]. The MS ring is composed of the transmembrane protein FliF and is part of a rotor [71]. FliG, FliM, and FliN form the C ring on the cytoplasmic face of the MS ring. The C ring acts not only as a central part of the rotor for torque generation but also as a structural device to switch the direction of motor rotation in *E. coli* and *Salmonella* [71]. Diameters of the LP ring complex, the S ring, the M ring, and the C ring are ~25 nm, ~24.5 nm, ~30 nm, and ~45 nm, respectively, in *Salmonella*.

FliG consists of N-terminal ($FliG_N$), middle ($FliG_M$), and C-terminal ($FliG_C$) domains. $FliG_N$ directly associates with the C-terminal cytoplasmic domain of FliF ($FliF_C$) with a one-to-one stoichiometry [72]. Inter-molecular interactions between $FliG_N$ domains and between $FliG_M$ and $FliG_C$ are responsible for FliG polymerization on the cytoplasmic face of the MS ring [73–76]. $FliG_C$ is involved in the interaction with the stator protein MotA [77–79]. The middle domain of FliM ($FliM_M$) binds to $FliG_M$ with a one-to-one stoichiometry to form the C ring wall [80]. An EHPQR motif in $FliG_M$ and a GGXG motif in $FliM_M$ are responsible for the $FliG_M$–$FliM_M$ interaction. The C-terminal domain of FliM ($FliM_C$) shows significant sequence and structural similarities with FliN, and $FliM_C$ and FliN together form a doughnut-shaped hetero-tetramer consisting of one copies of $FliM_C$ and three copies of FliN, and this hetero-tetrameric block produces a continuous spiral density along the circumference at the bottom edge of the C ring [81]. *B. subtilis* has a *fliY* gene, which shows sequence similarity to both $FliM_C$ and FliN, instead of the *fliN* gene [82]. In *B. subtilis*, FliG, FliM and FliY form the C ring in a similar manner to *E. coli* and *Salmonella* C ring structures although the overall structure and dimensions of the *B. subtilis* C ring remain unclear. Interestingly, high-resolution single-molecule fluorescence imaging techniques have revealed rapid exchanges of FliM and FliN labelled with a fluorescent protein between the basal body and the cytoplasmic pool in *E. coli*, suggesting that the C ring is a highly dynamic structure [83–85].

The stator units are assembled on the FliG ring (Figure 1c), and so stator–rotor interactions occur about 20 nm away from the center of the C ring in *Salmonella*. The *Salmonella* flagellar motor can accommodate about 10 stator units [86]. A *fliF–fliG* deletion fusion significantly shortens the diameter of the C ring, because $FliF_C$ and $FliG_N$, which together form the inner lobe structure connecting the M and C rings, are missing. It has been shown that the average number of active stator units is two units less in the FliF–FliG deletion fusion motor than in the wild-type motor [87]. This suggests that the diameter of the C ring determines the number of active stator units that can be bound to the motor. This is supported by recent observations that a diameter of the C ring of the *C. jejuni* and *H. pylori* flagellar motors is larger than that of the *Salmonella* C ring, allowing these motors to accommodate more active stator units around the rotor to generate much higher torque [88].

5. Stator

5.1. Diversity of the Stator Unit

The transmembrane stator unit of the flagellar motor conducts ions and exerts force on the rotor. Based on the coupling ion and sequence similarity, the stator units are classified into three groups: H^+-coupled MotAB complex, Na^+-coupled PomAB complex, and Na^+-coupled MotPS complex [14]. The MotAB complex is composed of four copies of MotA and two copies of MotB and acts as a transmembrane H^+ channel [89,90]. The PomAB and MotPS complexes form a Na^+ channel in a way similar to the MotAB complex [91–93]. In addition to these stator proteins, bacteria such as *S. meliloti* and *V. alginolyticus* have additional motor proteins. *S. meliloti* possesses three extra motor proteins, namely MotC, MotD, and MotE. MotC stabilizes the periplasmic domain of MotB to facilitate proton translocation through a H^+ channel of the MotAB complex. MotD binds to FliM for fast rotation, and MotE is involved in folding and stability of MotC [94]. *V. alginolyticus* has MotX and MotY to form the T ring structure located beneath the P ring, and an interaction between PomB and MotX is required for stable localization of PomAB complex around the basal body [9,95].

V. alginolyticus and *V. parahaemolyticus* use a single polar flagellum for swimming in low viscous liquid and induce lateral flagella when these *Vibrio* cells encounter solid surfaces [96–98]. The polar flagellum utilizes the PomAB complex as a stator unit whereas the lateral flagella use the MotAB complex as a stator unit [99]. *B. subtilis* possesses two distinct H^+-type MotAB and Na^+-type MotPS complexes to drive flagellar motor rotation, and these two types of stator units are exchanged in response to changes in external pH, external Na^+ concentration and viscosity [92,93,100]. Like *B. subtilis*, *Shewanella oneidensis* also utilizes two distinct H^+-type MotAB and Na^+-type PomAB complexes in response to changes in the environmental Na^+ concentration [101].

The MotPS complex of *Bacillus alcalophilus* conducts K^+ and Rb^+ in addition to Na^+ [102]. *Bacillus clausii* has only MotAB complex as a stator unit, and this MotAB complex exhibits the H^+ channel activity at neutral pH and the Na^+ channel activity at extremely high pH [103]. The MotAB complex of a spirochete *Leptospira biflexa* has the ability to conduct both H^+ and Na^+ in an external pH-dependent manner in a way similar to the MotAB complex of *B. clausii* [104]. These observations suggest that the stator function of these species would be optimized for environmental conditions of their habitats.

5.2. Topology of the Stator Complex

MotA, PomA and MotP possess four transmembrane helices (TM1, TM2, TM3, and TM4) and a relatively large cytoplasmic loop between TM2 and TM3 and a C-terminal cytoplasmic tail (Figure 2a). MotB, PomB and MotS possess an N-terminal cytoplasmic tail, a single transmembrane helix, and a relatively large C-terminal periplasmic domain containing a conserved peptidoglycan-binding (PGB) motif for anchoring the stator units to the rigid PG layer (Figure 2a). A plausible atomic model of the transmembrane H^+ channel of the MotAB stator complex derived from *E. coli* has been proposed [105]. The MotAB stator complex has two H^+ pathways formed by MotA-TM3, MotA-TM4 and MotB-TM (Figure 2b). A highly conserved Asp-32 residue lies near the cytoplasmic end of MotB-TM and plays an important role in the H^+ relay mechanism [106]. This Asp residue is located on the surface of MotB-TM facing MotA-TM3 and MotA-TM4 [90]. A plug segment in the flexible linker of MotB connecting MotB-TM and the PGB domain binds to the H^+ channel to suppress massive H^+ flow through the channel until the MotAB complex associates with the motor. It has been proposed that an interaction between MotA and FliG may induce a detachment of the plug segment form the H^+ channel to couple the H^+ flow through the channel to torque generation [107,108].

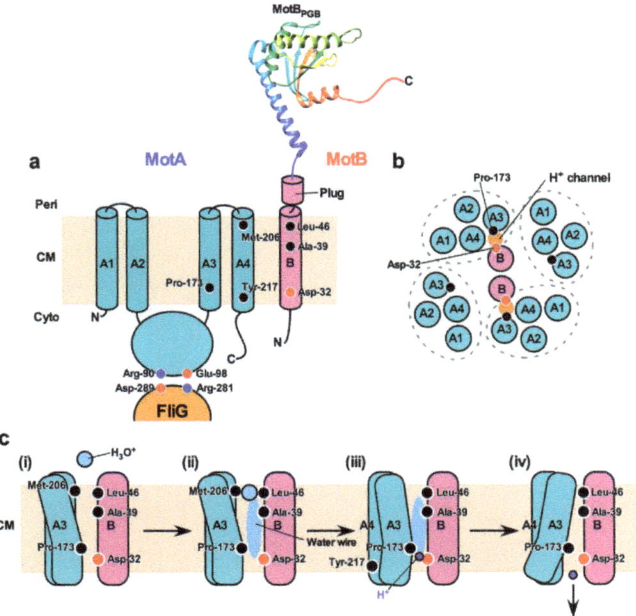

Figure 2. H^+ translocation mechanism of the flagellar motor. (**a**) Topology of the *E. coli* MotA and MotB and a crystal structure of the peptidoglycan-binding domain of MotB ($MotB_{PGB}$, PDB code: 2ZVY). Highly conserved Arg-90 and Glu-98 residues in the cytoplasmic loop between transmembrane helices 2 (A2) and 3 (A3) interact with conserved Asp-289 and Arg-281 residues of FliG, respectively, to drive motor rotation. Asp-32 of MotB provides a binding site for H^+. Pro-173, Met-206 and Tyr-217 of MotA and Ala-39 and Leu-46 of MotB are involved in the H^+ relay mechanism. Cyto, cytoplasm; CM, cytoplasmic membrane; Peri, periplasm. (**b**) Arrangement of transmembrane segments of MotA and MotB. The MotAB complex has two proton channels. Four MotA subunits are positioned with their TM3 (A3) and TM4 (A4) segments adjacent to the MotB dimer, and their TM1 (A1) and TM2 (A2) segments on the outside. (**c**) A plausible model for H^+ translocation through MotAB stator complex (see text for details).

5.3. H^+ Translocation Mechanism

The maximum rotation rate of the H^+-driven flagellar motors of *E. coli* and *Salmonella* is reduced with a decrease in the intracellular pH. In contrast, a change in external pH does not affect the maximum motor speed at all. These observations suggest that the intracellular H^+ concentration affects the rate of the H^+ flow through the MotAB complex [109,110].

Asp-33 of *Salmonella* MotB, which corresponds to Asp-32 in *E. coli* MotB, is critical for the binding of H^+ from the cell exterior, and its protonation and deprotonation cycle is directly linked to a torque generation step caused by stator–rotor interactions [111]. The *motB(D33E)* mutation results in a considerable decrease in the rate of H^+-coupled conformational change of the MotAB complex [112]. Furthermore, the *motB(D33E)* mutation causes not only large speed fluctuations but also frequent pausing of motor rotation at low load. However, neither speed fluctuation nor pausing is seen at high load [112]. These observations suggest that the protonation and deprotonation cycle of Asp-33 of MotB may occur in a load-dependent manner. The dissociation of H^+ from this Asp-33 residue to the cytoplasm is linked to conformational changes of a cytoplasmic loop of MotA, which is responsible for the interaction with FliG. Molecular dynamics (MD) simulation has predicted that the binding of H^+ to this Asp residue induces a conformational change of the proton channel to facilitate H^+ release to the cytoplasm [105]. Two highly conserved residues, Pro-173 of MotA-TM3 and Tyr-217 of MotA-TM4, are involved in such H^+-coupled conformation changes of the H^+ channel [113–115].

Based on MD simulation of the H^+ channel of the *E. coli* MotAB complex, the H^+ translocation through the channel is postulated to be mediated by water molecules aligned along a H^+ pathway (i.e., water wire). Leu-46 of MotB is assumed to act as a gate for hydronium ion (H_3O^+) and then to transfers H^+ to MotB-Asp32 via the water wire [105]. Mutations at position of Ala-39 of MotB, which resides on the same side as Asp-32 in the H^+ pathway, impair motility and are partially suppressed by extragenic mutations at Met-206 of MotA [116]. This Met-206 residue is located near the periplasmic end of TM4 and faces the H^+ pathway [105,117]. The *motA(M206I)* mutation reduces the H^+ channel activity, thereby reducing motility [118]. Taken all together, the H^+ translocation mechanism is postulated to be as follows: (i) H^+ permeates a H^+ channel in the H_3O^+ state through Leu-46 of MotB, (ii) Met-206 of MotA and Ala-39 of MotB are involved in the transfer of H^+ along the water wire, (iii) H^+ binds to Asp-32 of MotB, and (iv) the dissociation of H^+ from Asp-32 of MotB to the cytoplasm is facilitated by a conformational change of the H^+ channel through Pro-173 and Tyr-217 of MotA (Figure 2c). As a result, the cytoplasmic loop of MotA can interact with FliG to drive flagellar motor rotation [119].

6. Torque Generation

6.1. Rotation Mechanism

Highly conserved Arg-90 and Glu-98 residues of MotA, which are located in the cytoplasmic loop between TM2 and TM3 of MotA, interact with highly conserved Asp-289 and Arg-281 residues of FliG, respectively (Figure 2a) [77–79,120]. These two electrostatic interactions are responsible for efficient stator assembly around the rotor, and the interaction between Glu-98 of MotA and Arg-281 of FliG is likely to be involved in torque generation [79]. H^+ translocation through the transmembrane H^+ channel of the MotAB complex allows the cytoplasmic loop of MotA to associate with and dissociate from FliG to drive flagellar motor rotation [119]. However, the energy coupling mechanism of the flagellar motor remains unknown.

6.2. Torque-Speed Relationship

Precise measurements of motor rotation are important to elucidate the torque-generation mechanism of the flagellar motor. Direct evidence that the bacterial flagellum is a rotary motor is obtained by tethered cell assay (Figure 3a), in which the cell body rotates by tethering the filament to a glass surface [121]. The tethered cell assay is a simple method to measure the rotation of the flagellar motor to give fundamental knowledges on the motor mechanism. However, the maximum speed of tethered cells is limited below 20 Hz, because a cell body (~2 μm in length) is extremely large load against the flagellar motor (~45 nm in diameter). To measure the rotational speeds of the *E. coli* flagellar motor over a wide range of external load, bead assay was developed by the Howard Berg laboratory (Figure 3b) [11,122,123]. A bead is attached to a partially sheared sticky flagellar filament lacking domain D3 of flagellin as a probe, and then the rotation of the bead is recoded by a quadrant photodiode or a high-speed camera with high temporal and special resolutions. Therefore, bead assays enable us to investigate output properties of the flagellar motor over the wide range of external load by changing the bead size and medium viscosity. Viscous drags on a bead (γ_b) and a truncated filament (γ_f) are obtained from the bead diameter and the flagellar morphology (filament length and thickness, and helical pitch and radius), respectively, based on a hydrodynamic theory [124,125], and so motor torque (M) can be estimated by $M = (\gamma_b + \gamma_f) \cdot 2\pi f$, where f is the rotation rate.

Figure 3c shows a schematic diagram of the torque versus speed relationship of the flagellar motor, namely torque-speed curve. The torque-speed curve of the flagellar motor consists of two regimes: a high-load, low-speed regime and a low-load, high-speed regime [11]. As external load is decreased, torque decreases gradually up to a certain speed and then falls rapidly to zero. The rotation rate of the flagellar motor is proportional to IMF over a wide range of external load (Figure 3d) [126,127]. Both deuterium oxide and temperature affect the rotation rate of the *E. coli* motor operating in the low-load, high-speed regime but not in the high-load, low-speed regime (Figure 3d), suggesting that a steep decline of torque seen in the low-load, high-speed regime is limited by the rate of H^+-coupled conformational changes of the MotAB complex [11,123]. Torque at high load is dependent on the number of active stator units in the motor, whereas the maximum motor speed near zero load is independent of the stator number [122,128,129]. However, recent two biophysical analyses have revealed that the maximum speed near zero load increases with an increase in the number of active stator units in the motor [130,131], suggesting that both torque and speed would be proportional to not only IMF but also to the stator number over a wide range of external load.

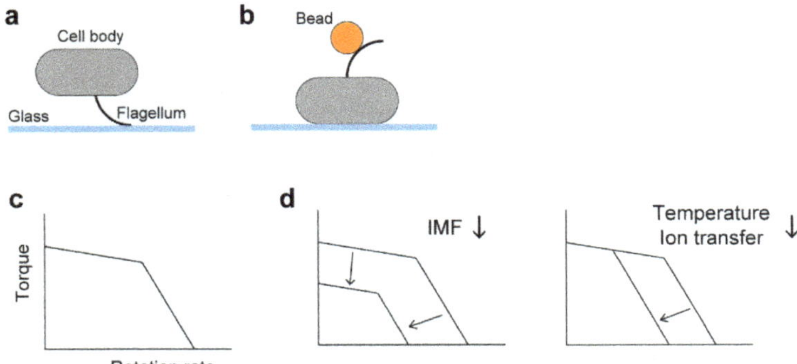

Figure 3. Characterization of the rotation of the flagellar motor. (**a**) Tethered cell assay. (**b**) Bead assay; gold nanoparticles (60–100 nm in diameter) and polystyrene beads (0.2–2.0 μm in diameter) are used. (**c**) A schematic of the torque-speed curve. (**d**) Effects of factors relevant to motor dynamics on the torque-speed curve. Dependence of the curve on the number of stator units is described in the section of *Duty ratio*.

6.3. Stepwise Rotation

Discretely stepwise movements have been observed in many molecular motors. For example, kinesin, which is an ATP-driven linear motor, moves along a microtubule with steps of 8 nm interval [132]; myosin V on an actin filament shows stepwise movements with 36 nm intervals with 90° random rotation either CCW or CW [133]; and F_1-ATPase, which is the ATP-driven rotary motor, shows a 120° step, which is further divided into 80° and 40° substeps [134]. Such stepwise movements reflect the elementary process of mechanochemical energy coupling, e.g., 80° and 40° substeps in F_1-ATPase are coupled with ATP binding and Pi release, respectively, and thus kinetics and dynamics of the step events are important for understanding the motor mechanism. When the flagellar motor labelled with a small bead (diameter: ~100 nm) contains only a single stator unit around a rotor and spins at a few Hz, stepping motions of the motor has been observed. The flagellar motor containing a single stator unit rotates with 26 steps per revolution in both CCW and CW directions [135,136]. Since the number of steps per revolution is consistent with the rotational symmetry of the FliG ring, it is suggested that torque is generated through cyclic association–dissociation of MotA with every FliG subunit along the circumference of the rotor and that such an elementary process is symmetric in CCW and CW rotation. However, it remains unknown how the protonation–deprotonation cycle of Asp-32 of MotB is linked to the cyclic association–dissociation of MotA with FliG.

6.4. Duty Ratio

The duty ratio is defined as a fraction of time that a stator unit is bound to a rotor in the mechanochemical cycle of the flagellar motor. The duty ratio is one of the fundamental properties of molecular motors and is an important parameter for understanding the operation mechanism. The duty ratio of the flagellar motor has been discussed based on the dependency of the rotation rate on the number of active stator units in the motor [122,128–131]. At high load where torque generation against load is a rate limiting step, the rotation rate is proportional to the number of active stator units in the motor regardless of the value of the duty ratio: If the duty ratio is large (~1), the rotation rate is proportional to the sum of the applied torque because multiple stator units work together at the same time; If the duty ratio is small (<< 1), each stator unit works independently and so the probability of torque generation by the motor per a certain period of time is increased with an increase in the number of active stator units in the motor. As a result, the rotational speed of the flagellar motor is proportional to the number of active stator units in the motor. At low load where kinetic processes (e.g., proton translocation and conformational change) are rate limiting steps, the relationship between the rotation rate and the number of active stator units would depend on the duty ratio: If the duty ratio is close to 1, total torque does not affect the rotation rate, and so the rotation rate of the motor does not depend on the number of active stator units in the motor; if the duty ratio is small, the probability of torque generation by stator-rotor interactions is increased with an increment in the stator number. Ryu et al. have shown that the stator number dependence of the rotational speed of the *E. coli* flagellar motor becomes smaller when external load becomes lower [122]. Furthermore, Yuan and Berg have shown that the maximum speed of the *E. coli* motor is independent of the number of active stator units in the motor [128]. Recently, Wang et al. have reported that the maximum speed of the *E. coli* motor near zero load is constant although the number of active stator units varies [129]. These three studies have suggested that the duty ratio of the flagellar motor seems to be large. Assuming that the flagellar motor has a high duty ratio, theoretical studies can reproduce the output properties of the flagellar motor such as a torque-speed curve [137–141]. In contrast, a recent study using a hybrid motor containing both H^+-type and Na^+-type stator units in *E. coli* cells has shown that the maximum speed of the hybrid flagellar motor near zero load varies with the number of active stator units in the motor [130]. This observation is supported by recent observation that the zero-torque speed of the *Salmonella* flagellar motor depends on the number of active stator units in the motor [131]. These suggest that the duty ratio of the flagellar motor operating at low load is smaller than the previous thought. By removing the high duty ratio constraint from the theoretical model, it is also possible to reproduce the stator-number-dependent rotational speed close to zero load. This physical model also predicts that the duty ratio will become larger with increase in the number of active stator units when the motor operates at low load and that a high duty ratio will be required for the motor to processivity generate much larger torque at high load [142]. Thus, the duty ratio of the flagellar motor is currently controversial, and hence further experimental verification over a wide range of external load will be necessary.

7. Switching of Direction of Flagellar Motor Rotation

7.1. Conformational Changes for Reversal of Motor Rotation

E. coli and *Salmonella* cells sense temporal changes in chemical concentrations of attractants and repellents via transmembrane chemoreceptors (methyl-accepting chemotaxis proteins, MCP) localized near the cell pole [143]. The binding of repellent to MCP induces auto-phosphorylation of CheA via the adopter protein CheW, and then CheA-P transfers a phosphate to the response regulator CheY. The binding of the phosphorylated form of CheY (CheY-P) to FliM and FliN induces the structural remodeling of the C ring responsible for the switching of direction of flagellar motor rotation from CCW to CW. The relationship between the switching frequency and CheY-P concentration shows a sigmoid curve with a Hill coefficient of ~10 [144]. This switching Hill coefficient value is larger than the Hill coefficient estimated from the binding affinity of CheY-P for the motor [145,146]. This suggests that CheY-P-dependent structural remodeling of the C ring occurs in a highly cooperative manner.

Since the elementary process of torque generation by stator-rotor interactions is symmetric in CCW and CW rotation, $FliG_C$, which contains highly conserved Arg-281 and Asp-289 residues involved in the interaction with MotA, is postulated to rotate 180° relative to MotA [136]. $FliG_C$ has a highly flexible MFXF motif between $FliG_{CN}$ and $FliG_{CC}$ subdomains and so the MFXF motif allows $FliG_{CC}$ to rotate 180° relative to $FliG_{CN}$ to reorient Arg-281 and Asp-289 residues in $FliG_{CC}$ to achieve the symmetric elementary process of torque generation in both CCW and CW rotations (Figure 4) [147,148].

$Helix_{MC}$ is a helical linker connecting $FliG_M$ and $FliG_N$ and plays an important role in directional switching of the flagellar motor [149]. A deletion of three residues in the N-terminal end of $Helix_{MC}$ (Pro-Ala-Ala, PAA) locks the flagellar motor in the CW state even in the absence of CheY-P [149]. The PAA deletion causes conformational rearrangements of the $FliG_M$–$FliM_M$ interface to induce a detachment of $Helix_{MC}$ from the interface. Furthermore, this PAA deletion induces a 90° rotation of $FliG_{CC}$ relative to $FliG_{CN}$ through the MFXF motif in solution [75,76,149]. This is supported by in vivo site-directed crosslinking experiments [150]. Recent cryoEM image analyses have shown that inter-subunit spacing between C ring proteins are closer in the C ring of the CW motor than in that of the CCW motor [87], suggesting that the binding of CheY-P to FliM and FliN significantly affects inter-molecular interactions between the C ring proteins. Therefore, it is possible that the binding of CheY-P to FliM and FliN changes inter-molecular $FliM_M$–$FliM_M$, $FliM_C$–FliN and $FliG_M$–$FliM_M$ interactions in the C ring to induces the dissociation of $Helix_{MC}$ from the $FliG_M$–$FliM_M$ interface, thereby affecting inter-molecular $FliG_M$–$FliG_{CN}$ interactions to allow $FliG_{CC}$ to rotate 180° relative to $FliG_{CN}$ through a conformational change of the MFXF motif (Figure 4).

In *E. coli* and *Salmonella*, the binding of repellent to MCP elevates the cytoplasmic CheY-P level, thereby increasing the probability that the motor spins in CW direction. In contrast, the chemotaxis signaling pathway and response are known to diverse among bacterial species. In *B. subtilis*, CheY-P acts in the opposite way to induce CCW rotation. The binding of attractant to MCP of *B. subtilis* facilitates phosphorylation of CheY, and CheY-P binds to FliM to switch motor rotation from CW to CCW [151]. *Rhodobacter sphaeroides* possesses six CheY proteins, $CheY_1$ to $CheY_6$. The decreased attractant concentration increases the cytoplasmic $CheY_3$-P, $CheY_4$-P, and $CheY_6$-P levels, and the binding of $CheY_6$-P to FliM stops motor rotation with the support of $CheY_3$-P and $CheY_4$-P [152].

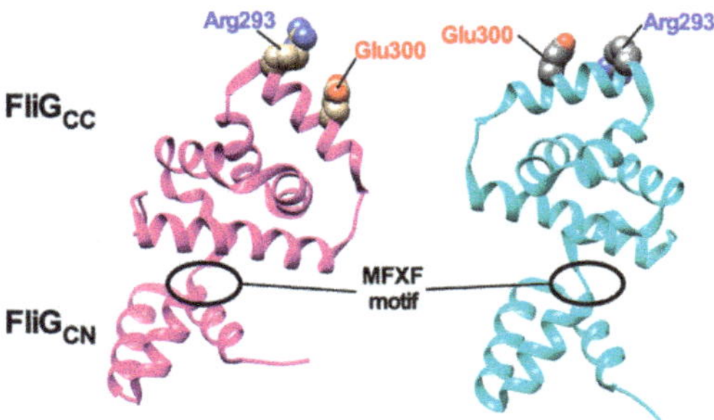

Figure 4. Structural comparisons between 3USY (cyan) and 3USW (magenta) structures of *Helicobacter pylori* FliG. Conformational rearrangements of the conserved MFXF motif induces a 180° rotation of $FliG_{CC}$ relative to $FliG_{CN}$ to reorient Arg-293 and Glu-300 residues, which correspond to Arg-281 and Asp-289 of *E. coli* FliG, respectively.

7.2. Conformational Spread for Cooperative Switching

Cooperative flagellar switching can be reproduced by an Ising-type model assuming allosteric cooperativity of the conformational change in C ring subunits [153]. The model assumes four states for each subunit, determined by whether a subunit conformation is placed in either the CCW or CW state with or without CheY-P bound. Assuming that homogeneous states of adjacent subunits (e.g., CCW-CCW-CCW or CW-CW-CW) are more stable than heterogeneous ones (e.g., CCW-CW-CCW or CW-CCW-CW), the directional switching is mediated by conformational changes in C ring subunits that extend from subunit to subunit via inter-molecular interactions between nearest adjacent subunits (Figure 5) [153]. The model prediction was verified by simultaneous measurements of motor rotation and a turnover of CheY labelled with a green fluorescent protein (GFP) between the motor and the cytoplasmic pool, showing that, in spite of the switch complex contains ~34 FliM subunits, the binding of about 13 CheY-P molecules can reverse the motor [154].

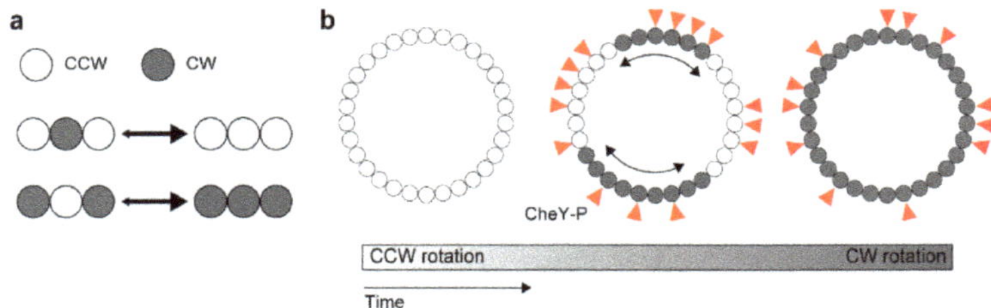

Figure 5. Model for cooperative switching between counterclockwise (CCW) and clockwise (CW) rotations. (**a**) Interaction between adjacent rotor subunits. (**b**) Conformational spread upon CheY-P binding.

The switching rate increases until the motor speed reaches ~150 Hz, and then decreases with further increase in the rotation rate [155,156]. The conformational spread model also explains a speed (load) dependent switching frequency by assuming the effect of mechanical force on the switching rate, which each stator unit applies force on the FliG subunit in the C ring [140]. The conventional Ising-type conformational spread model, which is an equilibrium model sufficient for detailed balance, shows exponentially decayed distributions of the duration time for CCW or CW rotation. Such exponential duration-time distributions have been observed experimentally, suggesting the equilibrium switching system. Recently, Wang et al. have measured the CCW and CW durations at various conditions of load, PMF, and the number of active stators and have shown non-exponential shaped distributions in a torque-dependent manner. The results suggest that the flagellar switch could be a non-equilibrium system rather than an equilibrium system under certain conditions, and that motor torque is a key factor for breaking detailed balance. Furthermore, the directional switching of the flagellar motors working under non-equilibrium conditions (e.g., at high load) can occur at lower CheY-P level compared to those placed under equilibrium conditions, suggesting that the binding affinity of the flagellar motor for CheY-P is enhanced by applied force [157]. Thus, the switching of direction of flagellar motor rotation is controlled not only by the chemotactic signaling pathway but also by the mechanical force [140,157].

8. Stator Assembly

The PGB domains of MotB ($MotB_{PGB}$) and PomB ($PomB_{PGB}$) bind to the PG layer to allow the MotAB and PomAB complexes to become an active stator unit around a rotor [93,158,159]. The N-terminal portions of $MotB_{PGB}$ and $PomB_{PGB}$ adopt a compact conformation in their crystal structure, but are structurally flexible to allow them to adopt an extended conformation as well (Figure 6). Structure-based mutational analyses of $MotB_{PGB}$ and $PomB_{PGB}$ have suggested that a 5 nm extension of the PGB domain from the transmembrane ion channel is required for the binding of $MotB_{PGB}$ and $PomB_{PGB}$ to the PG layer (Figure 6) [158,160]. Recently, such a 5 nm extension process of the PGB domain of MotS ($MotS_{PGB}$) of *B. subtilis* has been directly visualized by high-speed atomic force microscopy [93]. The 5 nm extension of $MotS_{PGB}$ is divided into at least two steps [93]. The first 2.5 nm extension step is caused by a detachment of a flexible linker connecting $MotS_{PGB}$ with MotS-TM from the transmembrane Na^+ channel of the MotPS complex, and the second 2.5 nm extension step results from an order-to-disorder transition of the N-terminal portion of $MotS_{PGB}$. Consistently, the *motB(L119P)* mutation in $MotB_{PGB}$ induces an extended conformation of the N-terminal portion of $MotB_{PGB}$ [159].

Interestingly, the *motB(L119P)* mutation increases not only the PGB binding activity of $MotB_{PGB}$ [159] but also the proton channel activity of the MotAB complex [107]. Therefore, it seems likely that proper positioning of an inactive MotAB complex around the rotor via stator–rotor interactions triggers a detachment of the flexible linker from the H^+ channel, followed by a structural transition of the N-terminal portion of $MotB_{PGB}$ from the compact to extended forms to become an active stator unit in the motor [159] (Figure 6).

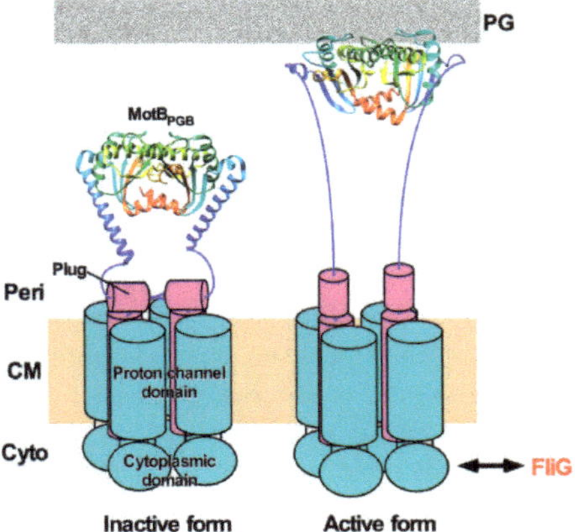

Figure 6. Activation mechanism of the H^+-type MotAB complex. The MotAB complex consists of at least three structural parts: a cytoplasmic domain, a transmembrane ion channel and a peptidoglycan-binding domain [$MotB_{PGB}$, PDB codes: 2ZVY (left panel) and 5Y40 (right panel). When the MotAB complex adopts a compact conformation, a plug segment of MotB binds to a transmembrane H^+ channel to suppress massive H^+ flow (left). When the MotAB complex encounters a rotor, electrostatic interactions between the cytoplasmic domain of MotA and FliG trigger the dissociation of the plug segment from the channel, followed by partial unfolding of the N-terminal portion of $MotB_{PGB}$ to allow $MotB_{PGB}$ to bind to the peptidoglycan (PG) layer. As a result, the MotAB complex becomes an active H^+-type stator unit to drive flagellar motor rotation (right).

The flagellar motor can accommodate about 10 stator units around a rotor in *E. coli* and *Salmonella* when the motor operates at high load [161]. High-resolution single molecule imaging techniques have revealed exchanges of the MotAB complex labelled with GFP between the basal body and the membrane pool during rotation at a rate constant of 0.04 s^{-1}, indicating that the dual time of a given stator unit is about 0.5 min. This suggests that the interaction of $MotB_{PGB}$ with the PG layer is highly dynamic, thereby allowing the MotAB complex to alternate in attachment to and detachment from the motor during motor rotation [162]. Interestingly, when external load becomes low enough, only a few stator units work around the rotor to drive motor rotation [163–165]. This suggests that such a dynamic assembly–disassembly process of the stator complex occurs in a load-dependent manner.

The number of active stator units can be estimated by resurrection experiments, in which time traces of the rotational speed of a single flagellar motor usually show stepwise speed increments and decrements. Each increment reflects the incorporation of a single MotAB complex around the rotor to become an active stator unit in the motor whereas each decrement unit reflects the disassembly of the MotAB stator complex from the rotor [163–165]. A deletion of a flexible linker connecting MotB-TM and $MotB_{PGB}$ results in a rapid decrease in the number of active stator units in the motor compared to the wild-type motor, suggesting that this flexible linker of MotB modulates the binding affinity of $MotB_{PGB}$ for the PG layer in a load-dependent manner [166]. Certain mutations in the cytoplasmic loop of MotA, which interacts with FliG, significantly affect the mechano-sensitivity of the MotAB complex, thereby causing distinct load-dependent assembly and disassembly dynamics compared to the wild-type. This suggests that the cytoplasmic loop of MotA may sense a change in external load through the interaction with FliG to control the number of active stator units around the rotor [86].

How does the cytoplasmic loop of MotA transmit the mechanical signal to $MotB_{PGB}$ associated with the PG layer to coordinate the number of active stator units in the motor in response to changes in external load? Nord et al. have reported that the dissociation rate of the MotAB stator complex becomes slower with an increase in applied force, thereby increasing the bound lifetime of each active stator unit incorporated into the motor, and that an abrupt relief from the stall makes the dissociation rate much faster, thereby decreasing the bound lifetime [130]. As a result, the average number of active stator units in the motor is maintained about 10 in the high-load, low speed regime whereas the stator number is decreased from 10 to a few when external load becomes quite low. Recently, it has been shown that a turnover process of the stator unit is divided into two distinct, slow (the rate constant of ~0.008 s^{-1}) and fast (~0.2 s^{-1}) steps [167]. Although the slow step called "hidden state" is not yet clarified, the fast step is assumed to reflect a rapid conformational change of $MotB_{PGB}$ to become an active stator unit.

The assembly and disassembly dynamics of the stator complex are also affected by changes in the extracellular ion concentration [93,101,118,168,169]. The Na^+-coupled PomAB and MotPS complexes can be assembled into a motor when the external Na^+ concentration is high enough [93,101,168]. How do the PomAB and MotPS complex sense external Na^+? High-speed atomic force microscopy with high special and temporal resolutions has revealed that $MotS_{PGB}$ adopts a folded conformation in the presence of 150 mM NaCl, but becomes denatured when the external Na^+ concentration is less than 150 mM NaCl. These direct observations suggest that $MotS_{PGB}$ functions as a Na^+ sensor to efficiently promote the assembly and disassembly of the MotPS complex with the motor in response to changes in external Na^+ concentration [93].

9. Conclusions and Perspectives

The flagellum of *E. coli* and *Salmonella* is a supramolecular rotary motor powered by an inward-directed H^+ translocation through a transmembrane H^+ channel of the MotAB stator complex and can spin in both CCW and CW directions without changing the direction of H^+ flow. The flagellar motor is conserved among bacterial species, but the flagellar structure has adopted to function in various environments of the habitant of bacteria [9,10]. The structure of the rod, hook and filament and their mechanical properties are understood at near atomic resolution. Because structural information on the rotor and stator is still limited, it remains unknown how the transmembrane stator complex conducts ions and exerts force on the rotor, how the rotor switches between the CCW and CW states in a highly cooperative manner, and how the stator complex senses external ion concentration to become an active stator unit around the rotor. To clarify these remaining questions, high-resolution structural analyses of the rotor and stator would be required.

The elementary process of the flagellar motor is visualized to be composed of a step and a dwell [135,136]. Since the dissociation rate of the stator unit becomes much faster at low load than at high load, the number of active stator units in the motor is decreased from 10 to a few when external loads become low enough [130]. Although the duty ratio of the flagellar motor seems to be small, the flagellar motor containing only a few stator units can processively generate torque for high-speed rotation near zero load. However, it remains unknown how the H^+ translocation process is linked to a torque generation step by stator-rotor interactions and how cyclic association–dissociation of MotA with every FliG subunit along the circumference of the rotor allow the motor to spin at about 300 revolutions per second in a highly processive manner. Much more precise measurements of the rotational speed of the flagellar motor near zero load would be essential to advance our mechanistic understanding of the energy coupling mechanism of the flagellar motor.

Funding: This work was supported in part by Grants-in-Aid for Scientific Research from the Japan Society for the Promotion of Science (JSPS KAKENHI Grant Numbers 18K07100 to S.N. and 19H03182 to T.M.).

Acknowledgments: We acknowledge Keiichi Namba, Seishi Kudo and Yusuke V. Morimoto for continuous support and encouragement.

Conflicts of Interest: The authors declare no conflict of interest.

References

1. Josenhans, C.; Suerbaum, S. The role of motility as a virulence factor in bacteria. *Int. J. Med. Microbiol.* **2002**, *291*, 605–614. [CrossRef] [PubMed]
2. Lertsethtakarn, P.; Ottemann, K.M.; Hendrixson, D.R. Motility and chemotaxis in *Campylobacter* and *Helicobacter*. *Annu. Rev. Microbiol.* **2011**, *65*, 389–410. [CrossRef] [PubMed]
3. Visick, K.L. An intricate network of regulators controls biofilm formation and colonization by *Vibrio fischeri*. *Mol. Microbiol.* **2009**, *74*, 782–789. [CrossRef] [PubMed]
4. Scharf, B.E.; Hynes, M.F.; Alexandre, G.M. Chemotaxis signaling systems in model beneficial plant–bacteria associations. *Plant Mol. Biol.* **2016**, *90*, 549–559. [CrossRef] [PubMed]
5. Ishikawa, T.; Yoshida, N.; Ueno, H.; Wiedeman, M.; Imai, Y.; Yamaguchi, T. Energy transport in a concentrated suspension of bacteria. *Phys. Rev. Lett.* **2011**, *107*, 028102. [CrossRef] [PubMed]
6. Jarrell, K.F.; McBride, M.J. The surprisingly diverse ways that prokaryotes move. *Nat. Rev. Microbiol.* **2008**, *6*, 466–476. [CrossRef] [PubMed]
7. Charon, N.W.; Goldstein, S.F. Genetics of motility and chemotaxis of a fascinating group of bacteria: The spirochetes. *Annu. Rev. Genet.* **2002**, *36*, 47–73. [CrossRef]
8. Minamino, T.; Namba, K. Self-assembly and type III protein export of the bacterial flagellum. *J. Mol. Microbiol. Biotechnol.* **2004**, *7*, 5–17. [CrossRef]
9. Terashima, H.; Kawamoto, A.; Morimoto, Y.V.; Imada, K.; Minamino, T. Structural differences in the bacterial flagellar motor among bacterial species. *Biophys. Physicobiology* **2017**, *14*, 191–198. [CrossRef]
10. Minamino, T.; Imada, K. The bacterial flagellar motor and its structural diversity. *Trends Microbiol.* **2015**, *23*, 267–274. [CrossRef]
11. Chen, X.; Berg, H.C. Torque-speed relationship of the flagellar rotary motor of *Escherichia coli*. *Biophys. J.* **2000**, *78*, 1036–1041. [CrossRef]
12. Magariyama, Y.; Sugiyama, S.; Muramoto, K.; Maekawa, Y.; Kawagishi, I.; Imae, Y.; Kudo, S. Very fast flagellar rotation. *Nature* **1994**, *371*, 752. [CrossRef] [PubMed]
13. Morimoto, Y.V.; Minamino, T. Structure and function of the bi-directional bacterial flagellar motor. *Biomolecules* **2014**, *4*, 217–234. [CrossRef] [PubMed]
14. Minamino, T.; Terahara, N.; Kojima, S.; Namba, K. Autonomous control mechanism of stator assembly in the bacterial flagellar motor in response to changes in the environment. *Mol. Microbiol.* **2018**, *109*, 723–734. [CrossRef] [PubMed]
15. Berg, H.C. The rotary motor of bacterial flagella. *Annu. Rev. Biochem.* **2003**, *72*, 19–54. [CrossRef] [PubMed]
16. Namba, K.; Vonderviszt, F. Molecular architecture of bacterial flagellum. *Q. Rev. Biophys.* **1997**, *30*, 1–65. [CrossRef]
17. Bonifield, H.R.; Hughes, K.T. Flagellar phase variation in *Salmonella enterica* is mediated by a posttranscriptional control mechanism. *J. Bacteriol.* **2003**, *185*, 3567–3574. [CrossRef]
18. Samatey, F.A.; Imada, K.; Nagashima, S.; Vonderviszt, F.; Kumasaka, T.; Yamamoto, M.; Namba, K. Structure of the bacterial flagellar protofilament and implications for a switch for supercoiling. *Nature* **2001**, *410*, 331–337. [CrossRef]
19. Yonekura, K.; Maki-Yonekura, S.; Namba, K. Complete atomic model of the bacterial flagellar filament by electron cryomicroscopy. *Nature* **2003**, *424*, 643–650. [CrossRef]
20. Maki-Yonekura, S.; Yonekura, K.; Namba, K. Conformational change of flagellin for polymorphic supercoiling of the flagellar filament. *Nat. Struct. Mol. Biol.* **2010**, *17*, 417–422. [CrossRef]
21. Beatson, S.A.; Minamino, T.; Pallen, M.J. Variation in bacterial flagellins: from sequence to structure. *Trends Microbiol.* **2006**, *14*, 151–155. [CrossRef] [PubMed]
22. Asakura, S. Polymerization of flagellin and polymorphism of flagella. *Adv. Biophys.* **1970**, *1*, 99–155. [PubMed]

23. Kamiya, R.; Asakura, S.; Wakabayashi, K.; Namba, K. Transition of bacterial flagella from helical to straight forms with different subunit arrangements. *J. Mol. Biol.* **1979**, *131*, 725–742. [CrossRef]
24. Yamashita, I.; Hasegawa, K.; Suzuki, H.; Vonderviszt, F.; Mimori-Kiyosue, Y.; Namba, K. Structure and switching of bacterial flagellar filaments studied by X-ray fiber diffraction. *Nat. Struct. Biol.* **1998**, *5*, 125–132. [CrossRef] [PubMed]
25. Wang, F.; Burrage, A.M.; Postel, S.; Clark, R.E.; Orlova, A.; Sundberg, E.J.; Kearns, D.B.; Egelman, E.H. A structural model of flagellar filament switching across multiple bacterial species. *Nat. Commun.* **2017**, *8*, 960. [CrossRef] [PubMed]
26. Faulds-Pain, A.; Birchall, C.; Aldridge, C.; Smith, W.D.; Grimaldi, G.; Nakamura, S.; Miyata, T.; Gray, J.; Li, G.; Tang, J.X.; et al. Flagellin redundancy in *Caulobacter crescentus* and its implications for flagellar filament assembly. *J. Bacteriol.* **2011**, *193*, 2695–2707. [CrossRef] [PubMed]
27. Scharf, B.; Schuster-Wolff-Bühring, H.; Rachel, R.; Schmitt, R. Mutational Analysis of the Rhizobium lupini H13-3 andSinorhizobium meliloti Flagellin Genes: Importance of Flagellin A for Flagellar Filament Structure and Transcriptional Regulation. *J. Bacteriol.* **2001**, *183*, 5334–5342. [CrossRef] [PubMed]
28. Tambalo, D.D.; Bustard, D.E.; Del Bel, K.L.; Koval, S.F.; Khan, M.F.; Hynes, M.F. Characterization and functional analysis of seven flagellin genes in *Rhizobium leguminosarum* bv. viciae. Characterization of R. leguminosarum flagellins. *BMC Microbiol.* **2010**, *10*, 219. [CrossRef]
29. Mohari, B.; Thompson, M.A.; Trinidad, J.C.; Setayeshgar, S.; Fuqua, C. Multiple Flagellin Proteins Have Distinct and Synergistic Roles in Agrobacterium tumefaciens Motility. *J. Bacteriol.* **2018**, *200*, e00327-18. [CrossRef]
30. Quelas, J.I.; Althabegoiti, M.J.; Jimenez-Sanchez, C.; Melgarejo, A.A.; Marconi, V.I.; Mongiardini, E.J.; Trejo, S.A.; Mengucci, F.; Ortega-Calvo, J.-J.; Lodeiro, A.R. Swimming performance of *Bradyrhizobium diazoefficiens* is an emergent property of its two flagellar systems. *Sci. Rep.* **2016**, *6*, 23841. [CrossRef]
31. Inoue, T.; Barker, C.S.; Matsunami, H.; Aizawa, S.-I.; Samatey, F.A. The FlaG regulator is involved in length control of the polar flagella of *Campylobacter jejuni*. *Microbiology* **2018**, *164*, 740–750. [CrossRef] [PubMed]
32. Kühn, M.J.; Schmidt, F.K.; Farthing, N.E.; Rossmann, F.M.; Helm, B.; Wilson, L.G.; Eckhardt, B.; Thormann, K.M. Spatial arrangement of several flagellins within bacterial flagella improves motility in different environments. *Nat. Commun.* **2018**, *9*, 5369. [CrossRef] [PubMed]
33. Lambert, C.; Evans, K.J.; Till, R.; Hobley, L.; Capeness, M.; Rendulic, S.; Schuster, S.C.; Aizawa, S.-I.; Sockett, R.E. Characterizing the flagellar filament and the role of motility in bacterial prey-penetration by *Bdellovibrio bacteriovorus*. *Mol. Microbiol.* **2006**, *60*, 274–286. [CrossRef] [PubMed]
34. Kim, S.Y.; Thanh, X.T.T.; Jeong, K.; Kim, S.B.; Pan, S.O.; Jung, C.H.; Hong, S.H.; Lee, S.E.; Rhee, J.H. Contribution of six flagellin genes to the flagellum biogenesis of *Vibrio vulnificus* and in vivo invasion. *Infect. Immun.* **2014**, *82*, 29–42. [CrossRef] [PubMed]
35. Ikeda, J.S.; Schmitt, C.K.; Darnell, S.C.; Watson, P.R.; Bispham, J.; Wallis, T.S.; Weinstein, D.L.; Metcalf, E.S.; Adams, P.; O'Connor, C.D.; et al. Flagellar phase variation of *Salmonella enterica* serovar Typhimurium contributes to virulence in the murine typhoid infection model but does not influence *Salmonella*-induced enteropathogenesis. *Infect. Immun.* **2001**, *69*, 3021–3030. [CrossRef] [PubMed]
36. Samatey, F.A.; Matsunami, H.; Imada, K.; Nagashima, S.; Shaikh, T.R.; Thomas, D.R.; Chen, J.Z.; Derosier, D.J.; Kitao, A.; Namba, K. Structure of the bacterial flagellar hook and implication for the molecular universal joint mechanism. *Nature* **2004**, *431*, 1062–1068. [CrossRef]
37. Fujii, T.; Kato, T.; Namba, K. Specific arrangement of α-helical coiled coils in the core domain of the bacterial flagellar hook for the universal joint function. *Structure* **2009**, *17*, 1485–1493. [CrossRef]
38. Sakai, T.; Inoue, Y.; Terahara, N.; Namba, K.; Minamino, T. A triangular loop of domain D1 of FlgE is essential for hook assembly but not for the mechanical function. *Biochem. Biophys. Res. Commun.* **2018**, *495*, 1789–1794. [CrossRef]
39. Fujii, T.; Matsunami, H.; Inoue, Y.; Namba, K. Evidence for the hook supercoiling mechanism of the bacterial flagellum. *Biophys. Physicobiology* **2018**, *15*, 28–32. [CrossRef]
40. Matsunami, H.; Barker, C.S.; Yoon, Y.-H.; Wolf, M.; Samatey, F.A. Complete structure of the bacterial flagellar hook reveals extensive set of stabilizing interactions. *Nat. Commun.* **2016**, *7*, 13425. [CrossRef]
41. Brown, M.T.; Steel, B.C.; Silvestrin, C.; Wilkinson, D.A.; Delalez, N.J.; Lumb, C.N.; Obara, B.; Armitage, J.P.; Berry, R.M. Flagellar hook flexibility is essential for bundle formation in swimming *Escherichia coli* cells. *J. Bacteriol.* **2012**, *194*, 3495–3501. [CrossRef] [PubMed]

42. Hiraoka, K.D.; Morimoto, Y.V.; Inoue, Y.; Fujii, T.; Miyata, T.; Makino, F.; Minamino, T.; Namba, K. Straight and rigid flagellar hook made by insertion of the FlgG specific sequence into FlgE. *Sci. Rep.* **2017**, *7*, 46723. [CrossRef] [PubMed]
43. Spöring, I.; Martinez, V.A.; Hotz, C.; Schwarz-Linek, J.; Grady, K.L.; Nava-Sedeño, J.M.; Vissers, T.; Singer, H.M.; Rohde, M.; Bourquin, C.; et al. Hook length of the bacterial flagellum is optimized for maximal stability of the flagellar bundle. *PLoS Biol.* **2018**, *16*, e2006989. [CrossRef] [PubMed]
44. Minamino, T. Hierarchical protein export mechanism of the bacterial flagellar type III protein export apparatus. *FEMS Microbiol. Lett.* **2018**, *365*, fny117. [CrossRef] [PubMed]
45. Son, K.; Guasto, J.S.; Stocker, R. Bacteria can exploit a flagellar buckling instability to change direction. *Nat. Phys.* **2013**, *9*, 494–498. [CrossRef]
46. Homma, M.; Kutsukake, K.; Hasebe, M.; Iino, T.; Macnab, R.M. FlgB, FlgC, FlgF and FlgG: A family of structurally related proteins in the flagellar basal body of *Salmonella typhimurium*. *J. Mol. Biol.* **1990**, *211*, 465–477. [CrossRef]
47. Kubori, T.; Shimamoto, N.; Yamaguchi, S.; Namba, K.; Aizawa, S.-I. Morphological pathway of flagellar assembly in *Salmonella typhimurium*. *J. Mol. Biol.* **1992**, *226*, 433–446. [CrossRef]
48. Minamino, T.; Yamaguchi, S.; Macnab, R.M. Interaction between FliE and FlgB, a proximal rod component of the flagellar basal body of *Salmonella*. *J. Bacteriol.* **2000**, *182*, 3029–3036. [CrossRef]
49. Fujii, T.; Kato, T.; Hiraoka, K.D.; Miyata, T.; Minamino, T.; Chevance, F.F.V.; Hughes, K.T.; Namba, K. Identical folds used for distinct mechanical functions of the bacterial flagellar rod and hook. *Nat. Commun.* **2017**, *8*, 14276. [CrossRef]
50. Minamino, T. Protein export through the bacterial flagellar type III export pathway. *Biochim. Biophys. Acta* **2014**, *1843*, 1642–1648. [CrossRef]
51. Minamino, T.; Macnab, R.M. Components of the *Salmonella* flagellar export apparatus and classification of export substrates. *J. Bacteriol.* **1999**, *181*, 1388–1394. [PubMed]
52. Fukumura, T.; Makino, F.; Dietsche, T.; Kinoshita, M.; Kato, T.; Wagner, S.; Namba, K.; Imada, K.; Minamino, T. Assembly and stoichiometry of the core structure of the bacterial flagellar type III export gate complex. *PLoS Biol.* **2017**, *15*, e2002281. [CrossRef] [PubMed]
53. Minamino, T.; Macnab, R.M. Interactions among components of the *Salmonella flagellar* export apparatus and its substrates. *Mol. Microbiol.* **2000**, *35*, 1052–1064. [CrossRef] [PubMed]
54. Minamino, T.; Morimoto, Y.V.; Hara, N.; Namba, K. An energy transduction mechanism used in bacterial flagellar type III protein export. *Nat. Commun.* **2011**, *2*, 475. [CrossRef] [PubMed]
55. Morimoto, Y.V.; Kami-ike, N.; Miyata, T.; Kawamoto, A.; Kato, T.; Namba, K.; Minamino, T. High-resolution pH imaging of living bacterial cells to detect local pH differences. *mBio* **2016**, *7*, e01911-16. [CrossRef] [PubMed]
56. Fabiani, F.D.; Renault, T.T.; Peters, B.; Dietsche, T.; Gálvez, E.J.C.; Guse, A.; Freier, K.; Charpentier, E.; Strowig, T.; Franz-Wachtel, M.; et al. A flagellum-specific chaperone facilitates assembly of the core type III export apparatus of the bacterial flagellum. *PLoS Biol.* **2017**, *15*, e2002267. [CrossRef] [PubMed]
57. Kuhlen, L.; Abrusci, P.; Johnson, S.; Gault, J.; Deme, J.; Caesar, J.; Dietsche, T.; Mebrhatu, M.T.; Ganief, T.; Macek, B.; et al. Structure of the core of the type III secretion system export apparatus. *Nat. Struct. Mol. Biol.* **2018**, *25*, 583–590. [CrossRef] [PubMed]
58. Dietsche, T.; Tesfazgi Mebrhatu, M.; Brunner, M.J.; Abrusci, P.; Yan, J.; Franz-Wachtel, M.; Schärfe, C.; Zilkenat, S.; Grin, I.; Galán, J.E.; et al. Structural and functional characterization of the bacterial type III secretion export apparatus. *PLoS Pathog.* **2016**, *12*, e1006071. [CrossRef]
59. Abrusci, P.; Vergara-Irigaray, M.; Johnson, S.; Beeby, M.D.; Hendrixson, D.R.; Roversi, P.; Friede, M.E.; Deane, J.E.; Jensen, G.J.; Tang, C.M.; et al. Architecture of the major component of the type III secretion system export apparatus. *Nat. Struct. Mol. Biol.* **2013**, *20*, 99–104. [CrossRef]
60. Kawamoto, A.; Morimoto, Y.V.; Miyata, T.; Minamino, T.; Hughes, K.T.; Kato, T.; Namba, K. Common and distinct structural features of *Salmonella* injectisome and flagellar basal body. *Sci. Rep.* **2013**, *3*, 3369. [CrossRef]
61. Terahara, N.; Inoue, Y.; Kodera, N.; Morimoto, Y.V.; Uchihashi, T.; Imada, K.; Ando, T.; Namba, K.; Minamino, T. Insight into structural remodeling of the FlhA ring responsible for bacterial flagellar type III protein export. *Sci. Adv.* **2018**, *4*, eaao7054. [CrossRef] [PubMed]

62. Minamino, T.; Morimoto, Y.V.; Hara, N.; Aldridge, P.D.; Namba, K. The bacterial flagellar type III export gate complex is a dual fuel engine that can use both H^+ and Na^+ for flagellar protein export. *PLoS Pathog.* **2016**, *12*, e1005495. [CrossRef] [PubMed]
63. Imada, K.; Minamino, T.; Tahara, A.; Namba, K. Structural similarity between the flagellar type III ATPase FliI and F1-ATPase subunits. *Proc. Natl. Acad. Sci. USA* **2007**, *104*, 485–490. [CrossRef] [PubMed]
64. Ibuki, T.; Imada, K.; Minamino, T.; Kato, T.; Miyata, T.; Namba, K. Common architecture of the flagellar type III protein export apparatus and F- and V-type ATPases. *Nat. Struct. Mol. Biol.* **2011**, *18*, 277–282. [CrossRef] [PubMed]
65. Imada, K.; Minamino, T.; Uchida, Y.; Kinoshita, M.; Namba, K. Insight into the flagella type III export revealed by the complex structure of the type III ATPase and its regulator. *Proc. Natl. Acad. Sci. USA* **2016**, *113*, 3633–3638. [CrossRef] [PubMed]
66. Gonzalez-Pedrajo, B.; Minamino, T.; Kihara, M.; Namba, K. Interactions between C ring proteins and export apparatus components: A possible mechanism for facilitating type III protein export. *Mol. Microbiol.* **2006**, *60*, 984–998. [CrossRef] [PubMed]
67. Minamino, T.; Yoshimura, S.D.J.; Morimoto, Y.V.; GonzÃ¡lez-Pedrajo, B.; Kami-ike, N.; Namba, K. Roles of the extreme N-terminal region of FliH for efficient localization of the FliH–FliI complex to the bacterial flagellar type III export apparatus: Localization of *Salmonella* FliH–FliI complex. *Mol. Microbiol.* **2009**, *74*, 1471–1483. [CrossRef] [PubMed]
68. Hara, N.; Morimoto, Y.V.; Kawamoto, A.; Namba, K.; Minamino, T. Interaction of the extreme N-terminal region of FliH with FlhA is required for efficient bacterial flagellar protein export. *J. Bacteriol.* **2012**, *194*, 5353–5360. [CrossRef]
69. Bai, F.; Morimoto, Y.V.; Yoshimura, S.D.J.; Hara, N.; Kami-ike, N.; Namba, K.; Minamino, T. Assembly dynamics and the roles of FliI ATPase of the bacterial flagellar export apparatus. *Sci. Rep.* **2014**, *4*, 6528. [CrossRef]
70. Minamino, T.; Morimoto, Y.V.; Kinoshita, M.; Aldridge, P.D.; Namba, K. The bacterial flagellar protein export apparatus processively transports flagellar proteins even with extremely infrequent ATP hydrolysis. *Sci. Rep.* **2014**, *4*, 7579. [CrossRef]
71. Minamino, T.; Imada, K.; Namba, K. Molecular motors of the bacterial flagella. *Curr. Opin. Struct. Biol.* **2008**, *18*, 693–701. [CrossRef] [PubMed]
72. Lynch, M.J.; Levenson, R.; Kim, E.A.; Sircar, R.; Blair, D.F.; Dahlquist, F.W.; Crane, B.R. Co-folding of a FliF-FliG split domain forms the basis of the MS:C ring interface within the bacterial flagellar motor. *Structure* **2017**, *25*, 317–328. [CrossRef] [PubMed]
73. Baker, M.A.B.; Hynson, R.M.G.; Ganuelas, L.A.; Mohammadi, N.S.; Liew, C.W.; Rey, A.A.; Duff, A.P.; Whitten, A.E.; Jeffries, C.M.; Delalez, N.J.; et al. Domain-swap polymerization drives the self-assembly of the bacterial flagellar motor. *Nat. Struct. Mol. Biol.* **2016**, *23*, 197–203. [CrossRef] [PubMed]
74. Kim, E.A.; Panushka, J.; Meyer, T.; Ide, N.; Carlisle, R.; Baker, S.; Blair, D.F. Biogenesis of the flagellar switch complex in *Escherichia coli*: Formation of sub-complexes independently of the basal-body MS-ring. *J. Mol. Biol.* **2017**, *429*, 2353–2359. [CrossRef] [PubMed]
75. Kinoshita, M.; Furukawa, Y.; Uchiyama, S.; Imada, K.; Namba, K.; Minamino, T. Insight into adaptive remodeling of the rotor ring complex of the bacterial flagellar motor. *Biochem. Biophys. Res. Commun.* **2018**, *496*, 12–17. [CrossRef] [PubMed]
76. Kinoshita, M.; Namba, K.; Minamino, T. Effect of a clockwise-locked deletion in FliG on the FliG ring structure of the bacterial flagellar motor. *Genes Cells* **2018**, *23*, 241–247. [CrossRef] [PubMed]
77. Zhou, J.; Lloyd, S.A.; Blair, D.F. Electrostatic interactions between rotor and stator in the bacterial flagellar motor. *Proc. Natl. Acad. Sci. USA* **1998**, *95*, 6436–6441. [CrossRef]
78. Morimoto, Y.V.; Nakamura, S.; Kami-ike, N.; Namba, K.; Minamino, T. Charged residues in the cytoplasmic loop of MotA are required for stator assembly into the bacterial flagellar motor. *Mol. Microbiol.* **2010**, *78*, 1117–1129. [CrossRef]
79. Morimoto, Y.V.; Nakamura, S.; Hiraoka, K.D.; Namba, K.; Minamino, T. Distinct roles of highly conserved charged residues at the MotA-FliG interface in bacterial flagellar motor rotation. *J. Bacteriol.* **2013**, *195*, 474–481. [CrossRef]
80. Paul, K.; Gonzalez-Bonet, G.; Bilwes, A.M.; Crane, B.R.; Blair, D. Architecture of the flagellar rotor. *EMBO J.* **2011**, *30*, 2962–2971. [CrossRef]

81. McDowell, M.A.; Marcoux, J.; McVicker, G.; Johnson, S.; Fong, Y.H.; Stevens, R.; Bowman, L.A.H.; Degiacomi, M.T.; Yan, J.; Wise, A.; et al. Characterisation of *Shigella* Spa33 and *Thermotoga* FliM/N reveals a new model for C-ring assembly in T3SS: Uniform C-ring assembly by NF- and flagellar-T3SS. *Mol. Microbiol.* **2016**, *99*, 749–766. [CrossRef] [PubMed]
82. Ward, E.; Kim, E.A.; Panushka, J.; Botelho, T.; Meyer, T.; Kearns, D.B.; Ordal, G.; Blair, D.F. Organization of the flagellar switch complex of *Bacillus subtilis*. *J. Bacteriol.* **2019**, *201*, e00626-18. [CrossRef] [PubMed]
83. Delalez, N.J.; Wadhams, G.H.; Rosser, G.; Xue, Q.; Brown, M.T.; Dobbie, I.M.; Berry, R.M.; Leake, M.C.; Armitage, J.P. Signal-dependent turnover of the bacterial flagellar switch protein FliM. *Proc. Natl. Acad. Sci. USA* **2010**, *107*, 11347–11351. [CrossRef] [PubMed]
84. Delalez, N.J.; Berry, R.M.; Armitage, J.P. Stoichiometry and turnover of the bacterial flagellar switch protein FliN. *mBio* **2014**, *5*, e01216-14. [CrossRef] [PubMed]
85. Branch, R.W.; Sayegh, M.N.; Shen, C.; Nathan, V.S.J.; Berg, H.C. Adaptive remodelling by FliN in the bacterial rotary motor. *J. Mol. Biol.* **2014**, *426*, 3314–3324. [CrossRef] [PubMed]
86. Pourjaberi, S.N.S.; Terahara, N.; Namba, K.; Minamino, T. The role of a cytoplasmic loop of MotA in load-dependent assembly and disassembly dynamics of the MotA/B stator complex in the bacterial flagellar motor. *Mol. Microbiol.* **2017**, *106*, 646–658. [CrossRef] [PubMed]
87. Sakai, T.; Miyata, T.; Terahara, N.; Mori, K.; Inoue, Y.; Morimoto, Y.V.; Kato, T.; Namba, K.; Minamino, T. Novel insights into conformational rearrangements of the bacterial flagellar switch complex. *mBio* **2019**, *10*, e00079-19. [CrossRef] [PubMed]
88. Beeby, M.; Ribardo, D.A.; Brennan, C.A.; Ruby, E.G.; Jensen, G.J.; Hendrixson, D.R. Diverse high-torque bacterial flagellar motors assemble wider stator rings using a conserved protein scaffold. *Proc. Natl. Acad. Sci. USA* **2016**, *113*, E1917–E1926. [CrossRef] [PubMed]
89. Kojima, S.; Blair, D.F. Solubilization and purification of the MotA/MotB complex of *Escherichia coli*. *Biochemistry* **2004**, *43*, 26–34. [CrossRef] [PubMed]
90. Braun, T.F.; Al-Mawsawi, L.Q.; Kojima, S.; Blair, D.F. Arrangement of core membrane segments in the MotA/MotB proton-channel complex of *Escherichia coli*. *Biochemistry* **2004**, *43*, 35–45. [CrossRef] [PubMed]
91. Sato, K.; Homma, M. Multimeric structure of PomA, a component of the Na^+ -driven polar flagellar motor of *Vibrio alginolyticus*. *J. Biol. Chem.* **2000**, *275*, 20223–20228. [CrossRef] [PubMed]
92. Ito, M.; Hicks, D.B.; Henkin, T.M.; Guffanti, A.A.; Powers, B.D.; Zvi, L.; Uematsu, K.; Krulwich, T.A. MotPS is the stator-force generator for motility of alkaliphilic *Bacillus*, and its homologue is a second functional Mot in *Bacillus subtilis*: Alkaliphile MotPS and its *B. subtilis* homologue. *Mol. Microbiol.* **2004**, *53*, 1035–1049. [CrossRef] [PubMed]
93. Terahara, N.; Kodera, N.; Uchihashi, T.; Ando, T.; Namba, K.; Minamino, T. Na^+-induced structural transition of MotPS for stator assembly of the *Bacillus* flagellar motor. *Sci. Adv.* **2017**, *3*, eaao4119. [CrossRef] [PubMed]
94. Eggenhofer, E.; Haslbeck, M.; Scharf, B. MotE serves as a new chaperone specific for the periplasmic motility protein, MotC, in *Sinorhizobium meliloti*. *Mol. Microbiol.* **2004**, *52*, 701–712. [CrossRef] [PubMed]
95. Terashima, H.; Fukuoka, H.; Yakushi, T.; Kojima, S.; Homma, M. The *Vibrio* motor proteins, MotX and MotY, are associated with the basal body of Na^+-driven flagella and required for stator formation. *Mol. Microbiol.* **2006**, *62*, 1170–1180. [CrossRef] [PubMed]
96. McCarter, L.; Silverman, M. Surface-induced swarmer cell differentiation of *Vibrio parahaemoiyticus*. *Mol. Microbiol.* **1990**, *4*, 1057–1062. [CrossRef] [PubMed]
97. McCarter, L.; Hilmen, M.; Silverman, M. Flagellar dynamometer controls swarmer cell differentiation of *V. parahaemolyticus*. *Cell* **1988**, *54*, 345–351. [CrossRef]
98. Kawagishi, I.; Imagawa, M.; Imae, Y.; McCarter, L.; Homma, M. The sodium-driven polar flagellar motor of marine *Vibrio* as the mechanosensor that regulates lateral flagellar expression. *Mol. Microbiol.* **1996**, *20*, 693–699. [CrossRef]
99. Kawagishi, I.; Maekawa, Y.; Atsumi, T.; Homma, M.; Imae, Y. Isolation of the polar and lateral flagellum-defective mutants in *Vibrio alginolyticus* and identification of their flagellar driving energy sources. *J. Bacteriol.* **1995**, *177*, 5158–5160. [CrossRef]
100. Ito, M.; Terahara, N.; Fujinami, S.; Krulwich, T.A. Properties of motility in *Bacillus subtilis* powered by the H^+-coupled MotAB flagellar stator, Na^+-coupled MotPS or hybrid stators MotAS or MotPB. *J. Mol. Biol.* **2005**, *352*, 396–408. [CrossRef]

101. Paulick, A.; Koerdt, A.; Lassak, J.; Huntley, S.; Wilms, I.; Narberhaus, F.; Thormann, K.M. Two different stator systems drive a single polar flagellum in *Shewanella oneidensis* MR-1. *Mol. Microbiol.* **2009**, *71*, 836–850. [CrossRef] [PubMed]
102. Terahara, N.; Sano, M.; Ito, M. A *Bacillus* flagellar motor that can use both Na^+ and K^+ as a coupling ion is converted by a single mutation to use only Na^+. *PLoS ONE* **2012**, *7*, e46248. [CrossRef] [PubMed]
103. Terahara, N.; Krulwich, T.A.; Ito, M. Mutations alter the sodium versus proton use of a *Bacillus clausii* flagellar motor and confer dual ion use on *Bacillus subtilis* motors. *Proc. Natl. Acad. Sci. USA* **2008**, *105*, 14359–14364. [CrossRef] [PubMed]
104. Islam, M.S.; Morimoto, Y.V.; Kudo, S.; Nakamura, S. H^+ and Na^+ are involved in flagellar rotation of the spirochete *Leptospira*. *Biochem. Biophys. Res. Commun.* **2015**, *466*, 196–200. [CrossRef] [PubMed]
105. Nishihara, Y.; Kitao, A. Gate-controlled proton diffusion and protonation-induced ratchet motion in the stator of the bacterial flagellar motor. *Proc. Natl. Acad. Sci. USA* **2015**, *112*, 7737–7742. [CrossRef] [PubMed]
106. Zhou, J.; Sharp, L.L.; Tang, H.L.; Lloyd, S.A.; Billings, S.; Braun, T.F.; Blair, D.F. Function of protonatable residues in the flagellar motor of *Escherichia coli*: A sritical role for Asp 32 of MotB. *J. Bacteriol.* **1998**, *180*, 2729–2735.
107. Morimoto, Y.V.; Che, Y.-S.; Minamino, T.; Namba, K. Proton-conductivity assay of plugged and unplugged MotA/B proton channel by cytoplasmic pHluorin expressed in *Salmonella*. *FEBS Lett.* **2018**, *584*, 1268–1272. [CrossRef]
108. Hosking, E.R.; Vogt, C.; Bakker, E.P.; Manson, M.D. The *Escherichia coli* MotAB proton channel unplugged. *J. Mol. Biol.* **2006**, *364*, 921–937. [CrossRef]
109. Minamino, T.; Imae, Y.; Oosawa, F.; Kobayashi, Y.; Oosawa, K. Effect of intracellular pH on rotational speed of bacterial flagellar motors. *J. Bacteriol.* **2003**, *185*, 1190–1194. [CrossRef]
110. Nakamura, S.; Kami-ike, N.; Yokota, J.P.; Kudo, S.; Minamino, T.; Namba, K. Effect of intracellular pH on the torque-speed relationship of bacterial proton-driven flagellar motor. *J. Mol. Biol.* **2009**, *386*, 332–338. [CrossRef]
111. Che, Y.-S.; Nakamura, S.; Kojima, S.; Kami-ike, N.; Namba, K.; Minamino, T. Suppressor analysis of the MotB(D33E) mutation to probe bacterial flagellar motor dynamics coupled with proton translocation. *J. Bacteriol.* **2008**, *190*, 6660–6667. [CrossRef] [PubMed]
112. Che, Y.-S.; Nakamura, S.; Morimoto, Y.V.; Kami-Ike, N.; Namba, K.; Minamino, T. Load-sensitive coupling of proton translocation and torque generation in the bacterial flagellar motor. *Mol. Microbiol.* **2014**, *91*, 175–184. [CrossRef] [PubMed]
113. Nakamura, S.; Morimoto, Y.V.; Kami-ike, N.; Minamino, T.; Namba, K. Role of a conserved prolyl residue (Pro173) of MotA in the mechanochemical reaction cycle of the proton-driven flagellar motor of *Salmonella*. *J. Mol. Biol.* **2009**, *393*, 300–307. [CrossRef] [PubMed]
114. Kim, E.A.; Price-Carter, M.; Carlquist, W.C.; Blair, D.F. Membrane segment organization in the stator complex of the flagellar motor: Implications for proton flow and proton-induced conformational change. *Biochemistry* **2008**, *47*, 11332–11339. [CrossRef] [PubMed]
115. Braun, T.F.; Poulson, S.; Gully, J.B.; Empey, J.C.; Van Way, S.; Putnam, A.; Blair, D.F. Function of proline residues of MotA in torque generation by the flagellar motor of *Escherichia coli*. *J. Bacteriol.* **1999**, *181*, 3542–3551.
116. Sudo, Y.; Terashima, H.; Abe-Yoshizumi, R.; Kojima, S.; Homma, M. Comparative study of the ion flux pathway in stator units of proton- and sodium-driven flagellar motors. *Biophysics* **2009**, *5*, 45–52. [CrossRef] [PubMed]
117. Sharp, L.L.; Zhou, J.; Blair, D.F. Features of MotA proton channel structure revealed by tryptophan-scanning mutagenesis. *Proc. Natl. Acad. Sci. USA* **1995**, *92*, 7946–7950. [CrossRef]
118. Suzuki, Y.; Morimoto, Y.V.; Oono, K.; Hayashi, F.; Oosawa, K.; Kudo, S.; Nakamura, S. Effect of the MotA(M206I) mutation on torque generation and stator assembly in the *Salmonella* H^+-driven flagellar motor. *J. Bacteriol.* **2019**, *201*, e00727-18. [CrossRef]
119. Kojima, S.; Blair, D.F. Conformational change in the stator of the bacterial flagellar motor. *Biochemistry* **2001**, *40*, 13041–13050. [CrossRef]
120. Zhou, J.; Blair, D.F. Residues of the cytoplasmic domain of MotA essential for torque generation in the bacterial flagellar motor. *J. Mol. Biol.* **1997**, *273*, 428–439. [CrossRef]
121. Silverman, M.; Simon, M. Flagellar rotation and the mechanism of bacterial motility. *Nature* **1974**, *249*, 73–74. [CrossRef] [PubMed]

122. Ryu, W.S.; Berry, R.M.; Berg, H.C. Torque-generating units of the flagellar motor of *Escherichia coli* have a high duty ratio. *Nature* **2000**, *403*, 444–447. [CrossRef] [PubMed]
123. Chen, X.; Berg, H.C. Solvent-isotope and pH effects on flagellar rotation in *Escherichia coli*. *Biophys. J.* **2000**, *78*, 2280–2284. [CrossRef]
124. Magariyama, Y.; Sugiyama, S.; Muramoto, K.; Kawagishi, I.; Imae, Y.; Kudo, S. Simultaneous measurement of bacterial flagellar rotation rate and swimming speed. *Biophys. J.* **1995**, *69*, 2154–2162. [CrossRef]
125. Sowa, Y.; Hotta, H.; Homma, M.; Ishijima, A. Torque–speed relationship of the Na^+-driven flagellar motor of *Vibrio alginolyticus*. *J. Mol. Biol.* **2003**, *327*, 1043–1051. [CrossRef]
126. Gabel, C.V.; Berg, H.C. The speed of the flagellar rotary motor of *Escherichia coli* varies linearly with protonmotive force. *Proc. Natl. Acad. Sci. USA* **2003**, *100*, 8748–8751. [CrossRef] [PubMed]
127. Lo, C.-J.; Sowa, Y.; Pilizota, T.; Berry, R.M. Mechanism and kinetics of a sodium-driven bacterial flagellar motor. *Proc. Natl. Acad. Sci. USA* **2013**, *110*, E2544–E2551. [CrossRef] [PubMed]
128. Yuan, J.; Berg, H.C. Resurrection of the flagellar rotary motor near zero load. *Proc. Natl. Acad. Sci. USA* **2008**, *105*, 1182–1185. [CrossRef]
129. Wang, B.; Zhang, R.; Yuan, J. Limiting (zero-load) speed of the rotary motor of *Escherichia coli* is independent of the number of torque-generating units. *Proc. Natl. Acad. Sci. USA* **2017**, *114*, 12478–12482. [CrossRef]
130. Nord, A.L.; Sowa, Y.; Steel, B.C.; Lo, C.-J.; Berry, R.M. Speed of the bacterial flagellar motor near zero load depends on the number of stator units. *Proc. Natl. Acad. Sci. USA* **2017**, *114*, 11603–11608. [CrossRef]
131. Sato, K.; Nakamura, S.; Kudo, S.; Toyabe, S. Evaluation of the duty ratio of the bacterial flagellar motor by dynamic load control. *Biophys. J.* **2019**, *116*, 1952–1959. [CrossRef] [PubMed]
132. Nishiyama, M.; Higuchi, H.; Ishii, Y.; Taniguchi, Y.; Yanagida, T. Single molecule processes on the stepwise movement of ATP-driven molecular motors. *Biosystems* **2003**, *71*, 145–156. [CrossRef]
133. Komori, Y.; Iwane, A.H.; Yanagida, T. Myosin-V makes two brownian 90° rotations per 36-nm step. *Nat. Struct. Mol. Biol.* **2007**, *14*, 968–973. [CrossRef] [PubMed]
134. Adachi, K.; Oiwa, K.; Nishizaka, T.; Furuike, S.; Noji, H.; Itoh, H.; Yoshida, M.; Kinosita, K. Coupling of rotation and catalysis in F_1-ATPase revealed by single-molecule imaging and manipulation. *Cell* **2007**, *130*, 309–321. [CrossRef] [PubMed]
135. Sowa, Y.; Rowe, A.D.; Leake, M.C.; Yakushi, T.; Homma, M.; Ishijima, A.; Berry, R.M. Direct observation of steps in rotation of the bacterial flagellar motor. *Nature* **2005**, *437*, 916–919. [CrossRef] [PubMed]
136. Nakamura, S.; Kami-ike, N.; Yokota, J.P.; Minamino, T.; Namba, K. Evidence for symmetry in the elementary process of bidirectional torque generation by the bacterial flagellar motor. *Proc. Natl. Acad. Sci. USA* **2010**, *107*, 17616–17620. [CrossRef] [PubMed]
137. Xing, J.; Bai, F.; Berry, R.; Oster, G. Torque-speed relationship of the bacterial flagellar motor. *Proc. Natl. Acad. Sci. USA* **2006**, *103*, 1260–1265. [CrossRef]
138. Bai, F.; Lo, C.-J.; Berry, R.M.; Xing, J. Model studies of the dynamics of bacterial flagellar motors. *Biophys. J.* **2009**, *96*, 3154–3167. [CrossRef]
139. Meacci, G.; Tu, Y. Dynamics of the bacterial flagellar motor with multiple stators. *Proc. Natl. Acad. Sci. USA* **2009**, *106*, 3746–3751. [CrossRef]
140. Bai, F.; Minamino, T.; Wu, Z.; Namba, K.; Xing, J. Coupling between switching regulation and torque generation in bacterial flagellar motor. *Phys. Rev. Lett.* **2012**, *108*, 178105. [CrossRef]
141. Mora, T.; Yu, H.; Sowa, Y.; Wingreen, N.S. Steps in the bacterial flagellar motor. *PLoS Comput. Biol.* **2009**, *5*, e1000540. [CrossRef] [PubMed]
142. Nirody, J.A.; Berry, R.M.; Oster, G. The limiting speed of the bacterial flagellar motor. *Biophys. J.* **2016**, *111*, 557–564. [CrossRef] [PubMed]
143. Briegel, A.; Ortega, D.R.; Tocheva, E.I.; Wuichet, K.; Li, Z.; Chen, S.; Muller, A.; Iancu, C.V.; Murphy, G.E.; Dobro, M.J.; et al. Universal architecture of bacterial chemoreceptor arrays. *Proc. Natl. Acad. Sci. USA* **2009**, *106*, 17181–17186. [CrossRef] [PubMed]
144. Cluzel, P. An ultrasensitive bacterial motor revealed by monitoring signaling proteins in single cells. *Science* **2000**, *287*, 1652–1655. [CrossRef] [PubMed]
145. Sourjik, V.; Berg, H.C. Binding of the *Escherichia coli* response regulator CheY to its target measured in vivo by fluorescence resonance energy transfer. *Proc. Natl. Acad. Sci. USA* **2002**, *99*, 12669–12674. [CrossRef] [PubMed]

146. Sagi, Y.; Khan, S.; Eisenbach, M. Binding of the chemotaxis response regulator CheY to the isolated, intact switch complex of the bacterial flagellar motor: Lack of cooperativity. *J. Biol. Chem.* **2003**, *278*, 25867–25871. [CrossRef]
147. Lam, K.-H.; Ip, W.-S.; Lam, Y.-W.; Chan, S.-O.; Ling, T.K.-W.; Au, S.W.-N. Multiple Conformations of the FliG C-Terminal Domain Provide Insight into Flagellar Motor Switching. *Structure* **2012**, *20*, 315–325. [CrossRef]
148. Miyanoiri, Y.; Hijikata, A.; Nishino, Y.; Gohara, M.; Onoue, Y.; Kojima, S.; Kojima, C.; Shirai, T.; Kainosho, M.; Homma, M. Structural and functional analysis of the C-terminal region of FliG, an essential motor component of *Vibrio* Na^+-driven flagella. *Structure* **2017**, *25*, 1540–1548. [CrossRef]
149. Minamino, T.; Imada, K.; Kinoshita, M.; Nakamura, S.; Morimoto, Y.V.; Namba, K. Structural insight into the rotational switching mechanism of the bacterial flagellar motor. *PLoS Biol.* **2011**, *9*, e1000616. [CrossRef]
150. Paul, K.; Brunstetter, D.; Titen, S.; Blair, D.F. A molecular mechanism of direction switching in the flagellar motor of *Escherichia coli*. *Proc. Natl. Acad. Sci. USA* **2011**, *108*, 17171–17176. [CrossRef]
151. Szurmant, H.; Ordal, G.W. Diversity in Chemotaxis Mechanisms among the Bacteria and Archaea. *Microbiol. Mol. Biol. Rev.* **2004**, *68*, 301–319. [CrossRef] [PubMed]
152. Porter, S.L.; Wadhams, G.H.; Armitage, J.P. *Rhodobacter sphaeroides*: Complexity in chemotactic signalling. *Trends Microbiol.* **2008**, *16*, 251–260. [CrossRef] [PubMed]
153. Bai, F.; Branch, R.W.; Nicolau, D.V.; Pilizota, T.; Steel, B.C.; Maini, P.K.; Berry, R.M. Conformational spread as a mechanism for cooperativity in the bacterial flagellar switch. *Science* **2010**, *327*, 685–689. [CrossRef] [PubMed]
154. Fukuoka, H.; Sagawa, T.; Inoue, Y.; Takahashi, H.; Ishijima, A. Direct imaging of intracellular signaling components that regulate bacterial chemotaxis. *Sci. Signal.* **2014**, *7*, ra32. [CrossRef] [PubMed]
155. Fahrner, K.A.; Ryu, W.S.; Berg, H.C. Biomechanics: Bacterial flagellar switching under load. *Nature* **2003**, *423*, 938. [CrossRef] [PubMed]
156. Yuan, J.; Fahrner, K.A.; Berg, H.C. Switching of the bacterial flagellar motor near zero load. *J. Mol. Biol.* **2009**, *390*, 394–400. [CrossRef] [PubMed]
157. Wang, F.; Shi, H.; He, R.; Wang, R.; Zhang, R.; Yuan, J. Non-equilibrium effect in the allosteric regulation of the bacterial flagellar switch. *Nat. Phys.* **2017**, *13*, 710–714. [CrossRef]
158. Kojima, S.; Imada, K.; Sakuma, M.; Sudo, Y.; Kojima, C.; Minamino, T.; Homma, M.; Namba, K. Stator assembly and activation mechanism of the flagellar motor by the periplasmic region of MotB. *Mol. Microbiol.* **2009**, *73*, 710–718. [CrossRef] [PubMed]
159. Kojima, S.; Takao, M.; Almira, G.; Kawahara, I.; Sakuma, M.; Homma, M.; Kojima, C.; Imada, K. The helix rearrangement in the periplasmic domain of the flagellar stator B subunit activates peptidoglycan binding and ion influx. *Structure* **2018**, *26*, 590–598. [CrossRef]
160. Zhu, S.; Takao, M.; Li, N.; Sakuma, M.; Nishino, Y.; Homma, M.; Kojima, S.; Imada, K. Conformational change in the periplasmic region of the flagellar stator coupled with the assembly around the rotor. *Proc. Natl. Acad. Sci. USA* **2014**, *111*, 13523–13528. [CrossRef]
161. Reid, S.W.; Leake, M.C.; Chandler, J.H.; Lo, C.-J.; Armitage, J.P.; Berry, R.M. The maximum number of torque-generating units in the flagellar motor of *Escherichia coli* is at least 11. *Proc. Natl. Acad. Sci. USA* **2006**, *103*, 8066–8071. [CrossRef] [PubMed]
162. Leake, M.C.; Chandler, J.H.; Wadhams, G.H.; Bai, F.; Berry, R.M.; Armitage, J.P. Stoichiometry and turnover in single, functioning membrane protein complexes. *Nature* **2006**, *443*, 355–358. [CrossRef] [PubMed]
163. Lele, P.P.; Hosu, B.G.; Berg, H.C. Dynamics of mechanosensing in the bacterial flagellar motor. *Proc. Natl. Acad. Sci. USA* **2013**, *110*, 11839–11844. [CrossRef] [PubMed]
164. Tipping, M.J.; Delalez, N.J.; Lim, R.; Berry, R.M.; Armitage, J.P. Load-dependent assembly of the bacterial flagellar motor. *mBio* **2013**, *4*, e00551-13. [CrossRef] [PubMed]
165. Terahara, N.; Noguchi, Y.; Nakamura, S.; Kami-ike, N.; Ito, M.; Namba, K.; Minamino, T. Load- and polysaccharide-dependent activation of the Na^+-type MotPS stator in the *Bacillus subtilis* flagellar motor. *Sci. Rep.* **2017**, *7*, 46081. [CrossRef] [PubMed]
166. Castillo, D.J.; Nakamura, S.; Morimoto, Y.V.; Che, Y.-S.; Kami-ike, N.; Kudo, S.; Minamino, T.; Namba, K. The C-terminal periplasmic domain of MotB is responsible for load-dependent control of the number of stators of the bacterial flagellar motor. *Biophysics* **2013**, *9*, 173–181. [CrossRef]
167. Shi, H.; Ma, S.; Zhang, R.; Yuan, J. A hidden state in the turnover of a functioning membrane protein complex. *Sci. Adv.* **2019**, *5*, eaau6885. [CrossRef]

168. Fukuoka, H.; Wada, T.; Kojima, S.; Ishijima, A.; Homma, M. Sodium-dependent dynamic assembly of membrane complexes in sodium-driven flagellar motors. *Mol. Microbiol.* **2009**, *71*, 825–835. [CrossRef]
169. Tipping, M.J.; Steel, B.C.; Delalez, N.J.; Berry, R.M.; Armitage, J.P. Quantification of flagellar motor stator dynamics through in vivo proton-motive force control: PMF-dependent motor dynamics. *Mol. Microbiol.* **2013**, *87*, 338–347. [CrossRef]

Review

Regulation of the Single Polar Flagellar Biogenesis

Seiji Kojima *, Hiroyuki Terashima and Michio Homma

Division of Biological Science, Graduate School of Science, Nagoya University, Chikusa-Ku, Nagoya 464-8602, Japan; terashima.hiroyuki@h.mbox.nagoya-u.ac.jp (H.T.); g44416a@cc.nagoya-u.ac.jp (M.H.)

* Correspondence: z47616a@cc.nagoya-u.ac.jp; Tel.: +81-52-789-2993

Received: 17 March 2020; Accepted: 30 March 2020; Published: 1 April 2020

Abstract: Some bacterial species, such as the marine bacterium *Vibrio alginolyticus,* have a single polar flagellum that allows it to swim in liquid environments. Two regulators, FlhF and FlhG, function antagonistically to generate only one flagellum at the cell pole. FlhF, a signal recognition particle (SRP)-type guanosine triphosphate (GTP)ase, works as a positive regulator for flagellar biogenesis and determines the location of flagellar assembly at the pole, whereas FlhG, a MinD-type ATPase, works as a negative regulator that inhibits flagellar formation. FlhF intrinsically localizes at the cell pole, and guanosine triphosphate (GTP) binding to FlhF is critical for its polar localization and flagellation. FlhG also localizes at the cell pole via the polar landmark protein HubP to directly inhibit FlhF function at the cell pole, and this localization depends on ATP binding to FlhG. However, the detailed regulatory mechanisms involved, played by FlhF and FlhG as the major factors, remain largely unknown. This article reviews recent studies that highlight the post-translational regulation mechanism that allows the synthesis of only a single flagellum at the cell pole.

Keywords: polar flagellum; FlhF; FlhG; HubP; FlaK; SflA; protein localization; ATPase; GTPase

---

1. Introduction

Motility is a fundamental function of bacteria required for survival in response to environmental changes. When bacteria encounter deleterious conditions, they must move out from such environments and seek more favorable ones. To achieve this cell movement (swimming in a liquid [1] or swarming on a surface [2]), many motile bacteria use the flagellum, their motility machinery [1,3]. The bacterial flagellum is structurally, functionally, and evolutionally distinct from its eukaryotic counterpart. The flagellum is extended from the cell body and consists of three parts: the filament (helical propeller), the hook (universal joint), and the basal body (rotary motor) (Figure 1a). Reversible rotation of the long helical filament thrusts the cell body forward or backward, and is driven by the membrane-embedded rotary motor at its base [4–6]. The energy source of the flagellar motor is the electrochemical gradient of specific ions (in most cases H^+ or Na^+) across the inner membrane. The motor is composed of a rotary part (the rotor) and energy-converting multiple stator units that surround the rotor. The motor torque is generated by rotor–stator interactions that couple to the ion influx through the stator channel [7,8]. Genetic, biochemical, and biophysical studies have unveiled the many protein components involved in flagellar function, their locations in the motor, and the rotational properties of the motor. However, the molecular mechanism of energy conversion during flagellar rotation has not yet been elucidated. For reviews of the rotary mechanism of the flagellar motor, please see references recently published elsewhere [9,10].

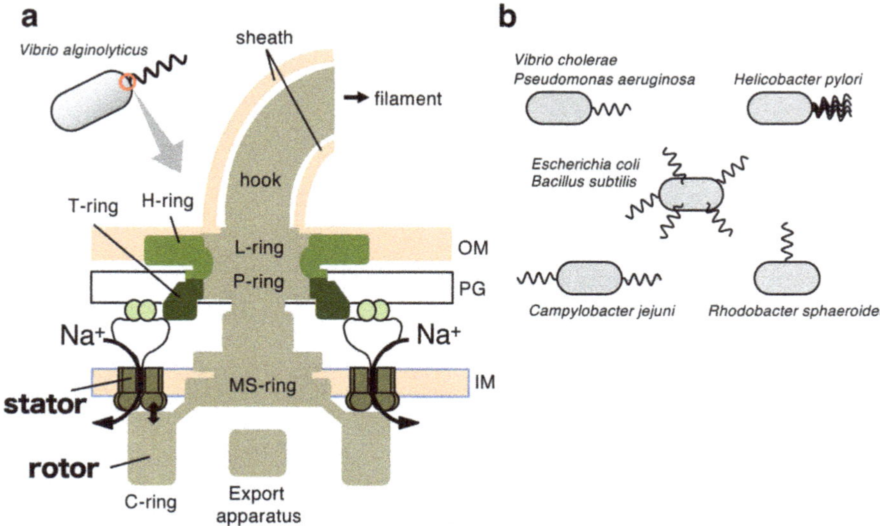

Figure 1. The flagellum is a bacterial motility organ. (**a**) Schematic of the polar flagellar motor of *Vibrio alginolyticus*. The rotor-stator interaction that couples with sodium ion influx through the stator channel generates motor torque. OM, outer membrane; IM, inner membrane; PG, peptidoglycan layer. (**b**) The number and location of flagella vary among bacterial species.

To achieve the best motility performance in a wide variety of habitats, bacteria have developed their own flagellar systems located at specific position(s) on the bacterial cells with suitable numbers, and the numbers and positions of flagella vary widely among bacterial species (Figure 1b) [11]. *Escherichia coli*, *Bacillus subtilis*, and *Salmonella enterica* have multiple peritrichous flagella and they form flagellar bundles to swim forward [9]. *Campylobacter jejuni* has bipolar flagella (one at each pole) [12], *Helicobacter pylori* has multiple polar flagella at one pole [13], and *Rhodobacter sphaeroides* has a single medially located flagellum [14]. *Pseudomonas aeruginosa* [15] and *Vibrio cholerae* [16] have a single polar flagellum at the cell pole. Surprisingly, the spirochete *Borrelia burgdorferi* has 7-11 periplasmic flagella [17,18] and *Leptospira biflexa* [19] possess two periplasmic flagella near each end of the cell body. In some species, two distinct types of flagella are produced. Well-known examples are *Vibrio alginolyticus* and *Vibrio parahaemolyticus*, which have a single sheathed polar flagellum suitable for swimming motility when grown in a liquid environment, but when they are grown on a surface, numerous lateral (peritrichous) flagella suitable for swarming motility are induced in response to the increased viscosity of the surrounding environment [20,21]. *Shewanella putrefaciens* CN-32 also possesses a two flagellar system, and the primary system generates a single polar flagellum, whereas a secondary flagellum is formed at a lateral position in subpopulations cultivated in complex medium [22]. In all cases, motility is impaired by mutations that cause defects in spatial and/or numerical control of flagella, indicating that the number and placement of flagella are precisely regulated to optimize motility under the environmental conditions of each bacterium [11].

It has been known that some species of bacteria have a single flagellum at their cell pole for motility. Then, one simple question arises—how do they generate only a single flagellum at the cell pole? In this regard, the marine bacterium *Vibrio alginolyticus*, which generates a single flagellum at its cell pole in a liquid environment [23], is a good model organism because genetic, biochemical, and structural analyses of its polar flagellar system have been extensively studied [4]. Moreover, the filament of the polar flagellum is covered with a membranous sheath contiguous with the outer membrane of the bacterial cell [24], and because of its thickness it can be easily observed using high intensity dark-field microscopy [25]. Using this bacterium, our group has reported that two key factors, FlhF and FlhG, play major roles in the regulation of polar flagellar number and placement. Here, we summarize our recent studies that have characterized the mechanisms by which FlhF and FlhG work to generate only a single polar flagellum in *Vibrio alginolyticus*, together with historical and recent insights obtained from other species. We also introduce other regulatory factors, HubP and SflA, which are involved in the biogenesis of polar flagella. It should be noted that the lateral flagella of *Vibrio alginolyticus*, which are induced in a viscous environment [20,26], are not the focus of this review. The unsheathed, thinner filament of lateral flagella is composed of components distinct from those of the polar flagella. All of our studies described in this review used bacterial strains that do not produce lateral flagella (VIO5 [27] and its derivatives).

2. FlhF and FlhG Regulate Flagellar Number and Placement

FlhF and FlhG have been reported as factors that regulate the number and position of polar flagella in *Pseudomonas aeruginsa* and *Pseudomonas putida* [28,29], *Campylobacter jejuni* [12], *Shewanella putrefaciens* CN-32 [30], *Shewanella oneidensis* [31], *Vibrio cholerae* [16], and *Vibrio alginolyticus* [32]. In the year 2000, FlhF was first identified as a factor involved in the starvation survival of *Pseudomonas putida* [29]. Sequence analysis of the transposon mutant (MK107) that was impaired in stationary phase survival appeared to have a transposon Tn*5* insertion in *flhF*. Because *flhF* is flanked by known flagellar genes, it was implicated in flagellar biogenesis. Indeed, MK107 did not spread in soft agar plates although its active motility was observed using phase-contrast microscopy. Electron microscopy showed that unlike the wild-type strain, the flagella of MK107 were randomly distributed on the cell surface. Overproduction of FlhF from the plasmid caused the production of a large number of polar flagella. Therefore, FlhF promotes flagellar biosynthesis at the cell pole, and positively regulates the number of flagella [29]. Meanwhile, genome sequence analysis of *Pseudomonas aeruginosa* revealed that *fleN* (*flhG*) is also implicated in flagellar biogenesis. *fleN* is flanked by *flhF* and *fliA*, and genetic inactivation of *fleN* by inserting a gentamicin cassette resulted in the productions of multiple flagella at the cell pole. Therefore, FleN negatively regulates the number of polar flagella [33].

In 2006, we reported that FlhF and FlhG also regulate the number of polar flagella in *Vibrio alginolyticus* [32]. During the screening of mutants defective in polar flagellar motility in soft agar plates, a mutant (KK148) was accidently isolated that had a large number of flagella at one pole (Figure 2a). KK148 formed a reduced motility ring compared to the wild-type strain VIO5, and an abnormal swimming behavior, caused by entangled multiple flagella, was observed for most cells using dark-field microscopy. Analysis of the *V. alginolyticus* genome revealed that *flhF* is flanked by *flhA* and *flhG* in the polar flagellar gene locus, and the mutation of KK148 was mapped to *flhG* (Gln109Amber, Figure 2b). Deletion studies revealed that the loss of FlhF resulted in a nonflagellated phenotype, whereas the loss of FlhG caused hyperflagellation (Figure 2b). Conversely, the overproduction of FlhF generated multiple polar flagella, but the overproduction of FlhG inhibited polar flagellation (Figure 2b). Therefore, similar to *Pseudomonas* spp., FlhF and FlhG function as positive and negative regulators of the number of flagella in *V. alginolyticus* [34], respectively. Because the deletion of both *flhF* and *flhG* confers a non-flagellated phenotype for most cells and FlhG mutants still form flagella at their cell pole, FlhF determines the flagellar placement at the cell pole [34].

Phenotypic analysis revealed that FlhF and FlhG work antagonistically to generate a single polar flagellum at the cell pole, but how do they achieve that? To determine the process involved, green fluorescent protein (GFP) was fused to the C-terminus of FlhF or FlhG, and their subcellular localization was observed [34]. Neither FlhF nor FlhG have a transmembrane segment and are thus expected to be cytoplasmic proteins. The fluorescent signal of FlhF-GFP was observed throughout the cytoplasm and most cells showed a fluorescent dot at the flagellated cell pole. This polar localization was observed more strongly in the absence of FlhG, in which multiple polar flagella were generated. Meanwhile, FlhG-GFP also diffused in the cytoplasm and its polar localization, which appeared to be independent of FlhF, was observed in ≈30% of cells. Because FlhG was immunoprecipitated by an anti-FlhF antibody from the cytoplasmic fraction, the first model proposed was that FlhF localization at the cell pole determines the polar localization and production of a flagellum, FlhG interacts with FlhF to prevent FlhF from localizing at the cell pole, and thus FlhG negatively regulates the number of flagella in *V. alginolyticus* (Figure 2c) [34].

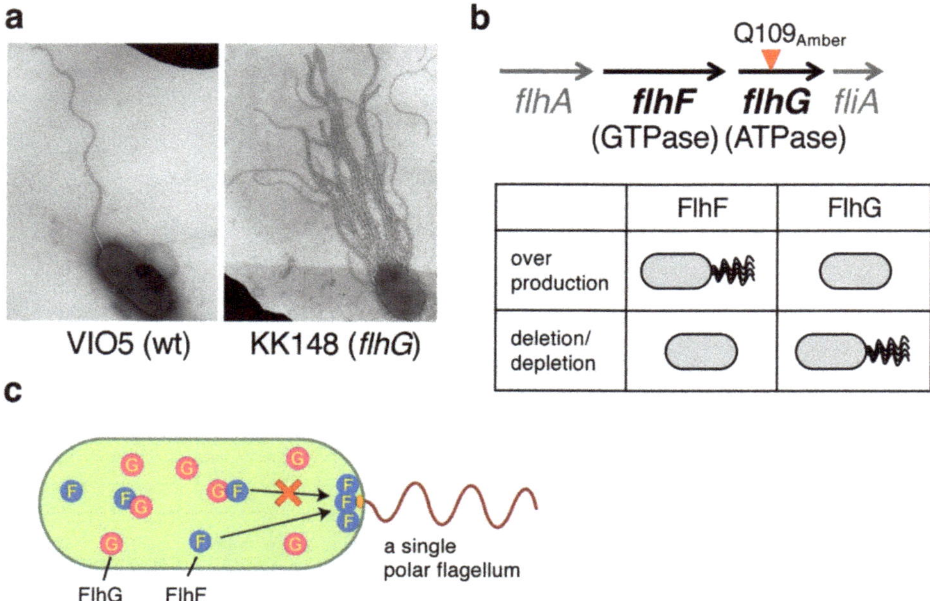

Figure 2. FlhF and FlhG regulate the number of polar flagella in *Vibrio alginolyticus*. (**a**) Electron micrographs of *V. alginolyticus* strain VIO5 (wild-type for polar flagellation) and KK148 (*flhG* mutant of VIO5). (**b**) The mutation site of KK148 (upper panel) and flagellar organisation of FlhF and FlhG variants of *Vibrio alginolyticus* (lower panel). The mutation was mapped on *flhG* (Q109Amber), which forms an operon with *flhF*. FlhF and FlhG work antagonistically. Their overproduction or deletion/depletion confers opposite phenotypes in *Vibrio alginolyticus*. (**c**) Model for the regulation of polar flagella by FlhF and FlhG proposed in [34]. In this model, FlhF localizes at the cell pole and determines the location of flagellation. FlhG interacts with FlhF to prevent its polar localization, and thereby negatively regulates the number of flagella.

It should be noted that the peritrichously flagellated bacterium *Bacillus subtilis* also produces both FlhF and FlhG. The deletion of *flhF* does not affect motility, but the basal body is formed at random positions compared to the wild-type strain. The deletion of *flhG* does not affect motility but causes the aggregation of basal bodies in the cell. FlhF and FlhG in *B. subtilis* seem to be important for the optimized spatial positioning of flagella with a grid-like pattern [35].

3. FlhF Is a SRP-Type GTPase

Sequence analysis revealed that FlhF is one of the three members of the signal recognition particle (SRP)-type guanosine triphosphate (GTP)ase subfamily of SIMIBI (signal recognition particle, MinD and BioD)-class nucleotide binding proteins [36]. The other two members of that family are the signal sequence binding protein Ffh and the SRP receptor FtsY, whose structures have already been solved (Figure 3a) [37]. As shown in Figure 3b, FlhF is composed of a basic N-terminal domain (B domain) followed by a conserved NG domain (regulatory N domain and GTPase G domain) [38]. Structural and biochemical analyses of FlhF have been carried out for *Bacillus subtilis* FlhF (hereafter, *Bs*FlhF). In 2007, the crystal structure of the *Bs*FlhF NG domain homodimer in complex with guanosine triphosphate (GTP) was solved [39], and later in 2011, the *Bs*FlhF homodimer in a complex with the peptide containing N-terminal 23 residues of *Bs*FlhG was reported (Figure 3c) [40]. That structure shares homology with Ffh and FtsY within the NG domain. Ffh and FtsY form a GTP-dependent heterodimer via their NG domains (Figure 3a), and its GTPase activity is coupled to their function in targeting a ribosome-nascent chain complex to the Sec machinery on the cytoplasmic membrane [37].

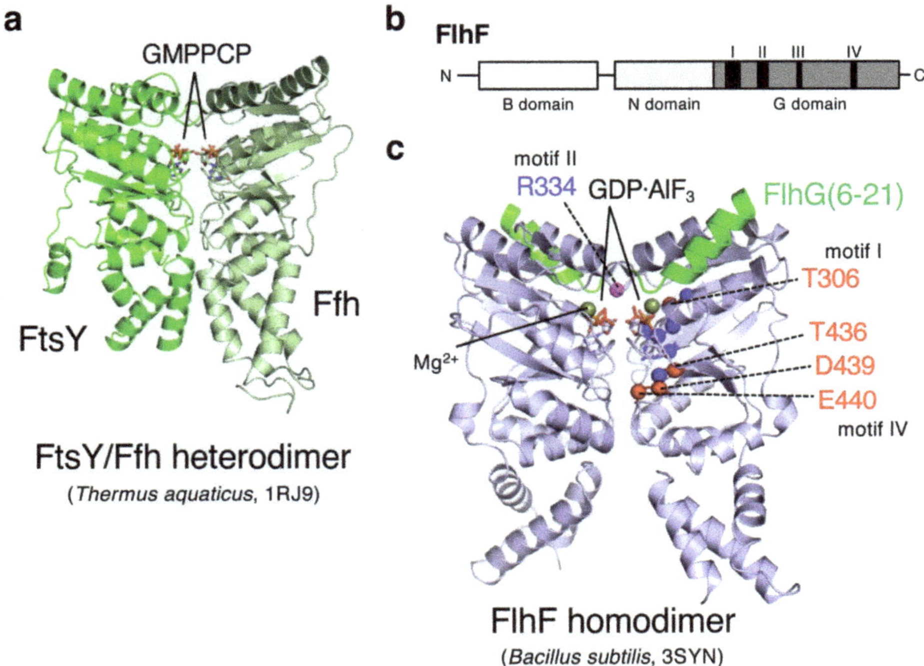

Figure 3. FlhF is a signal recognition particle (SRP)-type guanosine triphosphate (GTP)ase. (**a**) Crystal structure of the FtsY/Ffh heterodimer from *Thermus aquaticus* (PDB ID 1RJ9 [37]). The non-hydrolyzable guanosine triphosphate (GTP) analog β,γ-methyleneguanosine 5′-triphosphate (GMPPCP) stabilizes the heterodimer, and complex formation aligns the two molecules of this GTP analog in the composite active site. (**b**) Domain structure of FlhF proteins. *Bacillus subtilis* FlhF consists of 366 amino acids (41 kDa) with a smaller B domain than *Vibrio alginolyticus* FlhF (505 amino acids, 57 kDa). FlhF is composed of the function-unknown B domain, the regulatory N domain, and the G domain that contains the GTPase motif (I-IV). (**c**) Crystal structure of the NG domain homodimer from *Bacillus subtilis* FlhF in complex with the peptide containing N-terminal 23 residues of FlhG (PDB ID 3SYN [40]). FlhF is shown in light blue, and the FlhG peptide is shown in green. Guanosine diphosphate (GDP) and aluminum fluoride are shown as stick representations, and Mg^{2+} ions are shown in dark green. Residues mutated in corresponding *V. alginolyticus* proteins are highlighted by blue (function reduced) or red (abolished) balls with the residue number of *Vibrio* protein. The putative catalytic site (R334 of *Vibrio* FlhF) is also indicated. For simplicity, the above residues are highlighted only in one protomer. In *Vibrio* FlhF, alanine substitution of the catalytic residue (R334A) did not affect its function, indicating that GTP binding, but not hydrolysis, is essential.

Because of the structural similarity, it would be plausible that the FlhF homodimer functions in the numerous/spatial regulation of flagella similar to the way the Ffh/FtsY heterodimer does. To test that idea, mutational analyses of *Vibrio alginolyticus* FlhF (hereafter *Va*FlhF) have been performed. Site-specific mutations were introduced into conserved GTPase motifs (I, III, and IV; Figure 3b,c). The results showed that two of those mutations abolish the FlhF polar localization, flagellation, and thereby motility (T306A and D439A) [41]. Other mutants showed a correlation between the levels of polar localization and the ability to produce flagella. Later on, a random mutagenesis of full-length *flhF* was performed but mutations that abolished polar localization, flagellation, and motility were isolated only on the GTPase motif IV (Figure 3c, T436M and E440K) [42]. These results indicate that the GTPase motif of FlhF is functionally important, and to facilitate polar flagellation, the polar localization of FlhF is required.

The GTPase activity of FlhF was first reported for the *Bacillus subtilis* protein [40]. The results showed that *Bs*FlhF alone had only a low basal GTPase activity, but it was stimulated by FlhG (firstly named YlxH in *B. subtilis*). Subsequent analysis revealed that an N-terminal region of FlhG, which is not conserved in the MinD/ParA ATPase family, was responsible for the FlhF stimulation [40]. The crystal structure of the FlhF/FlhG-peptide (23 amino acids) complex showed that the FlhG peptide formed a helix and bound near the catalytic site of FlhF (Figure 3c) [40]. Later on, stimulation of the GTPase activity of *Campylobacter* FlhF (*Cj*FlhF) and *Va*FlhF by their cognate FlhG proteins was reported [43,44]. Consistent with *Bs*FlhF, purified *Va*FlhF existed as a homodimer in the presence of GTP but as a monomer in the presence of GDP [44]. These results suggest that the GTP-bound *Va*FlhF homodimer functions as an active form at the cell pole to promote flagellation, similar to the GTP-bound FtsY/Ffh complex at the Sec machinery. If so, a defect in the catalytic site would result in the accumulation of GTP-bound FlhF at the cell pole and cause hyperflagellation. This idea was supported by evidence that such a mutation in *Cj*FlhF (R324A) abolished the GTPase activity and increased the hyperflagellated population of *Campylobacter* cells [43,45]. On the other hand, the corresponding mutation (R334A, Figure 3c) in *Va*FlhF abolished its GTPase activity but still allowed it to localize at the cell pole and led to the normal polar flagellation of *V. alginolyticus* cells [44]. Similar results were reported for *Vibrio cholerae* FlhF (*Vc*FlhF), showing that substitutions of putative catalytic residues had little effect on *Vc*FlhF function, which indicated that GTP binding, but not hydrolysis, is critical for *Vibrio* FlhF function [16]. The varied phenotypes of FlhF mutants among species may reflect diverse flagellation patterns (mono- or bi-polar flagellation) or flagellar function (e.g., rotational speed or motor power) [11,38].

How FlhF promotes polar flagellation remains unknown. One plausible idea is that FlhF acts on the initial step of assembly of the flagellar basal body. Flagellar assembly begins with the formation of basal rings (MS- and C-rings; Figure 1a) that house the flagellum-specific export apparatus, followed by the construction of axial structures (rod, hook, and associated outer ring structures), and then is completed with the filament and motor part assembly [46]. The MS-ring is believed to be one of the earliest structures assembled in a flagellum [47], and FlhF may facilitate flagellation by recruiting FliF, a membrane protein and MS-ring component, to the cell pole. This idea is supported by evidence that the polar localization of GFP-fused FliF is dependent on FlhF expression in *Vibrio cholerae* [16]. Mutational analysis revealed that the B and N domains are essential for recruitment of FliF to the cell pole. A nonfunctional *Vc*FlhF D367A mutant of the GTPase motif III was still able to recruit FliF to the cell pole, but it inhibited flagellar assembly, suggesting the involvement of *Vc*FlhF in the MS-ring formation. Further studies are required to address the role of FlhF in the promotion of flagellar assembly.

4. FlhG Negatively Regulates Polar Flagellar Gene Expression

Assembly of a flagellum requires the synthesis of enormous amounts of protein components, including the long flagellar filament, and thus this process consumes large quantities of energy. Therefore, bacteria developed an efficient construction strategy—assembly occurs in a stepwise fashion to build from inner to outer structures, and is tightly coupled with flagellar gene transcription to provide necessary components at each step [46]. To achieve assembly-coupled transcription, flagellar genes are organized into a transcriptional hierarchy that is comprised of three to four classes of genes, with classification varying on species (Figure 4) [48,49]. On top of this hierarchy, a master regulator, which is usually the sole member of class 1, controls the expression of downstream flagellar genes. In *Pseudomonas aeruginosa*, *Vibrio cholerae* and *Vibrio parahaemolyticus*, the regulators FleQ [50],

FlrA [51], and FlaK [52] have been identified as the master regulators, respectively. As one can imagine, hyperflagellation due to the flhG mutation demands a large amount of flagellar components, and indeed, a lack of FlhG caused the upregulation of flagellar gene expression [33,34]. In *Pseudomonas aeruginosa*, FlhG (named FleN in *Pseudomonas*) does not affect the expression of *fleQ* [33] but rather physically interacts with the FleQ protein to inhibit its transcriptional activity [53], whereas in *Vibrio cholerae*, FlhG represses the expression of *flrA* to downregulate flagellar gene expression [54]. Therefore, FlhG regulates flagellar biogenesis at the transcriptional level by negatively acting on the expression (for *flrA*) or activity (for FleQ) of the master regulator. However, our research group also found that the negative regulation of FlhG for flagellar biogenesis occurs at the post-translational level. Such a mechanism is reviewed in the next two sections.

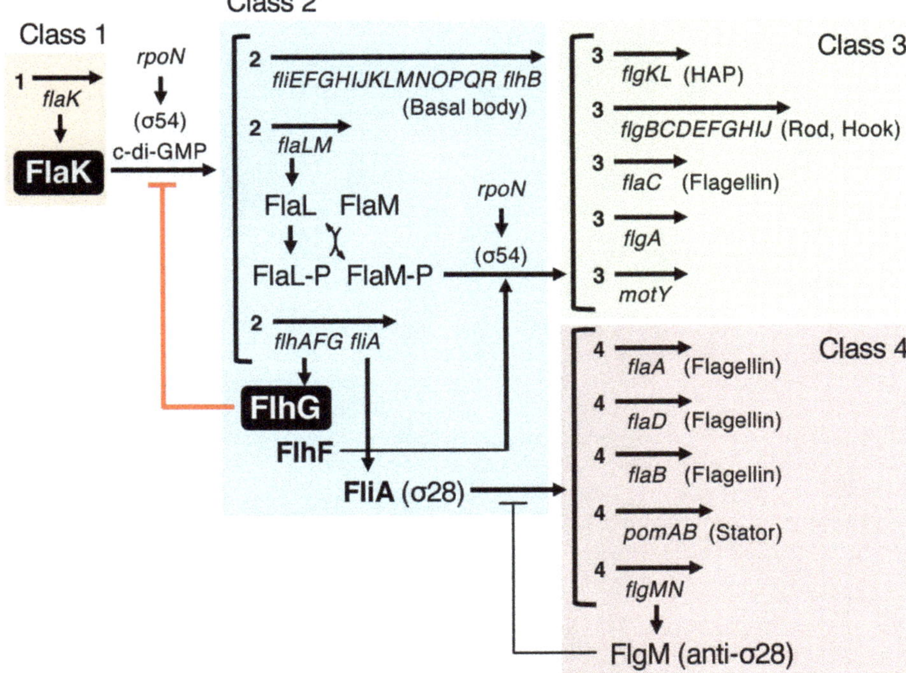

Figure 4. Proposed model for the regulation of *Vibrio alginolyticus* polar flagellar transcription hierarchy. This model is based on reports for *Vibrio cholerae* and *Vibrio parahaemolyticus* [54,55], whose flagellar genes are highly similar. The master regulator FlaK, which belongs to class 1 as a sole member, regulates downstream flagellar genes. FlaK activity is negatively regulated by FlhG, or FlhG may inhibit the transcription of *flaK*. The signaling molecule c-di-GMP also negatively regulates FlaK activity.

5. HubP, the Third Regulator of Polar Flagellar Biogenesis

HubP was first identified in *Vibrio cholerae* as a polar landmark protein that anchors three ParA-family proteins including FlhG [56]. HubP is conserved in *Vibrio* species [57], in *Shewanella* [30], and in some other gamma-proteobacteria, and shows similarity to FimV, a positive regulator for type IV pilus formation [58]. HubP is quite a large protein (1444 amino acids, ≈159 kDa for *V. alginolyticus* protein [57]), and has a single transmembrane segment with an N-terminal region placed in the periplasm (Figure 5a). The periplasmic LysM domain, which has been implicated in peptidoglycan binding, is important for the polar localization of HubP [56]. The large cytoplasmic C-terminal region contains 7–10 copies of the repeat sequence that interacts with ParA1, and FlhG is found to interact with the extreme C-terminus of HubP (Figure 5a,b) [56].

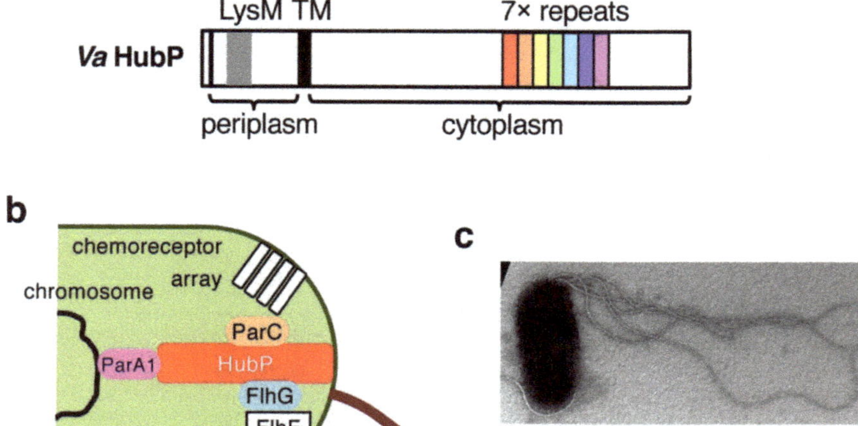

Figure 5. HubP, the third factor that regulates the number of polar flagella in *Vibrio alginolyticus*. (**a**) Domain structure of *V. alginolyticus* HubP, a single transmembrane protein of 1444 amino acids (≈159 kDa) with a large cytoplasmic region. The LysM domain in the N-terminal periplasmic region functions in anchoring HubP to the peptidoglycan layer. The cytoplasmic repeat sequence and its repetition numbers varies among *Vibrio* species. (**b**) HubP functions as the polar "hub". HubP localizes at the cell pole and anchors three ParA-like proteins at its large cytoplasmic platform. FlhF has an intrinsic property to localize at the cell pole, but FlhG polar localization is dependent on HubP. (**c**) Electron micrograph of a NMB303 cell, the *hubP* deletion strain of *V. alginolyticus*. It generates multiple sheathed flagella at the cell pole.

Although the deletion of *hubP* did not affect the polar flagellation of *V. cholerae* [56], the deletion of *hubP* in *V. alginolyticus* increased the number of polar flagella (Figure 5c) [57]. The level of hyperflagellation is stronger for the *flhG* mutant, which had more flagella per cell than the Δ*hubP* strain, and the additional deletion of *hubP* from the *flhG* mutant did not further increase the number of flagella [57]. These results indicate that HubP is also involved in regulating the number of polar flagella in *V. alginolyticus* to a certain level. How then is HubP involved in the biosynthesis of polar flagella? The *hubP* gene is not included in the polar flagellar gene cluster, and the endogenous chromosomal expression level of FlhG in the Δ*hubP* mutant is comparable to that in the wild-type strain [57], indicating that the expression of polar flagellar genes is not affected by the deletion of *hubP*. On the other hand, the polar localization of FlhG, but not FlhF, was abolished in the Δ*hubP* mutant as observed in *V. cholerae* [56,57]. These results suggest that FlhG localized at the cell pole negatively regulates flagellar biogenesis. If so, FlhG functions not only at the transcriptional level, but also at the post-translational level by localizing at the cell pole. In the next section, we describe the post-translational regulation of polar flagellar biogenesis by FlhG.

6. FlhG Is a MinD/ParA-Type ATPase

Sequence analysis revealed that FlhG (FleN) is classified as a MinD/ParA-type ATPase [38]. MinD is the ATPase component of the Min system that is involved in the spatial regulation of cell division [59]. It forms a homodimer in the presence of ATP and that homodimer binds to the membrane at the cell pole via the C-terminal amphipathic helix. MinD ATPase is then stimulated by MinE, and this hydrolysis induces the release of MinD from the membrane as a monomer. This ATP-dependent dimerization and polar localization is essential for the function of MinD. Meanwhile, FlhG is composed of a MinD/ParA-homologous domain and an N-terminal extension that stimulates FlhF GTPase [40]. All functionally important residues in MinD are conserved in FlhG, and both have a membrane binding sequence at their C-termini (Figure 6a). The crystal structures of *Geobacillus thermodenitrificans* FlhG [60] and *Pseudomonas aeruginosa* FleN (FlhG) [61] revealed that it is indeed a structurally close homolog of MinD (Figure 6b) [62]. Biochemical characterization of FlhG in *Geobacillus* has revealed that it has quite similar properties to MinD—an ATP-bound FlhG homodimer associates with the plasma membrane through its C-terminal amphipathic helix, and hydrolysis of ATP causes dissociation of FlhG from the membrane as a monomer [60]. Interestingly, deletion of *flhG* in *Campylobacter jejuni* caused more cell division at the polar region to form minicells [12]. Because *Campylobacter* species lack a Min system, FlhG may take over Min function to inhibit division at the cell pole.

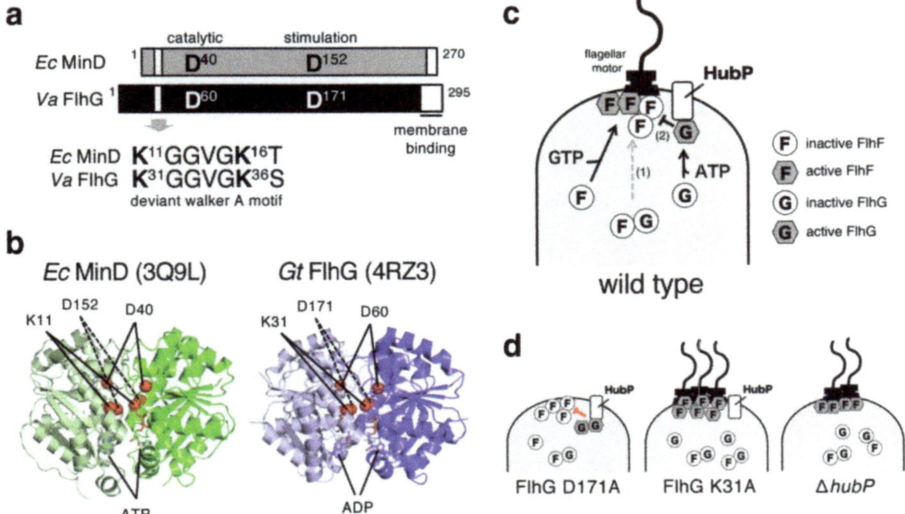

Figure 6. FlhG is a MinD/ParA-like ATPase that negatively regulates polar flagellar number. (**a**) Domain structures of *Escherichia coli* MinD (*Ec* MinD) and *V. alginolyticus* FlhG (*Va* FlhG). FlhG has a slightly longer N-terminal region than MinD, which functions in the stimulation of FlhF GTPase activity. (**b**) Crystal structures of *E. coli* MinD dimer in complex with ATP (PDB ID: 3Q9L [62]) and *Geobacillus thermodenitrificans* FlhG dimer in complex with adenosine diphosphate (ADP) (PDB ID: 4RZ3 [60]). ATP and ADP are shown by stick representations, and important conserved residues are colored red. (**c**) Model for the regulation of the number of polar flagella in *Vibrio alginolyticus*. GDP-bound FlhF and ADP-bound FlhG are in an inactive state, interact with each other and remain in the cytoplasm. When GTP is bound, FlhF becomes active and localizes at the cell pole to facilitate flagellation. Likewise, the ATP-bound active form of FlhG localizes to the cell pole via the landmark membrane protein HubP to inhibit FlhF activity. The inhibition of FlhF polar localization (1) and its activity (2) by FlhG optimizes the number of flagella into becoming a single one. (**d**) Working models to explain nonflagellated or hyperflagellated phenotypes of FlhG or HubP mutants of *V. alginolyticus*. FlhG D171A, a putative activated mutant, inhibits polar flagellation by more localization of FlhG at the cell pole. FlhG K31A, a nonfunctional mutant, causes hyperflagellation because it cannot localize at the cell pole. K31A also cannot inhibit FlaK so that more flagellar proteins are synthesized. The *ΔhubP* strain produces the wild-type level of flagellar proteins, but its polar flagellar number increases because FlhG cannot localize at the cell pole.

For *V. alginolyticus* FlhG, the purified protein alone exhibits a low basal ATPase activity, but it can be activated sevenfold by the D171A mutation [63]. The corresponding mutation, D152A of *E. coli* MinD, confers MinD insensitivity to MinE stimulation for ATPase activity [64]. As discussed in the previous section, FlhG at the cell pole inhibits flagellation, presumably by acting on the polar FlhF (Figure 6c). This active *Vibrio* FlhG D171A mutant localizes at the cell pole more strongly than wild-type FlhG and severely inhibits flagellation (Figure 6d) [63]. On the other hand, mutations at putative ATP binding residues in the deviant Walker A motif impair various properties of FlhG, such as its ATPase activity, polar localization, and negative regulator activity for flagellar biosynthesis, and thus confers the hyperflagellated phenotype (Figure 6d) [63].

Unexpectedly, a mutation at the catalytic residue (D60A) that abolishes ATPase activity but still allows ATP binding, only slightly affected FlhG function [63]. These results suggest that the ATP-dependent polar localization of FlhG is crucial for its negative regulator activity. Because the polar localization of FlhG is dependent on the landmark membrane protein HubP, we speculate that ATP-bound FlhG localizes at the cell pole via HubP and becomes active to directly inhibit FlhF at the cell pole (Figure 6c), and that the adenosine diphosphate (ADP)-bound inactive form interacts with cytoplasmic FlhF to interfere with its polar localization (Figure 6c) [57]. It should be noted that the number of polar flagella seem to not be determined primarily by the absolute amount of polar FlhF, as proposed in our first model (Figure 2b). The amount of FlhF at the cell pole was not increased by the deletion of hubP (hyperflagellation) and was not reduced by the overproduction of *flhG* (nonflagellation). Currently, we hypothesize that cytoplasmic FlhG works as a quantitative regulator that controls the amount of FlhF at the cell pole, and HubP-anchored polar FlhG works as a qualitative regulator that directly inhibits FlhF activity at the cell pole (Figure 6c,d) [57]. It should be noted that the FlhG mutant at the putative ATP binding site upregulated polar flagellar gene expression (thereby conferring the hyperflagellated phenotype) [63], which suggests that ATP binding is important for FlhG function to negatively regulate the master regulator FlaK.

As described above, MinD/ParA-family ATPases are known to form dimers in complex with ATP (Figure 6b), and that dimerization allows them to bind to the cell membrane where they exhibit their activities [65]. Recently, whether *Vibrio alginolyticus* FlhG undergoes ATP-dependent dimerization was examined [66]. The results showed that purified FlhG or FlhG in *Vibrio* cell lysates appeared to exist as a monomer in the presence of ATP or ADP, which suggests that ATP does not induce its dimerization. These results raise the possibility that monomeric FlhG can function *in vivo*, or alternatively, that an ATP-dependent FlhG dimer is unstable compared to other family member proteins and requires other factor(s) to stabilize the dimer structure [66]. In addition, mutations at the putative ATP binding or catalytic sites did not affect the elution profile of FlhG in size exclusion chromatography (eluted as a monomer regardless of the nucleotides), but the ATPase-active FlhG mutant (D171A) eluted slightly earlier in the presence of ATP but not ADP, presumably due to a subtle conformational change. Because the purified D171A mutant tends to aggregate in the presence of ATP, we speculate that ATP-bound active FlhG has a fragile conformation that causes its aggregation, but interactions with other proteins at the cell pole (most likely, HubP) prevent that aggregation and exhibit its function to inhibit FlhF activity [66].

It should be noted that *Geobacillus* and *Shewanella* FlhG have been shown to bind to the flagellar C-ring proteins FliM and FliY (FliN ortholog) in a nucleotide-independent manner [60]. Moreover, in the presence of ATP and lipids, *Geobacillus* FlhG (presumably in the dimer form) can activate FliM/FliY to assemble with another C-ring protein FliG in vitro [60]. These results raise the possibility that FlhG delivers C-ring proteins to the nascent flagellum, but this is puzzling because in this case FlhG functions as a positive regulator that promotes flagellar assembly. An alternative possibility is that FlhG binding to these proteins blocks the assembly of a nascent flagellum. Further analyses are required to clarify the enigmatic function of FlhG in the biogenesis of flagella.

7. SflA Represses Flagellar Biogenesis in the Absence of FlhF and FlhG

In *V. alginolyticus*, the deletion of both flhF and flhG from the strain VIO5 (wild-type for polar flagellum) resulted in a nonflagellated phenotype, but a very small fraction of the population produced several sheathed flagella at lateral positions [34]. The motile pseudo-revertants were isolated from the strain deleted for both *flhF* and *flhG* (Δ*flhFG*), which forms peritrichous flagella in the majority of cells [67]. Because these flagella were covered with a sheath and contained flagellins of the polar flagellum, the suppressor mutations increased the population of cells that produces multiple polar flagella at lateral positions. The mutation was mapped to a previously uncharacterized gene named *sflA* (suppressor of Δ*flhFG*) and the deletion of *sflA* from the Δ*flhFG* strain showed the suppression phenotype (Figure 7a) [68]. The *sflA* is specific for *Vibrio* species and is predicted to encode a single transmembrane protein (the mature protein contains 303 amino acids, ≈35 kDa) with its N-terminal region located at the periplasm. The cytoplasmic C-terminal region contains a DnaJ domain conserved in chaperone family proteins (Figure 7b) [69]. As with *hubP*, *sflA* is not included in known polar flagellar gene clusters, and therefore seems not to function specifically in flagellar biogenesis. The SflA protein was detected in the wild-type strain, but deletion of *sflA* from the wild-type strain did not affect polar flagellation and motility [68]. Overexpression of the C-terminal soluble region containing the DnaJ domain ($SflA_C$, Figure 7b) suppressed the lateral flagellation of the Δ*flhFG*Δ*sflA* strain [70]. SflA fused with fluorescent protein showed a HubP-dependent polar localization in the presence of FlhF and FlhG, but was observed at polar and lateral positions in Δ*flhFG* cells [70]. These observations suggest that SflA localizes with flagella and that $SflA_C$ represses the flagellar initiation in Δ*flhFG* cells by a currently unknown mechanism [70]. FlhF seems to be dominant over SflA in flagellation at the cell pole and voids the function of SflA. Recently, the crystal structure of SflA was solved for the N-terminal 131 residues ($SflA_{N1}$; Figure 7b,c) [71]. The core of $SflA_{N1}$ forms a domain-swapped dimer with a tetratricopeptide repeat (TPR)/Sel1-like repeat (SLR) motif, which is often found in domains responsible for protein–protein interactions in various proteins. $SflA_{N1}$ has a characteristic positively charged area at the surface, and alanine substitutions in that area reduced the SflA function of inhibiting flagellation in Δ*flhFG* cells, which suggests that $SflA_{N1}$ binds to an unknown partner protein and that the binding signal is transmitted to $SflA_C$ to suppress the formation of the sheathed flagellum at lateral positions (Figure 7c) [71].

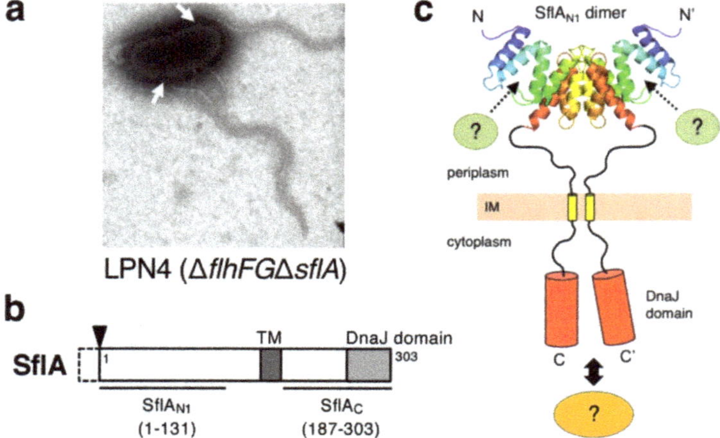

Figure 7. SflA represses lateral flagellation in the absence of both FlhF and FlhG in *Vibrio alginolyticus*. (**a**) Electron micrograph of a LPN4 cell, the *ΔflhFGΔsflA* mutant. The sheathed flagella were formed at lateral positions, as indicated by the white arrows. (**b**) Domain structure of SflA. SflA is synthesized as a precursor with an N-terminal signal sequence that is cleaved during maturation (the cleavage site is shown as a black arrowhead). (**c**) Model of the molecular architecture of the SflA dimer. Unknown binding partner proteins bind to the concave surface of the N-terminal tetratricopeptide repeat (TPR)/Sel1-like repeat (SLR) domain of SflA, as indicated by the broken arrows. The DnaJ domain is activated by the binding signal transmitted through the membrane, and interacts with an unknown partner protein to suppress the formation of the sheathed flagellum at peritrichous cell surfaces or promote it at the cell pole. IM, inner membrane.

8. Conclusions and Perspectives

To summarize all the insights presented in this review, we would like to propose a model for the biogenesis of a single polar flagellum in *Vibrio alginolyticus* (Figure 8). At least five factors are involved in this precise control: FlhF, FlhG, HubP, FlaK, and SflA. When cells are growing, the polar flagellar genes are transcribed in a cascade fashion. The class 1 master regulator FlaK activates expression of the class 2 genes, including *flhF*, *flhG*, and those for flagellar basal body components such as the MS-ring protein FliF. After it accumulates, FlhG negatively acts on the master regulator FlaK to shut off the expression of polar flagellar genes. This temporal regulatory mechanism prevents the unnecessary use of energy required for hyperflagellation. FlhF then forms a homodimer in complex with GTP and localizes at the cell pole. The polar localization of FlhF facilitates the accumulation of the MS-ring protein FliF at the cell pole, and thereby promotes the MS-ring formation there. Inactive ADP-bound FlhG binds to GDP-bound inactive FlhF and interferes with its polar localization in the cytoplasm, whereas ATP-bound FlhG is able to associate with HubP at the cell pole. HubP induces the structural change of FlhG to become an active form, in which the active site for ATP hydrolysis and binding interface for FlhF are constituted. This allows FlhG to directly inhibit the FlhF dimer, presumably by stimulating GTP hydrolysis. These four regulatory steps (indicated as "1" to "4" in Figure 8) together optimize FlhF activity at the cell pole to generate only a single polar flagellum. Meanwhile, initiation of sheathed flagellar assembly could occur at lateral positions once the MS-ring and flagellar specific export apparatus are assembled. However, in such a case, SflA inhibits the completion of flagellar assembly (indicated as "5" in Figure 8). Therefore, SflA also participates in regulation of flagellar biogenesis at the cell pole.

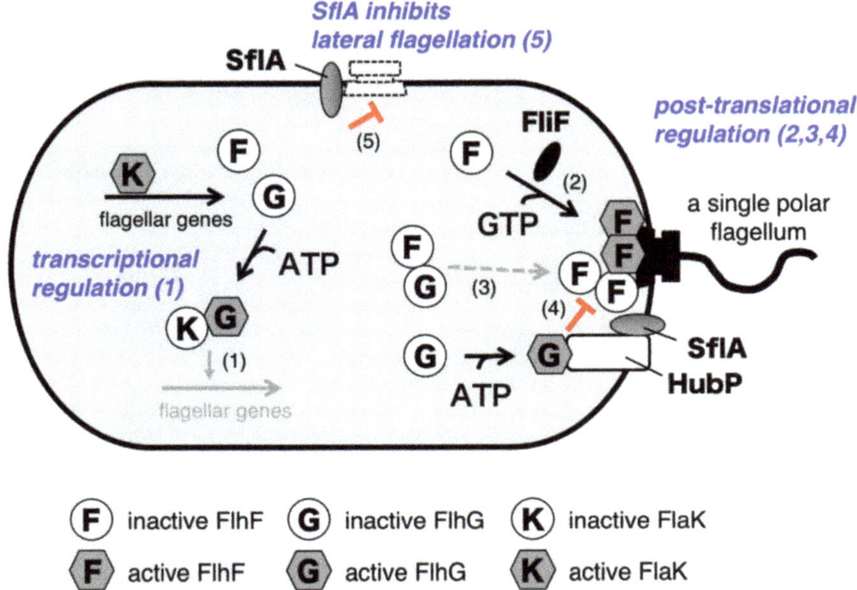

Figure 8. Model for the regulation of the biogenesis of a single polar flagellum in *Vibrio alginolyticus*. Five regulatory mechanisms are involved in this proposed model. When flagellar components are synthesized and accumulate, FlhG negatively acts on the master regulator FlaK to downregulate the expression of polar flagellar genes (1). FlhF forms a homodimer in complex with GTP and localizes at the cell pole. FlhF facilitates the accumulation of the MS-ring protein FliF at the cell pole and thereby promotes the MS-ring formation (2). Inactive FlhF and FlhG interact with each other and remain in the cytoplasm (3), and GTP-bound FlhG is activated by HubP at the cell pole and negatively acts on FlhF to inactivate the FlhF dimer at the cell pole (4). In addition to the transcriptional regulation, these post-translational mechanisms optimize polar FlhF activity that allows the cell to generate only a single polar flagellum. Meanwhile, SflA inhibits sheathed flagellar formation at lateral positions by negatively acting on its assembly step (5). FlhF activity is dominant over SflA at the cell pole, so that the effect of SflA, localized at the cell pole via HubP, is suppressed.

Many questions remain to be solved. For example, how does FlhF promote MS-ring assembly? The structural similarity suggests that GTP-bound FlhF homodimer may function similar to the GTP-dependent heterodimer of the signal recognition particle (SRP) and its receptor (SR). However, the *ΔflhFGΔsflA* strain can form sheathed flagella at lateral positions [68], and thus FlhF is not essential to insert nascent FliF, the MS-ring protein, in membranes. Therefore, FlhF seems to function in delivering or specifically inserting FliF to the cell pole. Alternatively, FlhF may facilitate the MS-ring assembly step at the cell pole. Further, how FlhG works at the cell pole is still largely unknown. It can function as a monomer or, alternatively, ATP-bound FlhG dimer may be stabilized at the cell pole to inhibit FlhF. Because the majority of *Bacillus* FlhG *in vivo* is reported as being highly mobile [60] but *Vibrio* FlhF is observed at the base of an assembled polar flagellum [34], FlhG may dissociate from the cell pole after the ATP hydrolysis, although FlhF somehow remains at the flagellated cell pole. To clarify the functions of FlhF and FlhG in flagellar biogenesis, their structural and biochemical properties must be understood on the basis of their temporal behavior at the cell pole. It should also be remembered that it remains enigmatic as to how SflA and FlaK recognize the flagellation state of cells. Altogether, it is fascinating that even simple organisms such as bacteria have such complex mechanisms to precisely control the temporal and subcellular positioning of biomolecules. The journey to elucidate the mechanism of how a single polar flagellum is generated by bacteria is a challenging one that will be exciting to explore.

Funding: This work was supported by Ministry of Education, Culture, Sports, Science and Technology (MEXT)/ Japan Society for the Promotion of Science (JSPS) KAKENHI grant numbers JP24115506, JP26115705, and JP16H04774 (to S.K.).

Acknowledgments: We thank Norihiro Takekawa for the preparation of the figures and for critically reading the manuscript. Some figures are adapted from references with permission ([57] for Figure 5c [66] for Figure 6a–c [68] for Figure 7a, and [71] for Figure 7c).

Conflicts of Interest: The authors declare no conflict of interest.

References

1. Berg, H.C. The Rotary Motor of Bacterial Flagella. *Annu. Rev. Biochem.* **2003**, *72*, 19–54. [CrossRef] [PubMed]
2. Kearns, D.B. A field guide to bacterial swarming motility. *Nat. Rev. Genet.* **2010**, *8*, 634–644. [CrossRef] [PubMed]
3. Macnab, R.M. How Bacteria Assemble Flagella. *Annu. Rev. Microbiol.* **2003**, *57*, 77–100. [CrossRef] [PubMed]
4. Li, N.; Kojima, S.; Homma, M. Sodium-driven motor of the polar flagellum in marine bacteria *Vibrio*. *Genes Cells* **2011**, *16*, 985–999. [CrossRef]
5. Sowa, Y.; Berry, R.M. Bacterial flagellar motor. *Q. Rev. Biophys.* **2008**, *41*, 103–132. [CrossRef]
6. Terashima, H.; Kojima, S.; Homma, M. Flagellar motility in bacteria: Structure and function of flagellar motor. *Int. Rev. Cell. Mol. Biol.* **2008**, *270*, 39–85.
7. Kojima, S. Dynamism and regulation of the stator, the energy conversion complex of the bacterial flagellar motor. *Curr. Opin. Microbiol.* **2015**, *28*, 66–71. [CrossRef]
8. Minamino, T.; Terahara, N.; Kojima, S.; Namba, K. Autonomous control mechanism of stator assembly in the bacterial flagellar motor in response to changes in the environment. *Mol. Microbiol.* **2018**, *109*, 723–734. [CrossRef]
9. Nakamura, S.; Minamino, T. Flagella-driven motility of bacteria. *Biomolecules* **2019**, *9*, 279. [CrossRef]
10. Subramanian, S.; Kearns, D.B. Functional Regulators of Bacterial Flagella. *Annu. Rev. Microbiol.* **2019**, *73*, 225–246. [CrossRef]
11. Kazmierczak, B.I.; Hendrixson, D.R. Spatial and numerical regulation of flagellar biosynthesis in polarly flagellated bacteria. *Mol. Microbiol.* **2013**, *88*, 655–663. [CrossRef] [PubMed]
12. Balaban, M.; Hendrixson, D.R. Polar flagellar biosynthesis and a regulator of flagellar number influence spatial parameters of cell division in *Campylobacter jejuni*. *PLOS Pathog.* **2011**, *7*, e1002420. [CrossRef]
13. Martinez, L.E.; Hardcastle, J.M.; Wang, J.; Pincus, Z.; Tsang, J.; Hoover, T.; Bansil, R.; Salama, N.R. *Helicobacter pylori* strains vary cell shape and flagellum number to maintain robust motility in viscous environments. *Mol. Microbiol.* **2015**, *99*, 88–110. [CrossRef] [PubMed]
14. Armitage, J.P.; Macnab, R.M. Unidirectional, intermittent rotation of the flagellum of *Rhodobacter sphaeroides*. *J. Bacteriol.* **1987**, *169*, 514–518. [CrossRef] [PubMed]
15. Schniederberend, M.; Williams, J.F.; Shine, E.E.; Shen, C.; Jain, R.; Emonet, T.; Kazmierczak, B.I. Modulation of flagellar rotation in surface-attached bacteria: A pathway for rapid surface-sensing after flagellar attachment. *PLOS Pathog.* **2019**, *15*, e1008149. [CrossRef] [PubMed]
16. Green, J.C.; Kahramanoglou, C.; Rahman, A.; Pender, A.; Charbonnel, N.; Fraser, G. Recruitment of the earliest component of the bacterial flagellum to the old cell division pole by a membrane-associated signal recognition particle family GTP-binding protein. *J. Mol. Boil.* **2009**, *391*, 679–690. [CrossRef]
17. Xu, H.; He, J.; Liu, J.; Motaleb, A. BB0326 is responsible for the formation of periplasmic flagellar collar and assembly of the stator complex in *Borrelia burgdorferi*. *Mol. Microbiol.* **2019**, *113*, 418–429. [CrossRef]
18. Zhang, K.; He, J.; Cantalano, C.; Guo, Y.; Liu, J.; Li, C. FlhF regulates the number and configuration of periplasmic flagella in *Borrelia burgdorferi*. *Mol. Microbiol.* **2020**, *00*, 1–18. [CrossRef]
19. Tahara, H.; Takabe, K.; Sasaki, Y.; Kasuga, K.; Kawamoto, A.; Koizumi, N.; Nakamura, S. The mechanism of two-phase motility in the spirochete *Leptospira*: swimming and crawling. *Sci. Adv.* **2018**, *4*, eaar7975. [CrossRef]

20. Kawagishi, I.; Imagawa, M.; Imae, Y.; McCarter, L.; Homma, M. The sodium-driven polar flagellar motor of marine *Vibrio* as the mechanosensor that regulates lateral flagellar expression. *Mol. Microbiol.* **1996**, *20*, 693–699. [CrossRef]
21. McCarter, L.; Hilmen, M.; Silverman, M. Flagellar dynamometer controls swarmer cell differentiation of *V. parahaemolyticus*. *Cell* **1988**, *54*, 345–351. [CrossRef]
22. Bubendorfer, S.; Held, S.; Windel, N.; Paulick, A.; Klingl, A.; Thormann, K.M. Specificity of motor components in the dual flagellar system of *Shewanella putrefaciens* CN-32. *Mol. Microbiol.* **2011**, *83*, 335–350. [CrossRef] [PubMed]
23. Kawagishi, I.; Maekawa, Y.; Atsumi, T.; Homma, M.; Imae, Y. Isolation of the polar and lateral flagellum-defective mutants in *Vibrio alginolyticus* and identification of their flagellar driving energy sources. *J. Bacteriol.* **1995**, *177*, 5158–5160. [CrossRef] [PubMed]
24. Furuno, M.; Sato, K.; Kawagishi, I.; Homma, M. Characterization of a flagellar sheath component, PF60, and its structural gene in marine *Vibrio*. *J. Biochem.* **2000**, *127*, 29–36. [CrossRef] [PubMed]
25. Muramoto, K.; Kawagishi, I.; Kudo, S.; Magariyama, Y.; Imae, Y.; Homma, M. High-speed rotation and speed stability of the sodium-driven flagellar motor in *Vibrio alginolyticus*. *J. Mol. Boil.* **1995**, *251*, 50–58. [CrossRef] [PubMed]
26. Kojima, M.; Kubo, R.; Yakushi, T.; Homma, M.; Kawagishi, I. The bidirectional polar and unidirectional lateral flagellar motors of *Vibrio alginolyticus* are controlled by a single CheY species. *Mol. Microbiol.* **2007**, *64*, 57–67. [CrossRef]
27. Okunishi, I.; Kawagishi, I.; Homma, M. Cloning and characterization of *motY*, a gene coding for a component of the sodium-driven flagellar motor in *Vibrio alginolyticus*. *J. Bacteriol.* **1996**, *178*, 2409–2415. [CrossRef]
28. Murray, T.S.; Kazmierczak, B.I. FlhF Is Required for swimming and swarming in *Pseudomonas aeruginosa*. *J. Bacteriol.* **2006**, *188*, 6995–7004. [CrossRef]
29. Pandza, S.; Baetens, M.; Park, C.H.; Au, T.; Keyhan, M.; Matin, A. The G-protein FlhF has a role in polar flagellar placement and general stress response induction in *Pseudomonas putida*. *Mol. Microbiol.* **2000**, *36*, 414–423. [CrossRef]
30. Rossmann, F.; Brenzinger, S.; Knauer, C.; Dörrich, A.K.; Bubendorfer, S.; Ruppert, U.; Bange, G.; Thormann, K.M. The role of FlhF and HubP as polar landmark proteins in *Shewanella putrefaciens* CN-32. *Mol. Microbiol.* **2015**, *98*, 727–742. [CrossRef]
31. Gao, T.; Shi, M.; Ju, L.; Gao, H. Investigation into FlhFG reveals distinct features of FlhF in regulating flagellum polarity in *Shewanella oneidensis*. *Mol. Microbiol.* **2015**, *98*, 571–585. [CrossRef] [PubMed]
32. Kusumoto, A.; Kamisaka, K.; Yakushi, T.; Terashima, H.; Shinohara, A.; Homma, M. Regulation of polar flagellar number by the *flhF* and *flhG* genes in *Vibrio alginolyticus*. *J. Biochem.* **2006**, *139*, 113–121. [CrossRef] [PubMed]
33. Dasgupta, N.; Arora, S.K.; Ramphal, R. *fleN*, a gene that regulates flagellar number in *Pseudomonas aeruginosa*. *J. Bacteriol.* **2000**, *182*, 357–364. [CrossRef] [PubMed]
34. Kusumoto, A.; Shinohara, A.; Terashima, H.; Kojima, S.; Yakushi, T.; Homma, M. Collaboration of FlhF and FlhG to regulate polar-flagella number and localization in *Vibrio alginolyticus*. *Microbiology* **2008**, *154*, 1390–1399. [CrossRef]
35. Guttenplan, S.B.; Shaw, S.; Kearns, D.B. The cell biology of peritrichous flagella in *Bacillus subtilis*. *Mol. Microbiol.* **2012**, *87*, 211–229. [CrossRef]
36. Bange, G.; Sinning, I. SIMIBI twins in protein targeting and localization. *Nat. Struct. Mol. Boil.* **2013**, *20*, 776–780. [CrossRef]
37. Egea, P.F.; Shan, S.-O.; Napetschnig, J.; Savage, D.F.; Walter, P.; Stroud, R.M. Substrate twinning activates the signal recognition particle and its receptor. *Nature* **2004**, *427*, 215–221. [CrossRef]
38. Schuhmacher, J.S.; Thormann, K.M.; Bange, G. How bacteria maintain location and number of flagella? *FEMS Microbiol. Rev.* **2015**, *39*, 812–822. [CrossRef]
39. Bange, G.; Petzold, G.; Wild, K.; Parlitz, R.O.; Sinning, I. The crystal structure of the third signal-recognition particle GTPase FlhF reveals a homodimer with bound GTP. *Proc. Natl. Acad. Sci. USA* **2007**, *104*, 13621–13625. [CrossRef]
40. Bange, G.; Kümmerer, N.; Grudnik, P.; Lindner, R.; Petzold, G.; Kressler, D.; Hurt, E.; Wild, K.; Sinning, I. Structural basis for the molecular evolution of SRP-GTPase activation by protein. *Nat. Struct. Mol. Boil.* **2011**, *18*, 1376–1380. [CrossRef]

41. Kusumoto, A.; Nishioka, N.; Kojima, S.; Homma, M. Mutational analysis of the GTP-binding motif of FlhF which regulates the number and placement of the polar flagellum in *Vibrio alginolyticus*. *J. Biochem.* **2009**, *146*, 643–650. [CrossRef] [PubMed]
42. Kondo, S.; Homma, M.; Kojima, S. Analysis of the GTPase motif of FlhF in the control of the number and location of polar flagella in *Vibrio alginolyticus*. *BPPB.* **2017**, *14*, 173–181. [CrossRef] [PubMed]
43. Gulbronson, C.J.; Ribardo, D.A.; Balaban, M.; Knauer, C.; Bange, G.; Hendrixson, D.R. FlhG employs diverse intrinsic domains and influences FlhF GTPase activity to numerically regulate polar flagellar biogenesis in *Campylobacter jejuni*. *Mol. Microbiol.* **2015**, *99*, 291–306. [CrossRef] [PubMed]
44. Kondo, S.; Imura, Y.; Mizuno, A.; Homma, M.; Kojima, S. Biochemical analysis of GTPase FlhF which controls the number and position of flagellar formation in marine *Vibrio*. *Sci. Rep.* **2018**, *8*, 12115. [CrossRef] [PubMed]
45. Balaban, M.; Joslin, S.N.; Hendrixson, D.R. FlhF and its GTPase activity are required for distinct processes in flagellar gene regulation and biosynthesis in *Campylobacter jejuni*. *J. Bacteriol.* **2009**, *191*, 6602–6611. [CrossRef]
46. Chevance, F.F.V.; Hughes, K.T. Coordinating assembly of a bacterial macromolecular machine. *Nat. Rev. Genet.* **2008**, *6*, 455–465. [CrossRef]
47. Kubori, T.; Yamaguchi, S.; Aizawa, S. Assembly of the switch complex onto the MS ring complex of *Salmonella typhimurium* does not require any other flagellar proteins. *J. Bacteriol.* **1997**, *179*, 813–817. [CrossRef]
48. Echazarreta, M.A.; Klose, K.E. Vibrio flagellar synthesis. *Front. Microbiol.* **2019**, *9*, 131. [CrossRef]
49. Kutsukake, K.; Ohya, Y.; Iino, T. Transcriptional analysis of the flagellar regulon of *Salmonella typhimurium*. *J. Bacteriol.* **1990**, *172*, 741–747. [CrossRef]
50. Hickman, J.W.; Harwood, C.S. Identification of FleQ from *Pseudomonas aeruginosa* as a c-di-GMP-responsive transcription factor. *Mol. Microbiol.* **2008**, *69*, 376–389. [CrossRef]
51. Klose, K.E.; Mekalanos, J.J. Differential Regulation of Multiple Flagellins in *Vibrio cholerae*. *J. Bacteriol.* **1998**, *180*, 303–316. [CrossRef] [PubMed]
52. Kim, Y.-K.; McCarter, L.L. Cross-Regulation in *Vibrio parahaemolyticus*: Compensatory activation of polar flagellar genes by the lateral flagellar regulator LafK. *J. Bacteriol.* **2004**, *186*, 4014–4018. [CrossRef] [PubMed]
53. Dasgupta, N.; Ramphal, R. Interaction of the antiactivator FleN with the transcriptional activator FleQ regulates flagellar number in *Pseudomonas aeruginosa*. *J. Bacteriol.* **2001**, *183*, 6636–6644. [CrossRef] [PubMed]
54. Correa, N.E.; Peng, F.; Klose, K.E. Roles of the Regulatory Proteins FlhF and FlhG in the *Vibrio cholerae* flagellar transcription hierarchy. *J. Bacteriol.* **2005**, *187*, 6324–6332. [CrossRef] [PubMed]
55. McCarter, L.L. Polar Flagellar Motility of the *Vibrionaceae*. *Microbiol. Mol. Boil. Rev.* **2001**, *65*, 445–462. [CrossRef] [PubMed]
56. Yamaichi, Y.; Bruckner, R.; Ringgaard, S.; Möll, A.; Cameron, D.E.; Briegel, A.; Jensen, G.J.; Davis, B.M.; Waldor, M.K. A multidomain hub anchors the chromosome segregation and chemotactic machinery to the bacterial pole. *Genes Dev.* **2012**, *26*, 2348–2360. [CrossRef]
57. Takekawa, N.; Kwon, S.; Nishioka, N.; Kojima, S.; Homma, M. HubP, a polar landmark protein, regulates flagellar number by assisting in the proper polar localization of FlhG in *Vibrio alginolyticus*. *J. Bacteriol.* **2016**, *198*, 3091–3098. [CrossRef]
58. Wehbi, H.; Portillo, E.; Harvey, H.; Shimkoff, A.E.; Scheurwater, E.M.; Howell, P.L.; Burrows, L.L. The Peptidoglycan-Binding protein FimV promotes assembly of the *Pseudomonas aeruginosa* Type IV pilus secretin. *J. Bacteriol.* **2010**, *193*, 540–550. [CrossRef]
59. Lutkenhaus, J. Assembly dynamics of the bacterial MinCDE system and spatial regulation of the Z ring. *Annu. Rev. Biochem.* **2007**, *76*, 539–562. [CrossRef]
60. Schuhmacher, J.S.; Rossmann, F.; Dempwolff, F.; Knauer, C.; Altegoer, F.; Steinchen, W.; Dörrich, A.K.; Klingl, A.; Stephan, M.; Linne, U.; et al. MinD-like ATPase FlhG effects location and number of bacterial flagella during C-ring assembly. *Proc. Natl. Acad. Sci. USA* **2015**, *112*, 3092–3097. [CrossRef]
61. Chanchal; Banerjee, P.; Jain, D. ATP-Induced structural remodeling in the antiactivator FleN enables formation of the functional dimeric form. *Structure* **2017**, *25*, 243–252. [CrossRef] [PubMed]
62. Wu, W.; Park, K.-T.; Holyoak, T.; Lutkenhaus, J. Determination of the structure of the MinD-ATP complex reveals the orientation of MinD on the membrane and the relative location of the binding sites for MinE and MinC. *Mol. Microbiol.* **2011**, *79*, 1515–1528. [CrossRef] [PubMed]
63. Ono, H.; Takashima, A.; Hirata, H.; Homma, M.; Kojima, S. The MinD homolog FlhG regulates the synthesis of the single polar flagellum of *Vibrio alginolyticus*. *Mol. Microbiol.* **2015**, *98*, 130–141. [CrossRef] [PubMed]

64. Ma, L.-Y.; King, G.F.; Rothfield, L. Mapping the MinE site involved in interaction with the MinD division site selection protein of *Escherichia coli*. *J. Bacteriol.* **2003**, *185*, 4948–4955. [CrossRef] [PubMed]
65. Lutkenhaus, J. The ParA/MinD family puts things in their place. *Trends Microbiol.* **2012**, *20*, 411–418. [CrossRef] [PubMed]
66. Kojima, S.; Imura, Y.; Hirata, H.; Homma, M. Characterization of the MinD/ParA-type ATPase FlhG in *Vibrio alginolyticus* and implications for function of its monomeric form. *Genes Cells* **2020**. [CrossRef] [PubMed]
67. Kojima, M.; Nishioka, N.; Kusumoto, A.; Yagasaki, J.; Fukuda, T.; Homma, M. Conversion of mono-polar to peritrichous flagellation in *Vibrio alginolyticus*. *Microbiol. Immunol.* **2011**, *55*, 76–83. [CrossRef] [PubMed]
68. Kitaoka, M.; Nishigaki, T.; Ihara, K.; Nishioka, N.; Kojima, S.; Homma, M. A novel *dnaJ* family gene, *sflA*, encodes an inhibitor of flagellation in marine *Vibrio* species. *J. Bacteriol.* **2012**, *195*, 816–822. [CrossRef]
69. Hennessy, F.; Cheetham, M.E.; Dirr, H.; Blatch, G.L. Analysis of the levels of conservation of the J domain among the various types of DnaJ-like proteins. *Cell Stress Chaperones* **2000**, *5*, 347–358. [CrossRef]
70. Inaba, S.; Nishigaki, T.; Takekawa, N.; Kojima, S.; Homma, M. Localization and domain characterization of the SflA regulator of flagellar formation in *Vibrio alginolyticus*. *Genes Cells* **2017**, *22*, 619–627. [CrossRef]
71. Sakuma, M.; Nishikawa, S.; Inaba, S.; Nishigaki, T.; Kojima, S.; Homma, M.; Imada, K. Structure of the periplasmic domain of SflA involved in spatial regulation of the flagellar biogenesis of *Vibrio* reveals a TPR/SLR-like fold. *J. Biochem.* **2019**, *166*, 197–204. [CrossRef] [PubMed]

 *biomolecules* 

Review

Protein Export via the Type III Secretion System of the Bacterial Flagellum

Manuel Halte and Marc Erhardt *

Institute for Biology–Bacterial Physiology, Humboldt-Universität zu Berlin, Philippstr. 13, 10115 Berlin, Germany; manuel.halte@hu-berlin.de
* Correspondence: marc.erhardt@hu-berlin.de

Abstract: The bacterial flagellum and the related virulence-associated injectisome system of pathogenic bacteria utilize a type III secretion system (T3SS) to export substrate proteins across the inner membrane in a proton motive force-dependent manner. The T3SS is composed of an export gate (FliPQR/FlhA/FlhB) located in the flagellar basal body and an associated soluble ATPase complex in the cytoplasm (FliHIJ). Here, we summarise recent insights into the structure, assembly and protein secretion mechanisms of the T3SS with a focus on energy transduction and protein transport across the cytoplasmic membrane.

Keywords: bacterial flagellum; flagellar assembly; type III protein export; ATPase; proton motive force; secretion model

Citation: Halte, M.; Erhardt, M. Protein Export via the Type III Secretion System of the Bacterial Flagellum. *Biomolecules* **2021**, *11*, 186. https://doi.org/10.3390/biom11020186

Academic Editors: Tohru Minamino and Keiichi Namba
Received: 30 December 2020
Accepted: 26 January 2021
Published: 29 January 2021

1. Introduction

Flagella are complex rotary nanomachines embedded in the cell envelope of many bacteria. In addition to functions in adhering to surfaces, flagella allow bacteria to move in their environment towards nutrients or to escape harmful molecules. They are present in both Gram-negative and Gram-positive bacteria, and are evolutionary related to the injectisome device, which various Gram-negative bacterial species use to inject effectors into eukaryotic target cells [1]. Both the flagellum and injectisome are complex nanomachines and made of around 20 different proteins, ranging from a copy number of very few to several thousand [2]. Structurally, the flagellum can be divided into three main parts: (i) a basal body embedded in the cell envelope; (ii) a flexible linking structure, the hook; and (iii) a long external filament, which functions as the propeller.

The basal body is composed of the rod (made of FliE, FlgB, FlgC, FlgF, and FlgG) and several protein rings: the MS ring (made of FliF) in the inner membrane (IM) and the C ring (made of FliG, FliM and FliN) in the cytoplasm, the periplasmic P ring located in the peptidoglycan (PG) layer (made of FlgI), and the L ring (made of FlgH) in the outer membrane (OM) [3]. The assembly and function of both the flagellum and injectisome relies on a conserved protein export apparatus located at the base of the basal body. The so-called type III secretion system (T3SS) found in the flagellum (flagellar T3SS; fT3SS) and injectisome (virulence-associated T3SS; vT3SS) is comprised of five conserved core export gate proteins, FlhA/SctV, FlhB/SctU, FliP/SctR, FliQ/SctS, and FliR/SctT, and is associated to three soluble proteins, FliH/SctL, FliI/SctN, and FliJ/SctO, respectively, for the fT3SS and vT3SS, forming the export apparatus (Figure 1) [4–6]. The T3SS is involved in substrate protein selection, i.e., docking of substrate proteins (and their cognate secretion chaperones) to the export apparatus, the subsequent unfolding of substrates and proton motive force-driven translocation across the inner membrane, and forms the central secretion pore through which the substrates are secreted. As the diameter of the secretion channel is only 2 nm, secreted proteins are likely in an (at least partially) unfolded state [7]. Additionally, the T3SS can discriminate between different substrate classes. Only upon completion of the hook–basal–body (HBB) complex, the T3SS switches substrate specificity from early type substrates (e.g., components of the rod or the hook subunits) to late-type substrates (e.g., flagellin or the filament cap), which enables a mechanism to prevent premature secretion of the many thousand flagellin subunits before the assembly of the hook is completed [8]. This switch in substrate specificity involves an interaction between a molecular ruler protein, FliK, and the export gate protein FlhB [9–11] in case of the flagellum, while a homologous ruler protein (SctU/SctP) is implicated in substrate selectivity switching of the injectisome [12].

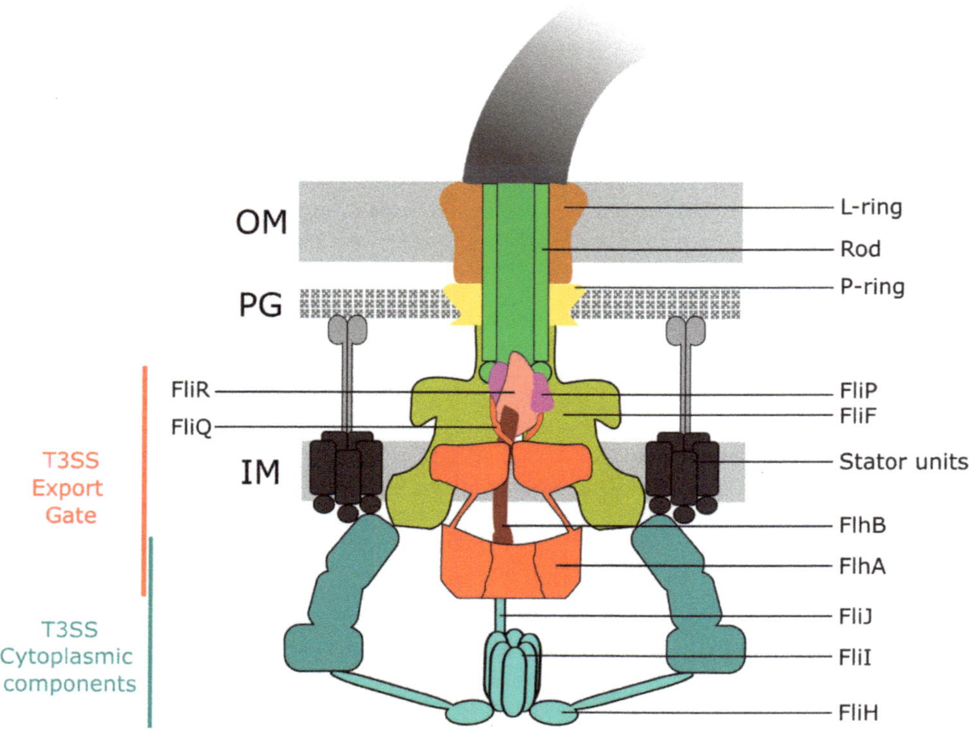

Figure 1. Schematic of the flagellar basal body complex. The basal body consists of the stator units, the periplasmic rod, the P- and L rings embedded in the peptidoglycan (PG) and outer membrane (OM), respectively, the MS ring in the inner membrane (IM), and the T3SS. The T3SS is composed of cytoplasmic components (the C ring and the ATPase complex consisting of FliI, FliH and FliJ), and the export gate (consisting of FliPQR, FlhA and FlhB) located at the base of the basal body within the MS ring.

Furthermore, protein secretion via the T3SS is remarkably fast. The T3SS is able to secrete several thousand amino acid per second [13], in comparison to the general secretion system (Sec system) that only secretes a few dozen amino acids per second [14,15]. In recent years, a model has emerged where primarily the proton motive force (PMF), i.e., the charge and proton gradient across the inner membrane, provides energy to drive protein translocation across the inner membrane via the T3SS [16]. The soluble ATPase complex made of FliHIJ is part of the export apparatus and thought to facilitate docking and unfolding of substrates [17]. Accordingly, the secretion of T3SS substrate proteins can be described as a three-step process: (i) docking of substrate proteins, which might be associated with their chaperones, to the export apparatus; (ii) unfolding of substrate proteins; and (iii) PMF-dependent injection of substrate proteins into the secretion channel. The secreted substrate proteins subsequently travel (presumably in an α-helical or partially unfolded conformation [18]) through the channel inside the flagellum to the tip, where they self-assemble.

This chapter highlights the protein components that make up the core T3SS of the flagellum, discusses potential mechanisms underlying the substrate energization processes, summarizes the various models that have been proposed to understand the secretion process and assembly of flagellin subunits into the growing filament and compares the protein export mechanism of the flagellar T3SS to the protein secretion mechanisms used by other bacterial secretion systems (Box 1).

2. Export Apparatus Structure and Assembly

The core export gate of the flagellar T3SS is located in the centre of the basal body and formed by FliPQR, FlhA and FlhB. Associated at the cytoplasmic face is the ATPase complex constituted of FliHIJ. FliPQR form a helical assembly with 5:4:1 stoichiometry ($FliP_5Q_4R_1$) embedded in the core of the basal body and located above the inner membrane (IM) [19]. A small bitopic membrane protein, FliO, was previously thought to be part of the export apparatus. However, it has recently been shown that FliO is not part of the assembled basal body complex and that FliO has no active role in the export process, but functions as a chaperone for productive assembly of the FliP-FliR complex [20] (Figure 2). It is interesting to note that, although FliPQR homologs are strongly conserved in the vT3SS, no homologs of FliO have been found thus far, which suggests a specific role for assembly of the fT3SS [21]. It has been postulated that FliO promotes stable FliP sub-assemblies in the IM, before forming a complex with FliR. Once this complex is formed, FliO would dissociate, and the complex would interact with FliQ to form the $FliP_5Q_4R_1$ helical assembly [19] before interacting with FlhB and finally FlhA. FlhB consists of an N-terminal, transmembrane (TM) domain ($FlhB_{TM}$) formed by four α-helices, and a C-terminal cytoplasmic domain ($FlhB_C$). $FlhB_C$ is responsible for the switching from early type substrates (e.g., the rod and hook subunits) to late-type substrates (e.g., the anti-sigma factor FlgM, the hook-filament junctions, the filament tip and flagellin). This switching requires a proteolytic autocleavage of $FlhB_C$ between Asn269 and Pro270 at a highly conserved NPTH motif splitting $FlhB_C$ into $FlhB_{CN}$ and $FlhB_{CC}$, in addition to an interaction with the molecular ruler FliK [12,22]. In the autocleavage deficient mutant FlhB(N269A), the T3SS is deficient for the secretion of late-type substrates [23]. Under normal conditions of flagellar assembly, FliK is secreted intermittently throughout the hook construction period: once the hook reached a length of ~55 nm, the N-terminus of FliK remains in the channel, thereby slowing-down its secretion. Presumably, this leaves the FliK C-terminal domain enough time to interact with FlhBc, inducing a conformational change in the export apparatus that results in the switch in substrate specificity [11]. Following the switch in secretion specificity, the anti-sigma factor FlgM is recognized as a late secretion substrate and secreted. Secretion of FlgM releases σ^{28} and allows for σ^{28}-dependent gene expressions from class III promoters, including *fliC* encoding for the flagellin FliC [8,24,25]. A recent study on the structure of the core flagellar export apparatus validated that FlhB is part of the export gate [26]. Interestingly, FlhB associates to the core secretion pore FliPQR by forming a loop ($FlhB_L$), containing the most conserved residues of FlhB. The four helices in the $FlhB_{TM}$ domain form two distinct hairpins that wrap around the cytoplasmic face of the FliPQR complex, inserting hydrophobic residues in the cavities between FliQs subunits. Kuhlen et al., initially hypothesised that $FlhB_L$ is involved in maintaining a closed export gate, but the FliPQR structure from a *flhB* deletion mutant is present in a closed conformation [26]. The wrapping of $FlhB_L$ might be involved in the opening of the export gate, either by moving away of the entrance with a hinge motion similar to a lid, or by staying in contact with the opening FliQs subunits and extending its structure in a similar way to a sphincter. Deletions of residues in $FlhB_L$ led to loss of motility, and the extensive interactions on the cytoplasmic face and on the surface of the FliPQR complex with FlhB demonstrates that FlhB is an integral component of the export gate complex directly involved in substrate protein secretion [26]. FlhA is a central component of the PMF-driven flagellar protein export machine and hypothesised to function as the proton/protein antiporter. FlhA can be separated into two regions: a hydrophobic N-terminal transmembrane (TM) region with eight predicted α-helical transmembrane domains ($FlhA_{TM}$) that interact with the MS ring [27], and a hydrophilic C-terminal cytoplasmic region ($FlhA_C$) that interacts with FliHIJ and FlhB, and with the substrate-chaperone complex (hook-filament junction protein FlgK/FlgN, filament capping protein FliD/FliT and flagellin FliC/FliS) [28–31]. The $FlhA_C$ region consists of four domains: D1 (residues 362 to 434 and residues 484 to 503), D2 (residues 435 to 483), D3 (residues 504 to 583), D4 (residues 584 to 682), and a flexible linker termed $FlhA_L$ (residues 328 to 361). The $FlhA_C$ regions form a nonameric, cytoplasmic ring beneath the export gate through D1-D3 and D3-D3 interactions [32], and is involved in the recognition of substrates, the binding of the substrate-chaperone complex and the actual secretion process [30,33].

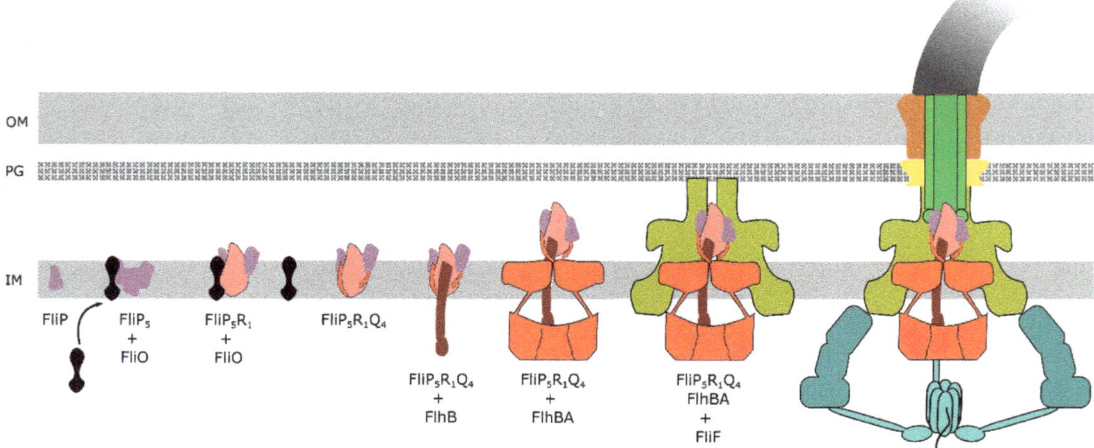

Figure 2. Step-wise export apparatus assembly model. Formation of the $FliP_5$ complex in the IM is presumably promoted by the integral-membrane chaperone FliO. FliO further facilitates the formation of a stable $FliP_5R_1$ complex. Once the $FliP_5R_1$ complex is formed, FliO is thought to dissociate from the complex. The $FliP_5R_1$ complex then constitutes the nucleus for the assembly of the subsequent subunits of the export gate ($FliQ_4$, FlhB and $FlhA_9$). Finally, the MS ring made of FliF forms around the core export apparatus $FliP_5FliQ_4FliR_1FlhB_1FlhA_9$ in the IM followed by the assembly of the C ring (made of FliG/M/N) and recruitment of the ATPase complex.

Upon assembly of the core secretion pore, the switch protein FlhB and the nonameric FlhA ring, the MS ring assembles in the inner membrane around the flagellar export gate. The MS ring functions as a scaffold and is formed by 34 subunits of FliF [34]. The FliG/M/N proteins (respectively 26–34, 34 and around 100 copies [35]) form the cytoplasmic C ring, which interacts tightly with the MS ring. Although it was proposed previously that 26 copies of FliG are present in the C ring [35], the stoichiometry of FliF and FliG are likely matched, suggesting that the FliG part of the C ring consists of 34 subunits [34]. The C ring primarily functions as the rotor the flagellum, which through interactions with the stator units formed by $MotA_5B_2$ drives rotation of the flagellum. The FliG subunits of the C ring interact with the rotating stator units (a pentamer of MotA that has been proposed to rotate around a dimer of MotB using energy derived from the ion gradient across the inner membrane), and enables bidirectional rotation of the flagellum by changing its conformation upon binding of the phosphorylated response regulator CheY during chemotaxis [36–38]. Additionally, the C ring plays a role as affinity site for substrates [39].

3. Structure and Function of the ATPase

As mentioned above, a soluble ATPase complex formed by FliHIJ is located at the cytoplasmic interface of the flagellar basal body. The ATPase FliI is found in two different complexes, (i) in the $FliI_6FliH_{12}FliJ$ complex at the base of the export apparatus, where FliI forms a hexameric ring, and (ii) in the freely diffusing $FliH_2FliI$ hetero-trimer [40–44]. The $FliI_6FliH_{12}FliJ$ complex is associated to the C ring through interactions between FliH and FliN/FlhA, and FliJ might play a role in the energy coupling mechanism for flagellar protein export by interacting with FlhA [28,45]. The structure of the complex SctV/SctO of *Chlamydia pneumoniae* (vT3SS homologs of FlhA/FliJ, respectively) indicates that interaction of SctO to the C-terminal region of SctV alters the binding site for substrate-chaperone complex and changes the conformation of SctVc. A potential rotation of FliJ (discussed in more details below) might then release the substrate-chaperone complex, allowing subsequent secretion of the substrate protein [46]. However, the exact role of the ATPase complex for protein secretion via the flagellar export apparatus remains poorly understood. Under physiological conditions, the role of the ATPase complex appears to be facilitating efficient secretion of substrate proteins via the T3SS. Interestingly, however, the FliHIJ ATPase complex has been shown to be dispensable for filament formation under certain conditions, such as overproduction of flagellar substrates or when the cell's available PMF is increased [16,47,48]. In support, several mutations in FliI that abolish or substantially reduce its ATPase activity are still able to assemble flagellar filaments, suggesting that any process energized by ATP hydrolysis is uncoupled from the actual protein secretion mechanism [49].

The ATPase FliI is a member of the Walker-type ATPase family. It shares similar structural characteristics with the α/β subunits of the F_OF_1 ATP synthase [50]. In contrast to the F_OF_1 ATP synthase where the α and β subunits forms a hetero-hexamer, FliI forms a homo-hexamer. Hexamer formation is needed for the enzyme to exert its full ATPase activity. FliJ binds to the $FliI_6$ ring and functions to stabilise the formation of the hexameric FliI ring, which then resembles the F1-$\alpha_3\beta_3\gamma$ complex [28,51]. FliH is divided into three regions: an N-terminal region $FliH_N$, a central region $FliH_M$ and the C-terminal region $FliH_C$. Only the first 10 amino acids of FliHc are critical for the export of substrates through the secretion channel by binding with the C ring through FliN-FliH interactions [45]. FliH promotes the interaction of $FliI_6FliJ$ with the C ring by interacting through its C-terminal region with FliI and N-terminal region with FliM-FliN [45]. Photocrosslinking experiments have additionally revealed an interaction between FliH and $FlhA_C$, but not with the other proteins of the basal body. It is assumed that the interaction of $FliH_N$ and $FlhA_C$ anchors the $FliI_6FliH_{12}FliJ$ ATPase complex to the export apparatus during the protein secretion process [52]. The second, soluble FliI complex ($FliH_2FliI$) has been shown to inhibit formation of the hexameric $FliI_6$ ring, and to bind to late export substrates in complex with their chaperones [40,41,43,44,53,54]. In this function, the $FliH_2FliI$ complex is thought to act as a dynamic carrier to deliver FliJ and the late export substrate-chaperone complexes to the docking platform of the T3SS export gate formed by $FlhA_C$ [43,55]. The $FliH_2FliI$ carrier binds to the hook-filament junction protein FlgK/FlgN and filament capping protein FliD/FliT substrate-chaperone complexes, but not to FliC/FliS [53,56,57]. It has been proposed that $FliH_2FliI$ may contribute to efficient interactions of the FlgK/FlgN and FliD/FliT substrate chaperone complexes, promoting an efficient assembly of the hook/cap structures before the assembly of the filament [58]. The interaction of the FliH C-terminal region with the FliI N-terminal region inhibits its ATPase activity, indicating that the N-terminal of FliI might be involved in the regulation of the ATPase activity. This inhibition might prevent consumption of ATP before the export of substrates. The presence of another factor (FliJ) is then required to start the activity of FliI [40].

It has been proposed that FliJ might also act partially as a rotor: the V_oV_1 and F_OF_1 ATPases use a rotational mechanism to hydrolyse ATP, and since the $FliI_6FliJ$ complex and the F1-$\alpha_3\beta_3\gamma$ complex are evolutionary related and share structural similarities, it appears possible that the two systems use a similar mechanism for distinct functions [59].

Accordingly, a rotational mechanism has recently been proposed to explain how the ATPase $FliI_6FliJ$ might assist the unfolding of substrates before their export into the channel. For this ATPase rotation mechanism, FliJ would function as the rotor and bind in the central cavity of the hexameric FliI ring, which itself is anchored to the C ring via FliH interactions and thereby would function as the stator. Further, each FliI monomer is predicted to bind an ATP molecule and harbour a FliJ binding site. When FliJ is bound to one FliI monomer, it cannot rotate. After hydrolysis of ATP in one of the FliI monomers, FliJ is released from its confinement and can freely rotate temporarily. The motion is governed by two parameters: (i) ATP hydrolysis rate and (ii) lifetime of the ADP-bound state. As a folded substrate protein arrives at the export gate, ATP hydrolysis would then induce rotation of the FliJ shaft, and the rotating FliJ interacts with the folded substrate (or substrate-chaperone complex), providing sufficient energy to help overcome the energy barrier and unfold the substrate protein (or strip-off the chaperone) before PMF-dependent injection into the secretion channel. Such a model is in agreement with previous observation that a certain level of protein export still occurs in the absence of the ATPase complex, suggesting that the complex is merely helping in the substrate protein export process [60]. Cryo-EM structure determination of the ATPase SctN (vT3SS FliI homolog) and its central stalk SctO (vT3SS FliJ homolog) of enteropathogenic *Escherichia coli* identified the presence of hydrophobic residues that contribute to the interaction between SctN and SctO, and which might facilitate rotation of SctO in a similar way to the F_1-ATPases. In such a mechanism, the SctO/FliJ stalk could stabilize the ATPase complex to facilitate binding of the chaperone. A transition from an ATP- to ADP-bound state of the ATPase might then cause the dissociation of the chaperone from its substrate protein and the ATPase complex, thereby facilitating subsequent substrate secretion [61].

It is now clear that the ATPase FliI of the flagellar T3SS only has a facilitating role in substrate protein secretion via the T3SS. The ATPase is not required for substrate translocation per se, as a Δ*fliHI* double mutant is still able to produce filaments with low probability. In contrast to the dispensability of the ATPase complex for the construction of flagella in vivo [16,47,51], however, the analysis of flagellar T3SS protein transport in an in vitro model using inverted cytoplasmic membrane vesicles (IMVs) showed confusingly that the secretion of early and late substrates was dependent on the presence of the cytoplasmic $FliH_2FliI$ complex, while abolishing the PMF did not affect secretion [17]. A potential explanation of these contradicting observations might be that the concentration of the $FliH_2FliI$ complex used in the in vitro experiments was kept constant, and higher than under in vivo conditions. Such conditions might allow substrates transport into the IMVs via the fT3SS in the absence of the PMF. Accordingly, further investigations are required to determine the molecular mechanism how the cytoplasmic ATPase complex contributes to export substrate targeting, unfolding and opening of the transmembrane export gate.

4. Translocation of Substrate Proteins

4.1. The Role of the PMF and ATPase

How translocation of substrate proteins via the T3SS is coupled to ATP hydrolysis by the soluble cytoplasmic ATPase and the PMF across the cytoplasmic membrane remains poorly understood. The PMF is made up of two components, the electric potential ($\Delta\Psi$) and the transmembrane proton gradient (ΔpH) and results from the translocation of protons by the electron transport chain across the IM. Importantly, the PMF is involved in several major biological processes including ATP synthesis by the F_OF_1-ATP synthase and for various transport processes. The PMF also plays a major role in energizing substrate protein secretion via T3SS. Disruption of the PMF in *Yersinia peptis* impaired secretion of Yop effector proteins [62]. Further, addition of an uncoupler that disrupts the PMF such as carbonyl cyanide *m*-chlorophenylhydrazone, abolished flagellar protein secretion in *Salmonella enterica* [16]. Interestingly, flagellation of mutant of the FliI ATPase complex can be restored to nearly wild-type levels by overexpression of late secretion substrates (e.g., using a mutant of the anti-σ^{28} factor FlgM) or by increasing the available PMF (e.g., by deleting the F_OF_1-ATP synthase as a major consumer of the PMF during ATP regeneration via electron transport phosphorylation) [48]. Similar observations were made in the vT3SS of *Pseudomonas aeruginosa*, where a cytoplasmic regulator and a cytoplasmic component control the access of effectors and the overall secretion by modulating the conversion efficiency of the PMF for protein export, adding yet another level of regulation [63].

While the overall importance of the PMF in energizing substrate protein secretion via T3SS is now rather clear, the underlying molecular mechanism of how the export apparatus couples the PMF to substrate protein translation remains unknown. Both components of the PMF (ΔpH and Δ_Ψ) might have different roles during substrate protein secretion, as it has been observed that only Δ_Ψ is necessary to promote protein translocation in wild-type cells, while both components are necessary in a mutant that allows flagellar substrate secretion in the absence of the T3SS ATPase FliI. This Δ*fliHI flhB*(P28T) bypass mutant secretes the early type substrates FliK and FlgD in equivalent amount as wild-type cells, adding evidence to the suggestion that the function of the ATPase FliI is not essential for the actual protein secretion process. Disruption of the PMF in the wild-type and the Δ*fliHI flhB*(P28T) bypass mutant abolished protein secretion and filament formation. An increase in ΔpH improved the secretion of the Δ*fliHI flhB*(P28T) bypass mutant, and use of deuterium oxide D_2O (heavy water isotope impacting the rate of proton translocation) led to a high decrease in FlgD secretion level only in the Δ*fliHI flhB*(P28T) bypass mutant, while the secretion levels of wild-type cells remained unchanged. Accordingly, the rate of proton translocation via the T3SS appears to limit protein export in the absence of FliHI [16,51].

Many mutations in the FlhA/FlhB components of the export gate have been identified that impact the formation of the flagellar filament and might help us to understand the mechanism of PMF-driven protein export. As mentioned above, the FlhB(P28T) mutation was shown to significantly improve the formation of flagella in absence of FliHI, but how this mutation actually bypasses the loss of the ATPase has remained obscure [47]. The recently solved structure of the export gate $FliP_5Q_4R_1FlhB_1$ revealed that the location of the FlhB N-terminal region (containing residue P28) at the cytoplasmic entrance of the gate might affect closure and opening of the gate. In support, in case of the vT3SS, a $SctU_{F28pBpa}$ mutant was shown to interact with SctS (vT3SS homolog of FliQ) by in vivo photocrosslinking. As the $FlhB_{TM}$ domain wraps around the export gate components FliPQR, the conformational change necessary for the opening of the export gate might happen from pulling forces imparted on helix 4 of the $FlhB_{TM}$ domain, linked to the other helixes through conserved buried charged residues, including FlhB D208. This force might be initiated in FlhA, the only component of the export apparatus demonstrated to have multiple conformations [64]. The FlhB D208A mutation in $FlhB_{TM}$ also disrupted motility, but was rescued by an overexpression of FlhA, highlighting the importance of a charged residues link between the two proteins [26].

4.2. The Role of FlhA

As mentioned previously, FlhA can be separated into two regions: a hydrophobic N-terminal transmembrane (TM) region with eight predicted α-helical transmembrane domains ($FlhA_{TM}$), and a hydrophilic C-terminal cytoplasmic region ($FlhA_C$) consisting of four domains: D1 (residues 362 to 434 and residues 484 to 503), D2 (residues 435 to 483), D3 (residues 504 to 583), D4 (residues 584 to 682). A flexible linker termed $FlhA_L$ connects the TM8 of the N-terminal region with the D1 domain of the C-terminal cytoplasmic region. D1-D3 and D3-D3 interactions within $FlhA_C$ subunits form a nonameric, cytoplasmic ring beneath the export gate [32].

As FlhA is a central component of PMF-driven flagellar protein export via the T3SS, several mutagenesis studies of conserved charged residues have been performed in an attempt to identify potential proton binding sites. This section discusses the structure and function of FlhA in T3SS protein export and highlights the mutations that affect substrate protein secretion, by affecting the interaction with substrate-chaperone complexes or by disrupting proton binding and the proton-driven conformational changes in FlhA.

Interactions of FlhA with FliHIJ and FlhB/FliPQR play an important role for the protein secretion process. $FlhA_L$ and a hydrophobic dimple located between D1 and D2 interact with FliJ and substrate-chaperone complexes, promoted by FliI and FliH. The interaction of FlhA with the ATPase complex might allow the export gate to efficiently utilize the PMF to facilitate flagellar protein export [28]. As the D2 domain is directly involved in the translocation of substrates, it has been proposed that FliH and FliI promote the interactions of FlhA to FliJ and substrate-chaperone complex to ensure an efficient energy coupling mechanism [55]. Based on mutational analysis, the D1 domain of FlhA might be involved in substrate entry into the secretion channel [65]. The flexible linker $FlhA_L$ is further required for $FlhA_C$ oligomerisation as alanine substitutions in the D1 domain (interacting with $FlhA_L$) and in $FlhA_L$ inhibit the formation of the $FlhA_C$ ring. These mutations reduce the efficiency of filament assembly after completion of the hook, as well as the binding affinity of FlgN, the chaperone of FlgKL [29]. Triple alanine substitution in the D1 domain resulted in a polyhook phenotype, indicating that the mutations potentially affected hook-length control. Other alanine substitutions mutants in $FlhA_L$ domain still assembled a complete hook structure, although hook length was not tightly regulated anymore, and secretion of FlgM (and accordingly of FlgK and FliC) was reduced (while early substrates secretion e.g., of FlgE/FlgD was not impaired). As hooks/basal bodies were still produced at the wild-type level, this suggests that these residues are involved in the interaction with late-type substrates and that the interactions of $FlhA_L$ with its neighbouring subunit may induce conformational changes in the $FlhA_C$ ring structure to initiate the export of filament-type proteins (e.g., FlgM/FlgK) once the hook has been completed [29]. The V404M mutation in the D1 domain of FlhA facilitates binding of FliI in absence of FliH to the export apparatus. $FlhA_C$(G368C) is a temperature-sensitive mutant which displays significantly reduced flagellar protein export at elevated temperatures but not at the permissive temperature of 30 °C [66–68]. Thermal stability experiments showed that this mutation affects denaturation of the C-terminal domain of $FlhA_C$, revealing the importance of FlhA conformational rearrangements for productive protein secretion [64,65]. Hara et al. identified the non-motile D208A mutation and suggested that this charged residue might be directly involved in the PMF-driven protein export. However, a more extensive mutagenesis study revealed that proton-binding at this position is non-essential for flagellar protein export, as suppressor mutations elsewhere in FlhA were able to rescue the motility defect of D208 mutants. In addition, significant transport activity was measured in D208 mutants using a more sensitive assay based on export of a hook protein fusion to beta-lactamase (FlgE-Bla) [69,70].

In the same study, three other charged residues (R147, R154 and D158) located in a small cytoplasmic loop between TM4 and TM5 ($FlhA_{CD1}$) were identified as necessary for FlhA function and potential proton binding sites [70]. This highly conserved cytoplasmic loop between TM4 and TM5 coordinates the secretion of early secretion substrates during hook assembly by interacting with the $FlhA_C$ domain. This suggests that substrate entry in the channel is regulated by this loop. A protonation-mimicking mutation in this loop (FlhA(D158N)) induces a large conformational change of the $FlhA_C$ domain, and it therefore has been proposed that during the secretion process the cytoplasmic $FlhA_C$ ring gets close to the loop between TM4 and TM5 through this proton-driven conformational change [70]. This protonation mimicking mutation triggers a global conformational change that also affects the large cytoplasmic domain of FlhA ($FlhA_C$, also termed $FlhA_{CD2}$ in order to discriminate it from the cytoplasmic loop between TM4 and TM5 termed $FlhA_{CD1}$), which interacts with secretion substrates. These data led to a model of how proton-driven conformational cycling of $FlhA_{CD1}$ might drive substrate protein secretion. In this model, the nine copies of FlhA form two cytoplasmic rings; the $FlhA_{CD1}$ ring positioned close to the membrane and the larger $FlhA_{CD2}$ ring more distal to the export gate. It is presumed that $FlhA_{CD2}$ can cycle between its more distal position from the membrane and a position more proximal to the membrane, where it is held in place through interactions with $FlhA_{CD1}$. The model further proposes that proton-driven conformational changes in

$FlhA_{CD1}$ modulate these interactions to drive the cyclical movements of the large $FlhA_{CD2}$ ring between the positions proximal and distal to the membrane (Figure 3). The proton-driven cyclical movements of the $FlhA_{CD2}$ ring are presumably linked to cycles of secretion substrate binding and release into the secretion channel [70].

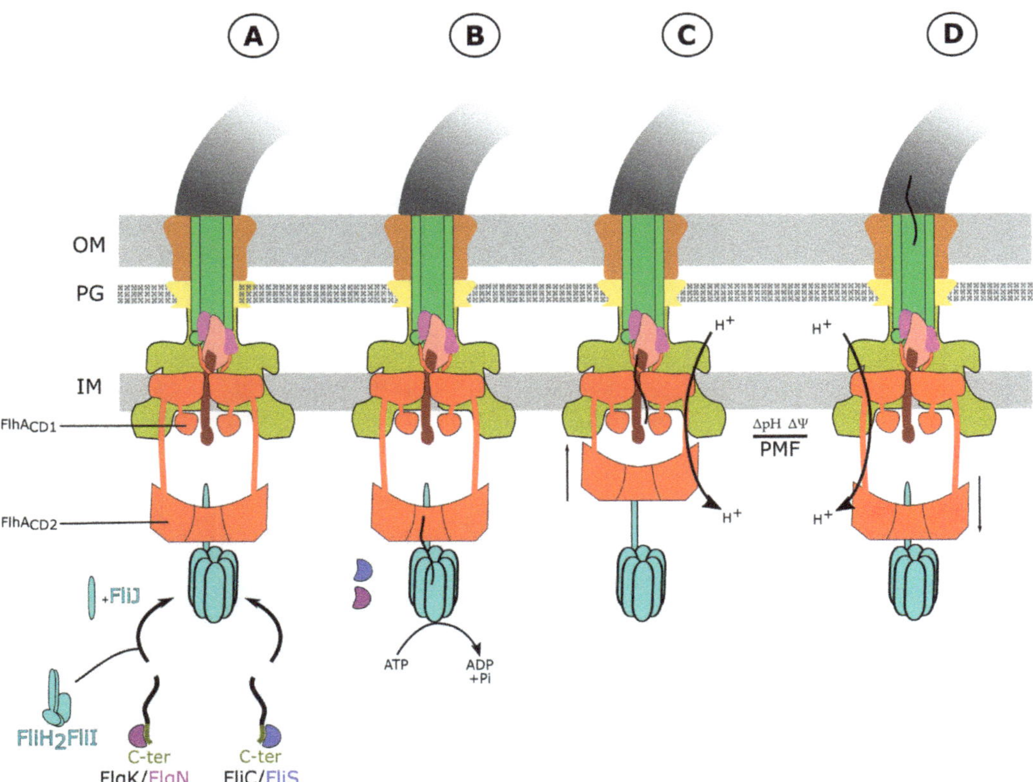

Figure 3. Putative model of the fT3SS secretion process. (**A**) Delivery of chaperones/substrates complexes to the $FliH_{12}FliI_6FliJ$ ATPase complex, followed by (**B**) removal of the chaperone/unfolding of the substrates mediated by ATP hydrolysis of the ATPase complex $FliH_{12}FliI_6FliJ$ and delivery to FlhA. (**C**) Proton-driven cyclical movements of the $FlhA_{CD2}$ ring cause an opening of the export gate and mediate cycles of secretion substrate binding and release into the secretion channel. (**D**) Diffusion of secreted substrates inside the secretion channel to the tip of the flagellar structure.

However, several crucial steps of the flagellar protein secretion process remain unclear. First, while the relation between FlhA and FlhB evidently is vital for an efficient PMF-dependent secretion of substrate proteins via the fT3SS, how exactly the T3SS distinguishes between early and late-type substrates is unknown. The proton-driven cyclical movements of $FlhA_{CD1}$ and $FlhA_{CD2}$ might impact FlhB, causing an opening of the export gate. Recent mutational analysis suggests that a structural rearrangement of the $FlhA_C$ ring is promoted by interaction between $FlhB_C$ and $FlhA_C$. A FlhA(A489E) suppressor mutant partially rescues the impaired late-type substrates secretion of FlhB(P270A) (a slow-cleaving FlhB autocleavage mutant, which forms polyhooks) and shortens the length of the polyhooks. These observations suggest that the FlhA(A489E) mutation assists the FlhB(P270A) mutant in switching to late-substrate secretion mode. Interestingly, the FlhA(A489E) mutation is located in the binding site for substrate-chaperone complexes and reduces the binding affinity for these complexes. Accordingly, it appears reasonable to conclude the FlhA(A489E) mutation mimics the conformation of the chaperone-bound $FlhA_C$ ring also in presence of the FlhB(P270A) mutation. In a *fliK*-null mutant, the FlhA(A489E) FlhB(P270A) double mutant is able to secrete early type substrates (e.g., FlgE) at a wild-type level, while late-type substrates secretion (e.g., FliC) is abolished. This result highlights the crucial role of the molecular ruler FliK for some structural rearrangement of the $FlhA_C$ ring that enables late substrate-chaperone docking. Accordingly, these mutational results support a model, where the molecular ruler FliK interacts with $FlhB_C$ as the hook reaches its final length of ~55nm. This interaction induces a conformational change in FlhB required for cleavage of $FlhB_C$. The cleaved $FlhB_C$ in turn interacts with $FlhA_C$ and causes a conformational change that promotes interaction of the $FlhA_C$ ring with the substrate-chaperone complexes, thereby enabling the secretion of late-type substrates [71].

Further, the role of the ATPase complex in the actual substrate secretion process remains poorly understood. It has been suggested that cycles of ATP hydrolysis may induce conformational changes in the export gate that result in opening of the gate, as the $FlhA_C$ ring is known to interact with FliJ and to be positioned between the export gate and the ATPase complex [26]. Based on cryo-electron tomogram imaging of basal body complexes in situ, the distance between the export gate and ATPase is large, and it is thus unclear how the ATPase complex would be able to mediate the gate opening. Kuhlen et al. proposed that the gate-opening conformational changes would presumably be mediated via interactions of the FliJ stalk of the ATPase complex with the cytoplasmic $FlhA_{CD2}$ ring [26]. Further, as outlined above, low levels of flagellar protein secretion is possible also in the absence of the ATPase complex and therefore, a spontaneous opening of the export gate complex must be possible. In such a model, where conformational changes in FlhB are driven via FliI-FliJ-FlhA interactions, the activity of the ATPase complex might only be required for the initial gate opening. It would appear reasonable to assume that, once opened, the export gate then remains open as long as substrate proteins are transported.

SctV, the vT3SS homolog of FlhA, also forms a nonameric ring and binds substrate-chaperone complexes before substrate secretion. The substrate-chaperone binding site of SctV is, however, not conserved in FlhA, as in place of the hydrophobic dimple found in FlhA, the SctV substrate-chaperone binding site is made of highly conserved charged residues glutamine and arginine located at the interface between the D3 and D4 domains [33]. As the fT3SS needs to switch substrate specificity from early to late-type substrate secretion, the vT3SS requires two specificity switches; the first from early to intermediate-type substrates, and the second from intermediate- to late-type substrates. The first switch might be similar to the switch of the fT3SS, as the second one requires the presence of a gatekeeper protein, SctW, which is absent in the fT3SS. This gatekeeper interacts with the C-terminal and membrane domains of SctV (the vT3SS homolog of FlhA), and this binding decreases the binding affinity of the chaperone/effector complexes to SctV, allowing secretion of translocators but preventing an early secretion of effectors before contact with the eukaryotic host cell. Substrates of the translocator class might also contain an N-terminal signal sequence that is recognized by the gatekeeper protein. Upon host cell contact, SctW dissociates from SctV, the binding affinity of effectors for SctV increases, which subsequently allows secretion of the effector class of vT3SS substrates [72].

5. Model for High-Speed Secretion of Flagellin

In the previous chapters, we discussed the organisation of the flagellar T3SS and how substrate proteins might be translocated across the inner membrane in an ATP- and PMF-dependent manner. However, the flagellum grows by polymerization of its building blocks at the distal end. Flagellin is pumped into the secretion channel, travels through the secretion channel and then polymerises on top of the completed hook at the distal end of the flagellum with the help of the filament cap FliD [73,74]. Around 20,000 subunits of flagellin assemble to form the flagellar filament, which can grow up to a 20 µm in length [25,75]. Accordingly, once translocated into the two nm narrow secretion channel, the many thousand building blocks of the flagellum must travel distances ranging from several dozen nanometer (e.g., in case of hook subunits) up to several micrometer (in case of flagellin subunits). This raised the question if transport beyond the inner membrane relied on additional energy sources. Several models have been proposed on how the bacteria might be able to transport thousands of building blocks over such long distances.

Initial observations of flagellar filament growth were made using electron microscopy-based measurements of filament lengths in vitro and in vivo. In vivo experiments showed an exponential decrease of the filament growth rate with increasing filament length [13], while in vitro measurements revealed a constant growth rate [76]. However, the flagellum grows by incorporating flagellin subunits at the tip of the filament and the in vitro measurements do not consider the transport of flagellin molecules inside the narrow secretion channel from the base to the tip. Accordingly, the authors concluded that the exponential decay was caused by a decrease in translocation efficiency [13]. Such a mechanism is supported by several theoretical and computational analyses: Keener presented a biophysical model for the growth of flagellar filaments where the monomers are translocated into the channel by the ATPase and then diffuse until reaching the tip where they assemble. Such a model resulted in similar qualitative results to the measurements made for the exponential decay. The quantitative differences with the experimental measurements were hypothesised by the fact that the movement of the monomer substrates is not driven solely by diffusion [77]. Using molecular dynamics simulation, Tanner et al. proposed a mathematical model of the translocation-elongation process which describes all of the properties included in the elongation process: the friction of the flagellin during interactions with the channel, the flagellin density and the flagellin translocation rate. This theoretical model was consistent with the exponential measurement of Iino et al., and the authors concluded that the flagellum growth rate decreases exponentially with length because of protein compression and friction between translocating flagellin and the flagellar channel [78].

An alternative, electrostatic model has been proposed to be responsible for secretion of effector proteins through the needle of the injectisome. In this model, secretion substrate proteins with charged residues would travel through the channel by electrostatic repulsion mediated through negative charged residues present on the inside of the channel wall. Since the negative charges would be present throughout the length of the needle, the electrostatic repulsion would enable the transport of the substrate proteins from the base of the channel to the tip. In this model, the insertion of the positive charged substrates inside the channel would be energized by the PMF-driven T3SS, which would push the substrate proteins into the needle channel [79]. However, this model requires the presence of negative electrostatic charges inside the secretion channel, and more recent study on the structure of the injectisome needle revealed that the inner linings of the channel wall is mainly neutral [80]. The flagellar channel is also mostly hydrophilic, probably to minimize hydrophobic interaction with the unfolded substrates that would hinder the diffusion to the distal end of the filament. Interestingly, among the lining residues, at least two amino acids are positively charged and/or polar for the flagellin FliC of *Salmonella*, and this number is generally up to four amino acids as shown by alignment of flagellin homologs. Most of the amino acids in the channel are polar non-charged, and although one lining residue of the flagellar filament channel in *Salmonella* is negatively charged, the presence of two other lining positively charged residues seem to indicate either a positive or neutral global charge inside the channel, tending to the idea that electrostatic repulsion could not propel substrate proteins through the channel [7,81].

The current model for the filament growth mechanism of the bacterial flagellum is based on diffusive motion of single flagellin molecules inside the filament channel. Evidence supporting such a mechanism was first published in 2012 when Turner et al. developed a clever technique to estimate the elongation rate of individual flagellar filaments. They used a flagellin variant harbouring a surface-exposed cysteine replacement mutation, which allowed to visualize filament fragments using maleimide-coupled fluorophores. After shearing of the filament, Turner et al. labelled the filament using maleimide dyes of two different colours and observed that the filament was able to re-grow, and that the length of the second fragment was independent of the length of the first fragment [82]. A length-independent growth mechanism was puzzling, however, and not consistent with the previous observations of an exponential decay of filament growth. Stern et al. predicted that PMF-dependent injection of partially folded, α-helical flagellin subunits followed by single-file diffusion inside the channel, would account for linear filament growth provided that the diffusion coefficient of flagellin molecules is sufficiently large [83].

An alternative model to explain the length-independent filament growth and resulting constant rate of flagellin transport was based on the observation that secretion substrates are captured by the C-terminal domain of FlhB through head-to-tail linkage of terminal helices [84]. This led to a model where a chain of head-to-tail linked unfolded flagellin subunits would span the complete length of the secretion channel from the export gate to the tip of the filament. Crystallization of the most distal flagellin subunit at the tip would exert a force to pull the next subunit from the export gate, maintaining a constant rate of flagellin subunit transport to the tip of the flagellum. While elegant, this model of inter-subunit chain formation is not compatible with the simultaneous secretion of non-chaining substrates such as the anti-σ factor FlgM or excess hook-associated proteins (FlgK, FlgL, FliD), which are known to be continuously secreted during the assembly process [85], but do not interact with FliC [86]. The chain model can also not explain the secretion of non-flagellar proteins fused to an N-terminal T3SS signal peptide (e.g., FlgM) and subsequently exported via the flagellar protein secretion system [87]. Finally, premature termination of translation is occurring frequently [88] and presumably result in a sub-population of C-terminal truncated secretion substrates, which would be unable to form inter-subunit chains. This was confirmed by co-expression of truncated flagellin unable to form inter-subunit chains and to assemble into the filament, which did not affect the filament elongation kinetics [89].

In 2017, Renault et al. provided further evidence in support of a model of flagellar filament growth that is dependent on PMF-driven injection of flagellin subunits into the empty secretion channel and one-dimensional diffusion inside the channel from the base to the distal tip [89]. Here, the site-specific labelling of flagellin subunits containing a surface-exposed cysteine residue using maleimide-coupled fluorochromes was optimized by exchanging dyes multiple times in situ during normal bacterial growth. The high-resolution multiple labelling approach revealed a length-dependent filament growth with an elongation speed that gradually decreased from ~100 nm·min^{-1} to ~20 nm·min^{-1} for 8 μm long filaments, which translates to an initial flagellin secretion rate of ~1700 amino acids per second. In an earlier study that investigated the growth rate of the filament, Iino used p-fluorophenylalanine (pFPA), which incorporates into the growing filament and changes the filament curvature. The incorporation of pFPA results in a curly filament that can be distinguished from the usual filament structure and enabled Iino to estimate the growth of filament fragments after addition of pFPA. Using this technique, Iino observed that the rate of elongation decreases exponentially with the increase in flagellar filament length, and attributed this to a decrease in the efficiency of flagellin transportation through the secretion channel. However, the decreased transport efficiency was not fully understood at that time, as the initial fast growth rate caused by the injection was not integrated in Iino equations [13]. The more recent injection-diffusion model considers the injection of the substrates into the channel, the diffusion coefficient of the monomer inside the channel, the length of the monomer and the increment in length of the flagellum. For more precise, quantitative measurements of the filament elongation rate, Renault et al. further employed a strain where the master regulator FlhDC expression is dependent of an inducible promoter. This enabled Renault et al. to control the timing of filament initiation and growth. Using the maleimide approach described above, the authors could determine the growth rate of several different filament segments over time. They found that the growth rate of a new, distal filament fragments was inversely proportional to the initial basal-fragment length. This approach also allowed to exclude broken filaments from the analysis [89]. In support of the injection-diffusion model, a length-dependent growth of the flagellar filament was also observed by Zhao et al. in *Vibrio alginolyticus* using real-time fluorescent labelling of the flagellar sheath, as well as in *E. coli* [90,91]. Here, real time fluorescence microscopy of filament growth confirmed that flagellar growth rate was inversely proportional to the length and further revealed that an insufficient cytoplasmic flagellin supply results in intermittent pauses during filament elongation in *E. coli* [90].

6. Conclusions

The T3SS of Gram-negative bacteria is a protein secretion machine that primarily uses a PMF-driven mechanism to translocate its substrate proteins across the inner membrane into a narrow secretion channel, and is essential for the assembly of complex nanomachines such as the flagellum and injectisome. The associated ATPase complex might have a facilitating role during the secretion process by activating the PMF-driven export gate or facilitating secretion substrate docking and unfolding. Among all of the secretion systems developed by bacteria, the T3SS displays striking features. It facilitates the secretion of substrate proteins from the cytoplasm at a remarkably high speed of several thousand amino acids per second in a one step process through the IM and OM, using both ATPase hydrolysis and the PMF. Although some principle aspects of the protein secretion process via the T3SS are now well-established, numerous questions remain on the underlying molecular mechanisms of the actual secretion and the energisation processes. In particular, it remains to be elucidated how the T3SS is able to transport substrate proteins at such a high rate while preventing the leakage of small molecules. How different classes of substrates are recognized and what constitutes the switch in secretion substrate specificity remain equally elusive. Considering its dispensability at least for secretion of flagellar proteins, the role of the cytoplasmic ATPase complex and its potential contribution to substrate targeting, unfolding and chaperone release continues to be a big mystery. Finally, the holy grail of the T3SS secretion mechanism will be a molecular understanding of how the PMF is actually coupled to the protein secretion process.

Box 1. Energy requirements of other bacterial secretion systems.

Bacteria have evolved different ways to secrete proteins into the external environment and/or into mammalian or plant hosts cells. Proteins are secreted into the periplasm or inserted in the inner membrane by the general secretion system (Sec) and the Twin-arginine translocation (Tat), which secrete unfolded and folded (co-factor containing) proteins, respectively. The Sec system is found in all domains of life, while the Tat system is found in bacteria and archaea. Protein translocation through the Sec system is powered by the ATPase SecA, which couples ATP hydrolysis to substrate protein insertion into the protein conducting channel formed by the translocase SecYEG. The PMF is required for a high substrate secretion rate, but it is currently unknown which component of the PMF is necessary. Usage of the PMF was proposed to favour the outward flow of the substrate polypeptide in a Brownian ratchet-type mechanism. It is still under debate how ATP hydrolysis and the PMF enable substrate protein translocation, however. Two main models have been proposed to explain it: the Brownian ratchet model for the ATP-driven reaction, potentially aided by the PMF, and a "push-and-slide" model, involving diffusion and powerstroke movements of the components driven by the ATPase SecA. There also seems to be a critical role of the phospholipids of the membrane. Notably, cardiolipin a specialized phospholipid, which is necessary for the stability of the complex, and for both ATP and PMF-driven protein translocation activity [92,93], indicating that an essential lipid-protein interface exists for the secretion process. Membranes lacking cardiolipin are unable anymore of PMF stimulated translocation, suggesting a direct role of cardiolipin on the diffusion of the protons. Contrary to the Sec system, substrate protein translocation through the Tat system (formed by TatA, TatB and TatC) is exclusively driven by the PMF. In vivo studies have shown that the $\Delta\psi$ component alone is sufficient to energize protein translocation. The assembly of TatA is also promoted by the PMF, but the mechanism how the PMF contributes to the assembly of the Tat system remains unknown. Currently, two different models exist to explain substrate protein translocation by the Tat system; the first is based on pore formation in the membrane, and the second on membrane weakening by TatA complexes, where short TM domains would locally reduce the membrane thickness after binding of the cargo proteins [94].

Box 1. *Cont.*

Among other protein secretion systems of Gram-negative bacteria, the T2SS and T5SS utilize a two-step protein secretion mechanism, as their substrates are first translocated into the periplasm in a Sec/Tat- and Sec-dependent manner, respectively, before getting translocated outside of the bacterial cell. The T2SS (e.g., responsible for secretion of the cholera toxin of *Vibrio cholerae* or the heat-labile enterotoxin of enterotoxigenic *Escherichia coli*) can be separated in four subassemblies: the OM complex, the IM platform, the secretion ATPase in the cytoplasm, and the pseudopilus in the periplasm. The secretion ATPase interacts with the IM platform at its cytoplasmic interface and the OM complex translocates the substrates through the outer membrane. The energy for the translocation is provided by the ATPase in the cytoplasm, as ATPase mutants with no activity are unable to secrete effectors. ATP hydrolysis by the ATPase induces the formation of the pseudopilus, that in turn pushes exoprotein substrates through the OM complex channel by alternating extension and retraction similar to a piston, and comparable to what can be observed for type 4 pilus assembly systems (T4PS) [95,96]. The high similarity between T4PS and T2SS, as well as to archaeal flagella, suggest an early common origin, potentially from archetypal structure that evolved in specialized T4PS and T2SS pseudopilus [97]. T5SS are autotransporters that can be classified into monomeric and trimeric autotransporters, and two-partner secretion systems (TPSS). Several types of T5SS exist, reviewed in Leo et al. [98]. Contrary to the others Gram-negative protein secretion systems, the energy requirement for the translocation of T5SS is not provided by ATP hydrolysis or the PMF, but by protein folding. It is presumed that the folding of the C-terminal domain of the secreted substrate protein in the extracellular space acts as a Brownian ratchet to move passively across the OM [99].

The T1SS and T4SS are one-step protein secretion systems that use energy derived from ATP hydrolysis to drive substrate protein transport from the cytoplasm through the IM and OM. T4SS form pili that are able to extend and retract, and are used for the transfer of DNA from a donor to a recipient cell by a process called conjugation, or to secrete proteins to the outside of the cell envelope. The cycles of pilus extension/retraction are powered by ATP hydrolysis of a dedicated ATPase, and pilin subunits are reinserted in the membrane during the retraction cycles. The pilin subunits are processed in the IM before incorporation into the growing pilus and assemble into the pilus at the base of the T4SS. In the case of the T4SS from F-like plasmids, it is now known that F-like pili are not composed solely of pilin, but also of phospholipids from the inner membrane (mainly phosphatidylglycerol 32:1 and phosphatidylglycerol 34:1, which are the two most common phosphatidylglycerol species found in the bacterial cell membrane) in a stoichiometric manner. The presence of these lipids inside of the F-like pili structure would then help the transfer of the ssDNA through the T4SS by making the channel moderately electronegative (without phosphatidylglycerol, the pilus lumen is overwhelmingly electropositive). The presence of phospholipids in the structure might also lower the energetic barrier for the extraction or re-insertion of pilus subunits from, or into, the inner membrane, thereby facilitating pilus extension or retraction, respectively. The lowered energy barrier may also facilitate pilus insertion into the recipient cell membrane for efficient cargo delivery [100]. Interestingly, the pilin subunit TraA is integrated in the inner membrane and extracted from it with a phosphatidylglycerol using an ATP and PMF dependent process, independent of the Sec pathway. Depletion of the PMF causes an inhibition of pilin processing, which in turn prevents pilus assembly and the conjugation process [101].

Funding: This project has received funding from the European Research Council (ERC) under the European Union's Horizon 2020 research and innovation programme (grant agreement No 864971).

Data Availability Statement: Not applicable.

Conflicts of Interest: The authors declare no conflict of interest.

References

1. Hueck, C.J. Type III protein secretion systems in bacterial pathogens of animals and plants. *Microbiol. Mol. Biol. Rev. MMBR* **1998**, *62*, 379–433. [CrossRef]
2. Cornelis, G.R. The type III secretion injectisome. *Nat. Rev. Microbiol.* **2006**, *4*, 811–825. [CrossRef]
3. Minamino, T.; Imada, K.; Namba, K. Mechanisms of type III protein export for bacterial flagellar assembly. *Mol. Biosyst.* **2008**, *4*, 1105–1115. [CrossRef]
4. Minamino, T.; Macnab, R.M. Components of the *Salmonella* flagellar export apparatus and classification of export substrates. *J. Bacteriol.* **1999**, *181*, 1388–1394. [CrossRef]
5. Minamino, T.; MacNab, R.M. Interactions among components of the *Salmonella* flagellar export apparatus and its substrates. *Mol. Microbiol.* **2000**, *35*, 1052–1064. [CrossRef]
6. Wagner, S.; Diepold, A. A Unified Nomenclature for Injectisome-Type Type III Secretion Systems. *Curr. Top. Microbiol. Immunol.* **2020**, *427*. [CrossRef]
7. Yonekura, K.; Maki-Yonekura, S.; Namba, K. Complete atomic model of the bacterial flagellar filament by electron cryomicroscopy. *Nature* **2003**, *424*, 643–650. [CrossRef] [PubMed]
8. Hughes, K.T.; Gillen, K.L.; Semon, M.J.; Karlinsey, J.E. Sensing structural intermediates in bacterial flagellar assembly by export of a negative regulator. *Science* **1993**, *262*, 1277–1280. [CrossRef] [PubMed]
9. Hirano, T.; Yamaguchi, S.; Oosawa, K.; Aizawa, S. Roles of FliK and FlhB in determination of flagellar hook length in *Salmonella* typhimurium. *J. Bacteriol.* **1994**, *176*, 5439–5449. [CrossRef] [PubMed]

10. Williams, A.W.; Yamaguchi, S.; Togashi, F.; Aizawa, S.I.; Kawagishi, I.; Macnab, R.M. Mutations in *fliK* and *flhB* affecting flagellar hook and filament assembly in *Salmonella* typhimurium. *J. Bacteriol.* **1996**, *178*, 2960–2970. [CrossRef] [PubMed]
11. Erhardt, M.; Singer, H.M.; Wee, D.H.; Keener, J.P.; Hughes, K.T. An infrequent molecular ruler controls flagellar hook length in *Salmonella enterica*. *EMBO J.* **2011**, *30*, 2948–2961. [CrossRef]
12. Sorg, I.; Wagner, S.; Amstutz, M.; Müller, S.A.; Broz, P.; Lussi, Y.; Engel, A.; Cornelis, G.R. YscU recognizes translocators as export substrates of the *Yersinia* injectisome. *EMBO J.* **2007**, *26*, 3015–3024. [CrossRef]
13. Iino, T. Assembly of *Salmonella* flagellin in vitro and in vivo. *J. Supramol. Struct.* **1974**, *2*, 372–384. [CrossRef]
14. Tomkiewicz, D.; Nouwen, N.; van Leeuwen, R.; Tans, S.; Driessen, A.J.M. SecA supports a constant rate of preprotein translocation. *J. Biol. Chem.* **2006**, *281*, 15709–15713. [CrossRef]
15. Fessl, T.; Watkins, D.; Oatley, P.; Allen, W.J.; Corey, R.A.; Horne, J.; Baldwin, S.A.; Radford, S.E.; Collinson, I.; Tuma, R. Dynamic action of the Sec machinery during initiation, protein translocation and termination. *eLife* **2018**, *7*, e35112. [CrossRef]
16. Paul, K.; Erhardt, M.; Hirano, T.; Blair, D.F.; Hughes, K.T. Energy source of flagellar type III secretion. *Nature* **2008**, *451*, 489–492. [CrossRef]
17. Terashima, H.; Kawamoto, A.; Tatsumi, C.; Namba, K.; Minamino, T.; Imada, K. In Vitro Reconstitution of Functional Type III Protein Export and Insights into Flagellar Assembly. *mBio* **2018**, *9*. [CrossRef]
18. Shibata, S.; Takahashi, N.; Chevance, F.F.V.; Karlinsey, J.E.; Hughes, K.T.; Aizawa, S.-I. FliK regulates flagellar hook length as an internal ruler. *Mol. Microbiol.* **2007**, *64*, 1404–1415. [CrossRef]
19. Kuhlen, L.; Abrusci, P.; Johnson, S.; Gault, J.; Deme, J.; Caesar, J.; Dietsche, T.; Mebrhatu, M.T.; Ganief, T.; Macek, B.; et al. Structure of the Core of the Type Three Secretion System Export Apparatus. *Nat. Struct. Mol. Biol.* **2018**, *25*, 583–590. [CrossRef]
20. Fabiani, F.D.; Renault, T.T.; Peters, B.; Dietsche, T.; Gálvez, E.J.C.; Guse, A.; Freier, K.; Charpentier, E.; Strowig, T.; Franz-Wachtel, M.; et al. A flagellum-specific chaperone facilitates assembly of the core type III export apparatus of the bacterial flagellum. *PLoS Biol.* **2017**, *15*. [CrossRef]
21. Liu, R.; Ochman, H. Stepwise formation of the bacterial flagellar system. *Proc. Natl. Acad. Sci. USA* **2007**, *104*, 7116–7121. [CrossRef] [PubMed]
22. Ferris, H.U.; Furukawa, Y.; Minamino, T.; Kroetz, M.B.; Kihara, M.; Namba, K.; Macnab, R.M. FlhB regulates ordered export of flagellar components via autocleavage mechanism. *J. Biol. Chem.* **2005**, *280*, 41236–41242. [CrossRef] [PubMed]
23. Fraser, G.M.; Hirano, T.; Ferris, H.U.; Devgan, L.L.; Kihara, M.; Macnab, R.M. Substrate specificity of type III flagellar protein export in *Salmonella* is controlled by subdomain interactions in FlhB. *Mol. Microbiol.* **2003**, *48*, 1043–1057. [CrossRef]
24. Ohnishi, K.; Kutsukake, K.; Suzuki, H.; Iino, T. Gene *fliA* encodes an alternative sigma factor specific for flagellar operons in *Salmonella* typhimurium. *Mol. Gen. Genet. MGG* **1990**, *221*, 139–147. [CrossRef]
25. Chevance, F.F.V.; Hughes, K.T. Coordinating assembly of a bacterial macromolecular machine. *Nat. Rev. Microbiol.* **2008**, *6*, 455–465. [CrossRef]
26. Kuhlen, L.; Johnson, S.; Zeitler, A.; Bäurle, S.; Deme, J.C.; Caesar, J.J.E.; Debo, R.; Fisher, J.; Wagner, S.; Lea, S.M. The substrate specificity switch FlhB assembles onto the export gate to regulate type three secretion. *Nat. Commun.* **2020**, *11*, 1–10. [CrossRef]
27. Kihara, M.; Minamino, T.; Yamaguchi, S.; Macnab, R.M. Intergenic Suppression between the Flagellar MS Ring Protein FliF of *Salmonella* and FlhA, a Membrane Component of Its Export Apparatus. *J. Bacteriol.* **2001**, *183*, 1655–1662. [CrossRef]
28. Ibuki, T.; Uchida, Y.; Hironaka, Y.; Namba, K.; Imada, K.; Minamino, T. Interaction between FliJ and FlhA, Components of the Bacterial Flagellar Type III Export Apparatus. *J. Bacteriol.* **2013**, *195*, 466–473. [CrossRef]
29. Terahara, N.; Inoue, Y.; Kodera, N.; Morimoto, Y.V.; Uchihashi, T.; Imada, K.; Ando, T.; Namba, K.; Minamino, T. Insight into structural remodeling of the FlhA ring responsible for bacterial flagellar type III protein export. *Sci. Adv.* **2018**, *4*. [CrossRef]
30. Kinoshita, M.; Hara, N.; Imada, K.; Namba, K.; Minamino, T. Interactions of bacterial flagellar chaperone–substrate complexes with FlhA contribute to co-ordinating assembly of the flagellar filament. *Mol. Microbiol.* **2013**, *90*, 1249–1261. [CrossRef]
31. Bange, G.; Kümmerer, N.; Engel, C.; Bozkurt, G.; Wild, K.; Sinning, I. FlhA provides the adaptor for coordinated delivery of late flagella building blocks to the type III secretion system. *Proc. Natl. Acad. Sci. USA* **2010**, *107*, 11295–11300. [CrossRef] [PubMed]
32. Abrusci, P.; Vergara-Irigaray, M.; Johnson, S.; Beeby, M.D.; Hendrixson, D.R.; Roversi, P.; Friede, M.E.; Deane, J.E.; Jensen, G.J.; Tang, C.M.; et al. Architecture of the major component of the type III secretion system export apparatus. *Nat. Struct. Mol. Biol.* **2013**, *20*, 99–104. [CrossRef] [PubMed]
33. Xing, Q.; Shi, K.; Portaliou, A.; Rossi, P.; Economou, A.; Kalodimos, C.G. Structures of chaperone-substrate complexes docked onto the export gate in a type III secretion system. *Nat. Commun.* **2018**, *9*. [CrossRef] [PubMed]
34. Johnson, S.; Fong, Y.H.; Deme, J.C.; Furlong, E.J.; Kuhlen, L.; Lea, S.M. Symmetry mismatch in the MS-ring of the bacterial flagellar rotor explains structural coordination of secretion and rotation. *Nat. Microbiol.* **2020**, *5*, 966–975. [CrossRef]
35. Lam, K.-H.; Ip, W.-S.; Lam, Y.-W.; Chan, S.-O.; Ling, T.K.-W.; Au, S.W.-N. Multiple Conformations of the FliG C-Terminal Domain Provide Insight into Flagellar Motor Switching. *Structure* **2012**, *20*, 315–325. [CrossRef]
36. Deme, J.C.; Johnson, S.; Vickery, O.; Aron, A.; Monkhouse, H.; Griffiths, T.; James, R.H.; Berks, B.C.; Coulton, J.W.; Stansfeld, P.J.; et al. Structures of the stator complex that drives rotation of the bacterial flagellum. *Nat. Microbiol.* **2020**, *5*, 1553–1564. [CrossRef]
37. Chang, Y.; Zhang, K.; Carroll, B.L.; Zhao, X.; Charon, N.W.; Norris, S.J.; Motaleb, M.A.; Li, C.; Liu, J. Molecular mechanism for rotational switching of the bacterial flagellar motor. *Nat. Struct. Mol. Biol.* **2020**, *27*, 1041–1047. [CrossRef]
38. Santiveri, M.; Roa-Eguiara, A.; Kühne, C.; Wadhwa, N.; Hu, H.; Berg, H.C.; Erhardt, M.; Taylor, N.M.I. Structure and Function of Stator Units of the Bacterial Flagellar Motor. *Cell* **2020**, *183*, 244–257.e16. [CrossRef]

39. Erhardt, M.; Hughes, K.T. C-ring requirement in flagellar type III secretion is bypassed by FlhDC upregulation. *Mol. Microbiol.* **2010**, *75*, 376–393. [CrossRef]
40. Minamino, T.; Macnab, R.M. FliH, a soluble component of the type III flagellar export apparatus of *Salmonella*, forms a complex with FliI and inhibits its ATPase activity. *Mol. Microbiol.* **2000**, *37*, 1494–1503. [CrossRef]
41. Chen, S.; Beeby, M.; Murphy, G.E.; Leadbetter, J.R.; Hendrixson, D.R.; Briegel, A.; Li, Z.; Shi, J.; Tocheva, E.I.; Müller, A.; et al. Structural diversity of bacterial flagellar motors. *EMBO J.* **2011**, *30*, 2972–2981. [CrossRef] [PubMed]
42. Imada, K.; Minamino, T.; Uchida, Y.; Kinoshita, M.; Namba, K. Insight into the flagella type III export revealed by the complex structure of the type III ATPase and its regulator. *Proc. Natl. Acad. Sci. USA* **2016**, *113*, 3633–3638. [CrossRef] [PubMed]
43. Bai, F.; Morimoto, Y.V.; Yoshimura, S.D.J.; Hara, N.; Kami-Ike, N.; Namba, K.; Minamino, T. Assembly dynamics and the roles of FliI ATPase of the bacterial flagellar export apparatus. *Sci. Rep.* **2014**, *4*, 6528. [CrossRef] [PubMed]
44. Kawamoto, A.; Morimoto, Y.V.; Miyata, T.; Minamino, T.; Hughes, K.T.; Kato, T.; Namba, K. Common and distinct structural features of *Salmonella* injectisome and flagellar basal body. *Sci. Rep.* **2013**, *3*. [CrossRef]
45. Minamino, T.; Yoshimura, S.D.J.; Morimoto, Y.V.; González-Pedrajo, B.; Kami-Ike, N.; Namba, K. Roles of the extreme N-terminal region of FliH for efficient localization of the FliH-FliI complex to the bacterial flagellar type III export apparatus. *Mol. Microbiol.* **2009**, *74*, 1471–1483. [CrossRef]
46. Jensen, J.L.; Yamini, S.; Rietsch, A.; Spiller, B.W. The structure of the Type III secretion system export gate with CdsO, an ATPase lever arm. *PLoS Pathog.* **2020**, *16*, e1008923. [CrossRef]
47. Minamino, T.; Namba, K. Distinct roles of the FliI ATPase and proton motive force in bacterial flagellar protein export. *Nature* **2008**, *451*, 485–488. [CrossRef]
48. Erhardt, M.; Mertens, M.E.; Fabiani, F.D.; Hughes, K.T. ATPase-Independent Type-III Protein Secretion in *Salmonella enterica*. *PLoS Genet.* **2014**, *10*, e1004800. [CrossRef]
49. Minamino, T.; Morimoto, Y.V.; Kinoshita, M.; Aldridge, P.D.; Namba, K. The bacterial flagellar protein export apparatus processively transports flagellar proteins even with extremely infrequent ATP hydrolysis. *Sci. Rep.* **2014**, *4*. [CrossRef]
50. Imada, K.; Minamino, T.; Tahara, A.; Namba, K. Structural similarity between the flagellar type III ATPase FliI and F1-ATPase subunits. *Proc. Natl. Acad. Sci. USA* **2007**, *104*, 485–490. [CrossRef]
51. Minamino, T.; Morimoto, Y.V.; Hara, N.; Namba, K. An energy transduction mechanism used in bacterial flagellar type III protein export. *Nat. Commun.* **2011**, *2*, 475. [CrossRef] [PubMed]
52. Hara, N.; Morimoto, Y.V.; Kawamoto, A.; Namba, K.; Minamino, T. Interaction of the Extreme N-Terminal Region of FliH with FlhA Is Required for Efficient Bacterial Flagellar Protein Export. *J. Bacteriol.* **2012**, *194*, 5353–5360. [CrossRef] [PubMed]
53. Minamino, T.; Kinoshita, M.; Imada, K.; Namba, K. Interaction between FliI ATPase and a flagellar chaperone FliT during bacterial flagellar protein export. *Mol. Microbiol.* **2012**, *83*, 168–178. [CrossRef] [PubMed]
54. Akeda, Y.; Galán, J.E. Chaperone release and unfolding of substrates in type III secretion. *Nature* **2005**, *437*, 911–915. [CrossRef] [PubMed]
55. Minamino, T.; Kinoshita, M.; Inoue, Y.; Morimoto, Y.V.; Ihara, K.; Koya, S.; Hara, N.; Nishioka, N.; Kojima, S.; Homma, M.; et al. FliH and FliI ensure efficient energy coupling of flagellar type III protein export in *Salmonella*. *MicrobiologyOpen* **2016**, *5*, 424–435. [CrossRef]
56. Sajó, R.; Liliom, K.; Muskotál, A.; Klein, A.; Závodszky, P.; Vonderviszt, F.; Dobó, J. Soluble components of the flagellar export apparatus, FliI, FliJ, and FliH, do not deliver flagellin, the major filament protein, from the cytosol to the export gate. *Biochim. Biophys. Acta* **2014**, *1843*, 2414–2423. [CrossRef]
57. Thomas, J.; Stafford, G.P.; Hughes, C. Docking of cytosolic chaperone-substrate complexes at the membrane ATPase during flagellar type III protein export. *Proc. Natl. Acad. Sci. USA* **2004**, *101*, 3945–3950. [CrossRef]
58. Inoue, Y.; Morimoto, Y.V.; Namba, K.; Minamino, T. Novel insights into the mechanism of well-ordered assembly of bacterial flagellar proteins in *Salmonella*. *Sci. Rep.* **2018**, *8*. [CrossRef]
59. Ibuki, T.; Imada, K.; Minamino, T.; Kato, T.; Miyata, T.; Namba, K. Common architecture of the flagellar type III protein export apparatus and F- and V-type ATPases. *Nat. Struct. Mol. Biol.* **2011**, *18*, 277–282. [CrossRef]
60. Kucera, J.; Terentjev, E.M. FliI 6 -FliJ molecular motor assists with unfolding in the type III secretion export apparatus. *Sci. Rep.* **2020**, *10*, 7127. [CrossRef]
61. Majewski, D.D.; Worrall, L.J.; Hong, C.; Atkinson, C.E.; Vuckovic, M.; Watanabe, N.; Yu, Z.; Strynadka, N.C.J. Cryo-EM structure of the homohexameric T3SS ATPase-central stalk complex reveals rotary ATPase-like asymmetry. *Nat. Commun.* **2019**, *10*, 626. [CrossRef] [PubMed]
62. Wilharm, G.; Lehmann, V.; Krauss, K.; Lehnert, B.; Richter, S.; Ruckdeschel, K.; Heesemann, J.; Trülzsch, K. *Yersinia enterocolitica* type III secretion depends on the proton motive force but not on the flagellar motor components MotA and MotB. *Infect. Immun.* **2004**, *72*, 4004–4009. [CrossRef] [PubMed]
63. Lee, P.-C.; Zmina, S.E.; Stopford, C.M.; Toska, J.; Rietsch, A. Control of type III secretion activity and substrate specificity by the cytoplasmic regulator PcrG. *Proc. Natl. Acad. Sci. USA* **2014**, *111*, E2027–E2036. [CrossRef] [PubMed]
64. Inoue, Y.; Ogawa, Y.; Kinoshita, M.; Terahara, N.; Shimada, M.; Kodera, N.; Ando, T.; Namba, K.; Kitao, A.; Imada, K.; et al. Structural Insights into the Substrate Specificity Switch Mechanism of the Type III Protein Export Apparatus. *Structure* **2019**, *27*, 965–976.e6. [CrossRef] [PubMed]

65. Shimada, M.; Saijo-Hamano, Y.; Furukawa, Y.; Minamino, T.; Imada, K.; Namba, K. Functional Defect and Restoration of Temperature-Sensitive Mutants of FlhA, a Subunit of the Flagellar Protein Export Apparatus. *J. Mol. Biol.* **2012**, *415*, 855–865. [CrossRef] [PubMed]
66. Minamino, T.; Shimada, M.; Okabe, M.; Saijo-Hamano, Y.; Imada, K.; Kihara, M.; Namba, K. Role of the C-terminal cytoplasmic domain of FlhA in bacterial flagellar type III protein export. *J. Bacteriol.* **2010**, *192*, 1929–1936. [CrossRef]
67. Saijo-Hamano, Y.; Minamino, T.; Macnab, R.M.; Namba, K. Structural and functional analysis of the C-terminal cytoplasmic domain of FlhA, an integral membrane component of the type III flagellar protein export apparatus in *Salmonella*. *J. Mol. Biol.* **2004**, *343*, 457–466. [CrossRef]
68. Saijo-Hamano, Y.; Imada, K.; Minamino, T.; Kihara, M.; Shimada, M.; Kitao, A.; Namba, K. Structure of the cytoplasmic domain of FlhA and implication for flagellar type III protein export. *Mol. Microbiol.* **2010**, *76*, 260–268. [CrossRef]
69. Hara, N.; Namba, K.; Minamino, T. Genetic Characterization of Conserved Charged Residues in the Bacterial Flagellar Type III Export Protein FlhA. *PLoS ONE* **2011**, *6*, e22417. [CrossRef]
70. Erhardt, M.; Wheatley, P.; Kim, E.A.; Hirano, T.; Zhang, Y.; Sarkar, M.K.; Hughes, K.T.; Blair, D.F. Mechanism of type-III protein secretion: Regulation of FlhA conformation by a functionally critical charged-residue cluster. *Mol. Microbiol.* **2017**, *104*, 234–249. [CrossRef]
71. Minamino, T.; Inoue, Y.; Kinoshita, M.; Namba, K. FliK-Driven Conformational Rearrangements of FlhA and FlhB Are Required for Export Switching of the Flagellar Protein Export Apparatus. *J. Bacteriol.* **2020**, *202*. [CrossRef] [PubMed]
72. Portaliou, A.G.; Tsolis, K.C.; Loos, M.S.; Balabanidou, V.; Rayo, J.; Tsirigotaki, A.; Crepin, V.F.; Frankel, G.; Kalodimos, C.G.; Karamanou, S.; et al. Hierarchical protein targeting and secretion is controlled by an affinity switch in the type III secretion system of enteropathogenic *Escherichia coli*. *EMBO J.* **2017**, *36*, 3517–3531. [CrossRef] [PubMed]
73. Maki-Yonekura, S.; Yonekura, K.; Namba, K. Domain movements of HAP2 in the cap–filament complex formation and growth process of the bacterial flagellum. *Proc. Natl. Acad. Sci. USA* **2003**, *100*, 15528–15533. [CrossRef] [PubMed]
74. Al-Otaibi, N.S.; Taylor, A.J.; Farrell, D.P.; Tzokov, S.B.; DiMaio, F.; Kelly, D.J.; Bergeron, J.R.C. The cryo-EM structure of the bacterial flagellum cap complex suggests a molecular mechanism for filament elongation. *Nat. Commun.* **2020**, *11*, 3210. [CrossRef] [PubMed]
75. Berg, H.C.; Anderson, R.A. Bacteria Swim by Rotating their Flagellar Filaments. *Nature* **1973**, *245*, 380–382. [CrossRef]
76. Hotani, H.; Asakura, S. Growth-saturation in vitro of *Salmonella* flagella. *J. Mol. Biol.* **1974**, *86*, 285–300. [CrossRef]
77. Keener, J.P. How *Salmonella* typhimurium measures the length of flagellar filaments. *Bull. Math. Biol.* **2006**, *68*, 1761–1778. [CrossRef]
78. Tanner, D.E.; Ma, W.; Chen, Z.; Schulten, K. Theoretical and Computational Investigation of Flagellin Translocation and Bacterial Flagellum Growth. *Biophys. J.* **2011**, *100*, 2548–2556. [CrossRef]
79. Rathinavelan, T.; Zhang, L.; Picking, W.L.; Weis, D.D.; De Guzman, R.N.; Im, W. A Repulsive Electrostatic Mechanism for Protein Export through the Type III Secretion Apparatus. *Biophys. J.* **2010**, *98*, 452–461. [CrossRef]
80. Demers, J.-P.; Habenstein, B.; Loquet, A.; Vasa, S.K.; Giller, K.; Becker, S.; Baker, D.; Lange, A.; Sgourakis, N.G. High-resolution structure of a *Shigella* type III secretion needle by solid-state NMR and cryo-electron microscopy. *Nat. Commun.* **2014**, *5*, 4976. [CrossRef]
81. Beatson, S.A.; Minamino, T.; Pallen, M.J. Variation in bacterial flagellins: From sequence to structure. *Trends Microbiol.* **2006**, *14*, 151–155. [CrossRef] [PubMed]
82. Turner, L.; Stern, A.S.; Berg, H.C. Growth of Flagellar Filaments of *Escherichia coli* Is Independent of Filament Length. *J. Bacteriol.* **2012**, *194*, 2437–2442. [CrossRef] [PubMed]
83. Stern, A.S.; Berg, H.C. Single-file diffusion of flagellin in flagellar filaments. *Biophys. J.* **2013**, *105*, 182–184. [CrossRef] [PubMed]
84. Evans, L.D.B.; Poulter, S.; Terentjev, E.M.; Hughes, C.; Fraser, G.M. A chain mechanism for flagellum growth. *Nature* **2013**, *504*, 287–290. [CrossRef]
85. Komoriya, K.; Shibano, N.; Higano, T.; Azuma, N.; Yamaguchi, S.; Aizawa, S.-I. Flagellar proteins and type III-exported virulence factors are the predominant proteins secreted into the culture media of *Salmonella* typhimurium. *Mol. Microbiol.* **1999**, *34*, 767–779. [CrossRef]
86. Furukawa, Y.; Imada, K.; Vonderviszt, F.; Matsunami, H.; Sano, K.; Kutsukake, K.; Namba, K. Interactions between bacterial flagellar axial proteins in their monomeric state in solution. *J. Mol. Biol.* **2002**, *318*, 889–900. [CrossRef]
87. Singer, H.M.; Erhardt, M.; Steiner, A.M.; Zhang, M.-M.; Yoshikami, D.; Bulaj, G.; Olivera, B.M.; Hughes, K.T. Selective Purification of Recombinant Neuroactive Peptides Using the Flagellar Type III Secretion System. *mBio* **2012**, *3*. [CrossRef]
88. Sin, C.; Chiarugi, D.; Valleriani, A. Quantitative assessment of ribosome drop-off in *E. coli*. *Nucleic Acids Res.* **2016**, *44*, 2528–2537. [CrossRef]
89. Renault, T.T.; Abraham, A.O.; Bergmiller, T.; Paradis, G.; Rainville, S.; Charpentier, E.; Guet, C.C.; Tu, Y.; Namba, K.; Keener, J.P.; et al. Bacterial flagella grow through an injection-diffusion mechanism. *eLife* **2017**, *6*. [CrossRef]
90. Zhao, Z.; Zhao, Y.; Zhuang, X.-Y.; Lo, W.-C.; Baker, M.A.B.; Lo, C.-J.; Bai, F. Frequent pauses in *Escherichia coli* flagella elongation revealed by single cell real-time fluorescence imaging. *Nat. Commun.* **2018**, *9*. [CrossRef]
91. Chen, M.; Zhao, Z.; Yang, J.; Peng, K.; Baker, M.A.; Bai, F.; Lo, C.-J. Length-dependent flagellar growth of *Vibrio alginolyticus* revealed by real time fluorescent imaging. *eLife* **2017**, *6*. [CrossRef] [PubMed]

92. Corey, R.A.; Pyle, E.; Allen, W.J.; Watkins, D.W.; Casiraghi, M.; Miroux, B.; Arechaga, I.; Politis, A.; Collinson, I. Specific cardiolipin–SecY interactions are required for proton-motive force stimulation of protein secretion. *Proc. Natl. Acad. Sci. USA* **2018**, *115*, 7967–7972. [CrossRef]
93. Collinson, I. The Dynamic ATP-Driven Mechanism of Bacterial Protein Translocation and the Critical Role of Phospholipids. *Front. Microbiol.* **2019**, *10*. [CrossRef]
94. Frain, K.M.; Robinson, C.; van Dijl, J.M. Transport of Folded Proteins by the Tat System. *Protein J.* **2019**, *38*, 377–388. [CrossRef] [PubMed]
95. Fulara, A.; Vandenberghe, I.; Read, R.J.; Devreese, B.; Savvides, S.N. Structure and oligomerization of the periplasmic domain of GspL from the type II secretion system of *Pseudomonas aeruginosa*. *Sci. Rep.* **2018**, *8*, 1–14. [CrossRef] [PubMed]
96. Nivaskumar, M.; Francetic, O. Type II secretion system: A magic beanstalk or a protein escalator. *Biochim. Biophys. Acta BBA Mol. Cell Res.* **2014**, *1843*, 1568–1577. [CrossRef] [PubMed]
97. Bergeron, J.R.C.; Sgourakis, N.G. Type IV Pilus: One Architectural Problem, Many Structural Solutions. *Structure* **2015**, *23*, 253–255. [CrossRef] [PubMed]
98. Leo, J.C.; Grin, I.; Linke, D. Type V secretion: Mechanism(s) of autotransport through the bacterial outer membrane. *Philos. Trans. R. Soc. B Biol. Sci.* **2012**, *367*, 1088–1101. [CrossRef]
99. Peterson, J.H.; Tian, P.; Ieva, R.; Dautin, N.; Bernstein, H.D. Secretion of a bacterial virulence factor is driven by the folding of a C-terminal segment. *Proc. Natl. Acad. Sci. USA* **2010**, *107*, 17739–17744. [CrossRef]
100. Costa, T.R.D.; Ilangovan, A.; Ukleja, M.; Redzej, A.; Santini, J.M.; Smith, T.K.; Egelman, E.H.; Waksman, G. Structure of the Bacterial Sex F Pilus Reveals an Assembly of a Stoichiometric Protein-Phospholipid Complex. *Cell* **2016**, *166*, 1436–1444.e10. [CrossRef]
101. Majdalani, N.; Ippen-Ihler, K. Membrane insertion of the F-pilin subunit is Sec independent but requires leader peptidase B and the proton motive force. *J. Bacteriol.* **1996**, *178*, 3742–3747. [CrossRef] [PubMed]

Review

Construction and Loss of Bacterial Flagellar Filaments

Xiang-Yu Zhuang and Chien-Jung Lo *

Department of Physics and Graduate Institute of Biophysics, National Central University, Taoyuan City 32001, Taiwan; xiangyu066@gmail.com
* Correspondence: cjlo@phy.ncu.edu.tw

Received: 31 July 2020; Accepted: 4 November 2020; Published: 9 November 2020

Abstract: The bacterial flagellar filament is an extracellular tubular protein structure that acts as a propeller for bacterial swimming motility. It is connected to the membrane-anchored rotary bacterial flagellar motor through a short hook. The bacterial flagellar filament consists of approximately 20,000 flagellins and can be several micrometers long. In this article, we reviewed the experimental works and models of flagellar filament construction and the recent findings of flagellar filament ejection during the cell cycle. The length-dependent decay of flagellar filament growth data supports the injection-diffusion model. The decay of flagellar growth rate is due to reduced transportation of long-distance diffusion and jamming. However, the filament is not a permeant structure. Several bacterial species actively abandon their flagella under starvation. Flagellum is disassembled when the rod is broken, resulting in an ejection of the filament with a partial rod and hook. The inner membrane component is then diffused on the membrane before further breakdown. These new findings open a new field of bacterial macro-molecule assembly, disassembly, and signal transduction.

Keywords: self-assembly; injection-diffusion model; flagellar ejection

1. Introduction

Since Antonie van Leeuwenhoek observed animalcules by using his single-lens microscope in the 18th century, we have entered a new era of microbiology. The motility of single-cell organisms is fascinating, but it took a long period of time to develop tools to determine the underlying mechanisms. Among these single-cell organisms, many bacterial species swim by using flagella consisting of a long extracellular filament, a hook, and a rotary bacterial flagellar motor anchored on the cell envelope (Figure 1) [1].

The flagellum consists of a thin helical flagellar filament that acts as a propeller, a reversible rotary molecular motor embedded on the envelope, and a hook that acts as a universal connection joint between the motor and the flagellar filament [2] (Figure 1). Flagellar distribution on the cell surface varies on different bacterial species. Peritrichous bacteria, such as *Escherichia coli* and *Salmonella enterica*, can produce multiple flagella distributed around the cell body. Monotrichous bacteria, such as *Vibrio alginolyticus*, have one single polar flagellum, and lophotrichous bacteria, such as *Vibrio fischeri*, have multiple flagella on one pole. The flagellar distribution affects bacterial swimming patterns as well as chemotaxis strategies. By switching motor rotation between counterclockwise to clockwise states, peritrichous *E. coli* can do a run-and-tumble swimming pattern [3]. Monotrichous *V. alginolyticus* can do forward-backward-turn through the flagellar flick [4,5].

The flagellar motor has a rotor and energy-conversion stator units that couple the ion-motive force and ion flux to the rotation. For example, *E. coli* and *Salmonella* use the proton (hydrogen nucleus), whereas marine *Vibrio* species use the sodium ion [6,7] (Figure 1). Recent reports have confirmed five MotA and two MotB proteins form one stator unit [8–10]. It is believed the ion flux passes through the stator unit ion channel coupled to the torque generation. Hence, a rotating flagellar motor can propel the cell body at a speed of 15–100 μm/s [11,12], and motor rotational speed is linear with proton motive force (PMF) [13]. The rotor comprises several stacked ring-link structures. The MS ring is composed of FliF, and the C ring is composed of FliG/FliM/FliN [14], which is located under the MS ring [15,16]. The rotary motor is driven by the interaction between the FliG and stator units to generate torque [17–19]. Moreover, a single stator unit can drive the rotor through conduction of at least 37 ions per revolution [20].

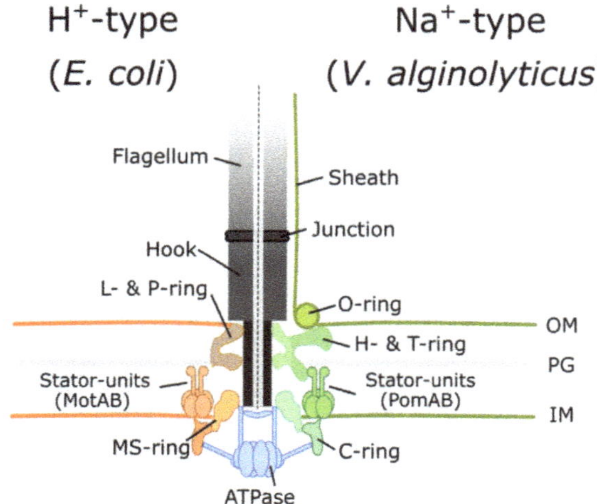

Figure 1. Bacterial flagellar motors are mainly classified into proton-driven and sodium-driven motors. Torque generation requires an interaction between stator units and FliG on the C-ring. In the proton-driven motor of *E. coli*, the stator is composed of MotA and MotB, whereas PomA and PomB are sodium-driven analogues in *V. alginolyticus*. The common elements for both motor types are LP, MS, and C rings. In *V. alginolyticus*, an additional sheath covers the flagellar filament.

In this article, we reviewed the current understanding of bacterial flagellar filament constructions and the mechanisms of flagellar loss. A flagellar filament is typically about 5–20 µm (2–10 times the cell body length) and is a hollow cylinder with outer and inner diameters of 20 and 2 nm, respectively [21]. This long extracellular component is self-assembled with several thousand flagellin monomers [21]. The construction of flagellar filament is considered to occur in an inside-out manner, that is, flagellins are delivered from the base of the flagellar type-III secretion system (fT3SS), which attaches to the basal body of the flagellum [2,14,22]. Using energy derived from ATP hydrolysis and PMF [23–25], the fT3SS pumps unfolded flagellins into the flagellar channel. These unfolded flagellins are transported to the distal end and are then folded as the new part of the flagellar filament.

Although the structure and mechanical functions of bacterial flagella are well researched, the dynamics of flagellar loss remains poorly understood. Recent works have revealed that flagella are impermanent cellular structures, and cells can actively abandon this large motility apparatus [26–30]. Whereas molecular triggering and the activation mechanism remain unknown, the novel finding of the active ejection of the flagellar filament provides a new direction of cellular adaptation to external stimuli.

2. Bacterial Flagellin Transportation

2.1. Architecture of the Type-III Secretion System

The fT3SS is involved in the construction of the flagellar axial structure consisting of the rod, hook, and filament and the virulence-associated T3SS (vT3SS) of the injection device used by gram-negative bacteria injects toxic effectors into target cells [25]. The fT3SS and vT3SS have similar cytoplasmic components and an inner membrane export apparatus, but the final destinations of secreted proteins are different. The fT3SS keeps the secreted proteins to form the flagellar filaments while the vT3SS delivers the effectors into the target cells.

The basal body of the flagellar rotary motor is composed of an axial rod and MS, C, and LP rings (Figure 1). The MS ring consists of 33 subunits of protein FliF [31], and FliP/FliQ/FliR is assembled and embedded in the cytoplasm first during the flagellum formation on the membrane [32–34]. The C ring consists of 26–34 FliG [35], 34–44 FliM [36], and > 100 FliN [37] subunits and is located under the MS ring. All flagellar axial parts self-assemble through protein export of the fT3SS [38]. The core integral-membrane components of the flagellar export apparatus (FlhA/B and FliP/Q/R) are similar to the virulence-associated vT3SS [39]. Furthermore, fT3SS and vT3SS are powered by ATP hydrolysis and PMF [23–25,40–42]. Hence, during flagellar assembly, FliF first forms the MS ring in the cytoplasmic membrane. Next, FliG, FliM, and FliN assemble the C ring on the cytoplasmic side. FliG is also directly associated with the MS-ring component FliF [43,44]. After the basal body is complete, the axial proteins are exported through fT3SS [12,22,45,46].

The T3SS needle complex or flagellum is composed of rings for supporting a needle filament or flagellar filament, which extend from the inner membrane, through the periplasmic space and peptidoglycan layer, to the outer membrane (Figure 2). It serves as the central channel for translocating proteins. Several additional proteins combine with the basal body to form a functioning needle complex [47,48]. SctI assembles between membranes forming an inner rod as a needle adaptor, which can anchor the needle filament or the flagellum [49–52]. Then, extending from the inner rod to the extracellular environment is the needle filament or flagellar filament composed of protein SctF or FliC, respectively [53]. Additionally, there are several proteins surrounding the cytoplasmic side of the membrane-spanning rings, which form the sorting platform complex [54] similar to the flagellar C ring complex [55].

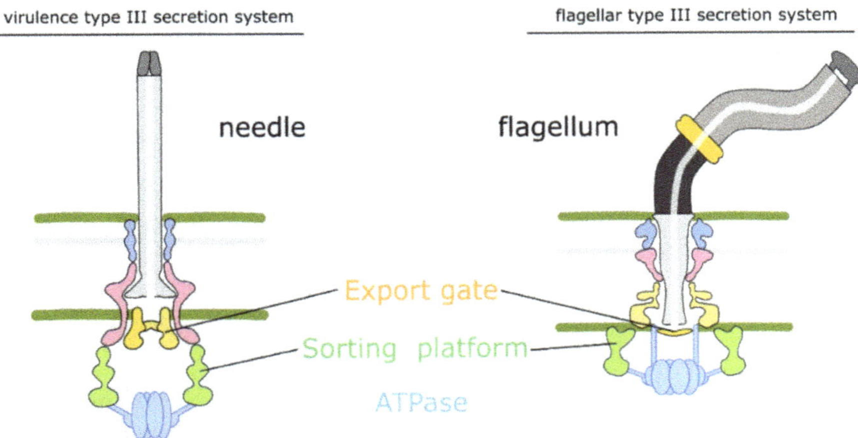

Figure 2. Schematic of vT3SS and fT3SS. The vT3SS and fT3SS are similar in terms of cytoplasmic components and the inner membrane export apparatus. ATPase associates with proteins to pump unfolded substrates into the export channel. Then, the sorting platform and export gate help line up the sequence of unfolded substrates. The flagellum consists of a rod, hook structure, and flagellar filament. The needle connects with rings to directly pass through the outer membrane.

2.2. Strcture of the Flagellin

The bacterial flagellar filament is a 5–20 μm long, thin, and hollow helical propeller with an outer diameter of 20 nm and an inner diameter of 2 nm. It is packed with flagellins of two conformations, L and R types [56,57]. Folded flagellin FliC consists of four domains (D0, D1, D2, and D3) and is shaped like "T" with vertical and horizontal lengths of approximately 140 and 110 Å, respectively. In total,

11 flagellins are packed into one round of the filament tube in which the D0/D1 domain forms the double tubular structure as the inner core and the D2/D3 domain forms the outer structure [21]. Most of the subunit interactions in the outer part are through polar-to-polar interaction, and hydrophobic interactions are small. By contrast, the inner tube is mostly hydrophobic for high stability of the flagellar filament. The narrow channel diameter would reduce the diffusion rate of the unfolded flagellins during transportation in the flagellar tube. More details would be discussed in Section 3.2.

2.3. Substrates Accumulation and Delivery

The fT3SS is similar to the vT3SS in that both have five proteins that form the export apparatus and connect with the MS ring [25] (Figure 2). The fT3SS plays a crucial role in bacterial flagellar motor system as the rod, hook, and filament subunits are transported via the fT3SS, which includes an ATPase complex (FliI, FliJ, and FliH), an export substrate-chaperon docking platform (FlhAc), and a transmembrane export gate (FlhAB and FliPQR). The ATPase complex (FliI and FliJ) connects with the C ring (FliG, FliM, and FliN) through an interaction between FliH and FliN. There are two components of fT3SS to collect substrates nearby the export gate to increase the binding efficiency [58]. One is the associated ATPase complex (FliHIJ), which is composed of an ATPase (FliI), a regulator of ATPase (FliH), and a central stalk protein (FliJ) [59–61]. The other one is the C ring, which is located under the MS ring. The ATPase complex can deliver proteins to the export gate after the cytoplasmic FliH2FliI complex recruit the substrate-chaperone complexes [61,62]. The C ring provides binding sites for substrate-chaperone complexes and promotes the accumulation of substrates near the export gate [63].

The ATPase and the T3SS export gate play an important role in cargo transfer across the inner membrane. In early studies, ATP has been considered as the main energy source for T3SS. Recent research studies have shown that the PMF provides the primary energy source [23,24,40,41,64,65]. PMF is the sum of electrical and chemical potential across the membrane. Bacteria use PMF for several import cellular functions such as ATP synthesis, active membrane transport, and flagellar motility. The stator proteins MotA and MotB couples the ion flux driven by PMF to the motor rotation [66]. Further investigation revealed that the electrical components of PMF contributes as the main energy source to flagellar protein export [67]. The ATPase may have dual roles for shuttling substrates to the export gate and enhancing the efficiency of the export apparatus. However, the exact mechanism of energy conversion for the protein translocation by the export apparatus remains unknown. Once proteins are translocated across the inner membrane, the substrate proteins diffuse through the narrow channel until they arrive at their site of assembly. There is no evidence of active transportation involved in the proteins' transportation inside the channel.

For example, for filament subunit FliC, the chaperone-subunit complex (FliS–FliC) binds with the cytoplasmic ATPase complex before loading into the export channel. Then, the chaperone–subunit complex goes through the export gate [62,68–70]. Finally, the flagellar filament subunits are translocated across the cell membrane into the export channel and folded to become part of the filament at the distal end. In gram-negative bacteria, there are two additional rings, L and P rings, surrounding the axial rod rotation. These rings act as bearings for the axial rod. In other words, they anchor to the outer membrane and peptidoglycan layer, respectively. Overall, proteins are delivered through the fT3SS apparatus by using energy from ATP hydrolysis and PMF [23–25,40–42]. Filament subunit transportation is discussed in detail in Section 3.

3. Flagellar Filament Construction

In this section, we outlined the crucial experimental and theoretical milestones in the study of the flagellar filament construction. A timeline of experiments and models on flagellar filament construction and loss is summarized in Figure 3.

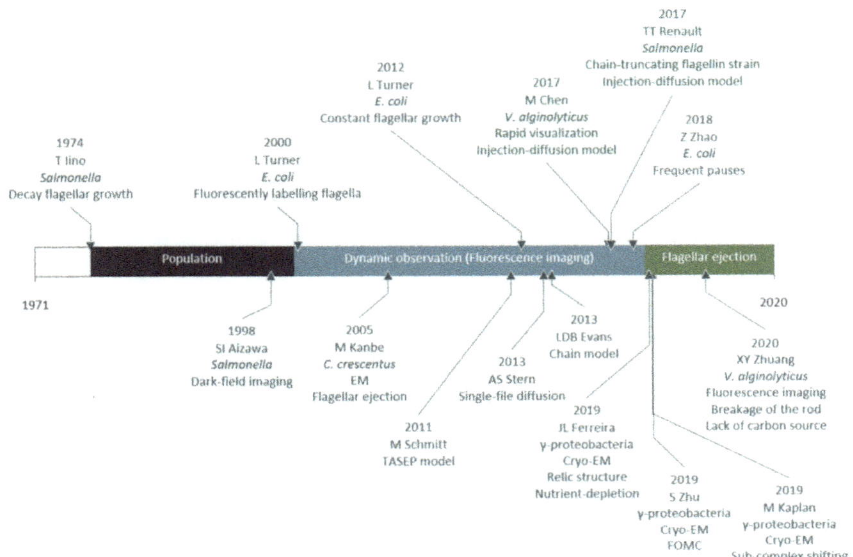

Figure 3. A timeline of experiments and models on flagellar filament construction and loss.

3.1. Milestones of Flagellar Filament Growth Measurements

The first attempt to probe the flagellar filament growth rate can be dated back to 1974. At that time, the only tool for measuring flagellar filament length precisely was electron microscopy. Iino compared the flagellar length histogram at two different time points in a *Salmonella typhimurium* growing culture [71]. Assuming that the length order of two samples are the same, he calculated the growth rate of the flagellar filament from the filament length histograms. The most important finding from his experiments was that flagellar growth rate is length dependent with exponential decay formulated as follows.

$$V = V_0 e^{-KL} \tag{1}$$

where V is the flagellar length growth rate, V_0 is the initial flagellar growth rate at L = 0, K is the decay rate, and L is the flagellar length. Iino found that the initial flagellar filament growth rate can be as high as 550 nm/min and that the fT3SS must secrete 18 FliC per second (Table 1).

In 1998, Aizawa and Kubori used dark-field microscopy to measure flagellar length distribution in different growth phases, and their results regarding flagellar growth rate were similar to those of Iino [72]. Both experiments show statistical filament growth from population data with limited temporal resolution. For measuring growth dynamics, development of flagellar filament fluorescent labelling is required [3].

In 2012, Turner et al. sequentially labeled *E. coli*'s flagellar filament with two colors of fluorophores and found that the average flagellar growth rate is independent of filament length [73]. The average filament growth rate was a constant (24 nm/min) but with high variation. These results motivate scientists in the field to revisit this long-standing question.

In 2017, based on improved in situ labeling and immunostaining, Renault et al. monitored the flagellar growth of *Salmonella* [74]. They provided a living-cell method for observing single cells dynamically growing individual flagella. Their results reported the length-dependent elongation mechanism with elongation speed decreasing gradually from approximately 100 (nm/min) to 20 (nm/min), thus, confirming the results of Iino [71].

Simultaneously, Chen et al. used the sheathed flagellar filament and fast easy sheath labeling of *V. alginolyticus* to largely improve temporal resolution of flagellar growth dynamics [75]. The sheath is a membrane-like structure and contiguous of the outer membrane [76–78]. The sheath can be easily labeled using lipophilic fluorescent dyes within a sub-second time scale [79,80]. Furthermore, the single polar flagellum system of *V. alginolyticus* reduces the difficulty faced during image analysis in separating entangled flagella. This report revealed that the flagellar growth process is highly length-dependent with an initial constant growth rate and then a decaying growth rate. The initial growth rate is approximately 50 nm/min (Table 1).

In 2018, Zhao et al. revisited *E. coli* flagellar growth by using biarsenical dyes to label flagellin with tetracysteine tag for single-cell observation [81] and showed that the flagellar growth of *E. coli* has frequent pauses due to insufficient flagellins [82]. Although a high fluctuation of the filament growth rate was observed similarly to that observed by Tuner. The average growth rate also decayed.

Table 1. Summary of flagellar filament growth in different species.

Species	V_0 (nm/min)	Secretion Rate (#/sec)	Microscopy	Reference
Salmonella	550 (Decay)	18.33	Electron microscopy	[71]
	Decay		Dark-field imaging	[72]
	100 (Decay)	3.33	Fluorescence imaging	[74]
E. coli	24 (Constant)	0.80	Fluorescence imaging	[73]
	27 (Decay, pauses)	0.90	Fluorescence imaging	[82]
V. alginolyticus	50 (Decay)	1.67	Fluorescence imaging	[75]

With improved fluorescent labelling techniques, flagellar filament growth was observed in live single cells and showed a length-dependent decay in the growth rate. Temporal and spatial resolutions can be further improved to reveal the initial flagellar growth rate and transportation details.

3.2. Models for Flagellin Transport and Filament Growth

Once the secreted flagellin passes across the inner membrane, it continuously travels through the flagellar filament that could be up to 10 μm long. The filament central channel is too narrow for folded flagellin to pass. The energy source and transportation mechanism are the main mysteries.

The simplest model for flagellin transportation is that partially unfolded flagellins diffuse through the channel in a single-file and then fold at the distal end. Considering this model, Schmitt and Stark used the totally asymmetric simple exclusion process (TASEP) models with open boundary to simulate flagellin transportation and flagellar growth [83]. TASEP has been applied successfully to study nonequilibrium steady states such as the motion of ribosomes along mRNA, molecular motors along microtubule filaments, or the traffic of the car on the highway. In the TASEP model, particle diffusion is described based on the forward/backward rate with lattice sites. The single-file feature is simulated using particles that can only move when the target site is empty. This Monte Carlo simulation has four parameters, namely loading rate, crystallization rate, forward rate, and backward rate. With a small negative drift, that is, the forward rate smaller than the backward rate, the simulation data can match the experimental result of Iino. However, it was unclear why the filament channel exhibits biased Brownian diffusion of flagellin transportation.

In 2013, Stern and Berg performed a single-file diffusion simulation with more realistic parameters [84]. They assumed that the flagellin is unfolded into an α-helical chain, and the pump extrudes one flagellin every two seconds into a one-dimensional lattice tube. The diffusion process is modeled using the diffusion constant D. They found that, by changing the flagellin diffusion constant, the flagellar growth could be varied from a constant growth rate to a length-dependent decay rate. Furthermore, they used diffusion constants 30–480 times smaller than the estimated diffusion constant for an α-helical flagellin subunit diffused freely in water.

In 2017, Chen et al. built an injection-diffusion model to explain the high-resolution experimental data from *V. alginolyticus* flagella growth [75] with an initial constant growth rate and then a decayed growth rate (Figure 4A,C). The main difference of this model from the previous diffusion mechanism model is the addition of the pumping force at the secretion side that can push flagellin into the channel

(Figure 4A). With Brownian dynamic simulation, they successfully reproduced the length-dependent growth rate and determined that the effective diffusion constant of flagellin in the channel is 1000 times smaller than that in the bulk water. With a reduction in the injection force, the initial constant growth rate region can be eliminated. Simultaneously, Renault et al. demonstrated the length-dependent *Salmonella* flagellar growth rate based on the injection-diffusion mechanism with analytical approaches [74].

To explain Tunner's 2012 result of constant flagellar growth rate, Evans et al. proposed a chain mechanism model that harnesses the entropic force of the unfolded flagellins for the flagellar growth (Figure 4A,B) [85]. This model required all the transporting flagellins in the flagellar channel to form a long chain and the folding force of the distal end flagellin to pull the entire chain (Figure 4A,B). They demonstrated that a subunit docked at the export apparatus can be captured using a free subunit with a head-to-tail linkage of N-terminals and C-terminals. The pulling force adjusts as the flagellar length increases to maintain the constant flagellar growth rate. Although the chain model is physically simple, it is incompatible with some properties of flagellins [74]. First, N and C terminals of flagellin are anti-parallel in the linked chain but parallel in the folded flagellar structure. Second, the channel is too narrow to accommodate the linking regions of chains. Third, Renault et al. demonstrated that truncation of flagellin N-terminal and C-terminal linking region do not affect the flagellar growth rate [74]. Thus, it is less likely that the chain model is the fundamental mechanism of flagellin transportation.

The flagellin is delivered in a 2-nm narrow channel that is different from the bulk water environments. A decrease in an effective diffusion constant is expected. The decays of the flagellar growth rate are due to reduced transportation of long-distance diffusion and increased jamming (Figure 4C). However, the underlying reason for high variation in the measured initial flagellar growth rate is unclear (Table 1). A high flagellar growth rate of 550 nm/min requires transporting 18 flagellins per second. A new high-spatial-resolution real-time imaging method is required for further studies to reveal the mysterious high-speed initial growth rate and transportation mechanism.

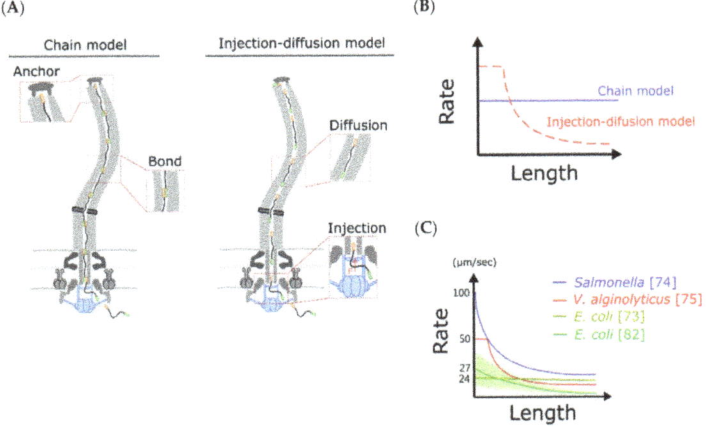

Figure 4. Potential mechanisms of the flagellar growth process. (**A**) Schematics of the chain model (left) and the injection-diffusion model (right) are shown. The chain model proposed that sequential flagellins are linked head-to-tail to form a chain, and the first flagellin anchors beneath the distal end of the flagellum to provide a pulling force. Therefore, constant force contributes to a constant growth rate. According to the injection-diffusion model, the secretion system applies a secretion force on an unfolded flagellin, and flagellins are delivered through diffusion after entering the channel. Hence, flagellins are crowded on the channel when the flagellum is getting longer. (**B**) The chain model predicts a constant growth rate, and the injection-diffusion model predicts a length-dependent growth rate. (**C**) The summary of flagellar growth rate of three bacteria using fluorescent-based techniques.

4. Loss of Flagella

Bacterial flagellar motility is a fascinating feature of single-cell organisms. The construction of whole flagellum requires >20,000 proteins. Therefore, it is generally believed that constructing flagellum is costive and flagella are valuable for bacteria. However, a highly motile planktonic bacterial phase is only a part of the bacterial life cycle. Bacteria must switch between different phases and manage their flagellar motility. We have studied the construction sequence of the bacterial flagellar motor [2], but little is known regarding flagellar disassembly.

The unique life cycle of freshwater bacterium *Caulobacter crescentus* provides an opportunity to learn the potential mechanism of flagella loss through ejection. *C. crescentus* transitions between two distinguishable cell states known as a motile swarmer cell state with a polar flagellum and a surface-attached stalked cell state. In 2004, Grünenfelder et al. demonstrated that, in *C. crescentus*, ejection started from the inside out and reported that flagellar ejection is trigged by ClpA [86]. During differentiation into a stalked cell state, the swarmer cell releases its single polar flagellum. The flagellum released is synchronized with cell differentiation, and the ejected flagellum has a hook and partial rod of approximately 18 nm [30]. The breaking point of the structure can be localized to the MS ring–rod junction. Thus, two potential flagellum-releasing models are available for *C. crescentus*–destruction of the MS ring or breakage between the MS ring and the rod [30]. Moreover, the assembly and loss of the polar flagellum in symbiont *V. fischeri* [87] and plant-associated *Methylobacteria* [88] have been reported, but the mechanism remains unclear.

In the beginning, scientists focused only on the released flagellar structure. However, in 2019, multiple groups simultaneously noted another important finding. These groups observed a flagellar outer membrane complex (FOMC) [26–28] in different bacterial species by using cryo-electron tomography. A common feature of the FOMC is containing L and P rings of the bacterial flagellar motor with a plug located inside the P ring and without the hook, the flagellar filament, and the MS ring. This finding raises a new question of whether the FOMC is a precursor or a relic of bacterial flagella.

To further investigate the FOMC and its relation to the fully assembled flagellar motor, these groups also imaged different mutants disrupting the flagellar construction sequence. The FOMC could not be detected in strains lacking the rod protein, FlgG (e.g., *Pseudomonas aeruginosa*) [27,28], or intrinsic flagellar type III secretion system protein, FlhA (e.g., *Shewanella putrefaciens*) [26]. These results strongly suggest that the FOMC cannot form in the absence of the distal rod and secretion system. These results support previous models that suggest that P and L rings form around the assembled rod. Therefore, the FOMC is not a precursor of the flagellar motor assembly process.

Evidence suggests that bacteria can somehow disassemble their flagella and leave FOMC or a relic structure in the membrane. Since the flagellar motor is used for swimming, Ferreira et al. demonstrated that the swimming speed and the number density of flagellar motors of *Plesiomonas shigelloides* and *V. fischeri* decrease at high cell density in the growth medium [26]. Later, Zhuang et al. used fluorescent labeled *V. alginolyticus* single polar flagellum and measured the percentage of flagellated bacteria (PFB) during the *V. alginolyticus* growth. They found that the PFB increase rapidly in the early exponential phase through widespread flagella production. The PFB peaks at approximately 76% in the mid-exponential phase. After entering the stationary phase, the PFB begins to decline due to cessation of flagella production in daughter cells. When the cells enter the prolonged stationary phase, the flagellated cell concentration suddenly drops, indicating that the bacteria actively abandon flagella [29]. In their study, the swimming speed of *V. alginolyticus* was strongly correlated to the PFB, which is consistent with the finding of Ferreira et al. [26].

To catch the flagella ejection event, Ferreira et al. presented a striking image from in situ cryo-EM, showing a flagellar filament with a hook and short distal rod breaking off from the motor [26]. More importantly, Zhuang et al. recorded time-lapse images showing the polar flagellum of *V. alginolyticus* being ejected from the cell pole [29]. These single-cell experimental results confirmed that these bacterial cells actively eject their flagella. However, the trigger is unknown.

Ferreira et al. first demonstrated that the depletion of nutrients triggers ejection [26]. Further studies on *V. alginolyticus* showed that a lack of carbon source promotes flagellar disassembly. Whether flagella ejection has a universal trigger or has a species-dependent trigger is unknown. Further investigation is required to determine the molecular-level triggers.

The mechanism of flagellar filament release by bacteria is not clear. In *C. crescentus*, the protease ClpAP is associated with FliF degradation. However, the depletion of ClpA and ClpX did not prevent flagellar ejection in *S. putrefaciens* [26]. Again, we can have a hint from in situ cryo-EM images. Kaplan et al. found inner-membrane complexes (C and MS rings) near FOMC (P and L rings). Zhuang et al., by using the single-molecular tracking of GFP-fused FliG of *V. alginolyticus*, found fast movement of FliG clusters on the inner membrane before flagellar filament loss. These results together suggest the "break the rod" model for flagellar filament release (Figure 5).

The FOMC are relics of ejected flagella but not flagellar assembly intermediates [26]. These findings raise the question of why bacteria do not keep the flagella as permanent cellular structures. A flagellum is composed of ~2×10^4 proteins, which is a significant fraction of the total of 3×10^6 proteins for a bacterial cell [29]. Building and rotating the flagella are energy-consuming for bacteria. However, several known mechanisms enable bacteria to stop flagellar rotation. For instance, with nutrient depletion, the activation of cyclic di-GMP signaling triggers YcgR, which is a c-di-GMP binding protein, to interact with the flagellar switch-complex proteins FliG and FliM, stopping flagellar rotation and acting as a "molecular brake" [89,90]. Additionally, reducing the ion motive force dissociates stator units from the flagellar motor in *E. coli* and *V. alginolyticus* [91]. The main purpose of active flagella loss under starvation is unclear.

Although flagella play a crucial role in bacterial motility, flagellins are also essential antigens that can stimulate both innate inflammatory response and adaptive immunity development [92–94]. Two specialized receptors on immune cells, cell surface Toll-like receptor 5 (TLR5) [95–97], and intracellular receptor Ipaf [94,98,99] are responsible for recognizing flagellins as a warning of a pathogenic bacterial invasion. Hence, flagella ejection may be a common feature of flagellated bacteria. Further investigation is required on the active response of flagella ejection.

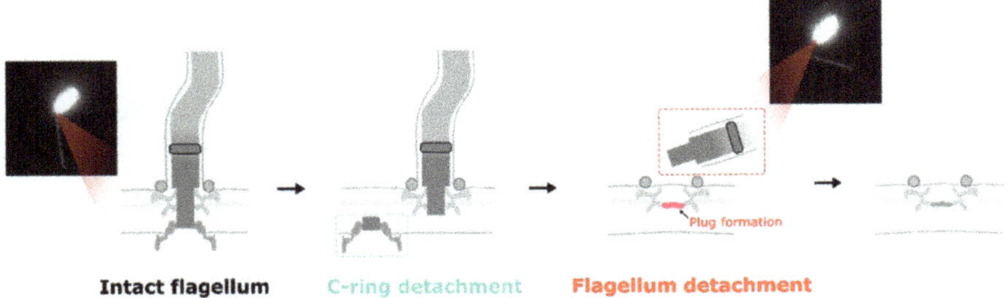

Figure 5. A model summarizing the disassembly process of a V. alginolyticus flagellum, showing that it begins with breaking the rod above the MS ring before FliG depolymerization. The C ring, with inner membrane components, then mobilizes on the cell membrane. Finally, the LP ring is likely sealed and the flagellum is ejected.

5. Concluding Remarks and Future Perspective

The self-assembly of tens of thousands of flagellins into an extracellular filament is an amazing process. Current data support the injection-diffusion mechanism for flagellin transportation and length-dependent decay in the flagellar growth rate. In the near future, investigating the efficiency of the secretion system, the initial flagellar growth rate, and the transportation mechanism will be noteworthy.

The active ejection of flagella causes flagellum loss and is different from the flagellar loss by shearing [100,101]. We speculate the presence of some trigger signals and a signal transduction pathway [102]. Moreover, a molecular level understanding is required to understand the "braking rod" process. Certainly, a significant amount of information needs to be explored regarding the flagellar motor, which is a phenomenal, tiny molecular machine.

Author Contributions: Conceptualization, X.-Y.Z. and C.-J.L; writing—original draft preparation, X.-Y.Z.; writing—review and editing, C.-J.L. All authors have read and agreed to the published version of the manuscript.

Funding: This work was supported by Ministry of Science and Technology, Republic of China, under contract No. MOST-107-2112-M-008-025-MY3 and MOST-109-2628-M-008-001-MY4.

Acknowledgments: We thank Keiichi Namba and Tohru Minamino for their comments on the manuscript.

Conflicts of Interest: The authors declare no conflict of interest.

References

1. Berg, H.C.; Anderson, R.A. Bacteria swim by rotating their flagellar filaments. *Nature* **1973**, *245*, 380–382. [CrossRef] [PubMed]
2. Macnab, R.M. How bacteria assemble flagella. *Annu. Rev. Microbiol.* **2003**, *57*, 77–100. [CrossRef] [PubMed]
3. Turner, L.; Ryu, W.S.; Berg, H.C. Real-time imaging of fluorescent flagellar filaments. *J. Bacteriol.* **2000**, *182*, 2793. [CrossRef] [PubMed]
4. Seishi, K.; Imai, N.; Nishitoba, M.; Sugiyama, S.; Magariyama, Y. Asymmetric swimming pattern of *Vibrio alginolyticus* cells with single polar flagella. *FEMS Microbiol. Lett.* **2005**, *242*, 221–225.
5. Taute, K.M.; Gude, S.; Tans, S.J.; Shimizu, T.S. High-throughput 3D tracking of bacteria on a standard phase contrast microscope. *Nat. Commun.* **2015**, *6*, 8776. [CrossRef] [PubMed]
6. Manson, M.D.; Tedesco, P.; Berg, H.C.; Harold, F.M.; Van Der Drift, C. A protonmotive force drives bacterial flagella. *Proc. Natl. Acad. Sci. USA* **1977**, *74*, 3060–3064. [CrossRef] [PubMed]
7. Hirota, N.; Kitada, M.; Imae, Y. Flagellar motors of alkalophilic bacillus are powered by an electrochemical potential gradient of Na^+. *FEBS Lett.* **1981**, *132*, 278–280. [CrossRef]
8. Hennell-James, R.; Deme, J.; Alcock, F.; Silale, A.; Lauber, F.; Berks, B.C.; Lea, S.M.; Kjaer, A. Structure of a proton-powered molecular motor that drives protein transport and gliding motility. *bioRxiv* **2020**. [CrossRef]
9. Deme, J.C.; Johnson, S.; Vickery, O.; Muellbauer, A.; Monkhouse, H.; Griffiths, T.; James, R.H.; Berks, B.C.; Coulton, J.W.; Stansfeld, P.J.; et al. Structures of the stator complex that drives rotation of the bacterial flagellum. *Nat. Microbiol.* **2020**, 1–12. [CrossRef] [PubMed]
10. Santiveri, M.; Roa-Eguiara, A.; Kühne, C.; Wadhwa, N.; Berg, H.C.; Erhardt, M.; Taylor, N.M.I. Structure and function of stator units of the bacterial flagellar motor. *Cell* **2020**, *183*, 244–257. [CrossRef]
11. Furuno, M.; Atsumi, T.; Yamada, T.; Kojima, S.; Nishioka, N.; Kawagishi, I.; Homma, M. Characterization of polar-flagellar-length mutants in *Vibrio alginolyticus*. *Microbiology* **1997**, *143*, 1615–1621. [CrossRef]
12. Xue, R.; Ma, Q.; Baker, M.A.; Bai, F. A delicate nanoscale motor made by nature—The bacterial flagellar motor. *Adv. Sci.* **2015**, *2*, 1500129. [CrossRef]
13. Gabel, C.V.; Berg, H.C. The speed of the flagellar rotary motor of *Escherichia coli* varies linearly with protonmotive force. *Proc. Natl. Acad. Sci. USA* **2003**, *100*, 8748–8751. [CrossRef]
14. Chevance, F.F.; Hughes, K.T. Coordinating assembly of a bacterial macromolecular machine. *Nat. Rev. Microbiol.* **2008**, *6*, 455–465. [CrossRef] [PubMed]
15. Murphy, G.E.; Leadbetter, J.R.; Jensen, G.J. In situ structure of the complete *Treponema primitia* flagellar motor. *Nature* **2006**, *442*, 1062–1064. [CrossRef] [PubMed]
16. Thomas, D.R.; Francis, N.R.; Xu, C.; DeRosier, D.J. The three-dimensional structure of the flagellar rotor from a clockwise-locked mutant of *Salmonella enterica serovar Typhimurium*. *J. Bacteriol.* **2006**, *188*, 7039–7048. [CrossRef] [PubMed]
17. Zhou, J.; Lloyd, S.A.; Blair, D.F. Electrostatic interactions between rotor and stator in the bacterial flagellar motor. *Proc. Natl. Acad. Sci. USA* **1998**, *95*, 6436–6441. [CrossRef]
18. Blair, D.F. Flagellar movement driven by proton translocation. *FEBS Lett.* **2003**, *544*, 86–95. [CrossRef]
19. Takekawa, N.; Kojima, S.; Homma, M. Contribution of many charged residues at the stator-rotor interface of the Na^+-driven flagellar motor to torque generation in *Vibrio alginolyticus*. *J. Bacteriol.* **2014**, *196*, 1377–1385. [CrossRef]

20. Lo, C.J.; Sowa, Y.; Pilizota, T.; Berry, R.M. Mechanism and kinetics of a sodium-driven bacterial flagellar motor. *Proc. Natl. Acad. Sci. USA* **2013**, *110*, E2544–E2551. [CrossRef]
21. Yonekura, K.; Maki-Yonekura, S.; Namba, K. Complete atomic model of the bacterial flagellar filament by electron cryomicroscopy. *Nature* **2003**, *424*, 643–650. [CrossRef]
22. Minamino, T.; Imada, K.; Namba, K. Mechanisms of type III protein export for bacterial flagellar assembly. *Mol. Biosyst.* **2008**, *4*, 1105–1115. [CrossRef]
23. Minamino, T.; Namba, K. Distinct roles of the FliI ATPase and proton motive force in bacterial flagellar protein export. *Nature* **2008**, *451*, 485–488. [CrossRef]
24. Paul, K.; Erhardt, M.; Hirano, T.; Blair, D.F.; Hughes, K.T. Energy source of flagellar type III secretion. *Nature* **2008**, *451*, 489–492. [CrossRef]
25. Lee, P.C.; Rietsch, A. Fueling type III secretion. *Trends Microbiol.* **2015**, *23*, 296–300. [CrossRef]
26. Ferreira, J.L.; Gao, F.Z.; Rossmann, F.M.; Nans, A.; Brenzinger, S.; Hosseini, R.; Wilson, A.; Briegel, A.; Thormann, K.M.; Rosenthal, P.B.; et al. γ-proteobacteria eject their polar flagella under nutrient depletion, retaining flagellar motor relic structures. *PLoS Biol.* **2019**, *17*, e3000165. [CrossRef]
27. Kaplan, M.; Subramanian, P.; Ghosal, D.; Oikonomou, C.M.; Pirbadian, S.; Starwalt-Lee, R.; Mageswaran, S.K.; Ortega, D.R.; Gralnick, J.A.; El-Naggar, M.Y.; et al. In situ imaging of the bacterial flagellar motor disassembly and assembly processes. *EMBO J.* **2019**, *38*, e100957. [CrossRef]
28. Zhu, S.; Schniederberend, M.; Zhitnitsky, D.; Jain, R.; Galán, J.E.; Kazmierczak, B.I.; Liu, J. In situ structures of polar and lateral flagella revealed by cryo-electron tomography. *J. Bacteriol.* **2019**, *201*, e00117–e00119. [CrossRef]
29. Zhuang, X.; Guo, S.; Li, Z.; Zhao, Z.; Kojima, S.; Homma, M.; Wang, P.; Lo, C.; Bai, F. Live-cell fluorescence imaging reveals dynamic production and loss of bacterial flagella. *Mol. Microbiol.* **2020**, *114*, 279–291. [CrossRef] [PubMed]
30. Kanbe, M.; Shibata, S.; Umino, Y.; Jenal, U.; Aizawa, S.I. Protease susceptibility of the *Caulobacter crescentus* flagellar hook-basal body: A possible mechanism of flagellar ejection during cell differentiation. *Microbiology* **2005**, *151*, 433–438. [CrossRef]
31. Johnson, S.; Fong, Y.H.; Deme, J.; Furlong, E.; Kuhlen, L.; Lea, S.M. Structure of the bacterial flagellar rotor MS-ring: A minimum inventory/maximum diversity system. *bioRxiv* **2019**, 718072. [CrossRef]
32. Fabiani, F.D.; Renault, T.T.; Peters, B.; Dietsche, T.; Gálvez, E.J.C.; Guse, A.; Freier, K.; Charpentier, E.; Strowig, T.; Franz-Wachtel, M.; et al. A flagellum-specific chaperone facilitates assembly of the core type III export apparatus of the bacterial flagellum. *PLoS Biol.* **2017**, *15*, e2002267. [CrossRef]
33. Fukumura, T.; Makino, F.; Dietsche, T.; Kinoshita, M.; Kato, T.; Wagner, S.; Namba, K.; Imada, K.; Minamino, T. Assembly and stoichiometry of the core structure of the bacterial flagellar type III export gate complex. *PLoS Biol.* **2017**, *15*, e2002281. [CrossRef]
34. Kuhlen, L.; Abrusci, P.; Johnson, S.; Gault, J.; Deme, J.; Caesar, J.; Dietsche, T.; Mebrhatu, M.T.; Ganief, T.; Macek, B.; et al. Structure of the core of the type III secretion system export apparatus. *Nat. Struct. Mol. Biol.* **2018**, *25*, 583–590. [CrossRef]
35. Suzuki, H.; Yonekura, K.; Namba, K. Structure of the rotor of the bacterial flagellar motor revealed by electron cryomicroscopy and single-particle image analysis. *J. Mol. Biol.* **2004**, *337*, 105–113. [CrossRef]
36. Lele, P.P.; Branch, R.W.; Nathan, V.S.; Berg, H.C. Mechanism for adaptive remodeling of the bacterial flagellar switch. *Proc. Natl. Acad. Sci. USA* **2012**, *109*, 20018–20022. [CrossRef]
37. Zhao, R.; Pathak, N.; Jaffe, H.; Reese, T.S.; Khan, S. FliN is a major structural protein of the C-ring in the *Salmonella typhimurium* flagellar basal body. *J. Mol. Biol.* **1996**, *261*, 195–208. [CrossRef]
38. Macnab, R.M. Type III flagellar protein export and flagellar assembly. *Biochim. Biophys. Acta* **2004**, *1694*, 207–217. [CrossRef]
39. Cornelis, G.R. The type III secretion injectisome. *Nat. Rev. Microbiol.* **2006**, *4*, 811–825. [CrossRef] [PubMed]
40. Galperin, M.Y.; Dibrov, P.A.; Glagolev, A.N. $\Delta\mu$H+ is required for flagellar growth in *Escherichia coli*. *FEBS Lett.* **1982**, *143*, 319–322. [CrossRef]
41. Wilharm, G.; Lehmann, V.; Neumayer, W.; Trcek, J.; Heesemann, J. *Yersinia enterocolitica* type III secretion: Evidence for the ability to transport proteins that are folded prior to secretion. *BMC Microbiol.* **2004**, *4*, 27.
42. Lee, P.C.; Stopford, C.M.; Svenson, A.G.; Rietsch, A. Control of effector export by the *Pseudomonas aeruginosa* type III secretion proteins PcrG and PcrV. *Mol. Microbiol.* **2010**, *75*, 924–941.

43. Ogawa, R.; Abe-Yoshizumi, R.; Kishi, T.; Homma, M.; Kojima, S. Interaction of the C-terminal tail of FliF with FliG from the Na+-driven flagellar motor of *Vibrio alginolyticus*. *J. Bacteriol.* **2015**, *197*, 63.
44. Lynch, M.J.; Levenson, R.; Kim, E.A.; Sircar, R.; Blair, D.F.; Dahlquist, F.W.; Crane, B.R. Co-folding of a FliF-FliG split domain forms the basis of the MS:C ring interface within the bacterial flagellar motor. *Structure* **2017**, *25*, 317–328. [PubMed]
45. Erhardt, M.; Namba, K.; Hughes, K.T. Bacterial nanomachines: The flagellum and type III injectisome. *Cold Spring Harb. Perspect. Biol.* **2010**, *2*, a000299.
46. Galan, J.E.; Lara-Tejero, M.; Marlovits, T.C.; Wagner, S. Bacterial type III secretion systems: Specialized nanomachines for protein delivery into target cells. *Annu. Rev. Microbiol.* **2014**, *68*, 415–438.
47. Miletic, S.; Goessweiner-Mohr, N.; Marlovits, T.C. The structure of the type III secretion system needle complex. In *Bacterial Type III Protein Secretion Systems*; Wagner, S., Galan, J.E., Eds.; Springer: Cham, Switzerland, 2020; pp. 67–90.
48. Minamino, T.; Kawamoto, A.; Kinoshita, M.; Namba, K. Molecular organization and assembly of the export apparatus of flagellar type III secretion systems. In *Bacterial Type III Protein Secretion Systems*; Wagner, S., Galan, J.E., Eds.; Springer: Cham, Switzerland, 2020; pp. 91–107.
49. Marlovits, T.C.; Kubori, T.; Sukhan, A.; Thomas, D.R.; Galán, J.E.; Unger, V.M. Structural insights into the assembly of the type III secretion needle complex. *Science* **2004**, *306*, 1040–1042.
50. Marlovits, T.C.; Kubori, T.; Lara-Tejero, M.; Thomas, D.; Unger, V.M.; Galán, J.E. Assembly of the inner rod determines needle length in the type III secretion injectisome. *Nature* **2006**, *441*, 637–640.
51. Hu, J.; Worrall, L.J.; Hong, C.; Vuckovic, M.; Atkinson, C.E.; Caveney, N.; Yu, Z.; Strynadka, N.C.J. Cryo-EM analysis of the T3S injectisome reveals the structure of the needle and open secretin. *Nat. Commun.* **2018**, *9*, 3840.
52. Torres-Vargas, C.E.; Kronenberger, T.; Roos, N.; Dietsche, T.; Poso, A.; Wagner, S. The inner rod of virulence-associated type III secretion systems constitutes a needle adapter of one helical turn that is deeply integrated into the system's export apparatus. *Mol. Microbiol.* **2019**, *112*, 918–931.
53. Kubori, T.; Sukhan, A.; Aizawa, S.I.; Galán, J.E. Molecular characterization and assembly of the needle complex of the *Salmonella typhimurium* type III protein secretion system. *Proc. Natl. Acad. Sci. USA* **2000**, *97*, 10225–10230. [CrossRef]
54. Lara-Tejero, M.; Kato, J.; Wagner, S.; Liu, X.; Galán, J.E. A sorting platform determines the order of protein secretion in bacterial type III systems. *Science* **2011**, *331*, 1188–1191. [CrossRef]
55. Francis, N.R.; Sosinsky, G.E.; Thomas, D.; DeRosier, D.J. Isolation, characterization and structure of bacterial flagellar motors containing the switch complex. *J. Mol. Biol.* **1994**, *235*, 1261–1270. [CrossRef] [PubMed]
56. Kamiya, R.; Asakura, S.; Wakabayashi, K.; Namba, K. Transition of bacterial flagella from helical to straight forms with different subunit arrangements. *J. Mol. Biol.* **1979**, *131*, 725–742. [CrossRef]
57. Yamashita, I.; Hasegawa, K.; Suzuki, H.; Vonderviszt, F.; Mimori-Kiyosue, Y.; Namba, K. Structure and switching of bacterial flagellar filaments studied by X-ray fiber diffraction. *Nat. Struct. Biol.* **1998**, *5*, 125–132.
58. Renault, T.T.; Guse, A.; Erhardt, M. Export mechanisms and energy transduction in type-III secretion machines. In *Bacterial Type III Protein Secretion Systems*; Wagner, S., Galan, J.E., Eds.; Springer: Cham, Switzerland, 2020; pp. 143–159.
59. Minamino, T.; Macnab, R.M. Components of the *Salmonella flagellar* export apparatus and classification of export substrates. *J. Bacteriol.* **1999**, *181*, 1388–1394. [CrossRef] [PubMed]
60. Minamino, T.; MacNab, R.M. FliH, a soluble component of the type III flagellar export apparatus of Salmonella, forms a complex with FliI and inhibits its ATPase activity. *Mol. Microbiol.* **2000**, *37*, 1494–1503. [CrossRef] [PubMed]
61. Ibuki, T.; Imada, K.; Minamino, T.; Kato, T.; Miyata, T.; Namba, K. Common architecture of the flagellar type III protein export apparatus and F- and V-type ATPases. *Nat. Struct. Mol. Biol.* **2011**, *18*, 277–282. [CrossRef]
62. Bai, F.; Morimoto, Y.V.; Yoshimura, S.D.; Hara, N.; Kami-Ike, N.; Namba, K.; Minamino, T. Assembly dynamics and the roles of FliI ATPase of the bacterial flagellar export apparatus. *Sci. Rep.* **2014**, *4*, 6528. [CrossRef]
63. González-Pedrajo, B.; Minamino, T.; Kihara, M.; Namba, K. Interactions between C ring proteins and export apparatus components: A possible mechanism for facilitating type III protein export. *Mol. Microbiol.* **2006**, *60*, 984–998. [CrossRef]
64. Erhardt, M.; Mertens, M.E.; Fabiani, F.D.; Hughes, K.T. ATPase-independent type-III protein secretion in *Salmonella enterica*. *PLoS Genet.* **2014**, *10*, e1004800. [CrossRef]

65. Lee, P.C.; Zmina, S.E.; Stopford, C.M.; Toska, J.; Rietsch, A. Control of type III secretion activity and substrate specificity by the cytoplasmic regulator PcrG. *Proc. Natl. Acad. Sci. USA* **2014**, *111*, E2027–E2036. [CrossRef]
66. Blair, D.F.; Berg, H.C. The MotA protein of E. coli is a proton-conducting component of the flagellar motor. *Cell* **1990**, *60*, 439–449. [CrossRef]
67. Minamino, T.; Morimoto, Y.V.; Hara, N.; Namba, K. An energy transduction mechanism used in bacterial flagellar type III protein export. *Nat. Commun.* **2011**, *2*, 475. [CrossRef]
68. Bange, G.; Kümmerer, N.; Engel, C.; Bozkurt, G.; Wild, K.; Sinning, I. FlhA provides the adaptor for coordinated delivery of late flagella building blocks to the type III secretion system. *Proc. Natl. Acad. Sci. USA* **2010**, *107*, 11295. [CrossRef]
69. Kinoshita, M.; Hara, N.; Imada, K.; Namba, K.; Minamino, T. Interactions of bacterial flagellar chaperone–substrate complexes with FlhA contribute to co-ordinating assembly of the flagellar filament. *Mol. Microbiol.* **2013**, *90*, 1249–1261. [CrossRef]
70. Abrusci, P.; Vergara-Irigaray, M.; Johnson, S.; Beeby, M.D.; Hendrixson, D.R.; Roversi, P.; Friede, M.E.; Deane, J.E.; Jensen, G.J.; Tang, C.M.; et al. Architecture of the major component of the type III secretion system export apparatus. *Nat. Struct. Mol. Biol.* **2013**, *20*, 99–104. [CrossRef]
71. Iino, T. Assembly of Salmonella flagellin in vitro and in vivo. *J. Supramol. Struct.* **1974**, *2*, 372–384.
72. Aizawa, S.I.; Kubori, T. Bacterial flagellation and cell division. *Genes Cells* **1998**, *3*, 625–634. [CrossRef]
73. Turner, L.; Stern, A.S.; Berg, H.C. Growth of flagellar filaments of *Escherichia coli* is independent of filament length. *J. Bacteriol.* **2012**, *194*, 2437. [CrossRef]
74. Renault, T.T.; Abraham, A.O.; Bergmiller, T.; Paradis, G.; Rainville, S.; Charpentier, E.; Guet, C.C.; Tu, Y.; Namba, K.; Keener, J.P.; et al. Bacterial flagella grow through an injection-diffusion mechanism. *eLife* **2017**, *6*, e23136. [CrossRef]
75. Chen, M.; Zhao, Z.; Yang, J.; Peng, K.; Baker, M.A.B.; Bai, F.; Lo, C.J. Length-dependent flagellar growth of *Vibrio alginolyticus* revealed by real time fluorescent imaging. *eLife* **2017**, *6*, e22140. [CrossRef] [PubMed]
76. Glauert, A.M.; Kerridge, D.; Horne, R.W. The fine structure and mode of attachment of the sheathed flagellum of *Vibrio metchnikovii*. *J. Cell. Biol.* **1963**, *18*, 327–336. [CrossRef]
77. Allen, R.D.; Baumann, P. Structure and arrangement of flagella in species of the genus Beneckea and *Photobacterium fischeri*. *J. Bacteriol.* **1971**, *107*, 295–302. [CrossRef]
78. McCarter, L.L. Polar flagellar motility of the Vibrionaceae. *Microbiol. Mol. Biol. Rev.* **2001**, *65*, 445–462. [CrossRef]
79. Grossart, H.P.; Steward, G.F.; Martinez, J.; Azam, F. A simple, rapid method for demonstrating bacterial flagella. *Appl. Environ. Microbiol.* **2000**, *66*, 3632. [CrossRef]
80. Wu, Y.; Yeh, F.L.; Mao, F.; Chapman, E.R. Biophysical characterization of styryl dye-membrane interactions. *Biophys. J.* **2009**, *97*, 101–109. [CrossRef]
81. Copeland, M.F.; Flickinger, S.T.; Tuson, H.H.; Weibel, D.B. Studying the dynamics of flagella in multicellular communities of *Escherichia coli* by using biarsenical dyes. *Appl. Environ. Microbiol.* **2010**, *76*, 1241–1250. [CrossRef]
82. Zhao, Z.; Zhao, Y.; Zhuang, X.Y.; Lo, W.C.; Baker, M.A.B.; Lo, C.J.; Bai, F. Frequent pauses in *Escherichia coli* flagella elongation revealed by single cell real-time fluorescence imaging. *Nat. Commun.* **2018**, *9*, 1885. [CrossRef]
83. Schmitt, M.; Stark, H. Modelling bacterial flagellar growth. *Europhys. Lett.* **2011**, *96*, 28001. [CrossRef]
84. Stern, A.S.; Berg, H.C. Single-file diffusion of flagellin in flagellar filaments. *Biophys. J.* **2013**, *105*, 182–184. [CrossRef]
85. Evans, L.D.B.; Poulter, S.; Terentjev, E.M.; Hughes, C.; Fraser, G.M. A chain mechanism for flagellum growth. *Nature* **2013**, *504*, 287–290. [CrossRef]
86. Grunenfelder, B.; Tawfilis, S.; Gehrig, S.; Østerås, M.; Eglin, D.; Jenal, U. Identification of the protease and the turnover signal responsible for cell cycle-dependent degradation of the *Caulobacter* FliF motor protein. *J. Bacteriol.* **2004**, *186*, 4960–4971. [CrossRef]
87. Ruby, E.G.; Asato, L.M. Growth and flagellation of *Vibrio fischeri* during initiation of the sepiolid squid light organ symbiosis. *Arch. Microbiol.* **1993**, *159*, 160–167. [CrossRef]
88. Doerges, L.; Kutschera, U. Assembly and loss of the polar flagellum in plant-associated methylobacteria. *Naturwissenschaften* **2014**, *101*, 339–346. [CrossRef]
89. Boehm, A.; Kaiser, M.; Li, H.; Spangler, C.; Kasper, C.A.; Ackermann, M.; Kaever, V.; Sourjik, V.; Roth, V.; Jenal, U. Second messenger-mediated adjustment of bacterial swimming velocity. *Cell* **2010**, *141*, 107–116. [CrossRef]

90. Paul, K.; Nieto, V.; Carlquist, W.C.; Blair, D.F.; Harshey, R.M. The c-di-GMP binding protein YcgR controls flagellar motor direction and speed to affect chemotaxis by a "backstop brake" mechanism. *Mol. Cell.* **2010**, *38*, 128–139. [CrossRef]
91. Fukuoka, H.; Wada, T.; Kojima, S.; Ishijima, A.; Homma, M. Sodium-dependent dynamic assembly of membrane complexes in sodium-driven flagellar motors. *Mol. Microbiol.* **2009**, *71*, 825–835. [CrossRef]
92. Honko, A.N.; Mizel, S.B. Effects of flagellin on innate and adaptive immunity. *Immunol. Res.* **2005**, *33*, 83–101. [CrossRef]
93. Salazar-Gonzalez, R.M.; McSorley, S.J. *Salmonella flagellin*, a microbial target of the innate and adaptive immune system. *Immunol. Lett.* **2005**, *101*, 117–122. [CrossRef]
94. Miao, E.A.; Andersen-Nissen, E.; Warren, S.E.; Aderem, A. TLR5 and Ipaf: Dual sensors of bacterial flagellin in the innate immune system. *Semin. Immunopathol.* **2007**, *29*, 275–288. [CrossRef]
95. Hayashi, F.; Smith, K.D.; Ozinsky, A.; Hawn, T.R.; Yi, E.C.; Goodlett, D.R.; Eng, J.K.; Akira, S.; Underhill, D.M.; Aderem, A. The innate immune response to bacterial flagellin is mediated by Toll-like receptor 5. *Nature* **2001**, *410*, 1099–1103. [CrossRef] [PubMed]
96. Smith, K.D.; Ozinsky, A. Toll-like receptor-5 and the innate immune response to bacterial flagellin. *Curr. Top. Microbiol. Immunol.* **2002**, *270*, 93–108.
97. Smith, K.D.; Andersen-Nissen, E.; Hayashi, F.; Strobe, K.; Bergman, M.A.; Barrett, S.L.; Cookson, B.T.; Aderem, A. Toll-like receptor 5 recognizes a conserved site on flagellin required for protofilament formation and bacterial motility. *Nat. Immunol.* **2003**, *4*, 1247–1253. [CrossRef]
98. Franchi, L.; Amer, A.; Body-Malapel, M.; Kanneganti, T.D.; Ozoren, N.; Jagirdar, R.; Inohara, N.; Vandenabeele, P.; Bertin, J.; Coyle, A.; et al. Cytosolic flagellin requires Ipaf for activation of caspase-1 and interleukin 1beta in salmonella-infected macrophages. *Nat. Immunol.* **2006**, *7*, 576–582. [CrossRef]
99. Miao, E.A.; Alpuche-Aranda, C.M.; Dors, M.; Clark, A.E.; Bader, M.W.; Miller, S.I.; Aderem, A. Cytoplasmic flagellin activates caspase-1 and secretion of interleukin 1β via Ipaf. *Nat. Immunol.* **2006**, *7*, 569–575. [CrossRef]
100. Vogler, A.P.; Homma, M.; Irikura, V.M.; Macnab, R.M. *Salmonella typhimurium* mutants defective in flagellar filament regrowth and sequence similarity of FliI to F0F1, vacuolar, and archaebacterial ATPase subunits. *J. Bacteriol.* **1991**, *173*, 3564–3572. [CrossRef]
101. Rosu, V.; Hughes, K.T. Sigma28-dependent transcription in *Salmonella enterica* is independent of flagellar shearing. *J. Bacteriol.* **2006**, *188*, 5196–5203. [CrossRef]
102. Zhu, S.; Gao, B. Bacterial flagella loss under starvation. *Trends Microbiol.* **2020**, *28*, 785–788. [CrossRef]

Publisher's Note: MDPI stays neutral with regard to jurisdictional claims in published maps and institutional affiliations.

Review

Structural Conservation and Adaptation of the Bacterial Flagella Motor

Brittany L. Carroll [1,2] and Jun Liu [1,2,*]

1 Department of Microbial Pathogenesis, Yale School of Medicine, New Haven, CT 06536, USA; Brittany.Carroll@yale.edu
2 Microbial Sciences Institute, Yale University, West Haven, CT 06516, USA
* Correspondence: jliu@yale.edu

Received: 5 October 2020; Accepted: 27 October 2020; Published: 29 October 2020

Abstract: Many bacteria require flagella for the ability to move, survive, and cause infection. The flagellum is a complex nanomachine that has evolved to increase the fitness of each bacterium to diverse environments. Over several decades, molecular, biochemical, and structural insights into the flagella have led to a comprehensive understanding of the structure and function of this fascinating nanomachine. Notably, X-ray crystallography, cryo-electron microscopy (cryo-EM), and cryo-electron tomography (cryo-ET) have elucidated the flagella and their components to unprecedented resolution, gleaning insights into their structural conservation and adaptation. In this review, we focus on recent structural studies that have led to a mechanistic understanding of flagellar assembly, function, and evolution.

Keywords: bacterial flagellum; cryo-electron tomography; cryo-electron microscopy; molecular motor; structure and function; torque generation; evolution

---

1. Introduction

The flagellum, a complex nanomachine, propels bacteria through media and along surfaces, using an ion gradient across the cytoplasmic membrane (for review [1]). All flagella share basic structural elements, including the filament, hook, and motor (Figure 1A). The filament acts as the propeller guiding the bacterium through space, while the hook acts as a joint transmitting energy from the motor to the filament [2–6]. The motor, or basal body is homologous to the non-flagellar type III secretion system (T3SS) (for review [7]). The filament can present either externally (Figure 1B,C) or periplasmically (Figure 1D). External flagella extend through the outer membrane into the media surrounding the bacterium and can further be categorized as lateral, peritrichous, and polar [8], while periplasmic flagella reside within the periplasmic space and are essential for spirochete motility [9].

The flagella of *Salmonella enterica* (henceforth called *Salmonella*) and *Escherichia coli* possess the best-studied motors, consisting of the membrane/supramembrane (MS) ring, cytoplasmic (C) ring, peptidoglycan (P) ring, lipopolysaccharide (L) ring, rod, stator, and export apparatus. The MS ring (FliF) acts a base upon which the motor sits, and the C ring (FliG, FliM, and FliN) controls the rotation sense [10–16]. The stator generates torque through ion gradients, mainly H^+ (MotA and MotB) and sometimes Na^+ (PomA and PomB), which drives the rotation of the C ring [14,15,17–19]. The rod (FlgB, FlgC, and FlgF, and FlgG) acts as a drive shaft, connecting the MS ring to the hook [20–22], and the L (FlgH) and P (FlgI) rings act as the bushings, providing support to the rotating rod [23]. The export gate complex, (FlhA, FlhB, FliP, FliQ, and FliR) and ATPase complex (FliH, FliI, and FliJ) [24–26] are responsible for the temporal and spatial assembly, ensuring that a functional flagellum is built [27]. Advances in structural biology techniques, specifically cryo-electron microscopy (cryo-EM) and cryo-electron tomography (cryo-ET), have led to the investigation of flagella from many other species, resulting in the identification of conserved and specifically adapted structural features.

Cryo-ET uniquely allows for the visualization of flagellar structures in situ, without the necessity of isolation and purification of the complexes. In this review, we summarize the plethora of structural work that has widened our view of the assembly, adaptation, and evolution of bacterial flagella.

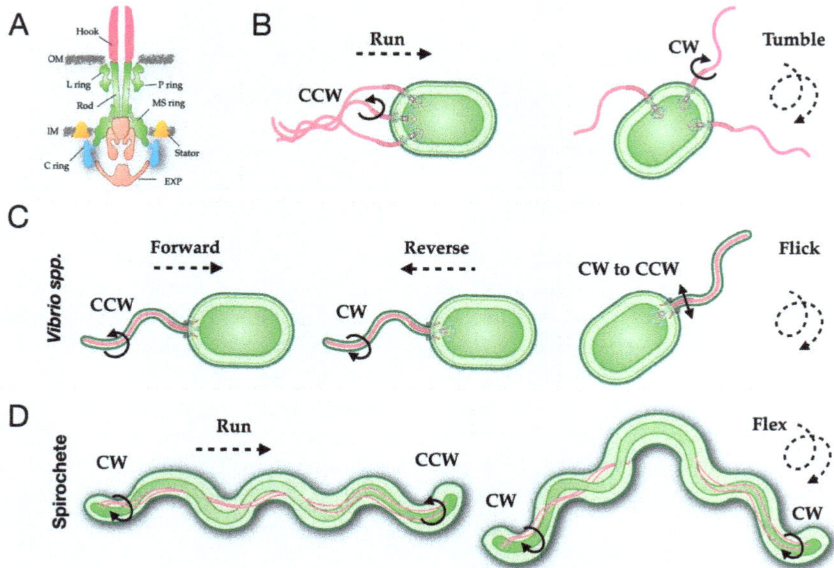

Figure 1. Bacterial flagella control distinct motility. The flagellar motor is a complex nanomachine that drives filament rotation. (**A**) Cartoon model of the flagellar motor. (**B**) In the two-step model used by many species, such as *E. coli* and *Salmonella*, the cell body is propelled forward, or runs, during counterclockwise (looking from the motor to the filament, CCW) rotation, and the filaments form an organized bundle. To change direction, the cell tumbles by rotating the filament in the clockwise (CW) direction, unwinding the bundle. (**C**) *Vibrio* spp. use a three-step method, with CCW rotation moving the cell body forward, CW rotation moving the cell body in reverse, and a flicking motion when CW-to-CCW randomly change direction. (**D**) Spirochetes, with periplasmic flagella at both poles, require a unique two-step method. During the run, the flagella rotate CCW and CW at opposite poles, such that one pole "pulls" while the other "pushes". Both poles rotate in the CW direction while the cell tumbles to change direction.

2. The Bacterial Flagellar Structure

Structural studies have illustrated how the flagellum is assembled and the unique features that have evolved in different species. X-ray crystallography is particularly powerful in unveiling many atomic structures of individual flagellar proteins as well as small subcomplexes (Table 1). These atomic models provide invaluable insight into the individual proteins and protein–protein interactions involved in flagellar assembly and aid in designing functional studies. Recently, cryo-EM has been increasingly utilized to provide both medium- and high-resolution structures of many flagellar subcomplexes, elucidating variable symmetry and complexity of the motor (Table 2). However, the flagellum as an intact organelle is far too complex and flexible for X-ray crystallography and cryo-EM. Cryo-ET coupled with subtomogram averaging [28] has the unique capacity to reveal the entirety of bacterial flagella in multiple species, depicting the relative arrangement of the rings and other protein complexes of the flagella in situ (Table 3). In this section, we review the structural information that not only is conserved but also provides a basis for understanding the functions.

Table 1. Crystal structures of flagellar proteins. A list of the flagellar protein structures deposited in the PDB.

Protein(s)	Species	PDB ID	Refs
		Axial	
Flagellin FliC	*Bacillus cereus*	5Z7Q	[29]
	Salmonella typhimurium	1IO1	[30]
	Sphingaomonas sp	2ZBI, 3K8V, 3K8W	[31]
	Burkholderia psuedomallei	4CFI	[32]
	Pseudomonas aeruginosa	4NX9	[33]
FliS	*Aquifex aeolicus*	1ORY, 1ORJ	[34]
	Bacillus cereus	5XEF	[35]
	Helicobacter pylori	3IQC	[36]
FliT	*Salmonella typhimurium*	5GNA	
	Yersinia enterocolitica	3NKZ	
FljB	*Salmonella typhimurium*	6RGV	[37]
FcpA	*Leptospira biflexa*	6NQY	[38]
FcpB	*Leptospira interrogans*	6NQZ	[38]
Flagellin–FliS	*Bacillus subtilis*	5MAW, 6GOW	[39]
FliC–FliS fusion	*Aquifex aeolicus*	4IWB	[40]
FlgD	*Helicobacter pylori*	4ZZF, 4ZZK, 5K5Y	[41,42]
	Salmonella typhimurium	6IEE, 6IEF	
FlgE	*Campylobacter jejuni*	5AZ4	[43]
	Caulobacter crescentus	5AY6	[43]
	Helicobacter pylori	5NPY	[44]
	Salmonella typhimurium	1WLG	[45]
	Treponema denticola	6NDT, 6NDW, 6NDV, 6NDX	[46]
FlgK	*Campylobacter jejuni*	5XBJ	[47]
	Bacillus cereus	5ZIY	[48]
FlgL	*Xanthomonas campestris*	5ZIZ, 5ZJ0	[48]
	Legionella pneumophila	5YTI	
FliD (HAP2)	*Pseudomonas aeruginosa*	5FHY	[49]
	Helicobacter pylori	6IWY	[50]
FlgG	*Salmonella typhimurium*	6JF2	[51]
FlgJ	*Salmonella typhimurium*	5DN4, 5DN5	[52]
	Basal Body		
FlgA	*Salmonella typhimurium*	3VKI, 3VJP, 3TEE	[53]
FliF–FliG	*Helicobater pylori*	5WUJ	[54]
$FliF_c$–$FliG_N$	*Thermotaoga maritima*	5TDY	[55]
FliG	*Aquifex aeolicus*	3HJL	[56]
	Helicobacter Pylori	3USY, 3USW	[57]
	Thermotoga maritima	1LKV, 1QC7, 3AJC	[58–60]
FliM	*Helicobacter pylori*	4GC8	[61]
	Thermotoga maritima	2HP7	[62]
	Helicobacter pylori	5XRW	[63]
FliN	*Thermotaoga maritima*	1YAB, 1O6A	[64]
FliY	*Thermotoga maritima*	4HYN	[65]
FliG–FliM	*Helicobacter pylori*	4FQ0	[61]
	Thermotaoga maritima	3SOH, 4FHR, 4QRM	[66–68]
FliM–FliN	*Salmonella typhimurium*	4XYB	[69]
FliM–FliN–FliH	*Salmonella typhimurium*	4XYC	[70]
FliM-SpeE	*Helicobacter pylori*	5X0Z	[71]
CheY	*Escherichia coli*	1U8T, 1ZDM, 2B1J, 2ID7, 2ID9, 2IDM, 6TG7	[72,73] [74]
	Thermotoga maritima	4IGA	[75]
CheY3	*Vibrio cholerae*	3TO5, 4H60, 4HNQ, 4JP1, 4LX8	[76]
CheY4	*Vibrio cholerae*	4HNR, 4HNS	[76]
CheY–FliM	*Escherichia coli*	1F4V	[72]
FlhG	*Geobacillus thermodenitrificans*	4RZ2, 4RZ3	[77]
MotB	*Salmonella typhimurium*	5Y3Z, 5Y40, 2ZVY, 2ZVZ, 2ZOV	[78,79]
$PomB_c$	*Vibrio alginolyticus*	3WPW, 3WPX	[80]
MotY	*Vibrio alginolyticus*	2ZF8	[81]
FliL	*Vibrio alginolyticus*	6AHQ, 6AHP	[82]
FlgT	*Vibrio alginolyticus*	3W1E	[83]

Table 1. *Cont.*

Protein(s)	Species	PDB ID	Refs
	Export Apparatus		
FlhA	*Bacillus subtilis*	3MIX	[84]
	Salmonella typhimurium	6CH1, 6AI0, 6AI1, 6AI2, 6AI3	[85,86]
FlhA FliT–FliD complex	*Salmonella typhimurium*	6CH2	[85]
FlhA FliS–FliC complex	*Salmonella typhimurium*	6CH3	[85]
FlhB	*Aquifex aeolicus*	3B1S	[87]
	Salmonella typhimurium	3B0Z	[87]
FlhF	*Bacillus subtilis*	2PX0, 2PX3	[88]
FliI	*Salmonella typhimurium*	2DPY	[89]
FliJ	*Salmonella typhimurium*	3AJW	[90]
FlgN	*Pseudomonas aeruginosa*	2FUP	
	Salmonella typhimurium	5B3D	[91]
FliH–FliI	*Salmonella typhimurium*	5B0O	[92]

Table 2. Cryo-EM structures for flagellar subcomplexes. A list of the cryo-EM maps and models deposited in the EMDB and PDB.

Protein(s)	Species	PDB ID	EMDB ID	Refs
		Axial		
Flagellin	*Campylobacter jejuni*		5007	[93]
	Salmonella typhimurium	1UCU, 3A5X	1641	[94,95]
	Bacillus subtilis	5WJT, 5WJU, 5WJV, 5WJW, 5WJX, 5WJY, 5WJZ	8447, 8848, 8849, 8850, 8851, 8852, 8853	[96]
	Pseudomonas aeruginosa	5WK5, 5WK6	8855, 8856	[96]
	Leptospira biflexa	6PWB	20504	[38]
	Salmonella typhimurium	6JY0	9896	[97]
	Kurthia spp.	6T17	10362	[98]
FlgE	*Helicobacter pylori*	5JXL	8179	[99]
	Caulobacter crescentus	2BGY	1132	[100]
	Salmonella typhimurium	2BGZ, 3A69, 6JZT, 6KFK, 6K3I	1132, 1647, 9974, 9909	[100,101] [51,102,103]
	Salmonella enterica	6K9Q	9952	[104]
FliD (HAP2)	*Escherichia coli*		1873	[105]
FlgG	*Salmonella typhimurium*	6JZR	6683	[51]
		Basal Body		
FliF	*Salmonella typhimurium*		1887	[106]
	Salmonella typhimurium	6SCN, 6SD1, 6SD2, 6SD3, 6SD4, 6SD5, 6TRE	10143, 10145, 10146, 10147, 10148, 10149, 10560, 6715	[107,108]
FliF–FliG	*Salmonella typhimurium*		6716	[108]
MotA	*Aquifex aeolicus*		3417	[109]
MotA/B	*Campylobacter jejuni*	6YKM, 6YKP, 6YKR	10828, 10829, 10830	[110]
	Clostridium sporogenes	6YSF	10895, 10897	[111]
	Bacillus subtilis	6YSL	10899	[111]
PomA/PomB	*Vibrio mimicus*		10901	[111]
		Export Apparatus		
FliPQR	*Salmonella typhimurium*	6R69, 6F2D	4733, 4173	[107]
	Vibrio mimicus	6S3S	10096	[112]
	Pseudomonas savastanoi	6S3R	10095	[112]
FliPQR–FlhB	*Vibrio mimicus*	6S3L	10093	[112]
SctRST	*Salmonella typhimurium*	6R6B	4734	[107]

Table 3. *In situ* flagellar motors visualized by cryo-ET. A list containing the cryo-ET maps of flagellar motors deposited in the EMDB. Note that not all cryo-ET maps are deposited.

Species	EMDB ID	Refs
Acetonema longum	5297	[113]
Arcobacter butzleri	3910	[114]
Borrelia burgdorferi	0525, 0534, 0536, 0537, 0538, 1644, 5298, 5627, 5628, 5629, 5630, 5631, 5632, 5633, 6088, 6089, 6090, 6091, 6092, 6093, 6094, 6095, 6096, 6097, 6098, 9123, 21885, 21884, 21886	[113,115–120]
Bdellovibrio bacteriovorus	3911	[114]
Campylbacter jejuni	3150, 3157, 3158, 3159, 3160, 3161, 5300, 10341, 10342, 10343, 10345, 10454, 10455, 10456, 10457	[113,121,122]
Caulobacter crescentus	5312, 10943, 10945, 10949, 10950, 10955, 10956, 10957	[113,123]
Escherichia coli	5311	[113]
Helicobacter pylori	8459	[57]
Heliobacter Hepaticus	5299	[113]
Hylemonella gracilis	5309	[113]
Hyphomonas neptunium	5313	[113]
Legionella pneumophila	0464	[124]
Leptospira biflexa	20503, 20504	[38]
Leptospira interrogans	5912, 5913, 5914	[6]
Plesiomonas shigelloides	4569, 10057	[125]
Pseudomonas aeruginosa	0465	[124]
Salmonella enterica	2520, 2521, 3154, 3813, 5310	[113,121,126]
Shewanella oneidensis	0467	[124]
Treponema primitia	1235	[127]
Vibrio cholerae	5308	[113]
Vibrio fischeri	3155, 3156, 3162	[121]
Vibrio alginolyticus	21819, 21837	[128]
Wolinella succinogenes	3912	[114]

2.1. The Rod, Hook, and Filament Extend from the Cell Body

The flagellar filament is comprised of 11 protofilaments, each with thousands of repeating units of flagellin (for review [129]). Although, variation of the filament is possible, such as in the case of *Campylobacter jejuni* with 7 protofilaments [93]. The flagellin protein (FliC) has four domains—D0, D1, D2, and D3 [30]—and the protofilaments can adopt both left- and right-handed helical rotations. The filament forms a left-handed supercoil when rotating CCW and a right-handed supercoil during CW rotation, together coined polymorphic switching [130–132]. The Namba group solved atomic models of locked right-handed and left-handed *Salmonella* filaments using cryo-EM, elucidating key interacting regions of the flagellin protein [94,95]. Recently, the *Bacillus subtilis* and *Pseudomonas aeruginosa* locked filaments were revealed by using cryo-EM as well [96]. Importantly, due to improved resolution, Wang et al. were able to predict point mutations involved in polymorphic switching, which will aid future work towards a better understanding of the filament rotation [96].

The hook, composed of ~120 copies of FlgE forming 11 protofilaments, has the critical job of joining the filament to the basal body, requiring a balance of rigidity and flexibility to allow the transfer of energy without breaking [133]. FlgE has 4 domains: D0 forms the channel, D1 forms the middle body, D2 forms the exposed surface, and Dc loops back in towards D0 [103]. Advances in cryo-EM enable high-resolution views of the hook as a bended structure during flagellar rotation [103,104] or the earlier structures that were limited to straight segments [45,94,99]. Different from the two-state model [134], these studies revealed 11 different subunit conformations, suggesting that each protofilament has unique interdomain interactions allowing for compression and extension as necessary during rotation. The super helical pitch of the hook is dependent upon the environment, with a helical pitch of 996 Å at pH 3.5 [103] and 1,290 Å at pH 8 [104], indicating that the environment also plays a role in the supercoiled form.

The rod is the most proximal region of the axial structure and acts as the drive shaft. It can be divided into two regions: the proximal rod contains six monomers of FlgB, FlgC, and FlgF and nine monomers of FliE, and the distal rod contains 26 copies of FlgG [135–137]. Biochemical characterization of the rod proteins suggests that FliE associates with the MS ring [25,137] and also with the proximal protein assembly of FlgB, FlgF, and FlgC [138]. A cryo-ET study looking at flagellar assembly in the spirochete *Borrelia burgdorferi* broke down the assembly of the proximal rod, distal rod, hook, and filament using various deletion mutants, confirming the previous cellular studies [119]. A recent crystal structure of the core fragment of FlgG from *Salmonella* docked into the cryo-EM maps of the distal rod [139] and hook [102] identified the importance of the L-stretch in stabilization of the rod-hook junction [51]. There was also striking similarity between FlgG and FlgE, highlighting the fluidity of the rod-hook junction evolution [51]. Importantly, the static structures of the rod, hook, and filament observed by X-ray and cryo-EM lack the payload stress that occurs during flagellar rotation; therefore, different forces acting on these structures during filament rotation may alter their configuration.

2.2. The Periplasmic P and L Rings Stabilize the Rod

Analogous to the bushing, the P (FlgI) and L (FlgH) rings are located within the periplasmic space and encircle and stabilize the rod [23]. The L ring was thought to catalyze the removal of the rod cap protein, FlgJ [140]. Bioinformatic analysis suggests that the P and L rings are highly conserved yet evolved separately, rather than via horizontal gene transfer [141]. Visualization of the PL rings from a dozen diverse bacterial species further supports their conservation among phyla [141,142]. Recent cryo-ET studies have found that the P and L rings form the outer membrane structures when the flagellum is absent [125,142,143]. These novel structures, identified by different groups, have been called outer-membrane partial flagellar structures, flagellar outer membrane complexes (FOMCs), or PL subcomplexes [125,141–143]. These complexes were suggested to be relics from which flagella have detached or been sheared, as the rod appears to be required for the assembly of the subcomplexes. Furthermore, Zhu et al. did not observe FOMCs in a *P. aeruginosa flgG* mutant, suggesting that the distal rod is necessary for the formation of the FOMCs [143]. Interestingly, the sheathed flagellum (discussed below) of *Vibrio* spp. also possesses the PL subcomplexes [141], and spirochetes and firmicutes lack the L ring and PL ring, respectively [144], raising the question of whether there are still unknown functions of the P and L rings.

2.3. The MS Ring is the Base of the Motor

The MS ring, comprised solely of FliF, sits mainly in the periplasmic space but is anchored to the inner membrane via N- and C-terminal transmembrane helices [16,145]. FliF is a multidomain protein with two transmembrane domains, the ring-building motif domains (RBM) RBM1, RMB2, RBM3a, RBM3b, the β-collar domain, and C-terminal domain [146]. The C-terminal domain of FliF interacts with the N-terminus of the C-ring protein FliG [54,55], and the export gate complex resides within the MS ring [126,147]. A recent cryo-EM structural analysis of the MS ring answered the outstanding question of symmetry mismatch between the MS ring (25-fold) [106,108] and C ring (34-fold) [106,148,149]. The Lea group found that symmetry within the MS ring due to FliF folding creates an inner and outer ring. The export gate complex interacts with the 21/22-fold inner RBM domains, and the outer ring with 33/34-fold symmetry matches that of the C ring [146]. The unique organization of FliF allows the MS ring to grasp the rotor and export gate, acting to stabilize the basal body.

2.4. The C Ring Acts as a Rotor Within the Cytosol

The C ring, a notable structure located in the cytosol, is essential for flagellar rotation and assembly. The overall structure and shape are conserved, while the diameter of the C ring can vary across species [113]. Cryo-EM and cryo-ET studies have shown that *Salmonella* and *E. coli* have C rings with ~34-fold symmetry [148,150], and bacterial species with larger motors, such as ε-proteobacteria [114,121] and spirochetes, possess C rings with higher symmetry [116]. The increased resolution of cryo-ET has confirmed and expanded upon the initial observations of the C ring diameter variation.

Insights into the C ring composition were inferred from the homologous non-flagellar type III secretion system (or injectosome, for review [7]) of *Shigella* [69,151]. Using sequence alignments, mass spectroscopy, and cryo-EM, McDowell et al. suggested that multimerization of a repeating heptamer [151] containing FliG, FliM, and FliN creates a C ring with a spiral base in lieu of the previously postulated hexamer [152,153]. This finding has further been substantiated by bioinformatics techniques, establishing an evolutionary precedent [154] and pseudo-atomic models built into cryo-ET maps [120,128]. FliG, comprised of three domains ($FliG_N$, $FliG_M$, and $FliG_C$), occupies the C ring adjacent to the MS ring and stator, with $FliG_N$ interacting with FliF of the MS ring [54,55,155], and $FliG_C$ interacting with MotA of the stator complex via charged residues [156–158]. FliM also contains three domains with similar nomenclature: $FliM_N$ binds to CheY-P [67,159], $FliM_M$ interacts with $FliG_M$ [160–162], and $FliM_C$ forms a heterodimer with FliN [69,154]. FliN is a single-domain protein that dimerizes with FliM or itself [163,164]. Numerous crystal structures of the C-ring proteins provide critical information on protein–protein interactions (Table 1).

Some species have FliY, a protein with strong sequence homology to FliN and weak homology to FliM [163]. Typically, FliY replaces FliN, but in *Leptospira* and ε-proteobacteria, both FliY and FliN are expressed and necessary for flagellation [63,165]. The crystal structure of the FliN and FliY complex showed that these proteins form a heterodimer [63]. Co-expression and purification showed that *Campylobacter jejuni* FliY interacts with both FliN and FliM, but interestingly, FliN and FliM do not interact in ε-proteobacteria, *Helicobacter pylori*, or *C. jejuni* [63,166,167]. Recently, a detailed study of the *C. jejuni* C ring composition established the distinct roles of FliY and FliN, as they appear to have evolved independently. The FliY and FliM interactions are important for stabilization of FliH, while FliN is necessary for stabilization of the C ring, suggesting that the C ring is composed of a FliG–FliM–FliN–FliY complex in *C. jejuni* [122]. Understanding C ring composition has proven very important in revealing the switching mechanism for controlling the rotational sense (discussed below).

2.5. Torque is Generated by the Stator Through Ion Gradients

The stator complex generates the torque required to rotate the C ring through a proton gradient, although some species use Na^+ ions [5,19,168,169]. Two membrane proteins, MotA and MotB, form the stator complex as the H^+ powered pump, while the Na^+-driven pump assembles from PomA and PomB [170]. The complexity of the stator complex is two-fold: (1) the stator complex undergoes conformational changes to gain functionality, and (2) the stator complex pool is known to be dynamic [171], leading to variations in stator assembly [172–174]. The dynamic nature of the stator complex makes trapping it with the motor during purification difficult. For these reasons, much of our knowledge of the conformational changes during stator assembly has been accumulated through biochemical experiments, although structural information is starting to accumulate [175].

Initially, cryo-EM structures of PomA/PomB and MotA uncovered the shape and organization of a stator subunit but lacked vital information about stator stoichiometry and rotor–stator interactions [109,171,176]. Freeze-fractured micrographs [177,178] and low-resolution cryo-ET [113,117,121,127,179,180] studies show the stator as a stud-like particle, with different species utilizing varying numbers of stators. Two recent high-resolution cryo-EM structures show that MotA:MotB and PomA:PomB exist in a 5:2 ratio [110,111]. Interestingly, one of these cryo-EM studies found very little conformational rearrangement of the stator complex during protonation in *C. jejuni*, using a protonation mimic mutant [110]. A cryo-ET study on *B. burgdorferi* greatly extended the resolution of the stator–C ring complex in situ, as the spirochete-specific collar of *B. burgdorferi* appears to stabilize the stator complexes around the C ring [116]. Mutations in MotB (D24N and D24E) result in non-motile and motile deficient spirochetes, respectively [116]. Furthermore, these mutations alter the number of stators assembled around the C ring; from these variations in stator number, C ring deformation increases with increased torque [116]. Cryo-ET partially resolved the elusive stator of *Vibrio alginolyticus* such that PomB appears to interact with MotX and MotY of the T ring, supporting the idea that the H and T ring help recruit the stators and allow for increased torque [181]. Evidently, bacteria have evolved sophisticated mechanisms to recruit stator complexes, perhaps to control torque necessary for different bacterial motility and behavior.

2.6. A Conserved Mechanism for Flagellar Rotational Switching

Bacteria move forward when the external flagella rotate in the CCW direction and tumble during CW rotation (Figure 1B) [5,56,160,169,182]. Notably, *Vibrio spp.* have a three-stroke swimming pattern, moving forward during CCW rotation, backward during CW rotation, and using a flicking motion upon CW-to-CCW rotation, analogous to the tumble (Figure 1C) [183]. Spirochete's periplasmic flagellar rotation is unique as forward movement occurs when flagella at one pole rotate CCW and the other CW, and tumbling occurs when flagella at both poles rotate in the CW direction (Figure 1D) [120,184–186]. The C ring controls the rotational sense in response to chemical attractions and repellents [12]. A chemotaxis system mediates the rotational sense via cooperative binding of phosphorylated CheY (CheY-P) to FliM, resulting in a CCW motor switching to the CW sense [67,159] (for review see [187]). A co-crystal structure of CheY-P bound to a truncation of $FliM_N$ provided direct evidence of this interaction [72]. The presence of CheY at the C ring has further been confirmed by two recent cryo-ET studies showing GFP-CheY at the outer periphery of the C ring. The first study used GFP-tagged CleD and CheY homolog in *Caulobacter cresecentus* [123], and the second used GFP-tagged CheY in *B. burgdorferi* [120].

The molecular mechanism of the C ring rotational switching has been extensively studied. High-resolution microscopy of fluorescently tagged FliM and FliN provided evidence of a high turnover rate of FliM and a slower but significant turnover of FliN [188–191]. Fluorescent studies of FliM suggest ~34 copies are in CW rotating motors and ~44 copies in CCW rotating motors [189]. It is still unknown what makes FliM appear more stable during CCW rotation. Cryo-EM studies of purified motors do not show the large change suggested by high-resolution light microscopy studies [191] but suggest a slight diameter difference between CCW and CW motors [192]. Two recent cryo-ET studies in *B. burgdorferi* and *V. alginolyticus* revealed the C ring conformational changes during rotational switching in situ [120,128]. These studies suggest that FliG–FliM–FliN stoichiometry remains consistent at 1:1:3 during switching, whereas there is a conformational change of the C ring subunits that leads to the different presentation of FliG to the stator. The stator complexes were resolved in the *B. burgdorferi* motor structure, showing direct evidence for a difference in FliG–MotA interactions between the two rotational senses [120]. Using cryo-EM coupled with functional assays, Santiveri et al. suggest that MotA of the stator unit in *C. jejuni* rotates, specifically in a clockwise direction during protonation [110]. Together, these studies support a new model for the C ring rotational switching, whereby the stator complex rotates in a CW manner, and the differences in the presentation of FliG to the stator complexes change the rotational sense of the C ring [110,111,120,128].

2.7. The Export Apparatus Secretes Flagellar Proteins for Assembly

The export apparatus is responsible for secreting proteins out of the cytoplasm and across the bacterial membranes to form a functional flagellum. Both proton motive force and ATP are utilized to unfold and translocate proteins across the cytoplasmic membrane. The export apparatus is composed of nine proteins: FlhA, FlhB, FliO, FliP, FliQ, FliR, FliH, FliI, and FliJ [24,25,118]. FlhA forms an ion channel [193–196] and has been shown biochemically and genetically to interact with multiple flagella-associated proteins [24,197–199]. FlhB, critical for substrate specificity, regulates the hook length and switching to flagellin secretion for filament assembly via an autocleavage event [200,201]. FliPQR forms the core complex, which is the channel that secretes the proteins [202]. The ATPase complex is formed by the ATPase (FliI), stalk protein (FliJ), and negative regulator (FliH) [7,118].

The first hints of structural and spatial information about the export apparatus came from freeze fracture experiments, establishing the presence of a pore [177,203]. Multiple cryo-ET studies proposed the location of the export apparatus [113,127,179,204]; however, Chen et al. were the first to study the structural detail export apparatus in depth [113]. By comparing flagella from many species, they showed that the export apparatus is highly conserved in shape and location, with a dome feature below the MS ring, a torus, and a spherical structure. A FliI deletion in *C. jejuni* resulted in intact flagella missing the spherical density, solidifying the location of the export apparatus, specifically the ATPase portion [113]. A recent cryo-ET study showed that the ATPase portion of the export apparatus is connected to the C ring via interactions with FliH and likely rotates with the C ring [118]. FliH is a negative regulator of FliI, but exactly how the assembled FliH–FliI complex is regulated is still unknown; the crystal structure, while revealing an intriguing FliH dimer, did not bare the assembled ATPase complex structure [92]. Deletion of *fliH* in *C. jejuni* led to loss of FliI density but still allowed for flagella assembly, providing direct evidence that the ATPase is non-essential for flagella assembly, consistent with biochemical results [122].

A proton channel in FlhA has been shown to be critical for powering the export of flagellar proteins [205–208]. FlhA is the largest protein of the export gate and contains three cytoplasmic domains, CD1 with the FHIPEP motif, a linker domain $FlhA_L$ and a C-terminal domain $FlhA_C$, as well as two transmembrane regions [193,209]. The C-terminal domain, which interacts with the chaperones and export substrate, has been crystallized and studied extensively but lacks structural information for the remaining regions [84–86]. Inferences of the FlhA structure can be drawn from a cryo-ET study of the *Salmonella* non-flagellar T3SS, in which a seahorse-shaped structure was resolved for InvA, the homolog to FlhA [210]. The FliPQR–FlhB complex has recently been resolved in multiple cryo-EM studies, whereby purified FliPQR and FliPQR–FlhB of the export gate complex from *Salmonella* revealed an unexpected topology and orientation of the complex, with no canonical transmembrane regions but rather with a helical structure that sits at the base of the basal body, mainly inside the periplasm [107,112,202]. These studies also confirmed, using native mass spectrometry, that both the flagellar and non-flagellar export gates have a $P_5Q_4R_1$ stoichiometry, and suggest that FlhB is important not only for the regulation of substrate export but also for the opening of the export gate, adding to the complexity [112,202]. A cryo-EM and cryo-ET study of the *Salmonella* non-flagellar T3SS showed that thinning of the membrane around the export apparatus allows the export gate to span the membrane by docking the high-resolution FliPQR structure [210]. The accumulation of information about the export apparatus points towards a complex highly conserved in sequence, structure, assembly, and function, although the molecular mechanism underlying protein secretion remains poorly understood [193].

3. Specific Examples of Adaptation within the Bacterial Flagellum

Since the first intact flagellar motor was visualized in *Treponema primitia*, a spirochete with periplasmic flagella, by cryo-ET [127,179], a thorough investigation of 11 bacterial species using cryo-ET by Chen et al. highlighted the vast differences among flagellar motors, leading to new insights into bacterial evolution [113]. The 'generic' model created by Chen et al., by averaging motors from 11 different species, suggests that the hook, rod, and L, P, and MS rings are highly conserved morphologically. The motors in *Salmonella* and *E. coli* are the best-known examples of the generic model (Figure 2). However, the flagellar motors in other species have evolved unique structural features, presumably to adapt to different environments [113]. In this section, we highlight evolutionary differences by specifically examining three subsets of bacteria: marine *Vibrio*, ε- proteobacteria, and spirochetes.

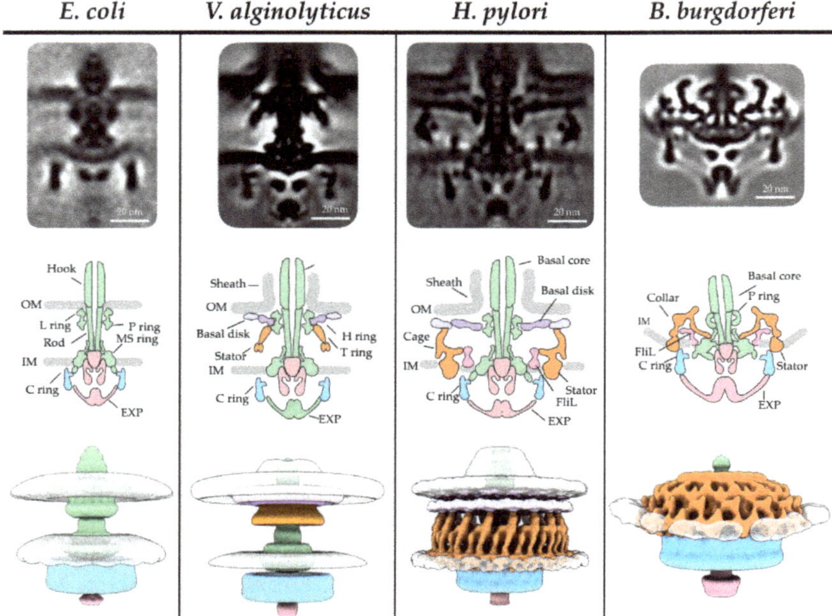

Figure 2. Intact flagellar motor structures reveal dramatic differences among species. Depicted for each species, from top to bottom, are the deposited class average of the motor, a cartoon model drawn from the class average, and a 3D reconstruction of the map. *E. coli* possess the simplest motor, resulting in a functional flagellum (EMDB 5311). *Vibrio spp.* have evolved additional rings that increase rotational speed. *H. pylori* (EMDB 8459), representing ε-proteobacteria, and *B. burgdorferi* (EMDB 0534), representing spirochetes, separately evolved structures that stabilize stators and increase rotor diameter, leading to greater torque generation.

3.1. *Vibrio Flagella Have Additional Rings in the Periplasm for Greater Torque Generation*

Vibrio species are marine bacteria that can cause gastroenteritis in humans via the consumption of contaminated water or seafood or via wound infections from swimming, with the well-known pathogen in this species being *Vibrio cholera*. The *Vibrio* single, polar, and sheathed flagellum has been studied in great detail biochemically (for example [211], for review of sheathed flagellum see [212]). Cryo-ET with STA revealed predominantly sheathed and, to a lesser extent, unsheathed flagella in wild-type *V. alginolyticus* (Figure 2). This allowed for the visualization of the membrane sheath and a novel O-ring structure [180]. A *V. alginolyticus flhG* mutant that assembles multiple polar flagella [213] was used to gain resolution due to more particles per cell pole, and as expected, the sheathed flagellum appears very different from the unsheathed flagella of *V. alginolyticus* and *E. coli*. The diameter at the base of the flagellum was larger due to the membranous sheath, and the loss of the outer membrane–L ring fusion led to more mobility of the basal body. Additional density, named the O ring, was observed outside of the outer membrane, creating a 90° kink in the outer membrane to form the sheath [180].

The *Vibrio spp.* motors also differ from *E. coli* and *Salmonella*, with the identification of the H (FlgT) and T (MotX and MotY) rings believed to have evolved to help the rotor spin faster, and stators that use Na^+ ion pumps in lieu of the more common H^+ ion pump composed of PomA/B [80,214,215]. The T ring was first identified via negative stain EM, whereby the Homma group showed that MotX and MotY form additional density associated with the LP rings and are required for PomA/B localization to the motor [214]. The H ring was later identified as FlgT and located above the T ring via negative stain EM [215]. A *V. fischeri ΔmotB* mutant showed that the stator interacts with the

T ring, allowing for the wider incorporation of the stator relative to *Salmonella* and thus increasing the torque generation [80,83,121]. Further use of the *V. alginolyticus flhG* mutation strain in the presence of Δ*motY* or Δ*motX* suggests that the majority of the T ring is composed of MotY, as the Δ*motY* mutation resulted in the loss of the T ring density, and the Δ*motX* motors appeared relatively unchanged at low resolution. Importantly, the *V. alginolyticus* map revealed 13-fold symmetry of MotY, corresponding to the 13-fold symmetry of MotB in *V. ficsheri* [180]. Cryo-ET of *V. ficheri* Δ*flgP* [121] and *V. alginolyticus* Δ*flgO* and Δ*flgT* [216] mutants suggests that FlgT, FlgO, and FlgP create the proximal, medial, and distal regions of the H ring, respectively. In the *V. ficheri* Δ*flgP*, the stators did not assemble, and in the *V. alginolyticus* Δ*flgT* the flagella were periplasmic. Taken together, these results suggest that the H and T rings, unique to Na^+ ion pump flagella, are required for proper flagellar assembly, stator association, and outer membrane penetration.

3.2. The ε-Proteobacteria Flagellum Cage Traps Additional Stators

H. pylori is a well-known gastrointestinal pathogen that can cause stomach ulcers and cancer. *H. pylori* cells possess unipolar, sheathed flagella which allow the microbe to swim through the stomach mucosal lining and are essential for host infection. The function of the sheath still remains unknown. One possibility is that it protects the filament from the low pH of the stomach. Cryo-ET of the *H. pylori* motor revealed a very large motor ~86 nm in diameter and ~81 nm in height (Figure 2) [57]. The motor consists of the basal body core structures along with a novel periplasmic "cage-like" structure. The cage structure had 18-fold symmetry, with the densities below occupied by the stators [57]. This scaffold likely evolved to secure the 18 stators for the high torque generation needed to swim though the viscous environment of the human stomach mucous [57]. *E. coli* require only 11 stators in their flagella, as identified by total internal reflection fluorescence microscopy (TIRF) [171,174]. Cryo-ET revealed similar stator scaffolds in *C. jejuni* [121] (a gut pathogen that causes food poisoning) and *Wolinella succinogenes* [114] (a cattle rumen commensal) motors, albeit with 17-fold symmetry, suggesting that these microbes possess 17 stators. In *C. jejuni*, deletion mutants Δ*flgP*, Δ*flgQ*, Δ*pflA*, *and* Δ*pflB*, were analyzed by cryo-ET to address questions of motor assembly and the composition of the basal and medial disks. It was determined that FlgP creates the basal disk, FlgQ and PflA create the medial disk, and PflB creates the proximal disk [121]. There is a notable difference in *C. jejuni*, where the medial ring is parallel to the proximal ring and basal disk, contrasting with the perpendicular medial ring in *H. pylori* and *W. succinogenes* [57,114]. These structural difference most likely arise due to the FlgQ sequence diversity [114]. Chaban et al. postulate that the energy demand for such a continuously high stator load may be offset by the nutrient-rich habitat, as all three species are part of the gut flora in animals.

3.3. The Periplasmic Flagella of Spirochetes Uses a Collar to Stabilize Stators

Spirochetes are a unique family of bacteria with distinct morphology and motility. Some of them are known to cause diseases such as leptospirosis (*Leptospira interrogans*), syphilis (*Treponema pallidum*), and Lyme disease (*B. burgdorferi*). The flagella of spirochetes are unique due to the placement of the filament in the periplasmic space; this location has implications for the unique motility, host infection, and cell morphology of spirochetes [217] (for review [9]). From the first visualized in situ structures of the periplasmic flagellar motors in *T. primitia* [127] and *B. burgdorferi* [117,204], it has been readily apparent that the periplasmic flagella have a larger C ring, stator ring, and MS ring than those of *Salmonella* external flagella [12,218]. A spirochete-specific structure, also known as the collar, was identified [127]. The collar structure is approximately 71 nm in diameter and 24 nm in height, meaning the assembly is larger than the C ring [204] (Figure 2).

The composition of the collar has recently been studied using *B. burgdorferi* as the model system [9,217]. To begin assigning *B. burgdorferi* proteins to the collar structure, all known flagellar proteins in *B. burgdorferi* were compared to those of externally flagellated genomes, and (BB0286) FlbB was identified as a potential hit. The Δ*flbB* mutant cells are rod-shaped and non-motile.

Visualization of the Δ*flbB* motors by cryo-ET revealed that the collar did not assemble [219]. Furthermore, when Δ*flbB* was complemented, *flbB* fused with green fluorescent protein (GFP) extra densities near the MS ring were resolved, suggesting that FlbB constitutes the base of the collar and that other proteins must be involved in collar formation [219]. To further identify collar proteins, the *T. pallidum* protein–protein interaction map was used to identify homologs with FlbB and interactors [220]. The protein of unknown function (BB0236) was identified and characterized via molecular and cryo-ET experiments. BB0236 was determined to directly interact with FlbB in pull-down assays. Like the FlbB deletion mutant, Δ*bb0236* resulted in non-motile, rod-shaped bacteria. Cryo-ET showed that BB0236 is necessary for collar formation as well as for FliL and stator assembly, suggesting that BB0236 is a chaperone protein that aids in the formation of the collar, and that the collar provides support for the assembly of the stator and FliL [221]. The most recently identified collar protein was determined by a blast search of the peptidoglycan binding loop of MotB. The gene product of *bb0326* was renamed FlcA. The Δ*flcA* mutant cells exhibited motility and morphology defects. Interestingly, cryo-ET demonstrated that the collar was assembled minus a region at the periphery, where FlcA resides. Density for FliL and FlbB was observed, suggesting that FlcA subsequently binds the collar. The stator was absent from the collar. FlcA was shown to interact with the stator protein MotB and the collar proteins FlbB and FliL, but not with BB0236 [222]. While the story of the spirochetal collar is still unfolding, cryo-ET combined with genetics has elegantly identified three proteins involved in collar assembly and shown the importance of the collar both for stabilization of the stator, by directly binding to MotB and the PG layer, and as a foundation for the stator assembly.

4. Conclusions and Perspectives

Bacterial flagella have evolved as highly versatile nanomachines that enable bacteria to navigate and survive diverse environments such as the mucous of the mammalian gut. Over the last decade, cryo-ET has enabled direct visualization of conservation and adaptation of the bacterial flagellum to niche environments. Cryo-EM and X-ray crystallography have led to near-atomic views of purified flagellar proteins and subcomplexes, such as the MS ring, C ring, and stator complexes. By combining these techniques, it is becoming feasible to establish nearly complete models of the flagellar motor, such as the one shown in Figure 3. High-resolution views of the intact flagellar motor not only significantly enhance our understanding of flagellar structure and assembly but also provide the basis to address fundamental questions about bacterial flagella: How does proton motive force drive the flagellar assembly and rotation? How does the flagellum switch its rotational direction? And how has the flagellum evolved as a highly diverse nanomachine?

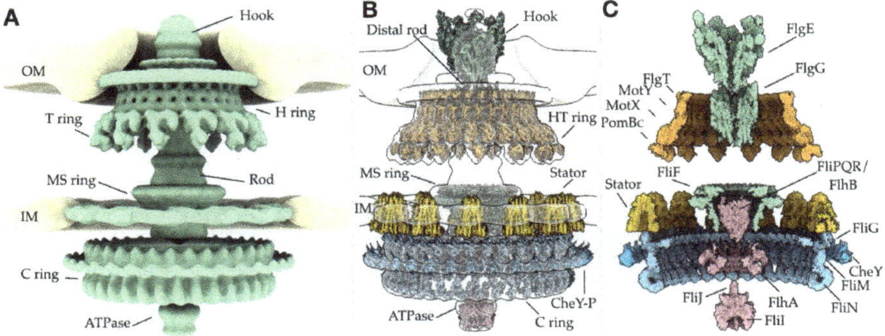

Figure 3. High-resolution cryo-EM and X-ray models placed in cryo-ET maps provide a basis for understanding flagellar assembly and function. (**A**). An assembled cryo-ET map of z motor trapped in the CW rotation (EMDB 3155, 21837, and [143]), depicting the general shape of the molecular components that assemble into the intact motor. (**B**). High-resolution cryo-EM and X-ray structures of the flagellar components are placed in the cryo-ET map (white). The motor is sliced in half to show the inner and outer structures. (**C**). Available high-resolution structures are shown in full. The models used for this reconstruction are: FlgE (PDB 6KFK), FlgG (PDB 6JZR), FlgT (PDB 3W1E), MotY (PDB 2ZF8), MotX (theoretical [181]), PomB$_C$ (PDB 3WPW), FliF (PDB 6SD5), FliPQR–FlhB (PDB 6S3L), FliG (PDB 3HJL and 4FHR), CheY (PDB 1F4V), FliM (PDB 4FHR and 4YXB), FliN (PDB 4YXB and 1YAB), FlhA (PDB 6CH1), FliI (PDB 2DPY), FliJ (PDB 3AJW), and stator (PDB 6YKM).

Author Contributions: B.L.C. wrote the draft, and B.L.C. and J.L. revised the manuscript. All authors have read and agreed to the published version of the manuscript.

Funding: The work in the Liu laboratory was supported in part by grants GM107629 and R01AI087946 from the National Institutes of Health.

Acknowledgments: We thank Jennifer Aronson for critical reading and suggestions.

Conflicts of Interest: The authors declare no conflict of interest.

References

1. Nakamura, S.; Minamino, T. Flagella-Driven Motility of Bacteria. *Biomolecules* **2019**, *9*, 279. [CrossRef]
2. Minamino, T.; Imada, K.; Namba, K. Molecular motors of the bacterial flagella. *Curr. Opin. Struct. Biol.* **2008**, *18*, 693–701. [CrossRef]
3. Pallen, M.J.; Matzke, N.J. From The Origin of Species to the origin of bacterial flagella. *Nat. Rev. Genet.* **2006**, *4*, 784–790. [CrossRef]
4. Terashima, H.; Kawamoto, A.; Morimoto, Y.V.; Imada, K.; Minamino, T. Structural differences in the bacterial flagellar motor among bacterial species. *Biophys. Physicobiology* **2017**, *14*, 191–198. [CrossRef]
5. Terashima, H.; Kojima, S.; Homma, M. Flagellar motility in bacteria structure and function of flagellar motor. *Int. Rev. Cell. Mol. Biol.* **2008**, *270*, 39–85. [CrossRef]
6. Zhao, X.; Norris, S.J.; Liu, J. Molecular Architecture of the Bacterial Flagellar Motor in Cells. *Biochemistry* **2014**, *53*, 4323–4333. [CrossRef]
7. Diepold, A.; Armitage, J.P. Type III secretion systems: The bacterial flagellum and the injectisome. *Philos. Trans. R. Soc. B Biol. Sci.* **2015**, *370*, 20150020. [CrossRef]
8. McCarter, L.L. Dual Flagellar Systems Enable Motility under Different Circumstances. *J. Mol. Microbiol. Biotechnol.* **2004**, *7*, 18–29. [CrossRef]
9. Chang, Y.; Liu, J. Architecture and Assembly of Periplasmic Flagellum. *Microbiol. Spectr.* **2019**, *7*, 10. [CrossRef]
10. Blair, D.F. Flagellar movement driven by proton translocation. *FEBS Lett.* **2003**, *545*, 86–95. [CrossRef]

11. Francis, N.R.; Irikura, V.M.; Yamaguchi, S.; DeRosier, D.J.; Macnab, R.M. Localization of the Salmonella typhimurium flagellar switch protein FliG to the cytoplasmic M-ring face of the basal body. *Proc. Natl. Acad. Sci. USA* **1992**, *89*, 6304–6308. [CrossRef]
12. Francis, N.R.; Sosinsky, G.E.; Thomas, D.; DeRosier, D.J. Isolation, Characterization and Structure of Bacterial Flagellar Motors Containing the Switch Complex. *J. Mol. Biol.* **1994**, *235*, 1261–1270. [CrossRef]
13. Homma, M.; Ohnishi, K.; Iino, T.; Macnab, R.M. Identification of flagellar hook and basal body gene products (FlaFV, FlaFVI, FlaFVII and FlaFVIII) in Salmonella typhimurium. *J. Bacteriol.* **1987**, *169*, 3617–3624. [CrossRef]
14. Sato, K.; Homma, M. Multimeric Structure of PomA, a Component of the Na+-driven Polar Flagellar Motor ofVibrio alginolyticus. *J. Biol. Chem.* **2000**, *275*, 20223–20228. [CrossRef]
15. Sato, K.; Homma, M. Functional Reconstitution of the Na+-driven Polar Flagellar Motor Component ofVibrio alginolyticus. *J. Biol. Chem.* **2000**, *275*, 5718–5722. [CrossRef]
16. Ueno, T.; Oosawa, K.; Aizawa, S.-I. M ring, S ring and proximal rod of the flagellar basal body of Salmonella typhimurium are composed of subunits of a single protein, FliF. *J. Mol. Biol.* **1992**, *227*, 672–677. [CrossRef]
17. Asai, Y.; Kojima, S.; Kato, H.; Nishioka, N.; Kawagishi, I.; Homma, M. Putative channel components for the fast-rotating sodium-driven flagellar motor of a marine bacterium. *J. Bacteriol.* **1997**, *179*, 5104–5110. [CrossRef]
18. Braun, T.F.; Al-Mawsawi, L.Q.; Kojima, A.S.; Blair, D.F. Arrangement of Core Membrane Segments in the MotA/MotB Proton-Channel Complex ofEscherichia coli. *Biochemistry* **2004**, *43*, 35–45. [CrossRef]
19. Kojima, S.; Blair, D.F. The Bacterial Flagellar Motor: Structure and Function of a Complex Molecular Machine. *Int. Rev. Cytol.* **2004**, *233*, 93–134. [CrossRef]
20. Homma, M.; Kutsukake, K.; Hasebe, M.; Iino, T.; Macnab, R.M. FlgB, FlgC, FlgF and FlgG. A family of structurally related proteins in the flagellar basal body of Salmonella typhimurium. *J. Mol. Biol.* **1990**, *211*, 465–477. [CrossRef]
21. Kubori, T.; Shimamoto, N.; Yamaguchi, S.; Namba, K.; Aizawa, S.-I. Morphological pathway of flagellar assembly in Salmonella typhimurium. *J. Mol. Biol.* **1992**, *226*, 433–446. [CrossRef]
22. Minamino, T.; Yamaguchi, S.; Macnab, R.M. Interaction between FliE and FlgB, a Proximal Rod Component of the Flagellar Basal Body ofSalmonella. *J. Bacteriol.* **2000**, *182*, 3029–3036. [CrossRef]
23. Karlinsey, J.; Pease, A.J.; Winkler, M.E.; Bailey, J.L.; Hughes, K.T. The flk gene of Salmonella typhimurium couples flagellar P- and L-ring assembly to flagellar morphogenesis. *J. Bacteriol.* **1997**, *179*, 2389–2400. [CrossRef]
24. Fukumura, T.; Makino, F.; Dietsche, T.; Kinoshita, M.; Kato, T.; Wagner, S.; Namba, K.; Imada, K.; Minamino, T. Assembly and stoichiometry of the core structure of the bacterial flagellar type III export gate complex. *PLoS Biol.* **2017**, *15*, e2002281. [CrossRef]
25. Minamino, T.; Macnab, R.M. Components of the Salmonella Flagellar Export Apparatus and Classification of Export Substrates. *J. Bacteriol.* **1999**, *181*, 1388–1394. [CrossRef]
26. Minamino, T.; Macnab, R.M. Interactions among components of the Salmonella flagellar export apparatus and its substrates. *Mol. Microbiol.* **2000**, *35*, 1052–1064. [CrossRef]
27. Minamino, T. Protein export through the bacterial flagellar type III export pathway. *Biochim. et Biophys. Acta (BBA)-Bioenerg.* **2014**, *1843*, 1642–1648. [CrossRef]
28. Lucic, V.; Rigort, A.; Baumeister, W. Cryo-electron tomography: The challenge of doing structural biology in situ. *J. Cell. Biol.* **2013**, *202*, 407–419. [CrossRef]
29. Kim, M.I.; Lee, C.; Park, J.; Jeon, B.-Y.; Hong, M. Crystal structure of Bacillus cereus flagellin and structure-guided fusion-protein designs. *Sci. Rep.* **2018**, *8*, 5814. [CrossRef]
30. Samatey, F.A.; Imada, K.; Nagashima, S.; Vonderviszt, F.; Kumasaka, T.; Yamamoto, M.; Namba, K. Structure of the bacterial flagellar protofilament and implications for a switch for supercoiling. *Nat. Cell Biol.* **2001**, *410*, 331–337. [CrossRef]
31. Maruyama, Y.; Momma, M.; Mikami, B.; Hashimoto, W.; Murata, K. Crystal Structure of a Novel Bacterial Cell-Surface Flagellin Binding to a Polysaccharide. *Biochemistry* **2008**, *47*, 1393–1402. [CrossRef]
32. Nithichanon, A.; Rinchai, D.; Gori, A.; Lassaux, P.; Peri, C.; Conchillio-Solé, O.; Ferrer-Navarro, M.; Gourlay, L.J.; Nardini, M.; Vila, J.; et al. Sequence- and Structure-Based Immunoreactive Epitope Discovery for Burkholderia pseudomallei Flagellin. *PLoS Neglected Trop. Dis.* **2015**, *9*, e0003917. [CrossRef]
33. Song, W.S.; Yoon, S.-I. Crystal structure of FliC flagellin from Pseudomonas aeruginosa and its implication in TLR5 binding and formation of the flagellar filament. *Biochem. Biophys. Res. Commun.* **2014**, *444*, 109–115. [CrossRef]

34. Evdokimov, A.G.; Phan, J.; Tropea, J.E.; Routzahn, K.M.; Peters, H.K.; Pokross, M.; Waugh, D.S. Similar modes of polypeptide recognition by export chaperones in flagellar biosynthesis and type III secretion. *Nat. Struct. Mol. Biol.* **2003**, *10*, 789–793. [CrossRef]
35. Lee, C.; Kim, M.I.; Park, J.; Jeon, B.-Y.; Yoon, S.-I.; Hong, M. Crystal structure of the flagellar chaperone FliS from Bacillus cereus and an invariant proline critical for FliS dimerization and flagellin recognition. *Biochem. Biophys. Res. Commun.* **2017**, *487*, 381–387. [CrossRef]
36. Lam, W.W.L.; Woo, E.J.; Kotaka, M.; Tam, W.K.; Leung, Y.C.; Ling, T.K.W.; Au, S.W.N. Molecular interaction of flagellar export chaperone FliS and cochaperone HP1076 in Helicobacter pylori. *FASEB J.* **2010**, *24*, 4020–4032. [CrossRef]
37. Horstmann, J.A.; Lunelli, M.; Cazzola, H.; Heidemann, J.; Kühne, C.; Steffen, P.; Szefs, S.; Rossi, C.; Lokareddy, R.K.; Wang, C.; et al. Methylation of Salmonella Typhimurium flagella promotes bacterial adhesion and host cell invasion. *Nat. Commun.* **2020**, *11*, 1–11. [CrossRef]
38. Gibson, K.H.; Trajtenberg, F.; Wunder, E.A.; Brady, M.R.; San Martin, F.; Mechaly, A.; Shang, Z.; Liu, J.; Picardeau, M.; Ko, A.; et al. An asymmetric sheath controls flagellar supercoiling and motility in the leptospira spirochete. *eLife* **2020**, *9*. [CrossRef]
39. Altegoer, F.; Mukherjee, S.; Steinchen, W.; Bedrunka, P.; Linne, U.; Kearns, D.B.; Bange, G. FliS/flagellin/FliW heterotrimer couples type III secretion and flagellin homeostasis. *Sci. Rep.* **2018**, *8*, 11552. [CrossRef]
40. Skorupka, K.; Han, S.K.; Nam, H.-J.; Kim, S.; Faham, S. Protein design by fusion: Implications for protein structure prediction and evolution. *Acta Crystallogr. Sect. D Biol. Crystallogr.* **2013**, *69*, 2451–2460. [CrossRef]
41. Kekez, I.; Cendron, L.; Stojanović, M.; Zanotti, G.; Matković-Čalogović, D. Structure and Stability of FlgD from the Pathogenic 26695 Strain of Helicobacter pylori. *Croat. Chem. Acta* **2016**, *89*, 1–7. [CrossRef]
42. Pulić, I.; Cendron, L.; Salamina, M.; De Laureto, P.P.; Matković-Čalogović, D.; Zanotti, G. Crystal structure of truncated FlgD from the human pathogen Helicobacter pylori. *J. Struct. Biol.* **2016**, *194*, 147–155. [CrossRef]
43. Yoon, Y.-H.; Barker, C.S.; Bulieris, P.V.; Matsunami, H.; Samatey, F.A. Structural insights into bacterial flagellar hooks similarities and specificities. *Sci. Rep.* **2016**, *6*, 35552. [CrossRef]
44. LoConte, V.; Kekez, I.; Matković-Čalogović, D.; Zanotti, G. Structural characterization of FlgE2 protein from Helicobacter pylori hook. *FEBS J.* **2017**, *284*, 4328–4342. [CrossRef]
45. Samatey, F.A.; Matsunami, H.; Imada, K.; Nagashima, S.; Shaikh, T.R.; Thomas, D.R.; Chen, J.Z.; DeRosier, D.J.; Kitao, A.; Namba, K. Structure of the bacterial flagellar hook and implication for the molecular universal joint mechanism. *Nat. Cell Biol.* **2004**, *431*, 1062–1068. [CrossRef]
46. Lynch, M.J.; Miller, M.; James, M.; Zhang, S.; Zhang, K.; Li, C.; Charon, N.W.; Pollack, L. Structure and chemistry of lysinoalanine crosslinking in the spirochaete flagella hook. *Nat. Chem. Biol.* **2019**, *15*, 959–965. [CrossRef]
47. Bulieris, P.V.; Shaikh, N.H.; Freddolino, P.L.; Samatey, F.A. Structure of FlgK reveals the divergence of the bacterial Hook-Filament Junction of Campylobacter. *Sci. Rep.* **2017**, *7*, 15743. [CrossRef]
48. Hong, H.J.; Kim, T.H.; Song, W.S.; Ko, H.-J.; Lee, G.-S.; Kang, S.G.; Kim, P.-H.; Yoon, S.-I. Crystal structure of FlgL and its implications for flagellar assembly. *Sci. Rep.* **2018**, *8*, 1–11. [CrossRef]
49. Postel, S.; Deredge, D.; Bonsor, D.A.; Yu, X.; Diederichs, K.; Helmsing, S.; Vromen, A.; Friedler, A.; Hust, M.; Egelman, E.H.; et al. Bacterial flagellar capping proteins adopt diverse oligomeric states. *eLife* **2016**, *5*, e18857. [CrossRef]
50. Cho, S.Y.; Song, W.S.; Oh, H.-B.; Kim, H.-U.; Jung, H.S.; Yoon, S.-I. Structural analysis of the flagellar capping protein FliD from Helicobacter pylori. *Biochem. Biophys. Res. Commun.* **2019**, *514*, 98–104. [CrossRef]
51. Saijo-Hamano, Y.; Matsunami, H.; Namba, K.; Imada, K. Architecture of the Bacterial Flagellar Distal Rod and Hook of Salmonella. *Biochemistry* **2019**, *9*, 260. [CrossRef]
52. Zaloba, P.; Bailey-Elkin, B.A.; Derksen, M.; Mark, B.L. Structural and Biochemical Insights into the Peptidoglycan Hydrolase Domain of FlgJ from Salmonella typhimurium. *PLoS ONE* **2016**, *11*, e0149204. [CrossRef]
53. Matsunami, H.; Yoon, Y.-H.; Meshcheryakov, V.A.; Namba, K.; Samatey, F.A. Structural flexibility of the periplasmic protein, FlgA, regulates flagellar P-ring assembly in Salmonella enterica. *Sci. Rep.* **2016**, *6*, 27399. [CrossRef]
54. Xue, C.; Lam, K.H.; Zhang, H.; Sun, K.; Lee, S.H.; Chen, X.; Au, S.W.N. Crystal structure of the FliF-FliG complex from Helicobacter pylori yields insight into the assembly of the motor MS-C ring in the bacterial flagellum. *J. Biol. Chem.* **2018**, *293*, 2066–2078. [CrossRef]

55. Lynch, M.J.; Levenson, R.; Kim, E.A.; Sircar, R.; Blair, D.F.; Dahlquist, F.W.; Crane, B.R. Co-Folding of a FliF-FliG Split Domain Forms the Basis of the MS:C Ring Interface within the Bacterial Flagellar Motor. *Struct.* **2017**, *25*, 317–328. [CrossRef]
56. Lee, L.K.; Ginsburg, M.A.; Crovace, C.; Donohoe, M.; Stock, D. Structure of the torque ring of the flagellar motor and the molecular basis for rotational switching. *Nat. Cell Biol.* **2010**, *466*, 996–1000. [CrossRef]
57. Qin, Z.; Lin, W.-T.; Zhu, S.; Franco, A.T.; Liu, J. Imaging the Motility and Chemotaxis Machineries in Helicobacter pylori by Cryo-Electron Tomography. *J. Bacteriol.* **2016**, *199*, e00695-16. [CrossRef]
58. Hou, Y.; Sun, W.; Zhang, C.; Wang, T.; Guo, X.; Wu, L.; Qin, L.; Liu, T. Meta-analysis of the correlation between Helicobacter pylori infection and autoimmune thyroid diseases. *Oncotarget* **2017**, *8*, 115691–115700. [CrossRef]
59. Li, T.H.; Qin, Y.; Sham, P.C.; Lau, K.; Chu, K.-M.; Leung, W.K. Alterations in Gastric Microbiota After H. Pylori Eradication and in Different Histological Stages of Gastric Carcinogenesis. *Sci. Rep.* **2017**, *7*, 44935. [CrossRef]
60. Liu, A.Q.; Xie, Z.; Chen, X.N.; Feng, J.; Chen, J.W.; Qin, F.J.; Ge, L.Y. Fas-associated factor 1 inhibits tumor growth by suppressing Helicobacter pylori-induced activation of NF-kappaB signaling in human gastric carcinoma. *Oncotarget* **2017**, *8*, 7999–8009. [CrossRef]
61. Lam, K.-H.; Lam, W.W.L.; Wong, J.Y.-K.; Chan, L.-C.; Kotaka, M.; Ling, T.K.-W.; Jin, D.-Y.; Ottemann, K.M.; Au, S.W.N. Structural basis of FliG-FliM interaction inHelicobacter pylori. *Mol. Microbiol.* **2013**, *88*, 798–812. [CrossRef]
62. Park, S.-Y.; Lowder, B.; Bilwes, A.M.; Blair, D.F.; Pollack, L. Structure of FliM provides insight into assembly of the switch complex in the bacterial flagella motor. *Proc. Natl. Acad. Sci. USA* **2006**, *103*, 11886–11891. [CrossRef]
63. Lam, K.-H.; Xue, C.; Sun, K.; Zhang, H.; Lam, W.W.L.; Zhu, Z.; Ng, J.T.Y.; Sause, W.E.; Lertsethtakarn, P.; Lau, K.; et al. Three SpoA-domain proteins interact in the creation of the flagellar type III secretion system in Helicobacter pylori. *J. Biol. Chem.* **2018**, *293*, 13961–13973. [CrossRef]
64. Brown, P.N.; Mathews, M.A.A.; Joss, L.A.; Hill, C.P.; Blair, D.F. Crystal Structure of the Flagellar Rotor Protein FliN from Thermotoga maritima. *J. Bacteriol.* **2005**, *187*, 2890–2902. [CrossRef]
65. Sircar, R.; Greenswag, A.R.; Bilwes, A.M.; Gonzalez-Bonet, G.; Crane, B.R. Structure and Activity of the Flagellar Rotor Protein FliY. *J. Biol. Chem.* **2013**, *288*, 13493–13502. [CrossRef]
66. Sircar, R.; Borbat, P.P.; Lynch, M.J.; Bhatnagar, J.; Beyersdorf, M.S.; Halkides, C.J.; Freed, J.H.; Crane, B.R. Assembly States of FliM and FliG within the Flagellar Switch Complex. *J. Mol. Biol.* **2015**, *427*, 867–886. [CrossRef]
67. Vartanian, A.S.; Paz, A.; Fortgang, E.A.; Abramson, J.; Dahlquist, F.W. Structure of Flagellar Motor Proteins in Complex Allows for Insights into Motor Structure and Switching. *J. Biol. Chem.* **2012**, *287*, 35779–35783. [CrossRef]
68. Paul, K.; Gonzalez-Bonet, G.; Bilwes, A.M.; Crane, B.R.; Blair, D.F. Architecture of the flagellar rotor. *EMBO J.* **2011**, *30*, 2962–2971. [CrossRef]
69. Notti, R.Q.; Bhattacharya, S.; Lilic, M.; Stebbins, C.E. A common assembly module in injectisome and flagellar type III secretion sorting platforms. *Nat. Commun.* **2015**, *6*, 7125. [CrossRef]
70. Couturier, C.; Silve, S.; Morales, R.; Pessegue, B.; Llopart, S.; Nair, A.; Bauer, A.; Scheiper, B.; Pöverlein, C.; Ganzhorn, A.; et al. Nanomolar inhibitors of Mycobacterium tuberculosis glutamine synthetase 1: Synthesis, biological evaluation and X-ray crystallographic studies. *Bioorganic Med. Chem. Lett.* **2015**, *25*, 1455–1459. [CrossRef]
71. Zhang, H.; Lam, K.-H.; Lam, W.W.L.; Wong, S.Y.Y.; Chan, V.S.F.; Au, S.W.N. A putative spermidine synthase interacts with flagellar switch protein FliM and regulates motility in Helicobacter pylori. *Mol. Microbiol.* **2017**, *106*, 690–703. [CrossRef]
72. Lee, S.Y.; Cho, H.; Pelton, J.G.; Yan, D.; Henderson, R.K.; King, D.S.; Huang, L.-S.; Kustu, S.; Berry, E.A.; Wemmer, D.E. Crystal structure of an activated response regulator bound to its target. *Nat. Genet.* **2001**, *8*, 52–56. [CrossRef]
73. Dyer, C.M.; Quillin, M.L.; Campos, A.; Lu, J.; McEvoy, M.M.; Hausrath, A.C.; Westbrook, E.M.; Matsumura, P.; Matthews, B.W.; Dahlquist, F.W. Structure of the Constitutively Active Double Mutant CheYD13K Y106W Alone and in Complex with a FliM Peptide. *J. Mol. Biol.* **2004**, *342*, 1325–1335. [CrossRef]
74. Dyer, C.M.; Dahlquist, F.W. Switched or Not?: The Structure of Unphosphorylated CheY Bound to the N Terminus of FliM. *J. Bacteriol.* **2006**, *188*, 7354–7363. [CrossRef]

75. Ahn, D.-R.; Song, H.; Kim, J.; Lee, S.; Park, S. The crystal structure of an activated Thermotoga maritima CheY with N-terminal region of FliM. *Int. J. Biol. Macromol.* **2013**, *54*, 76–83. [CrossRef]
76. Biswas, M.; Dey, S.; Khamrui, S.; Sen, U.; Dasgupta, J. Conformational Barrier of CheY3 and Inability of CheY4 to Bind FliM Control the Flagellar Motor Action in Vibrio cholerae. *PLoS ONE* **2013**, *8*, e73923. [CrossRef]
77. Schuhmacher, J.S.; Rossmann, F.M.; Dempwolff, F.; Knauer, C.; Altegoer, F.; Steinchen, W.; Dörrich, A.K.; Klingl, A.; Stephan, M.; Linne, U.; et al. MinD-like ATPase FlhG effects location and number of bacterial flagella during C-ring assembly. *Proc. Natl. Acad. Sci. USA* **2015**, *112*, 3092–3097. [CrossRef]
78. Kojima, S.; Takao, M.; Almira, G.; Kawahara, I.; Sakuma, M.; Homma, M.; Kojima, C.; Imada, K. The Helix Rearrangement in the Periplasmic Domain of the Flagellar Stator B Subunit Activates Peptidoglycan Binding and Ion Influx. *Struct.* **2018**, *26*, 590–598.e5. [CrossRef]
79. Kojima, S.; Imada, K.; Sakuma, M.; Sudo, Y.; Kojima, C.; Minamino, T.; Homma, M.; Namba, K. Stator assembly and activation mechanism of the flagellar motor by the periplasmic region of MotB. *Mol. Microbiol.* **2009**, *73*, 710–718. [CrossRef]
80. Zhu, S.; Takao, M.; Li, N.; Sakuma, M.; Nishino, Y.; Homma, M.; Kojima, S.; Imada, K. Conformational change in the periplamic region of the flagellar stator coupled with the assembly around the rotor. *Proc. Natl. Acad. Sci. USA* **2014**, *111*, 13523–13528. [CrossRef]
81. Kojima, S.; Shinohara, A.; Terashima, H.; Yakushi, T.; Sakuma, M.; Homma, M.; Namba, K.; Imada, K. Insights into the stator assembly of the Vibrio flagellar motor from the crystal structure of MotY. *Proc. Natl. Acad. Sci. USA* **2008**, *105*, 7696–7701. [CrossRef]
82. Takekawa, N.; Isumi, M.; Terashima, H.; Zhu, S.; Nishino, Y.; Sakuma, M.; Kojima, S.; Homma, M.; Imada, K. Structure of Vibrio FliL, a New Stomatin-like Protein That Assists the Bacterial Flagellar Motor Function. *mBio* **2019**, *10*, e00292-19. [CrossRef]
83. Terashima, H.; Li, N.; Sakuma, M.; Koike, M.; Kojima, S.; Homma, M.; Imada, K. Insight into the assembly mechanism in the supramolecular rings of the sodium-driven Vibrio flagellar motor from the structure of FlgT. *Proc. Natl. Acad. Sci. USA* **2013**, *110*, 6133–6138. [CrossRef]
84. Bange, G.; Kümmerer, N.; Engel, C.; Bozkurt, G.; Wild, K.; Sinning, I. FlhA provides the adaptor for coordinated delivery of late flagella building blocks to the type III secretion system. *Proc. Natl. Acad. Sci. USA* **2010**, *107*, 11295–11300. [CrossRef]
85. Xing, Q.; Shi, K.; Portaliou, A.; Rossi, P.; Economou, A.; Kalodimos, C.G. Structures of chaperone-substrate complexes docked onto the export gate in a type III secretion system. *Nat. Commun.* **2018**, *9*, 1–9. [CrossRef]
86. Inoue, Y.; Ogawa, Y.; Kinoshita, M.; Terahara, N.; Shimada, M.; Kodera, N.; Ando, T.; Namba, K.; Kitao, A.; Imada, K.; et al. Structural Insights into the Substrate Specificity Switch Mechanism of the Type III Protein Export Apparatus. *Struct.* **2019**, *27*, 965–976.e6. [CrossRef]
87. Meshcheryakov, V.A.; Kitao, A.; Matsunami, H.; Samatey, F.A. Inhibition of a type III secretion system by the deletion of a short loop in one of its membrane proteins. *Acta Crystallogr. Sect. D Biol. Crystallogr.* **2013**, *69*, 812–820. [CrossRef]
88. Bange, G.; Petzold, G.; Wild, K.; Parlitz, R.O.; Sinning, I. The crystal structure of the third signal-recognition particle GTPase FlhF reveals a homodimer with bound GTP. *Proc. Natl. Acad. Sci. USA* **2007**, *104*, 13621–13625. [CrossRef]
89. Imada, K.; Minamino, T.; Tahara, A.; Namba, K. Structural similarity between the flagellar type III ATPase FliI and F1-ATPase subunits. *Proc. Natl. Acad. Sci. USA* **2007**, *104*, 485–490. [CrossRef]
90. Ibuki, T.; Imada, K.; Minamino, T.; Kato, T.; Miyata, T.; Namba, K. Common architecture of the flagellar type III protein export apparatus and F- and V-type ATPases. *Nat. Struct. Mol. Biol.* **2011**, *18*, 277–282. [CrossRef]
91. Kinoshita, M.; Nakanishi, Y.; Furukawa, Y.; Namba, K.; Imada, K.; Minamino, T. Rearrangements of α-helical structures of FlgN chaperone control the binding affinity for its cognate substrates during flagellar type III export. *Mol. Microbiol.* **2016**, *101*, 656–670. [CrossRef]
92. Imada, K.; Minamino, T.; Uchida, Y.; Kinoshita, M.; Namba, K. Insight into the flagella type III export revealed by the complex structure of the type III ATPase and its regulator. *Proc. Natl. Acad. Sci. USA* **2016**, *113*, 3633–3638. [CrossRef]
93. Galkin, V.E.; Yu, X.; Bielnicki, J.; Heuser, J.; Ewing, C.P.; Guerry, P.; Egelman, E.H. Divergence of Quaternary Structures Among Bacterial Flagellar Filaments. *Science* **2008**, *320*, 382–385. [CrossRef]
94. Yonekura, K.; Maki-Yonekura, S.; Namba, K. Complete atomic model of the bacterial flagellar filament by electron cryomicroscopy. *Nat. Cell Biol.* **2003**, *424*, 643–650. [CrossRef]

95. Maki-Yonekura, S.; Yonekura, K.; Namba, K. Conformational change of flagellin for polymorphic supercoiling of the flagellar filament. *Nat. Struct. Mol. Biol.* **2010**, *17*, 417–422. [CrossRef]
96. Wang, F.; Burrage, A.M.; Postel, S.; Clark, R.E.; Orlova, A.; Sundberg, E.J.; Kearns, D.B.; Egelman, E.H. A structural model of flagellar filament switching across multiple bacterial species. *Nat. Commun.* **2017**, *8*, 960. [CrossRef]
97. Yamaguchi, T.; Toma, S.; Terahara, N.; Miyata, T.; Ashihara, M.; Minamino, T.; Namba, K.; Kato, T. Structural and Functional Comparison of Salmonella Flagellar Filaments Composed of FljB and FliC. *Biochemistry* **2020**, *10*, 246. [CrossRef]
98. Blum, T.B.; Filippidou, S.; Fatton, M.; Junier, P.; Abrahams, J.P. The wild-type flagellar filament of the Firmicute Kurthia at 2.8 Å resolution in vivo. *Sci. Rep.* **2019**, *9*, 14948. [CrossRef]
99. Matsunami, H.; Barker, C.S.; Yoon, Y.-H.; Wolf, M.; Samatey, F.A. Complete structure of the bacterial flagellar hook reveals extensive set of stabilizing interactions. *Nat. Commun.* **2016**, *7*, 13425. [CrossRef]
100. Shaikh, T.R.; Thomas, D.R.; Chen, J.Z.; Samatey, F.A.; Matsunami, H.; Imada, K.; Namba, K.; DeRosier, D.J. A partial atomic structure for the flagellar hook of Salmonella typhimurium. *Proc. Natl. Acad. Sci. USA* **2005**, *102*, 1023–1028. [CrossRef]
101. Horváth, P.; Kato, T.; Miyata, T.; Namba, K. Structure of Salmonella Flagellar Hook Reveals Intermolecular Domain Interactions for the Universal Joint Function. *Biochemistry* **2019**, *9*, 462. [CrossRef]
102. Fujii, T.; Kato, T.; Namba, K. Specific Arrangement of α-Helical Coiled Coils in the Core Domain of the Bacterial Flagellar Hook for the Universal Joint Function. *Structure* **2009**, *17*, 1485–1493. [CrossRef]
103. Shibata, S.; Matsunami, H.; Aizawa, S.-I.; Wolf, M. Torque transmission mechanism of the curved bacterial flagellar hook revealed by cryo-EM. *Nat. Struct. Mol. Biol.* **2019**, *26*, 941–945. [CrossRef]
104. Kato, T.; Makino, F.; Miyata, T.; Horváth, P.; Namba, K. Structure of the native supercoiled flagellar hook as a universal joint. *Nat. Commun.* **2019**, *10*, 5295–5298. [CrossRef]
105. Maki-Yonekura, S.; Yonekura, K.; Namba, K. Domain movements of HAP2 in the cap-filament complex formation and growth process of the bacterial flagellum. *Proc. Natl. Acad. Sci. USA* **2003**, *100*, 15528–15533. [CrossRef] [PubMed]
106. Thomas, D.R.; Francis, N.R.; Xu, C.; DeRosier, D.J. The Three-Dimensional Structure of the Flagellar Rotor from a Clockwise-Locked Mutant of Salmonella enterica Serovar Typhimurium. *J. Bacteriol.* **2006**, *188*, 7039–7048. [CrossRef]
107. Johnson, S.; Kuhlen, L.; Deme, J.C.; Abrusci, P.; Lea, S.M. The Structure of an Injectisome Export Gate Demonstrates Conservation of Architecture in the Core Export Gate between Flagellar and Virulence Type III Secretion Systems. *mBio* **2019**, *10*, e00818-19. [CrossRef]
108. Suzuki, H.; Yonekura, K.; Namba, K. Structure of the Rotor of the Bacterial Flagellar Motor Revealed by Electron Cryomicroscopy and Single-particle Image Analysis. *J. Mol. Biol.* **2004**, *337*, 105–113. [CrossRef] [PubMed]
109. Takekawa, N.; Terahara, N.; Kato, T.; Gohara, M.; Mayanagi, K.; Hijikata, A.; Onoue, Y.; Kojima, S.; Shirai, T.; Namba, K.; et al. The tetrameric MotA complex as the core of the flagellar motor stator from hyperthermophilic bacterium. *Sci. Rep.* **2016**, *6*, 31526. [CrossRef]
110. Santiveri, M.; Roa-Eguiara, A.; Kühne, C.; Wadhwa, N.; Berg, H.C.; Erhardt, M.; Taylor, N.M.I. Structure and Function of Stator Units of the Bacterial Flagellar Motor. *SSRN Electron. J.* **2020**, *183*, 244–257. [CrossRef]
111. Deme, J.C.; Johnson, S.; Vickery, O.; Muellbauer, A.; Monkhouse, H.; Griffiths, T.; James, R.H.; Berks, B.C.; Coulton, J.W.; Stansfeld, P.J.; et al. Structures of the stator complex that drives rotation of the bacterial flagellum. *Nat. Microbiol.* **2020**, 1–12. [CrossRef]
112. Kuhlen, L.; Johnson, S.; Zeitler, A.; Bäurle, S.; Deme, J.C.; Caesar, J.J.E.; Debo, R.; Fisher, J.; Wagner, S.; Lea, S.M. The substrate specificity switch FlhB assembles onto the export gate to regulate type three secretion. *Nat. Commun.* **2020**, *11*, 1296. [CrossRef] [PubMed]
113. Chen, S.; Beeby, M.; Murphy, G.E.; Leadbetter, J.R.; Hendrixson, D.R.; Briegel, A.; Li, Z.; Shi, J.; Tocheva, I.E.; Müller, A.; et al. Structural diversity of bacterial flagellar motors. *EMBO J.* **2011**, *30*, 2972–2981. [CrossRef] [PubMed]
114. Chaban, B.; Coleman, I.; Beeby, M. Evolution of higher torque in Campylobacter-type bacterial flagellar motors. *Sci. Rep.* **2018**, *8*, 1–11. [CrossRef]

115. Zhang, K.; Qin, Z.; Chang, Y.; Liu, J.; Malkowski, M.G.; Shipa, S.; Li, L.; Qiu, W.; Zhang, J.; Li, C. Analysis of a flagellar filament cap mutant reveals that HtrA serine protease degrades unfolded flagellin protein in the periplasm of Borrelia burgdorferi. *Mol. Microbiol.* **2019**, *111*, 1652–1670. [CrossRef]
116. Chang, Y.; Moon, K.H.; Zhao, X.; Norris, S.J.; Motaleb, A.; Liu, J. Structural insights into flagellar stator–rotor interactions. *eLife* **2019**, *8*. [CrossRef] [PubMed]
117. Kudryashev, M.; Cyrklaff, M.; Wallich, R.; Baumeister, W.; Frischknecht, F. Distinct in situ structures of the Borrelia flagellar motor. *J. Struct. Biol.* **2010**, *169*, 54–61. [CrossRef]
118. Qin, Z.; Tu, J.; Lin, T.; Norris, S.J.; Li, C.; Motaleb, A.; Liu, J. Cryo-electron tomography of periplasmic flagella in Borrelia burgdorferi reveals a distinct cytoplasmic ATPase complex. *PLoS Biol.* **2018**, *16*, e3000050. [CrossRef]
119. Zhao, X.; Zhang, K.; Boquoi, T.; Hu, B.; Motaleb, A.; Miller, K.A.; James, M.E.; Charon, N.W.; Manson, M.D.; Norris, S.J.; et al. Cryoelectron tomography reveals the sequential assembly of bacterial flagella in Borrelia burgdorferi. *Proc. Natl. Acad. Sci. USA* **2013**, *110*, 14390–14395. [CrossRef]
120. Chang, Y.; Zhang, K.; Carroll, B.; Zhao, X.; Charon, N.W.; Norris, S.J.; Motaleb, A.; Li, C.; Liu, J. Molecular mechanism for rotational switching of the bacterial flagellar motor. *Nat. Struct. Mol. Biol.* **2020**, 1–7. [CrossRef]
121. Beeby, M.; Ribardo, D.A.; Brennan, C.A.; Ruby, E.G.; Jensen, G.J.; Hendrixson, D.R. Diverse high-torque bacterial flagellar motors assemble wider stator rings using a conserved protein scaffold. *Proc. Natl. Acad. Sci. USA* **2016**, *113*, E1917–E1926. [CrossRef] [PubMed]
122. Henderson, L.D.; Matthews-Palmer, T.R.S.; Gulbronson, C.J.; Ribardo, D.A.; Beeby, M.; Hendrixson, D.R. Diversification of Campylobacter jejuni Flagellar C-Ring Composition Impacts Its Structure and Function in Motility, Flagellar Assembly, and Cellular Processes. *mBio* **2020**, *11*. [CrossRef]
123. Rossmann, F.M.; Hug, I.; Sangermani, M.; Jenal, U.; Beeby, M. In situ structure of the Caulobacter crescentus flagellar motor and visualization of binding of a CheY-homolog. *Mol. Microbiol.* **2020**, *114*, 443–453. [CrossRef]
124. Kaplan, M.; Ghosal, D.; Subramanian, P.; Oikonomou, C.M.; Kjaer, A.; Pirbadian, S.; Ortega, D.R.; Briegel, A.; El-Naggar, M.Y.; Jensen, G.J. The presence and absence of periplasmic rings in bacterial flagellar motors correlates with stator type. *eLife* **2019**, *8*. [CrossRef] [PubMed]
125. Ferreira, J.L.; Gao, F.Z.; Rossmann, F.M.; Nans, A.; Brenzinger, S.; Hosseini, R.; Wilson, A.; Briegel, A.; Thormann, K.M.; Rosenthal, P.B.; et al. gamma-proteobacteria eject their polar flagella under nutrient depletion, retaining flagellar motor relic structures. *PLoS Biol.* **2019**, *17*, e3000165. [CrossRef] [PubMed]
126. Kawamoto, A.; Morimoto, Y.V.; Miyata, T.; Minamino, T.; Hughes, K.T.; Kato, T.; Namba, K. Common and distinct structural features of Salmonella injectisome and flagellar basal body. *Sci. Rep.* **2013**, *3*, 3369. [CrossRef]
127. Murphy, G.E.; Leadbetter, J.R.; Jensen, G.J. In situ structure of the complete Treponema primitia flagellar motor. *Nat. Cell Biol.* **2006**, *442*, 1062–1064. [CrossRef]
128. Carroll, B.; Nishikino, T.; Guo, W.; Zhu, S.; Kojima, S.; Homma, M.; Liu, J. The flagellar motor of Vibrio alginolyticus undergoes major structural remodeling during rotational switching. *eLife* **2020**, *9*. [CrossRef] [PubMed]
129. Imada, K. Bacterial flagellar axial structure and its construction. *Biophys. Rev.* **2017**, *10*, 559–570. [CrossRef]
130. Turner, L.; Ryu, W.S.; Berg, H.C. Real-Time Imaging of Fluorescent Flagellar Filaments. *J. Bacteriol.* **2000**, *182*, 2793–2801. [CrossRef]
131. Macnab, R.M.; Ornston, M.K. Normal-to-curly flagellar transitions and their role in bacterial tumbling. Stabilization of an alternative quaternary structure by mechanical force. *J. Mol. Biol.* **1977**, *112*, 1–30. [CrossRef]
132. Macnab, R.M.; Koshland, D.E. The Gradient-Sensing Mechanism in Bacterial Chemotaxis. *Proc. Natl. Acad. Sci. USA* **1972**, *69*, 2509–2512. [CrossRef]
133. Berg, H.C.; Anderson, R.A. Bacteria Swim by Rotating their Flagellar Filaments. *Nat. Cell Biol.* **1973**, *245*, 380–382. [CrossRef] [PubMed]
134. Calladine, C.R. Construction of bacterial flagella. *Nat. Cell Biol.* **1975**, *255*, 121–124. [CrossRef] [PubMed]
135. Chevance, F.F.; Takahashi, N.; Karlinsey, J.E.; Gnerer, J.; Hirano, T.; Samudrala, R.; Aizawa, S.-I.; Hughes, K.T. The mechanism of outer membrane penetration by the eubacterial flagellum and implications for spirochete evolution. *Genes Dev.* **2007**, *21*, 2326–2335. [CrossRef] [PubMed]
136. Jones, C.J.; Macnab, R.M.; Okino, H.; Aizawa, S.-I. Stoichiometric analysis of the flagellar hook-(basal-body) complex of Salmonella typhimurium. *J. Mol. Biol.* **1990**, *212*, 377–387. [CrossRef]

137. Müller, V.; Jones, C.J.; Kawagishi, I.; Aizawa, S.; Macnab, R.M. Characterization of the fliE genes of Escherichia coli and Salmonella typhimurium and identification of the FliE protein as a component of the flagellar hook-basal body complex. *J. Bacteriol.* **1992**, *174*, 2298–2304. [CrossRef]
138. Osorio-Valeriano, M.; De La Mora, J.; Camarena, L.; Dreyfus, G. Biochemical Characterization of the Flagellar Rod Components of Rhodobacter sphaeroides: Properties and Interactions. *J. Bacteriol.* **2015**, *198*, 544–552. [CrossRef]
139. Fujii, T.; Kato, T.; Hiraoka, K.D.; Miyata, T.; Minamino, T.; Chevance, F.F.V.; Hughes, K.T.; Namba, K. Identical folds used for distinct mechanical functions of the bacterial flagellar rod and hook. *Nat. Commun.* **2017**, *8*, 14276. [CrossRef]
140. Cohen, E.J.; Hughes, K.T. Rod-to-Hook Transition for Extracellular Flagellum Assembly Is Catalyzed by the L-Ring-Dependent Rod Scaffold Removal. *J. Bacteriol.* **2014**, *196*, 2387–2395. [CrossRef]
141. Kaplan, M.; Sweredoski, M.J.; Rodrigues, J.P.G.L.M.; Tocheva, E.I.; Chang, Y.-W.; Ortega, D.R.; Beeby, M.; Jensen, G.J. Bacterial flagellar motor PL-ring disassembly subcomplexes are widespread and ancient. *Proc. Natl. Acad. Sci. USA* **2020**, *117*, 8941–8947. [CrossRef]
142. Kaplan, M.; Subramanian, P.; Ghosal, D.; Oikonomou, C.M.; Pirbadian, S.; Starwalt-Lee, R.; Mageswaran, S.K.; Ortega, D.R.; Gralnick, J.A.; El-Naggar, M.Y.; et al. In situ imaging of the bacterial flagellar motor disassembly and assembly processes. *EMBO J.* **2019**, *38*, e100957. [CrossRef] [PubMed]
143. Zhu, S.; Schniederberend, M.; Zhitnitsky, D.; Jain, R.; Galán, J.E.; Kazmierczak, B.I.; Liu, J. In Situ Structures of Polar and Lateral Flagella Revealed by Cryo-Electron Tomography. *J. Bacteriol.* **2019**, *201*. [CrossRef]
144. Liu, R.; Ochman, H. Stepwise formation of the bacterial flagellar system. *Proc. Natl. Acad. Sci. USA* **2007**, *104*, 7116–7121. [CrossRef] [PubMed]
145. Ueno, T.; Oosawa, K.; Aizawa, S.-I. Domain Structures of the MS Ring Component Protein (FliF) of the Flagellar Basal Body of Salmonella typhimurium. *J. Mol. Biol.* **1994**, *236*, 546–555. [CrossRef] [PubMed]
146. Johnson, S.; Fong, Y.H.; Deme, J.C.; Furlong, E.J.; Kuhlen, L.; Lea, S.M. Symmetry mismatch in the MS-ring of the bacterial flagellar rotor explains the structural coordination of secretion and rotation. *Nat. Microbiol.* **2020**, *5*, 966–975. [CrossRef]
147. Wagner, S.; Königsmaier, L.; Lara-Tejero, M.; Lefebre, M.; Marlovits, T.C.; Galán, J.E. Organization and coordinated assembly of the type III secretion export apparatus. *Proc. Natl. Acad. Sci. USA* **2010**, *107*, 17745–17750. [CrossRef]
148. Thomas, D.R.; Morgan, D.G.; DeRosier, D.J. Rotational symmetry of the C ring and a mechanism for the flagellar rotary motor. *Proc. Natl. Acad. Sci. USA* **1999**, *96*, 10134–10139. [CrossRef]
149. Young, H.S.; Dang, H.; Lai, Y.; DeRosier, D.J.; Khan, S. Variable Symmetry in Salmonella typhimurium Flagellar Motors. *Biophys. J.* **2003**, *84*, 571–577. [CrossRef]
150. Zhao, R.; Amsler, C.D.; Matsumura, P.; Khan, S. FliG and FliM distribution in the Salmonella typhimurium cell and flagellar basal bodies. *J. Bacteriol.* **1996**, *178*, 258–265. [CrossRef]
151. McDowell, M.A.; Marcoux, J.; McVicker, G.; Johnson, S.; Fong, Y.H.; Stevens, R.; Bowman, L.A.H.; Degiacomi, M.T.; Yan, J.; Wise, A.; et al. Characterisation ofShigella Spa33 andThermotoga FliM/N reveals a new model for C-ring assembly in T3SS. *Mol. Microbiol.* **2015**, *99*, 749–766. [CrossRef]
152. Sarkar, M.K.; Paul, K.; Blair, D.F. Subunit Organization and Reversal-associated Movements in the Flagellar Switch of Escherichia coli. *J. Biol. Chem.* **2009**, *285*, 675–684. [CrossRef] [PubMed]
153. Bai, F.; Branch, R.W.; Nicolau, D.V.; Pilizota, T.; Steel, B.C.; Maini, P.K.; Berry, R.M. Conformational Spread as a Mechanism for Cooperativity in the Bacterial Flagellar Switch. *Science* **2010**, *327*, 685–689. [CrossRef] [PubMed]
154. Dos Santos, R.N.; Khan, S.; Morcos, F. Characterization of C-ring component assembly in flagellar motors from amino acid coevolution. *R. Soc. Open Sci.* **2018**, *5*, 171854. [CrossRef] [PubMed]
155. Ogawa, R.; Abe-Yoshizumi, R.; Kishi, T.; Homma, M.; Kojima, S. Interaction of the C-Terminal Tail of FliF with FliG from the Na+-Driven Flagellar Motor of Vibrio alginolyticus. *J. Bacteriol.* **2014**, *197*, 63–72. [CrossRef] [PubMed]
156. Lloyd, S.A.; Blair, D.F. Charged residues of the rotor protein FliG essential for torque generation in the flagellar motor of Escherichia coli. *J. Mol. Biol.* **1997**, *266*, 733–744. [CrossRef] [PubMed]
157. Yakushi, T.; Yang, J.; Fukuoka, H.; Homma, M.; Blair, D.F. Roles of Charged Residues of Rotor and Stator in Flagellar Rotation: Comparative Study using H+-Driven and Na+-Driven Motors in Escherichia coli. *J. Bacteriol.* **2006**, *188*, 1466–1472. [CrossRef] [PubMed]

158. Takekawa, N.; Kojima, S.; Homma, M. Contribution of Many Charged Residues at the Stator-Rotor Interface of the Na+-Driven Flagellar Motor to Torque Generation in Vibrio alginolyticus. *J. Bacteriol.* **2014**, *196*, 1377–1385. [CrossRef]
159. Paul, K.; Brunstetter, D.; Titen, S.; Blair, D.F. A molecular mechanism of direction switching in the flagellar motor of Escherichia coli. *Proc. Natl. Acad. Sci. USA* **2011**, *108*, 17171–17176. [CrossRef]
160. Minamino, T.; Imada, K.; Kinoshita, M.; Nakamura, S.; Morimoto, Y.V.; Namba, K. Structural Insight into the Rotational Switching Mechanism of the Bacterial Flagellar Motor. *PLoS Biol.* **2011**, *9*, e1000616. [CrossRef]
161. Brown, P.N.; Hill, C.P.; Blair, D.F. Crystal structure of the middle and C-terminal domains of the flagellar rotor protein FliG. *EMBO J.* **2002**, *21*, 3225–3234. [CrossRef] [PubMed]
162. Brown, P.N.; Terrazas, M.; Paul, K.; Blair, D.F. Mutational Analysis of the Flagellar Protein FliG: Sites of Interaction with FliM and Implications for Organization of the Switch Complex. *J. Bacteriol.* **2006**, *189*, 305–312. [CrossRef] [PubMed]
163. Mathews, M.A.A.; Tang, H.L.; Blair, D.F. Domain Analysis of the FliM Protein ofEscherichia coli. *J. Bacteriol.* **1998**, *180*, 5580–5590. [CrossRef]
164. Paul, K.; Blair, D.F. Organization of FliN Subunits in the Flagellar Motor of Escherichia coli. *J. Bacteriol.* **2006**, *188*, 2502–2511. [CrossRef]
165. Lowenthal, A.C.; Hill, M.; Sycuro, L.K.; Mehmood, K.; Salama, N.R.; Ottemann, K.M. Functional Analysis of the Helicobacter pylori Flagellar Switch Proteins. *J. Bacteriol.* **2009**, *191*, 7147–7156. [CrossRef] [PubMed]
166. Häuser, R.; Ceol, A.; Rajagopala, S.V.; Mosca, R.; Siszler, G.; Wermke, N.; Sikorski, P.; Schwarz, F.; Schick, M.; Wuchty, S.; et al. A Second-generation Protein–Protein Interaction Network ofHelicobacter pylori. *Mol. Cell. Proteom.* **2014**, *13*, 1318–1329. [CrossRef] [PubMed]
167. Parrish, J.R.; Yu, J.; Liu, G.; Hines, J.A.; Chan, J.E.; Mangiola, B.A.; Zhang, H.; Pacifico, S.; Fotouhi, F.; DiRita, V.J.; et al. A proteome-wide protein interaction map for Campylobacter jejuni. *Genome Biol.* **2007**, *8*, 1–19. [CrossRef]
168. Li, N.; Kojima, S.; Homma, M. Sodium-driven motor of the polar flagellum in marine bacteria Vibrio. *Genes Cells* **2011**, *16*, 985–999. [CrossRef]
169. Berg, H.C. The Rotary Motor of Bacterial Flagella. *Annu. Rev. Biochem.* **2003**, *72*, 19–54. [CrossRef]
170. Kojima, S.; Blair, D.F. Solubilization and Purification of the MotA/MotB Complex ofEscherichia coli. *Biochemistry* **2004**, *43*, 26–34. [CrossRef]
171. Leake, M.C.; Chandler, J.H.; Wadhams, G.H.; Bai, F.; Berry, R.M.; Armitage, J.P. Stoichiometry and turnover in single, functioning membrane protein complexes. *Nat. Cell Biol.* **2006**, *443*, 355–358. [CrossRef]
172. Blair, D.F.; Berg, H.C. Restoration of torque in defective flagellar motors. *Science* **1988**, *242*, 1678–1681. [CrossRef] [PubMed]
173. Block, S.M.; Berg, H.C. Successive incorporation of force-generating units in the bacterial rotary motor. *Nat. Cell Biol.* **1984**, *309*, 470–472. [CrossRef] [PubMed]
174. Reid, S.W.; Leake, M.C.; Chandler, J.H.; Lo, C.-J.; Armitage, J.P.; Berry, R.M. The maximum number of torque-generating units in the flagellar motor of Escherichia coli is at least 11. *Proc. Natl. Acad. Sci. USA* **2006**, *103*, 8066–8071. [CrossRef]
175. Baker, A.E.; O'Toole, G.A. Bacteria, Rev Your Engines: Stator Dynamics Regulate Flagellar Motility. *J. Bacteriol.* **2017**, *199*, e00088-17. [CrossRef] [PubMed]
176. Yonekura, K.; Maki-Yonekura, S.; Homma, M. Structure of the Flagellar Motor Protein Complex PomAB: Implications for the Torque-Generating Conformation. *J. Bacteriol.* **2011**, *193*, 3863–3870. [CrossRef]
177. Coulton, J.W.; Murray, R.G. Cell envelope associations of Aquaspirillum serpens flagella. *J. Bacteriol.* **1978**, *136*, 1037–1049. [CrossRef]
178. Khan, S.; Dapice, M.; Reese, T.S. Effects of mot gene expression on the structure of the flagellar motor. *J. Mol. Biol.* **1988**, *202*, 575–584. [CrossRef]
179. Liu, J.; Howell, J.K.; Bradley, S.D.; Zheng, Y.; Zhou, Z.H.; Norris, S.J. Cellular Architecture of Treponema pallidum: Novel Flagellum, Periplasmic Cone, and Cell Envelope as Revealed by Cryo Electron Tomography. *J. Mol. Biol.* **2010**, *403*, 546–561. [CrossRef]
180. Zhu, S.; Nishikino, T.; Hu, B.; Kojima, S.; Homma, M.; Liu, J. Molecular architecture of the sheathed polar flagellum in Vibrio alginolyticus. *Proc. Natl. Acad. Sci. USA* **2017**, *114*, 10966–10971. [CrossRef]

181. Zhu, S.; Nishikino, T.; Takekawa, N.; Terashima, H.; Kojima, S.; Imada, K.; Homma, M.; Liu, J. In Situ Structure of the Vibrio Polar Flagellum Reveals a Distinct Outer Membrane Complex and Its Specific Interaction with the Stator. *J. Bacteriol.* **2019**, *202*. [CrossRef] [PubMed]
182. Chevance, F.F.V.; Hughes, K.T. Coordinating assembly of a bacterial macromolecular machine. *Nat. Rev. Genet.* **2008**, *6*, 455–465. [CrossRef]
183. Xie, L.; Altindal, T.; Chattopadhyay, S.; Wu, X.L. From the Cover: Bacterial flagellum as a propeller and as a rudder for efficient chemotaxis. *Proc. Natl. Acad. Sci. USA* **2011**, *108*, 2246–2251. [CrossRef]
184. Charon, N.W.; Cockburn, A.; Li, C.; Liu, J.; Miller, K.A.; Miller, M.R.; Motaleb, A.; Wolgemuth, C.W. The Unique Paradigm of Spirochete Motility and Chemotaxis. *Annu. Rev. Microbiol.* **2012**, *66*, 349–370. [CrossRef]
185. Charon, N.W.; Goldstein, S.F. Genetics of Motility and Chemotaxis of a Fascinating Group of Bacteria: The Spirochetes. *Annu. Rev. Genet.* **2002**, *36*, 47–73. [CrossRef] [PubMed]
186. Goldstein, S.F.; Buttle, K.F.; Charon, N.W. Structural analysis of the Leptospiraceae and Borrelia burgdorferi by high-voltage electron microscopy. *J. Bacteriol.* **1996**, *178*, 6539–6545. [CrossRef] [PubMed]
187. Minamino, T.; Kinoshita, M.; Namba, K. Directional Switching Mechanism of the Bacterial Flagellar Motor. *Comput. Struct. Biotechnol. J.* **2019**, *17*, 1075–1081. [CrossRef]
188. Branch, R.W.; Sayegh, M.N.; Shen, C.; Nathan, V.S.; Berg, H.C. Adaptive Remodelling by FliN in the Bacterial Rotary Motor. *J. Mol. Biol.* **2014**, *426*, 3314–3324. [CrossRef]
189. Lele, P.P.; Branch, R.W.; Nathan, V.S.J.; Berg, H.C. Mechanism for adaptive remodeling of the bacterial flagellar switch. *Proc. Natl. Acad. Sci. USA* **2012**, *109*, 20018–20022. [CrossRef]
190. Delalez, N.J.; Wadhams, G.H.; Rosser, G.; Xue, Q.; Brown, M.T.; Dobbie, I.M.; Berry, R.M.; Leake, M.C.; Armitage, J.P. Signal-dependent turnover of the bacterial flagellar switch protein FliM. *Proc. Natl. Acad. Sci. USA* **2010**, *107*, 11347–11351. [CrossRef]
191. Delalez, N.J.; Berry, R.M.; Armitage, J.P. Stoichiometry and Turnover of the Bacterial Flagellar Switch Protein FliN. *mBio* **2014**, *5*, e01216-14. [CrossRef] [PubMed]
192. Sakai, T.; Miyata, T.; Terahara, N.; Mori, K.; Inoue, Y.; Morimoto, Y.V.; Kato, T.; Namba, K.; Minamino, T. Novel Insights into Conformational Rearrangements of the Bacterial Flagellar Switch Complex. *mBio* **2019**, *10*, e00079-19. [CrossRef] [PubMed]
193. Erhardt, M.; Wheatley, P.; Kim, E.A.; Hirano, T.; Zhang, Y.; Sarkar, M.K.; Hughes, K.T.; Blair, D.F. Mechanism of type-III protein secretion: Regulation of FlhA conformation by a functionally critical charged-residue cluster. *Mol. Microbiol.* **2017**, *104*, 234–249. [CrossRef] [PubMed]
194. Minamino, T.; Morimoto, Y.V.; Hara, N.; Aldridge, P.D.; Namba, K. The Bacterial Flagellar Type III Export Gate Complex Is a Dual Fuel Engine That Can Use Both H+ and Na+ for Flagellar Protein Export. *PLoS Pathog.* **2016**, *12*, e1005495. [CrossRef]
195. Minamino, T.; Morimoto, Y.V.; Hara, N.; Namba, K. An energy transduction mechanism used in bacterial flagellar type III protein export. *Nat. Commun.* **2011**, *2*, 475. [CrossRef]
196. Morimoto, Y.V.; Kami-Ike, N.; Miyata, T.; Kawamoto, A.; Kato, T.; Namba, K.; Minamino, T. High-Resolution pH Imaging of Living Bacterial Cells To Detect Local pH Differences. *mBio* **2016**, *7*, e01911-16. [CrossRef]
197. Barker, C.S.; Samatey, F.A. Cross-Complementation Study of the Flagellar Type III Export Apparatus Membrane Protein FlhB. *PLoS ONE* **2012**, *7*, e44030. [CrossRef]
198. Hara, N.; Namba, K.; Minamino, T. Genetic Characterization of Conserved Charged Residues in the Bacterial Flagellar Type III Export Protein FlhA. *PLoS ONE* **2011**, *6*, e22417. [CrossRef]
199. Kihara, M.; Minamino, T.; Yamaguchi, S.; Macnab, R.M. Intergenic Suppression between the Flagellar MS Ring Protein FliF of Salmonella and FlhA, a Membrane Component of Its Export Apparatus. *J. Bacteriol.* **2001**, *183*, 1655–1662. [CrossRef]
200. Ferris, H.U.; Furukawa, Y.; Minamino, T.; Kroetz, M.B.; Kihara, M.; Namba, K.; Macnab, R.M. FlhB Regulates Ordered Export of Flagellar Components via Autocleavage Mechanism. *J. Biol. Chem.* **2005**, *280*, 41236–41242. [CrossRef]
201. Fraser, G.M.; Hirano, T.; Ferris, H.U.; Devgan, L.L.; Kihara, M.; Macnab, R.M. Substrate specificity of type III flagellar protein export in Salmonella is controlled by subdomain interactions in FlhB. *Mol. Microbiol.* **2003**, *48*, 1043–1057. [CrossRef] [PubMed]

202. Kuhlen, L.; Abrusci, P.; Johnson, S.; Gault, J.; Deme, J.; Caesar, J.; Dietsche, T.; Mebrhatu, M.T.; Ganief, T.; Macek, B.; et al. Structure of the core of the type III secretion system export apparatus. *Nat. Struct. Mol. Biol.* **2018**, *25*, 583–590. [CrossRef] [PubMed]
203. Katayama, E.; Shiraishi, T.; Oosawa, K.; Baba, N.; Aizawa, S. Geometry of the flagellar motor in the cytoplasmic membrane of Salmonella typhimurium as determined by stereo-photogrammetry of quick-freeze deep-etch replica images. *J. Mol. Biol.* **1996**, *255*, 458–475. [CrossRef] [PubMed]
204. Liu, J.; Lin, T.; Botkin, D.J.; McCrum, E.; Winkler, H.; Norris, S.J. Intact Flagellar Motor of Borrelia burgdorferi Revealed by Cryo-Electron Tomography: Evidence for Stator Ring Curvature and Rotor/C-Ring Assembly Flexion. *J. Bacteriol.* **2009**, *191*, 5026–5036. [CrossRef] [PubMed]
205. Galperin, M.; Dibrov, P.A.; Glagolev, A.N. delta mu H+ is required for flagellar growth in Escherichia coli. *FEBS Lett.* **1982**, *143*, 319–322. [CrossRef]
206. Ewilharm, G.; Lehmann, V.; Krauss, K.; Lehnert, B.; Richter, S.; Ruckdeschel, K.; Heesemann, J.; Trülzsch, K. Yersinia enterocolitica Type III Secretion Depends on the Proton Motive Force but Not on the Flagellar Motor Components MotA and MotB. *Infect. Immun.* **2004**, *72*, 4004–4009. [CrossRef]
207. Paul, K.; Erhardt, M.; Hirano, T.; Blair, D.F.; Hughes, K.T. Energy source of flagellar type III secretion. *Nat. Cell Biol.* **2008**, *451*, 489–492. [CrossRef]
208. Minamino, T.; Namba, K. Distinct roles of the FliI ATPase and proton motive force in bacterial flagellar protein export. *Nat. Cell Biol.* **2008**, *451*, 485–488. [CrossRef]
209. Barker, C.S.; Inoue, T.; Meshcheryakova, I.V.; Kitanobo, S.; Samatey, F.A. Function of the conserved FHIPEP domain of the flagellar type III export apparatus, protein FlhA. *Mol. Microbiol.* **2016**, *100*, 278–288. [CrossRef]
210. Butan, C.; Lara-Tejero, M.; Li, W.; Liu, J.; Galán, J.E. High-resolution view of the type III secretion export apparatus in situ reveals membrane remodeling and a secretion pathway. *Proc. Natl. Acad. Sci. USA* **2019**, *116*, 24786–24795. [CrossRef]
211. Echazarreta, M.A.; Klose, K.E. Vibrio Flagellar Synthesis. *Front. Cell. Infect. Microbiol.* **2019**, *9*, 131. [CrossRef] [PubMed]
212. Chu, J.; Liu, J.; Hoover, T.R. Phylogenetic Distribution, Ultrastructure, and Function of Bacterial Flagellar Sheaths. *Biomol.* **2020**, *10*, 363. [CrossRef] [PubMed]
213. Kusumoto, A.; Shinohara, A.; Terashima, H.; Kojima, S.; Yakushi, T.; Homma, M. Collaboration of FlhF and FlhG to regulate polar-flagella number and localization in Vibrio alginolyticus. *Microbiol.* **2008**, *154*, 1390–1399. [CrossRef] [PubMed]
214. Terashima, H.; Fukuoka, H.; Yakushi, T.; Kojima, S.; Homma, M. The Vibrio motor proteins, MotX and MotY, are associated with the basal body of Na+-driven flagella and required for stator formation. *Mol. Microbiol.* **2006**, *62*, 1170–1180. [CrossRef]
215. Terashima, H.; Koike, M.; Kojima, S.; Homma, M. The Flagellar Basal Body-Associated Protein FlgT Is Essential for a Novel Ring Structure in the Sodium-Driven Vibrio Motor. *J. Bacteriol.* **2010**, *192*, 5609–5615. [CrossRef]
216. Zhu, S.; Nishikino, T.; Kojima, S.; Homma, M.; Liu, J. The Vibrio H-Ring Facilitates the Outer Membrane Penetration of the Polar Sheathed Flagellum. *J. Bacteriol.* **2018**, *200*. [CrossRef]
217. Motaleb, A.; Corum, L.; Bono, J.L.; Elias, A.F.; Rosa, P.; Samuels, D.; Charon, N.W. Borrelia burgdorferi periplasmic flagella have both skeletal and motility functions. *Proc. Natl. Acad. Sci. USA* **2000**, *97*, 10899–10904. [CrossRef]
218. Sosinsky, G.E.; Francis, N.R.; Stallmeyer, M.; DeRosier, D.J. Substructure of the flagellar basal body of Salmonella typhimurium. *J. Mol. Biol.* **1992**, *223*, 171–184. [CrossRef]
219. Moon, K.H.; Zhao, X.; Manne, A.; Wang, J.; Yu, Z.; Liu, J.; Motaleb, A. Spirochetes flagellar collar protein FlbB has astounding effects in orientation of periplasmic flagella, bacterial shape, motility, and assembly of motors inBorrelia burgdorferi. *Mol. Microbiol.* **2016**, *102*, 336–348. [CrossRef]
220. Rajagopala, S.V.; Titz, B.; Goll, J.; Parrish, J.R.; Wohlbold, K.; McKevitt, M.T.; Palzkill, T.; Mori, H.; Jr, R.L.F.; Uetz, P. The protein network of bacterial motility. *Mol. Syst. Biol.* **2007**, *3*, 128. [CrossRef]

221. Moon, K.H.; Zhao, X.; Xu, H.; Liu, J.; Motaleb, A. A tetratricopeptide repeat domain protein has profound effects on assembly of periplasmic flagella, morphology and motility of the lyme disease spirochete Borrelia burgdorferi. *Mol. Microbiol.* **2018**, *110*, 634–647. [CrossRef] [PubMed]
222. Xu, H.; He, J.; Liu, J.; Motaleb, A. BB0326 is responsible for the formation of periplasmic flagellar collar and assembly of the stator complex in Borrelia burgdorferi. *Mol. Microbiol.* **2019**, *113*, 418–429. [CrossRef]

Review

The Architectural Dynamics of the Bacterial Flagellar Motor Switch

Shahid Khan

Molecular Biology Consortium, Lawrence Berkeley National Laboratory, Berkeley, CA 94720, USA; khan@mbc-als.org

Received: 31 March 2020; Accepted: 25 May 2020; Published: 29 May 2020

Abstract: The rotary bacterial flagellar motor is remarkable in biochemistry for its highly synchronized operation and amplification during switching of rotation sense. The motor is part of the flagellar basal body, a complex multi-protein assembly. Sensory and energy transduction depends on a core of six proteins that are adapted in different species to adjust torque and produce diverse switches. Motor response to chemotactic and environmental stimuli is driven by interactions of the core with small signal proteins. The initial protein interactions are propagated across a multi-subunit cytoplasmic ring to switch torque. Torque reversal triggers structural transitions in the flagellar filament to change motile behavior. Subtle variations in the core components invert or block switch operation. The mechanics of the flagellar switch have been studied with multiple approaches, from protein dynamics to single molecule and cell biophysics. The architecture, driven by recent advances in electron cryo-microscopy, is available for several species. Computational methods have correlated structure with genetic and biochemical databases. The design principles underlying the basis of switch ultra-sensitivity and its dependence on motor torque remain elusive, but tantalizing clues have emerged. This review aims to consolidate recent knowledge into a unified platform that can inspire new research strategies.

Keywords: rotary molecular motor; protein allostery; chemotactic signaling

---

1. The Problem Framed—Historical Background (1973–2003)

Bacterial motility has been a long-standing example of motion on a microscopic scale [1]. The modern era began with the realization that bacterial flagella rotate, as opposed to eukaryotic flagella that beat [2]. The fundamental issues that drive current research on the bacterial flagellar switch were framed in the first thirty years (1973–2003). The first stage, the "classical period", established that the energy source for motility was the chemiosmotic ion potential rather than ATP. Tethered cell assays demonstrated cell rotation driven by a single flagellum immobilized on glass coverslips. These assays showed that eubacterial motors rotate both counterclockwise (CCW) and clockwise (CW) and switch rotation sense without a detectible change in rotation speed. The CW and CCW rotation intervals were Poisson distributed. Chemo-effectors changed motor rotation bias with sub-second excitation followed by adaptation over seconds back to the pre-stimulus level. Motor rotation bias (CW/(CW + CCW)) measured in tethered cell assays coupled to flagellar filament polymorphic transitions could be correlated with the swim-tumble motility of free-swimming bacteria. This literature has been reviewed [3]. It established there was a fundamental difference in switch design and operation between bacterial and eukaryotic flagella (see [4] for a minireview). Advances in bacterial flagellar switch function and structure in the second half of these thirty years were based on the development of high-throughput genetic screens, sophisticated motor rotation assays, isolation and biochemical characterization of the intact switch and sub-complexes together with atomic structures as summarized in this section (Figure 1). Subsequent sections in this review consider progress in switch dynamics and architecture in the light of these advances.

The structural complexity of the flagellar switch: Swarm plate assays [5] provided high-throughput isolation of motility mutants that could be grouped into three categories (non-flagellate (*fla*), non-motile (*mot*) and non-chemotactic (*che*)). In 1986, a switch complex of three interacting proteins was proposed based on swarm plate assays of suppressor mutations [6]. Five years later, the gene sequences encoding the proteins were obtained by polymerase chain reaction (PCR), an early example of the use of PCR in bacterial motility research. The sequences indicated that all three proteins (renamed FliG, FliM and FliN) were cytoplasmic [7], and revealed mutation hotspots. The *che* mutations mostly localized to *fliM*; the *mot* mutations largely localized to *fliG* [8,9]. The clustering of the *fliM che* mutations to distinct regions that accentuated CW or CCW rotation suggested that the FliM structural determinants assigned the rotation state [8]. Genetic evidence for electrostatic residue interactions between the FliG C-terminal and MotA [10,11] implicated FliG in motor function. The *motA* and *motB* genes only had *mot* alleles in contrast to genes for the switch complex. Tethered cell motility was resurrected in similar, stepwise increments by *motA* or *motB* induction in the corresponding deletion strains, implying multiple, independently acting MotA and MotB stator complexes [12]. None of the protonatable *E. coli* FliG, FliM or FliN residues that were sites for *mot* substitutions was essential for motility [13]. These data indicated that the energizing proton flux did not traverse the switch complex.

The development of gentler protocols in 1992 led to the isolation and morphological identification of the switch complex, based on its impaired mutant structures, as an extended cytoplasmic component of the basal body [14], subsequently termed the C ring [15]. Further purification enabled its biochemical characterization [16]. FliN copy numbers (n) were 3–4 times the estimates for FliM ($n = 34 \pm 3$). Analysis of the sub-complexes concurrently established that the switch proteins self-associate and interact with each other [17]. Finally, structures that were morphologically identical to the C ring formed upon overproduction of the switch complex proteins together with the FliF MS ring [18]. 3D reconstructions in ice of the *S. enterica* basal body [15] combined developments in cryo-electron microscopy with single-particle image analysis (reviewed in [19,20]) to resolve C ring periodicity [21] and position individual domains, with FliG an early example [22]. The overproduced C rings had variable symmetry (n = 32–38) [23], but the dominant 34-fold symmetry was consistent with the biochemical estimates of FliM copies in the native C ring. This advance exploited the fact that *fliF*, *fliG*, *fliM* and *fliN* were "early" genes in the flagellar regulon [24], and built upon MS ring assembly by FliF overproduction [25].

Torque generation and switch activation: The torque, T, on a rotating spherical tethered cell of radius, a, is balanced by the hydrodynamic drag. $8\Pi\eta a^3 W$, where W is the angular velocity, and η the medium viscosity.

The torque velocity relation was examined over a limited load ($8\Pi\eta a^3$) range by changing η [26]. The relation had a biphasic form for CCW rotation. Visualization of the rotation of the "tethered" beads attached to flagellar stubs extended the range of the torque velocity relation to high rotation speeds. The torque was constant at low speed, and then decreased linearly above a threshold speed. Temperature and isotope effects in the linearly decreasing, but not the constant, torque regime implied the energizing proton transfer reactions limited the decreasing velocity [27]. Comparable results were obtained for the sodium *Vibrio alginolyticus* motor [28]. Application of an external force with optical traps [29] or electrorotation [30] allowed the study of the relation at negative as well as positive torque. The integration of optical trap and bead rotation assays revealed the load-dependent modulation of CW rotation interval [31], showing that the switching mechanism was not isolated from motor mechanics. Rapid switching events, damped out in tethered cells due to compliance of the hook structure connecting the basal body with the flagellar filament, were resolved earlier by laser dark-field microscopy measurements of filament rotation under conditions similar to free-swimming bacteria [32]. This study recorded slowed rotation and pausing events in addition to rapid reversals and importantly showed that these events increased in strains carrying switch complex mutations providing a direct window into switch mechanics not accessed by swarm plate assays.

The atomic structure of the FliG carboxy-terminal domain ($FliG_C$) [33] heralded the molecular era in structural analysis of the switch complex. Armadillo (ARM) folds, a ubiquitous architecture found in signal proteins, characterized both middle ($FliG_M$) and C-terminal FliG domains [34]. The structures revealed that charged FliG residues essential for torque generation [11] localized to a surface-exposed face of an α-helix. Suppressor residue substitutions for the MotB stator protein [35] clustered to the $FliG_M$–$FliG_C$ inter-domain loop that included a conserved glycine pair. The CW and CCW biasing substitutions mainly localized to $FliG_M$ and the inter-domain linker, but importantly also to the $FliG_C$ conserved MXVF loop and adjacent α-helices.

Ultra-sensitivity of the chemotactic motor response: The 1989 atomic structure of the chemotaxis signal protein CheY [36] identified a cluster of three aspartate residues as the probable phosphorylation site. This study was an early application of site-specific mutagenesis for structure determination in bacterial motility and chemotaxis. Subsequent studies established that CheY aspartyl phosphorylation by the receptor associated CheA kinase coupled receptor occupancy to the motor response. CheY is one of a large superfamily of response regulators with diverse functional roles (reviewed in [37]). Aspartyl phosphate is labile in contrast to the corresponding serine/threonine phosphates exploited in developmental circuits, but beryllium fluoride (BeF3), an acyl-phosphate analogue, binds stably [38]. Biochemical studies determined the FliM N-terminus ($FliM_N$) to be the CheY binding target at the flagellar switch [39,40]. Phospho-CheY and BeF3-CheY had a comparable affinity, with the activating structural transitions visualized in the BeF3-CheY.$FliM_N$ crystal structure [41]. 2D-NMR further showed the bound $FliM_N$ influenced phosphorylation site dynamics [42]. CheY did not associate with incomplete switch complexes formed by FliF MS rings with FliG [43]. A library of *cheY* mutant alleles was generated guided by the atomic structures (reviewed in [37]). The phospho-mimetic mutations 13DK and 13DK106YW have figured prominently in the study of switch physiology (e.g., [44]).

The motor rotation bias, reported by tethered beads, was a function of intracellular GFP-tagged CheY concentration, estimated by correlation intensity analysis in single *S. typhimurium* cells. The bacteria carried mutations that ensured CheY was phosphorylated. The bias changed sharply with CheY concentration (Hill coefficient, $H = 10.3$) [45]. Binding assays of CheY with overproduced complexes reported a similar difference in affinity for the phosphorylated and non-phosphorylated forms to that for $FliM_N$, but the binding was not cooperative ($H \sim 1$) [46]. Early models had formalized the switch as an equilibrium thermal isomerization machine [44,47]. An important advance over these models was the conformational spread model. that explicitly considered the multiple subunit stoichiometry, N. The individual subunits fluctuated between the CW and CCW states with adjacent subunits linked by a coupling energy term influenced by ligand (CheY) occupancy. The mean size of the contiguous CW or CCW domains increased with the coupling energy. Above a critical threshold, the entire ring flipped as a 1D Ising type switch to simulate the ultra-sensitive response with n = 34 [48].

In conclusion, a cluster of key publications between 1986 and 2003 (A) established a conceptual framework for the bacterial flagellar motor (BFM) switch and (B) introduced new methodologies pivotal for future advances in structure and dynamics. (A) The idea of the switch had developed from a process, rotation reversal, to a material entity, the switch complex, to a physical object, the C ring, The C ring was composed of a small set of core proteins that self-assembled into a large multi-subunit assembly attached to the MS-ring. The C ring did not conduct protons but interacted as the rotor module with the proton-conducting Mot complexes to generate torque. Mutant phenotypes linked component lesions to switch phenotypes, influenced by motor operation, that included, but were not restricted to, rotation reversal. The switch set-point was shown to be an ultrasensitive function of the activity of the CheY signal protein, in contrast to non-cooperative CheY binding to the C ring. The linkage between the highly cooperative output (motor rotation) and the non-cooperative input required long-range allosteric communication across subunits between the $FliM_N$ binding sites for CheY and the $FliG_C$ interface with stator complexes. A formal model was developed [48] while atomic protein structures [33,34,36,41] provided important clues into possible interactions. (B) The period witnessed the timely application of new NMR methodologies for structure determination of large macromolecules [49] to CheY complexes [42], as well as cryo-EM allied single-particle image processing of multi-subunit assemblies [19,20] to isolated basal bodies [15,23,50]. The *Thermatoga maritama* FliG structures [33,34] set an important precedent for X-ray crystallography of thermophile switch proteins. The first applications of live cell imaging with GFP biotechnology [51] to determine CheY bias modulation [45] or localization [43]—and together with single-molecule, force microscopy to characterize the load-dependent switching [31]—were to prove equally influential.

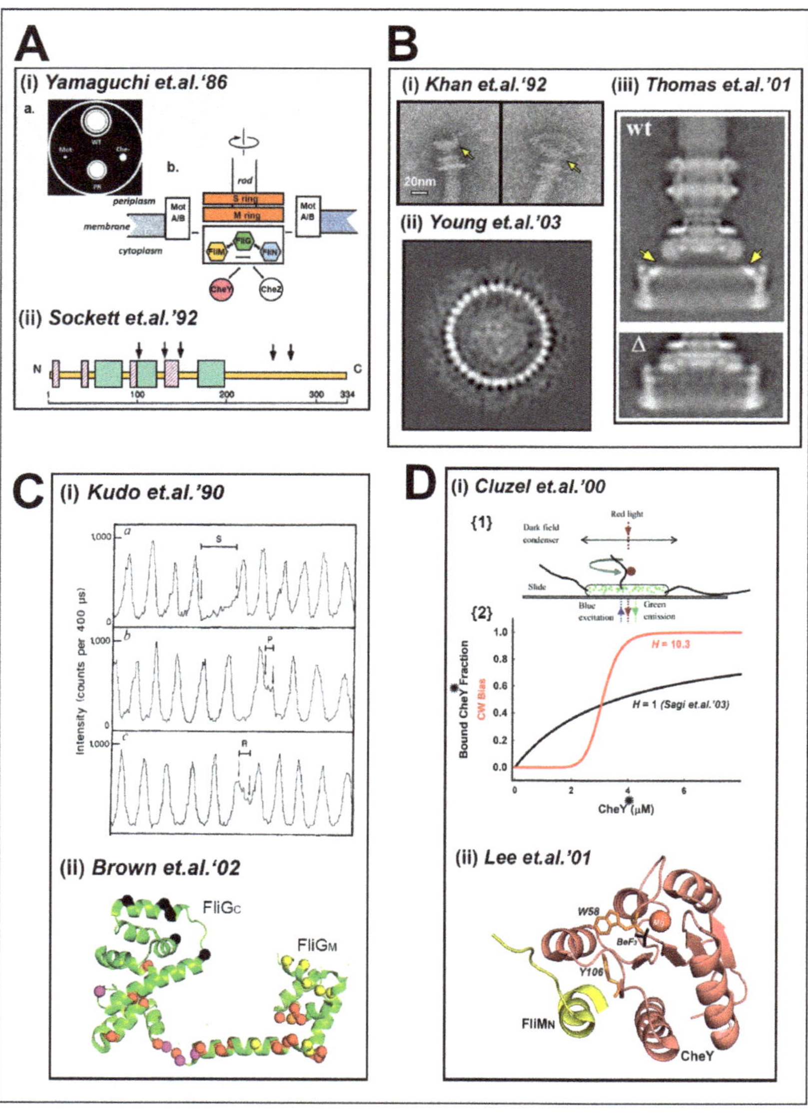

Figure 1. The bacterial flagellar motor (BFM) switch—landmarks. (**A**) Conceptualization: (i) The switch complex was proposed based on phenotypic characterization of *mot*, *che* and *fla* alleles and their

suppressor mutations in swarm plate assays. Its interactions with chemotaxis components and Mot proteins were also identified. {**a**} Schematic of a swarm plate—the native (WT) strain forms a swarm with chemotactic rings. Strains carrying *mot* mutations (Mot-) do not swarm while those with *che* mutations (Che-) have reduced swarms. Suppressor mutations yield pseudo-revertant strain (PR) with partially restored swarming. {**b**} Color codes are followed in subsequent Figures for the switch complex components (FliG (green), FliM (gold), FliN (cyan)), the CheY protein (salmon) and the MS-ring scaffold (orange) (adapted from [6]). (**ii**) Gene sequencing identified the mutations. The *fliM* gene (N–C terminal residue numbers) predominantly contained the *che* lesions, clustered into distinct CW (green) and CCW (magenta) regions. Arrows mark *mot* lesions (adapted from [8]). (**B**) Structural identification: (**i**) An extended cytoplasmic structure contiguous with the basal body MS-ring (yellow arrow) was isolated using gentler protocols and subsequently established as the switch complex by immuno-EM and biochemistry (from [14]). (**ii**) Assembly of switch complex by overproduction of plasmid-encoded components allowed biochemical characterization culminating in the determination of the C ring subunit stoichiometry (n = 33–34) (from [23]). (**iii**) Single-particle analysis resolved FliG domain substructure (yellow arrows) from differences in central sections from wild-type (WT) and ΔFliFFliG (Δ) 3D basal-body reconstructions (from [22] with permission). (**C**) Motor function and mechanism: (**i**) Temporally resolved measurement of filament rotation, as a sinusoidal variation of laser dark-field spot intensity, characterized aberrant phenotypes in switch complex mutant strains. Panels (top to bottom) show slow rotation (S), pausing (P) and reversal (R) episodes (reproduced from [32] with permission). (**ii**) The first atomic structure of a switch component (FliGc [33]) followed by the $FliG_{MC}$ structure localized much of the mutant library then available ((mot lesions (black); CW lesions (red); CCW lesions (yellow); CW or CCW, depending on the residue substitution, orange; and *motB* suppressors (purple)) to generate chemically explicit ideas for motor reversal (PDB: 1lkv (modified from [34])). (**D**) Switch chemotactic signal transduction: (**i**) {1}—Determination of switch "ultra-sensitivity" (Hill coefficient, $H = 10.3$) by simultaneous measurement of the CW bias of beads on flagellar stubs (red) and concentration of a fluorescent GFP-CheY fusion (green) locked in the active state (*) in engineered strains (reproduced from [45] with permission). {2}—Plots show non-cooperative binding of acetate-activated CheY to overproduced C rings [46] compared to the in-vivo change in CW bias. (**ii**) The atomic structure of beryllium-fluoride (BeF3 (black))-activated CheY (salmon) bound to the FliM N-terminal peptide (yellow) initiated structure guided mutagenesis to explain the switch ultra-sensitivity. Aromatic residue (W58, Y106 (orange)) motions were early diagnostics for activation. Magnesium ion (red) (PDB: 1f4v (modified from [41])).

Four fundamental issues could now be addressed. First, what was the nature of the coupling between the $FliG_C$ motor domain and the $FliM_N$ CheY binding target? Secondly, how were the dynamics of the C ring, a large multi-subunit assembly, synchronized for smooth rotation and rapid reversal? Thirdly, how did CheY activation trigger an ultrasensitive switch response? Fourthly, how was the switch regulated by the motor operation as dictated by the torque–velocity relations?

2. Switch Physiology and Mathematical Models

Motor dynamics are the direct outcome of the architectural dynamics of the molecular machinery. Knowledge of motor dynamics, and the associated development of mathematical models, has advanced concurrently with knowledge of the molecular architecture and dynamics. The spatiotemporal resolution and mechanical range of the rotation assays have continued to increase for characterization of speed fluctuations, the load dependence and stochastic properties of the switch machinery in unprecedented detail. These advances provide additional constraints that must be addressed by the study of the molecular mechanism. Motor dynamics have been reviewed recently [52]. An overview is given in this short section, summarized in Figure 2, to provide additional context for the main body of the review.

Focal back-plane interferometry with high spatiotemporal resolution (1°, 10^{-3} s) recorded incomplete, intermediate switching events to extend the temporally resolved measurement of switch transitions [53]. More recently, gold nanospheres (d = 60 nm) conjugated to genetically engineered, rigid

Salmonella flagellar hooks, imaged by dark-field and rapid (5000 Hz) CMOS cameras, have been exploited to measure motor rotation [54]. This study interpreted the large speed fluctuations recorded near the zero-torque speed (~400 Hz) that persisted over many revolutions as the association–dissociation of individual stator units. The highlight of such studies was that the resolution of the angular step periodicity per revolution was resolved in motors resurrected by single or a few stator units; first for a sodium powered chimeric motor [55] and then for the *S. enterica* proton motor [56]. In each case, 26 steps per revolution were reported for both CW and CCW rotation. However, while the step size was symmetrical for the *Salmonella* motor, the CW steps were smaller than the CCW steps in the chimeric motor. A study of the unidirectional *R. sphaeroides* motor reported a similar estimate of 27–28 discrete stopping angles [57]. The steps may be due to modulation of the elastic potential well by periodic contacts between the static and mobile motor modules rather than individual energy coupling events. A formal model of this idea successfully predicted the difference in CW versus CW step size [58].

The long-standing tethered cell measurements of the Poisson-distributed CCW and CW intervals had been the bedrock for theoretical models. Filament-associated bead rotation measurements in the phospho-mimetic *E. coli* CheY13DK mutant strain first revealed gamma distributions for both CW and CCW intervals [59] to challenge the view of switch transitions as a single Poisson process based on tethered cell data. Subsequent measurements of the rotation of beads attached to flagellar stubs [53] and nanoparticles attached to flagellar hooks [60] obtained exponential distributions in apparent contrast to [59], fueling speculation that filament polymorphic transitions [61] may have complicated the rotation of filament-associated latex beads (0.5 μm). The use of different-sized latex spheres extended the dependence of switching kinetics on load [62]. The study reported that at near zero-load speed both CCW<->CW and CW<->CCW transitions increased with the load. Furthermore, the torque–velocity curve for CW rotation was linear in contrast to CCW rotation [63]. The load dependence raised the possibility that it may underlie the apparent discrepancy between the gamma and Poisson interval distributions. The comprehensive analysis of the rotation of different-sized beads in the double mutant phospho-mimetic strain CheY13DK106YW showed this was indeed the case. Both exponential and peaked distributions were obtained, depending on the different load, proton potential and torque [64], to rule out speculation that the gamma distribution was an artefact.

The knowledge about the regulation of (CW/CCW) rotation bias by CheY has also advanced. Temporally resolved recordings of bead rotation over minutes re-evaluated motor individuality in terms of its rotation bias. Long-term recordings of variation in single bead rotation compared to population variation of multiple beads concluded that individuality was due to intracellular chemistry. The bias of individual motors had a bimodal distribution with cells with high switching frequencies binned into a central trough. The CW and CCW fractions were dominant, consistent with the ultrasensitive switch in rotation bias by CheY [65]. Single bead CW and CCW rotational intervals had earlier been reported to be asymmetrically distributed around the bias midpoint ((CW/(CW + CCW)) = 0.5), with the CCW but not CW intervals varying with rotation bias [66]. These different relations argued against the determination of transition rates by CheY or any other single parameter. The ultra-sensitivity has also been re-evaluated. The local GFP-CheY concentration around single flagellar motors was measured and correlated with (CW/CCW) rotation bias in a Δ*cheY E. coli* strain [67] for an estimate of the mean CheY rotor occupancy during CW rotation. GFP-tagged fusion proteins also provided evidence that the basal C ring components FliM and FliN undergo dynamic exchange [68,69]. An experiment was designed to correct for the FliM copy variation that would be obtained based on the exchange being adaptive. It estimated the Hill coefficient, H, for the ultra-sensitivity from bias changes due to attractant removal and addition in strains lacking receptor adaptation was as high as 21 [70].

The experiments motivated the development of the conformational spread model [48] and the formulation of a fundamentally distinct non-equilibrium model [71]. The initial interpretation for the peaked gamma distribution was based on multiple Poisson events [59] to refine, but not radically alter the conformational spread model. The asymmetry in the experimental torque–velocity curves and the load dependence of the switching kinetics could also be accounted for by an integrated model that combined torque generation models with the switch ultra-sensitivity as formalized by conformational spread [72]. Theoretical models, in general, seek to explain the interdependence between rotation bias, switching frequency and interval distributions. The experimental relations between these motor parameters were set as constraints to model the coupling between synchronous switching and signal amplification as a function of subunit number [73]. More recent simulations explained the correlation between local CheY concentration and single motor rotation bias as well as the utility of dynamic subunit exchange in the maintenance of switch ultra-sensitivity [74]. Detailed balance within the equilibrium conformational spread framework was followed in both studies [73,74]. Tu's non-equilibrium model, where the detailed balance is broken based on energy input, also explains the gamma distribution data [71,75]. The model is consistent with the expanded set of interval distributions obtained under different load and energization conditions as detailed in [64]. The energy input for switching need only be a small fraction of the total input that is dominantly utilized to power rotation.

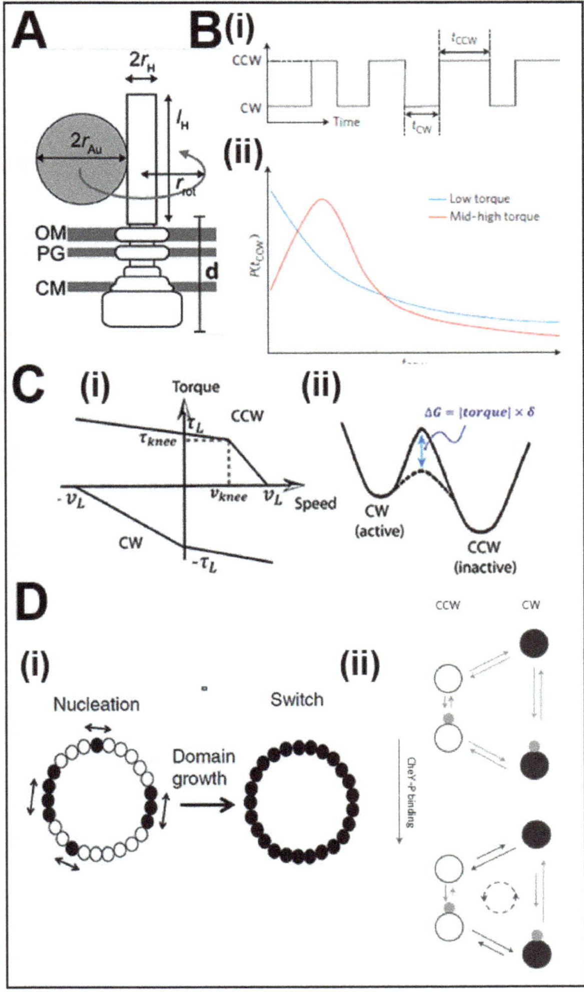

Figure 2. Advances in switch physiology. (**A**) Time-resolved motor rotation: Schematic of a motor rotation assay with a nanosphere conjugated to the hook connector contiguous with the rod and basal

body (reproduced from [54] with permission). (**B**) CW and CCW interval distributions: (**i**) Idealized time series of a motor alternating between CW and CCW rotation. (**ii**) Interval (τ_{CW}, τ_{CCW}) distributions measured under low and high torque (reproduced from [75] with permission). (**C**) Coupled energization and switching of rotation: (**i**) Torque velocity curves for CW and CCW rotation [63]. (**ii**) Free energy diagram of CW <-> CCW transitions and the mechanical work (blue arrow) contribution (reproduced from [72] with permission). (**D**) Models for non-Poisson interval distributions: (**i**) Conformational spread seeded from multiple CheY binding events (reproduced from [59] with permission). (**ii**) Breakdown of detailed balance from motor energy dissipation (reproduced from [64] with permission).

3. Architecture and Molecular Mechanism

Structural work on the issues outlined in the previous two sections can be grouped into four major areas based on the protein–protein interactions: the central processing unit ($FliM_M.FliG_N$), the trigger machinery (CheY/C ring), the torque reversal mechanism ($FliG_C/FliG_M$) and the torque transmission platform ($FliG_N/FliF_C$).

3.1. *The Central Processing Unit—The $FliM_M.FliG_M$ Complex*

X-ray crystallography led the effort guided by mutagenesis, in situ crosslinking and EM reconstruction to characterize the linkage between the domains responsible for torque generation and CheY signal reception. The atomic structure of the *T. maritima* FliG middle ($FliG_M$) and C-terminal ($FliG_C$) domains had an established ARM architecture for both domains and localized hotspots for *E. coli che* mutations to the sequence encoding the linker region between $FliG_M$ and $FliG_C$ [34]. The atomic structure of the *T. maritima* FliM middle domain ($FliM_M$), the central element in switch function, showed that the CCW and CW substitutions localized to residue positions at the intradomain contact interface, as deduced by subunit crosslinking in situ [76]. The CW substitutions aligned differently to the CCW substitutions, suggesting distinct $FliM_M$ orientations for CW versus CCW rotation (Figure 3A). In situ crosslinking had been introduced in earlier work on $FliG_M$ domain organization [77]. The in-situ crosslinking experiments probed the subunit organization of the switch proteins as appropriate controls established that cross-linking was negligible in non-flagellate strains [76,77]. The structures of $FliM_M.FliG_M$ complexes from *T. maritima* [78,79] and *H. pylori* [80] determined the functional importance of the FliM GXXG and FliG EHPQR loop motifs at the $FliM_M.FliG_M$ interface. Comparison of the *T. maritima* and *H. pylori* interfaces showed conservation of essential GXXG and EHPQR loop contacts and their stabilization by complex formation. Additional interfacial residues important for coupling were identified in *H. pylori*, supported by mutagenesis.

Protein dynamic simulations showed that the core architecture and dynamics of $FliM_M$ do not change upon complex formation and are conserved between *T. maritima* and *H. pylori*. $FliM_M$ conformational ensembles generated from the *T. maritima* structure had a bimodal distribution, while the $FliG_M$ domain movements also alternated between the bi-stable states strongly coupled to $FliM_M$ [81]. Targeted tryptophan substitutions had identified $FliG_M$ residue positions important for association with $FliM_M$ [82]. In addition to the EHPQR loop, these residues were in a FliGc ARM-C hydrophobic patch adjacent to the $FliG_{MC}$ GG linker as well as non-interfacial $FliG_M$ residue positions. Recent NMR measurements of the $FliG_M$ dynamics in the sodium *V. alginolyticus* motor have provided important validation, supported by mutagenesis, of the conformational plasticity of the $FliG_M$ domain between the bi-stable conformational states [83]. The EHPQR E144D residue substitution increased the switching frequency, a similar response to one obtained upon phenol repellent addition. It did not alter $FliG_M$ conformation. In contrast, residue substitutions in the GG hinge both locked the bias and reduced the $FliG_M$ dynamics, as assessed by NMR. The E144D and CCW-biased G214S motors were both CCW-locked in the Δ*che* strains, implying a close linkage between CheY occupancy and $FliG_M$ conformational fluctuations (Figure 3B).

In *H. pylori*, co-crystallization of putative spermidine synthase (SpeE) with $FliM_M$, combined with mutagenesis and motility assays, has shown how this protein affects motile speed and switching behavior. The $SpeE.FliM_N$ contact interface partially overlaps with the $FliG_M.FliM_M$ interface [84]. While several proteins important for cell metabolism have been identified in diverse bacteria to influence motile behavior via direct interaction with FliG (cited in [84]), SpeE is the first known to act at the $FliG_M.FliM_M$ interface. In conclusion, the crystal structure, allelic mutations, in situ crosslinking, and protein dynamic simulations all argue for the alternation of $FliM_M$ between bi-stable conformational states, inter-species conservation of its fold and interfacial $FliG_M$ coupling.

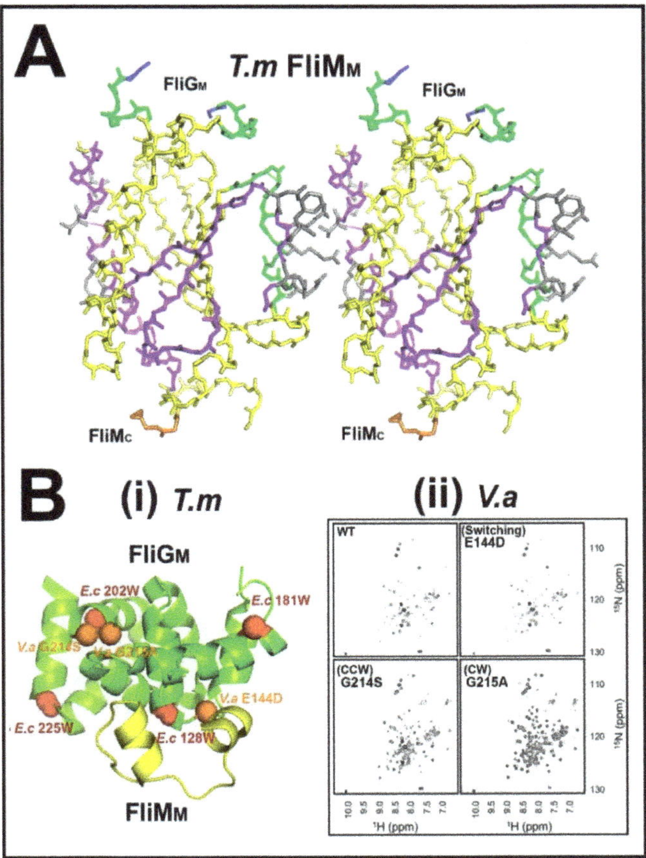

Figure 3. The central signal processing unit. (**A**) $FliM_M$ dimer model. The model is based on the *T. maritima* $FliM_M$ structure (gold)) guided by in situ cross-link data (grey). Localized CW (green) and CCW (magenta) biasing residue substitutions from bacterial homologs are mapped onto the colored structural elements. FliG GXXG interaction loop (blue). C-terminus (orange) (PDB: 2hp7 (modified from ([76])). (**B**) The central processing interface—$FliG_M$-$FliM_M$. (**i**) *T. maritima* structure (PDB: 4fhr) highlighted with homologous $FliG_M$ residues to those implicated by *E. coli* tryptophan mutagenesis (red [82]) and *V. alginolyticus* switch mutants G214S, G215A and E144D (orange [83]). (**ii**) 2D-NMR ^1H–^{15}N correlation spectra reveal that the native (WT) *V. alginolyticus* $FliG_M$ architecture is substantially altered by substitutions in the conserved glycine pair that alter rotation bias, but is unaffected by the EPQR E144D residue substitution that increases the switching frequency. Axes indicate the shift in the magnetic field, in parts per million (ppm), relative to a reference compound for resonance (reproduced from [83] with permission).

3.2. The Trigger Machinery—CheY and the Basal C Ring

Motor response to chemotactic signals can be parsed into CheY activation and binding to the C ring to bias $FliM_M$ conformational fluctuations. There is a 20-fold increase upon phosphorylation in CheY affinity for *E. coli* $FliM_N$ [41]. The molecular analysis of CheY activation by the phospho-mimetic CheY 13DK106YW double residue substitutions used to study motor response was initiated by co-crystallization of the *E. coli* protein alone and in complex with the FliM N-terminal peptide ($FliM_N$) [85]. Strikingly, while the CheY13DK106YW in the complex was in a conformation akin to BeF3 activated CheY, the free CheY13DK106YW adopted the inactive conformation. Subsequently, the atomic structure of the native CheY.$FliM_N$ complex revealed that the electron density of a critical hinge, the α4–β4 loop, in this complex partitioned between its active and inactive states [86]. Quantitative comparison of MD conformational ensembles from crystal structures of the CheY superfamily representatives has extracted common signatures, including the α4–β4 hinge, for allosteric communication between the phosphorylation and target binding sites [87]; the importance of the α4–β4 loop further emphasized by MD simulations of acetate-activated CheY [88]. MD simulations supported by oxygen-radical foot-printing solution measurements, a sensitive probe for sidechain solvent accessibility [89], have now shown that closure of this loop hinge buried the allosteric relay aromatic sidechains to both stabilize the global CheY fold and increase the local $FliM_N$ affinity.

The motor response is likely to depend on the intervening disordered linker between $FliM_N$ and $FliM_M$ that may, in principle, serve as a flexible tether to deliver $FliM_N$-bound CheY to second binding sites. Occupancy of these sites could control $FliM_M$ conformational fluctuations and inter-subunit coupling. Candidates have been identified in two species. There is NMR evidence for an interaction between *T. maritima* $FliM_N$-tethered CheY and $FliM_M$ [90]. Alternatively, in *E. coli*, pull-down assays have reported evidence for an association of the CheY-$FliM_N$ fusion with FliN [91]. This evidence is summarized in Figure 4A.

There is a close homology between FliN and $FliM_C$. The atomic structure of the *T. maritima* FliN homodimer [92] motivated mutagenesis and in situ crosslinking experiments to infer FliN tetrameric quaternary organization in the *E. coli* flagellar motor [93]. These experiments reported cross-link changes upon addition of chemotactic stimuli, consistent with domain motions, but the connectivity between $FliM_N$ and the $FliM_C{}^1.FliN^3$ module is not well-understood as the FliM inter-domain linkers have not been structurally characterized. The atomic structure of the $FliM_C$.FliN heterodimer is similar to the FliN homodimer [94], consistent with a $FliM_C{}^1.FliN^3$ tetramer identified by mass spectroscopy and structural homology with the Type III injectosome components [95]. The C-terminal tail of the FliH component of the flagellar export ATPase assembly has been co-crystallized and bound to the $FliM_C$.FliN heterodimer (Figure 4B). The co-crystal validated early predictions of the interactions of the C ring with the flagellar export apparatus [96], as well as the proposal based on cryo-tomographic (cryo-ET) reconstruction and FliH sequences from diverse species that FliH acts as a spacer to set C ring diameter [97].

The switch complex contains FliY instead of, or in addition to, FliM and FliN in numerous species [98]. The structure of *T. maritima* FliY reveals a middle domain with strong structural similarity to $FliM_M$, while the FliY C-terminal domain is similar to FliN [98]. This study also reported solution assays to show that *T. maritima* FliY homodimerizes via its N-terminal domain and does not have an increased affinity for activated versus inactive CheY, and that its middle domain does not bind FliG. FliY was first reported in the Gram-positive *B. subtilis* [99] where CheY phosphorylation enhances CCW, not CW rotation [100], in contrast to the Gram-negative γ-proteobacteria *E. coli* and *S. enterica*. The characterization of the *B. subtilis* rotor module by in situ crosslinking found that the $FliM_M FliG_M$ interface is conserved as in other bacteria with FliY proposed to form an external ring adjacent to the $FliM_M$ ring [101].

Thus, $FliM_C$ interacts with FliN to form the C ring base and can, in principle, transmit structural perturbations triggered by CheY-FliN association to $FliM_M$. Other bacteria, such as *T. maritima*, may utilize a different signal strategy and a distinct basal ring architecture.

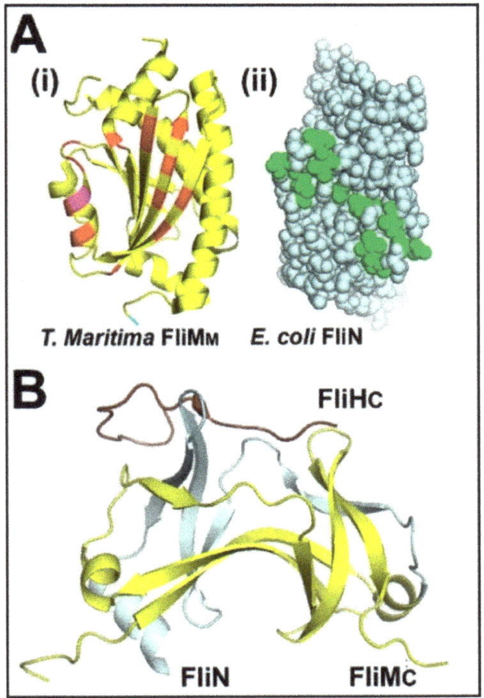

Figure 4. Chemotactic signal processing in the basal C ring. (**A**) Secondary CheY binding sites: (**i**) *T. maritima* $FliM_M$ residues showing NMR chemical shifts (red) in the presence of BeF3-activated CheY. Strongly shifted residue position (magenta) (PDB: 2hp7 (modified from [90])). (**ii**) *E. coli* FliN homodimer showing residues that impair binding of an activated CheY-$FliM_N$ fusion protein. Substitutions at these residue positions result in CCW rotation (PDB: 1yab (modified from [91])). (**B**) Basal C ring architecture: (i) Atomic structure of the $FliM_C$ (yellow)–FliN (cyan) heterodimer in complex with the FliHc terminal peptide (brown) fused with lysozyme (PDB: 4yxc (modified from [94])).

3.3. Bidirectional Torque Generation—$FliG_M$–$FliG_C$ Interactions

Vital clues driven by FliG multi-domain crystal structures have emerged since 2003 on the mechanism of rotation reversal, albeit from a limited set of organisms (*S. enterica*, *T. maritima*, *H. pylori* and *A. aeolicus*). Mutagenesis screens, experimental and computational analyses of protein dynamics and coevolutionary information have supplemented the crystal structures in important ways, as outlined below.

The variable orientations of the $FliG_C$ domain relative to $FliG_M$ inform on both rotation reversal in addition to assembly. A striking example has been a pair of *H. pylori* $FliG_{MC}$ structures that show $FliG_C$ in orthogonal orientations relative to $FliG_M$ [102]. The underlying motions have been characterized by experimental probes and atomistic protein dynamics simulations since the snapshots of the kinetic mechanism provided by the crystal structures, though valuable, are too sparse to determine reaction trajectories. $FliG_C$ reorientation was first deduced from in situ crosslinking in *E. coli* [77]. Reorientation of the *S. enterica* α-helical linker ($helix_{MC}$) adjacent to the GG pair was inferred from in vivo crosslinking experiments and simulations, to regulate the switching between the CW and CCW states. The crosslinking experiments found that the $helix_{MC}$ G174C residue substitution formed crosslinks in the CW-locked $FliG_{PEV}$ but not the native, dominantly CCW *S. enterica* strain [103]. Analyses of conformational ensembles simulated from the $FliG_{MC}$ structures had identified $helix_{MC}$ as a central hinge and predicted its melting regulated the $FliG_C$ orientation relative to $FliG_{MC}$ [81].

These studies, taken together, supported the large reorientations deduced from the superimposition of different crystal structures.

The simulations also reported that the *T. maritima* MFXF linker between $FliG_C$ ARM-C and $C\alpha_{1\text{-}6}$ amplifies thermal fluctuations of the coupled $FliM_M$–$FliG_M$ central processing unit [81]. This idea was supported and extended in important ways by NMR and MD simulations of *V. alginolyticus* $FliG_C$. The conformational dynamics of $FliG_C$-$C\alpha_{1\text{-}6}$ helix $\alpha 1$ in FliG A282T were reduced relative to the native protein to give detectible peaks in the NMR spectra. The A282T strain has a CW bias in contrast to the CCW bias of the native strain. MD simulations further showed the reduction was due to additional hydrogen bonding contacts between the ARM-C and $FliG_C$ α1-6 that constrained the flexibility of the intervening linker. $MFXF_{254}$ hinge orientation, monitored by residue F254, partitioned into conformational clusters that overlapped the subsets of the crystal structures representing either the CW or CCW rotation states based on other criteria (Figure 5A,B). A prescient early analysis of *E. coli* FliG residue substitutions in the $helix_{MC}$ linker, sensitized by a serendipitous E232G substitution a few positions upstream of the MFXF hinge, had reported these gave rise to diverse phenotypes with altered switching frequencies, pausing or altered bias [104].

The complete *Aquifex aeolicus* FliG structure revealed an ARM fold for the N-terminal domain ($FliG_N$) in addition to $FliG_M$ and $FliG_C$, separated by α-helical linkers [105]). Sequence similarity suggests these domains arose from gene duplication [106]. The structure reported inter-molecular stacking between the $FliG_M$ and FliG ARM_C domains. In contrast, the *S. enterica* $FliG_{PEV}$ [107] and *T. maritima* [78] crystal structures of the $FliG_{MC}$ complexes showed intramolecular stacking of these domains. The consensus view now is that the assembly of the FliG middle and C-terminal domain is mediated by intermolecular stacking. Solution pulsed dipolar ESR spectroscopy combined with residue substitutions first indicated that, in *T. maritima*, these domains self-assemble via the $FliG_M$–FliG ARM_C intermolecular stacking contact; then, direct assembly of FliM [108]. The central and membrane-proximal sections of the *S. enterica* C ring map was fit well by the ESR-derived model (Figure 5C). A domain-swap mechanism for FliG ring assembly in the *E. coli* motor was subsequently determined with SAXS analysis supported by in situ cross-linking [109]. Binding energy from FliG association with the FliF MS ring was speculated to alter the conformational equilibrium between the compact and extended conformation of the malleable $helix_{MC}$ to favor polymerization of a chained $FliG_{MC}$ ring in the flagellar motor but not in solution.

Coevolutionary information has emerged as an important high throughput tool to assess the design principles of bacterial multi-protein complexes at the single residue level. Residue coevolution supported the design of the switch machinery framed by experiments and simulations. First, the role of $FliM_M$ as the central relay was consistent with its strongly coevolved inter-subunit and $FliG_M$ interfacial contacts [110]. Two distinct *T. maritima* $FliM_M$ dimer configurations were obtained when dimer formation was simulated based on coevolved subunit couplings [111], although one orientation did not match either the CW and CCW dimers deduced from *S. enterica* residue substitutions [76], either because of limited sampling and/or because the bi-directional switch is not universal across species. Secondly, coevolution provided strong support for conservation of the $FliG_C$–$FliG_M$ stacking contact [109,112] relative to other contacts identified by cross-link data [77]. Thirdly, the coevolved coupling was mapped, in part, onto the identified dynamic couplings and modules. Notably, the coevolved $FliM_{Mc}$–$FliG_M$ contacts was mapped onto the dynamic couplings across the *T. maritima* $FliM_M$.$FliG_M$ interface [81] and $FliG_{MC}$, as a coevolved network with distinct nodes as allosteric sectors [110]. The nodes included the well-characterized EHPQR and PEV (in *S. enterica*) at the $FliM_M$.$FliG_M$ interface as well as α-helices ($helix_{MC}$, $C\alpha_{1\text{-}6}$ helix α_1). The melting of these α-helices has been noted above. The coevolution signal from $FliG_C$ ARM-C is sparse. This sub-domain may be the converter element that encodes different species-specific outputs from a conserved input signal (Figure 5D).

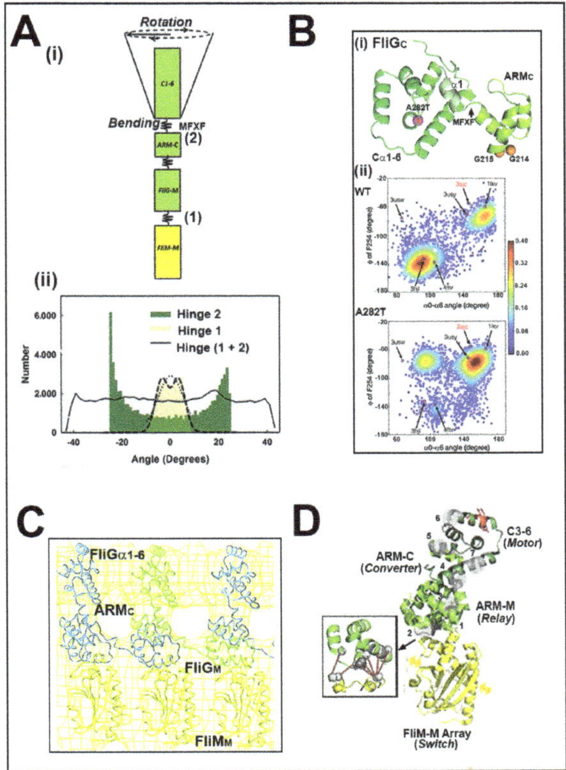

Figure 5. Mechanics of rotation reversal. (**A**) Mechanical amplification: (**i**) The *T. maritima* $FliM_MFliG_{MC}$ crystal structure (PDB: 4fhr. $FliM_M$ (gold), $FliG_{MC}$ (green)), modelled as a segmented rod with flexible hinges. Two-stage amplification of $FliM_M$ motions triggers large $FliG_C$ domain reorientations. (**ii**) Bimodal hinge planar angle distributions from the dominant principal collective motions 1–3 extracted from the generated conformational ensemble. Lines plot the 3D angular displacement distribution (solid) and unimodal fit (dotted). The distribution of the $FlIM_MFliG_M$ interfacial hinge (1 (gold)) has a narrow spread relative to the MFXF hinge distribution (2 (green)), indicating most of the amplification occurs at the latter hinge (from [81]). (**B**) MFXF hinge dynamics: (**i**) The representative structure of the *V. alginolyticus* FliGc A282T homology model from cluster analysis of the MD ensemble. The predicted dynamic helix α_1 (pale green), the $MFXF_{254}$ motif, and residues G214, G215 (orange) and T282 (magenta with an asterisk) are marked. 2D-NMR reports the CW-biasing A282T residue substitution melts helix α_1 and reorients $MFXF_{254}$. (**ii**) The 2D plot of F254 dihedral angle versus helix α_6 orientation relative to the $FliG_M.FliG_C$ interface obtained from the MD trajectories. The wild-type (WT) and mutant (A282T) conformational clusters map onto different subsets of the FliGc crystal structures. Labeled circles (magenta) mark structures from *T. maritima* (CW-locked. PDB: 3ajc; wild type. PDB: 1lkv; complexed with $FliM_M$ (PDB: 4fhr)), *Helicobacter pylori* (PDB: 3usy and 3usw) and *A. aeolicus* (full-length. PDB: 3hjl) (from [113] with permission). (**C**) The $FliG_M$-ARMc stacking interaction: The best-fit of the pulsed dipolar ESR spectroscopy model of the *T. maritima* $FliG_M$-ARMc inter-domain stack to the 3D *S. enterica* electron density map (from [108] with permission). (**D**) Residue coevolution model for rotation reversal: Strong, coevolved $FliM_M$ (yellow) contacts mediate conformational spread (arrows) in the $FliM_M$ ring. The FliG αC3-6 "motor domain" (dark green) is organized around the αC5 "torque helix" with charged residues (red) important for rotor–stator interactions. The primary nodes of the coevolved network form a relay of allosteric sectors (numbered grey patches) across ARM-M

and ARM-C (light green). ARM-C has sparse coevolved contacts implying a variable fold to generate different motor responses from a conserved $FliM_M$ switch transition Box. Coevolved interfacial pairs (residues (white spheres) and contacts (red)) link the $FliM_M$ GGXG motif with the $FliG_M$ EHPQ motif. The contacts and nodes are mapped onto PDB: 4fhr (from [110]).

In conclusion, the library of crystal structures now available, together with experimental and computational measurements of their dynamics, position FliM, FliG and FliN in the *S. enterica* C ring map and charts out their connectivity and conformational plasticity. $FliM_M$ has a central location in the C ring with subunit contacts designed for propagation of both the CW and CCW conformational states in the species studied thus far. $FliM_M$ forms a tightly coupled complex with $FliG_M$, with a central location in the C ring, connected via $FliG_M$ to $FliG_C$ adjacent to the membrane stator complexes, and via $FliM_M$ to $FliM_C$ in the basal C ring.

3.4. The Association of the Switch with the Mot Stators and the $FliF_C$ Scaffold

The interactions of the FliGc domain with the MotA.MotB stator complexes determine torque. The biochemical studies of rotor–stator interactions in *E. coli* [11] were soon extended to other bacterial motors, notably the sodium-driven *V. alginolyticus* motor [114]. The $FliG_C$ torque helix and adjacent segments in the *V. alginolyticus* motor contain more charged residues compared to *E. coli* [115]. Furthermore, substitutions at two residue positions selectively impair only CCW rotation [116]. Interestingly, the *V. alginolyticus* $FliG_C$ also determines the correct polar localization and assembly of the stator complexes [117]. The mutagenesis of the residues involved in rotor–stator interactions in the *E. coli* motor was extended in the related *S. enterica* [118]. Notably, these investigators coupled fluorescent localization of the GFP-MotB fusion proteins with motility assays to parse out charged residues at the $FliG_c$–MotA interface that influence stator assembly from those dedicated to torque generation. The first cryo-ET images documenting the alteration of C ring morphology by stator complexes have now been obtained in the spirochete *Borrelia burgdorferi*. Comparative study of *motB* mutant strains with defective versus restored proton conduction suggests that functional stators are required for full expression of this effect [119] (Figure 6A). Thus, torque affects switch architecture, in addition to CW/CCW rotation bias (Section 2), while the FliG rotor ting, in turn, affects assembly of the force-generating stator complexes.

The CCW and CW torque generated by stator interactions with the FliG ring must be transmitted via the intervening FliF MS-ring to the external components of the filament whose rotation is the physiologically relevant parameter. The *fliF* and *fliG* genes are adjacent in one operon for middle gene expression in the *S. enterica* flagellar regulon [120]. Strains with these fusions are motile and form C rings but have aberrant CW/CCW bias [121]. Their analysis has provided valuable clues on the effects of the FliF–FliG association on switch operation and C ring morphology.

The bias of the full-frame fusion was restored by suppressor mutations in $FliM_M$ and $FliG_N$, as well as by insertion of a flexible, nine-residue glycine-rich linker at the fusion site [121], emphasizing the influence of $FliG_N$ and $FliG_M$ conformational plasticity on the sign of the transmitted torque. The deletion the FliF–FliG fusion (ΔFliF.FliG) strain lacked a major part of $FliG_N$,and formed smaller C rings with lower (n = 31) subunit stoichiometry [122]. In contrast, the $FliG_M$ΔPAA deletion strain formed smaller C rings with altered packing as their subunit stoichiometry was unchanged. The bias of the ΔFliF.FliG strain could be also be compensated by residue substitutions localized to the $FliM_M$.$FliG_M$ interface ($FliG_{D124Y}$) and a likely $FliM_M$ dimerization interface (F188Y, V186A). Cryo-electron microscopy of isolated $FliG_{D124Y}$ basal bodies showed that D124Y substitution partly restored the C ring diameter towards wild-type values. These results demonstrate that either subunit number or spacing variation can change C ring size. While FliF.$FliG_N$ association primarily transmits torque, it also has downstream effects on the switch machinery for torque reversal. There seems to be a synergistic relationship between the supramolecular organization of the $FliG_N$ and $FliM_M$ rings with defects in one restored by compensating alterations in the other (Figure 6B).

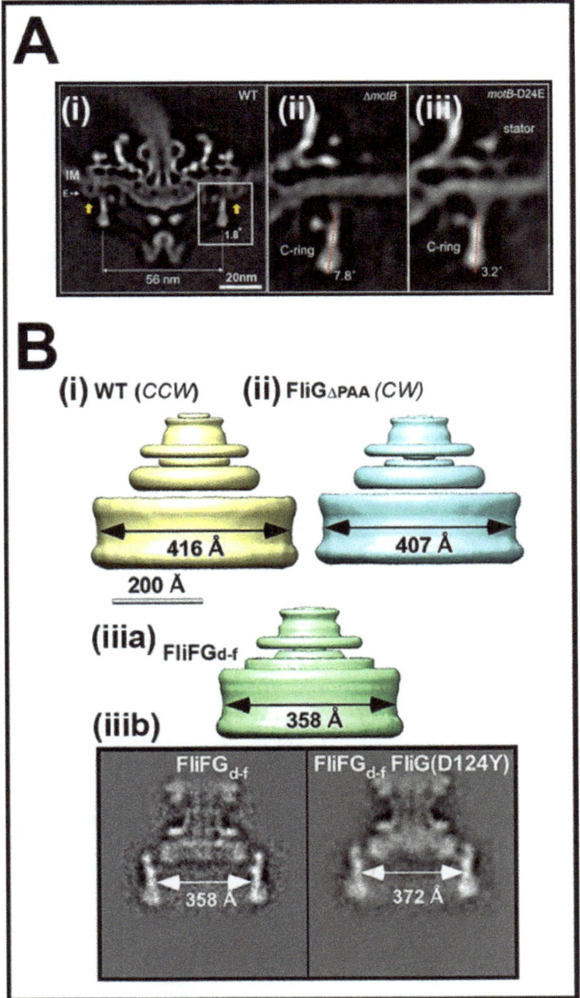

Figure 6. C ring modulation by interfacial membrane protein interactions. (**A**) C ring modulation by stator operation: (**i**) Central tomogram section of the wild-type *B. burgdorferi* flagellar basal structure. The C ring wall (box) is almost perpendicular (1.8°) to the Mot stator complexes (yellow arrows). The C ring has different orientations in (**ii**) ΔMotB (7.8°) and (**iii**) MotB-24DE (3.2°) mutants. The Δ*motB* strain does not assemble stators; the *motB*-24DE strain assembles stators and is motile (reproduced from [119] with permission). (**B**) FliF-FliG deletions alter C ring size. Size differences between *S. enterica* (i) CCW and (ii) CW-locked ($FliG_{\Delta PAA}$) C rings. (**iii.a**) The CW-biased *S. enterica* ΔFliF–FliG fusion assembles smaller C rings. (**iii.b**) 2D class averages from the ΔFliF–FliG fusion basal body; without and with the $FliG_M$ D124Y residue substitution that restores normal bias (reproduced from [122] with permission).

How does $FliG_N$ transmit torque generated at the FliG ring periphery to the axial rod and filament via FliF? The architecture of the *S. enterica* hook basal body revealed by the work of DeRosier and colleagues (see [50] and references therein) notably shows different symmetries for the internal and external modules; the cytoplasmic C ring (n = 33–36) and the external hook (n = 11), for example. The architectural design by which torque transmission is achieved and symmetry mismatch accommodated is starting to be understood. The structural adaptation for torque transmission was determined in *T. maritima* [123] and, subsequently, the sodium-powered *V. alginolyticus* motor [124]. NMR reported extensive conformational changes in the *T. maritima* $FliG_N$ ARM fold upon interaction with 46 FliF C-terminal residues. These changes were due to formation of the co-folded $FliG_N$ and FiF C-tail domain. $FliG_N$ in the crystallized co-folded domain alters conformation to closely match the $FliG_M$ fold [125], and possibly also conformational plasticity compatible with the compensatory interactions between the two domains in the regulation of C ring size. The crystal structure of the *H. pylori* co-folded domain corroborated the essential features reported for *T. maritima* [126].

The cytoplasmic FliF segment contiguous with the co-folded $FliG_N.FliF_{C\text{-tail}}$ domain is predicted to form a predominantly α-helical connector to the periplasmic C-terminal FliF ($FliF_C{}^{per}$) [112]. The recent 3D-cryoelectron microscopy reconstruction of overproduced *S. enterica* FliF rings has shown that $FliF_C{}^{per}$ forms a periplasmic scaffold with a split ring-building motif (RBM) that staples together an anti-parallel β-barrel to form the external periplasmic modules of the MS ring while the N-terminal FliF forms inner RBM modules structurally homologous to RBMs characterized for Type-III injectosomes [127]. Both the split RBM and the β-barrel were predicted by residue coevolution in the course of ongoing work on the full-frame FliF.FliG fusion ring [112]. Thus, the FliF flagellar motor scaffold has evolved to add a particularly stable, periplasmic C-terminal domain to the injectosome RBMs. The map further reveals that $FliF_C{}^{per}$ symmetry (n = 33–34), within the design tolerance reported for other biomolecular assemblies, matches the C ring symmetry to dispel the MS and C ring symmetry mismatch conundrum raised by initial estimates of a lower FliF subunit stoichiometry. Figure 7 summarizes these advances.

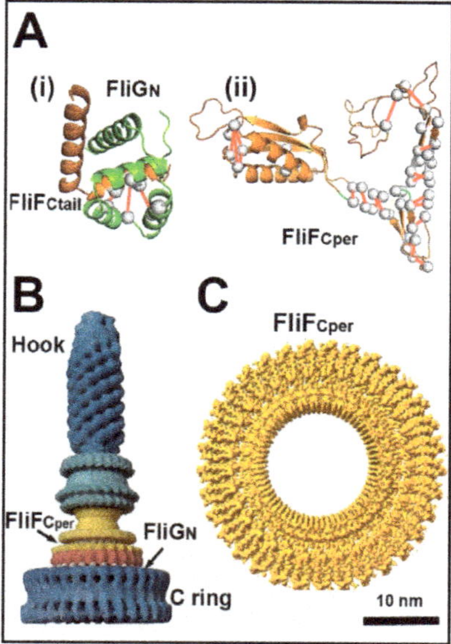

Figure 7. Transmembrane torque transmission. (**A**) (**i**) Conserved $FliG_N/FliF_{C\text{-tail}}$ coevolved contacts mapped onto the *T. maritima* crystal structure (PDB: 5tdy [125]). (**ii**) Predicted $FliF_C{}^{per}$ architecture obtained by residue coevolution. Red lines denote coevolved residue pairs as in Figure 5D (modified from [112]). (**B**) The 3D model of the *S. enterica* hook–basal body complex. The basal body part of the structure was from a 3D reconstruction as published [50]. The hook with attached FliD cap structure was done by single-particle methods using the entire hook with the cap as the single-particle (Dennis Thomas, unpublished results (with permission)). (**C**) The 3D map of the *S. enterica* $FliF_C$ torque transmitter module (EMD-10143, 3.1-angstrom resolution (modified from [127])). The *en-face* view resolves 33-subunits.

In summary, rotor–stator interactions employ a core set of charged residues and control large scale changes in stator assembly and C ring morphology either side of the interaction interface. The $FliF_C^{per}$ module presumably templates the assembly of the FliG ring via $FliF_{C\text{-tail}}$ that is part of the co-folded domain. $FliG_M$ then dictates FliM assembly via contacts with $FliM_M$. The $FliM_M$ ring drives reorientation of $FliG_C$ for bidirectional torque generation and remodels in response to lesions in the co-folded $FliG_N.FliF_{C\text{-tail}}$ domain, Thus, there is conformational coupling of $FliM_M$ to these distant domains via $FliG_M$. The correlation between C ring size and rotation state, driven by defined lesions, is comparable to the rotation–fluorescence intensity correlation reported for GFP-tagged motors. It provides the first structural clue for the difference reported for the CW versus CCW torque–velocity relations.

4. Current Challenges

This review draws overwhelmingly from the *E. coli*, *S. enterica* and sodium *V. alginolyticus* motors for the elucidation of their molecular mechanisms, with a notable contribution from thermophile and *H. pylori* crystal structures. The fundamental issues regarding switch synchrony and ultra-sensitivity had been posed by 2003. The present knowledge of the structural basis for torque generation and transmission does not answer these issues but takes their study to a new level. Other species with diverse phylogeny have revealed the diversity of the switch operation. The bacillus *B. subtilis* has inverted motile responses to CheY. *V. alginolyticus* is one of several bacteria that alter switch frequency rather than rotation bias in response to CheY activation. The flagellum of the α-proteobacterium *R. sphaeroides* stops and starts. The thermophile *A. aeolicus* has been reported to lack FliM, consistent with its mostly smooth-swim motile behavior [128]. A major new challenge as illustrated by the phylogenetic tree of $FliG_C$, the "motor" domain (Figure 8A), is to explain how such diversity can arise from a small core of protein components.

Cryo-electron tomography (cryo-ET) has been an important role for the appreciation of the diverse morphology of flagellar motors (reviewed in [129,130]), even though it is presently limited to thin bacteria or mini-cells. Crucially, the technique provides 3D-reconstructions of complete motors that capture rotor–stator interactions [131], the effect of the cell membrane upon rotor flexure [132] and novel cell wall motor components [133]. Mutagenesis guided Cryo-ET has, interestingly, established diversification of C ring architecture for various cellular functions within a single species (*Campylobacter jejuni*) [133], and has led to the realization that high-torque motors have larger diameter stator and C rings [134].

The emergence of informatics due to developments in high-throughput sequencing, mass spectroscopy and high-performance computing is the highlight of this era, as appreciated by the increase in protein sequences (135,850 (2003) -> 177,754,527 (2020) (www.ebi.ac.uk/uniprot)) and atomic structures (<20,000 (2003) -> >140,000 (2018) [135]). Computational strategies for protein dynamic simulations [87,136] have benefited from the expanded databases. Most core switch components belong to a larger superfamily. The remarkable diversity of the CheY response regulator superfamily has been reviewed recently [137]. Several response regulators are expressed concurrently within one species. Many species have multiple CheYs of which typically one interacts with the flagellar motor: a study on *V. cholerae* being an early example [138]. FliF, FliM, FliN and FliY are members of larger families that include Type III injectosome components (reviewed in [139]). FliG has distant homology to the MgtE transporter family [140]. The diversity of the basal C ring and the CheY atomic structures suggests that bacteria utilize multiple strategies for signal reception.

The *E. coli* and *S. enterica* bacteria remain the primary source for motor rotation assays in conjunction with native and chimeric *V. alginolyticus* sodium motors. Structural knowledge, while anchored in these bacteria, now has important contributions from other species driven by the developments in cryo-ET and informatics. The current understanding of the flagellar motor switch is based on a generic, integrated assembly with contributions from multiple species, as schematized in Figure 8B.

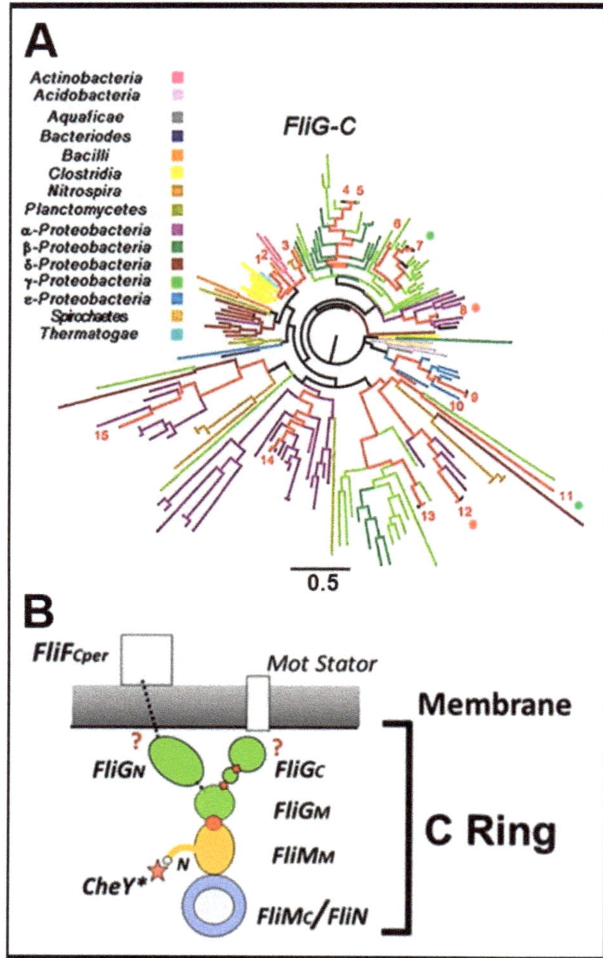

Figure 8. (**A**) The phylogenetic tree of the $FliG_C$ motor domain. 160 seed sequences (multiple sequences from 23 species). Species (red lines), with well-studied flagellar biochemistry, physiology or structure (1 = *Thermatoga maritima*, 2 = *Bacillus subtilis*, 3 = *Borrelia burgdorferi*, 4 = *Escherichia coli*, 5 = *Salmonella enterica*, 6 = *Vibrio cholerae*, 7 = *Vibrio alginolyticus1*, 8 = *Rhodobacter sphaeroides1*, 9 = *Helicobacter pylori*, 10 = *Aquifex aeolicus*, 11 = *Vibrio alginolyticus2*, 12 = *Rhodobacter sphaeroides2*, 13 = *Vibrio parahaemolyticus*, 14 = *Caulobacter crescentus*, 15 = *Rhizobium meliloti*). Asterisks (*R. sphaeroides* (red), *V. alginolyticus* (green)) mark duplicates. (from ([110])). (**B**) The signal pathways in the BFM switch for the response to chemotactic stimuli, motor torque and load. The N-terminal FliM peptide (N) anchors activated CheY (CheY*) by a flexible tether and primes it to bind additional sites on the basal C ring (FliM/FliN). The bound CheY* modulates thermal $FliM_N$ fluctuations for radial circumferential conformation spread and amplifies them via the conserved $FliM_M$.$FliG_M$ interface (red circle), and two hinges (red diamonds) bordering FliG ARM-C to reorient the peripheral FliGc Ca1-6 sub-domain. Inter-subunit stacking between the dynamic $FliG_M$ and ARM-C modules stabilizes the FliG ring and regulates rotation reversal. The Mot stator complexes step along Ca1-6 to generate torque and influence $FliG_{MC}$ dynamics. The conserved co-folded $FliG_N$/$FliF_{C\text{-tail}}$ domain ensures matching MS and C ring subunit stoichiometry. It is contiguous with the FliF C-terminal half ($FliF_{Cper}$) that forms a stable assembly scaffold. Co-folded domain lesions/fusions have long-range effects on C ring architecture. Thus, $FliG_M$ orchestrates bidirectional conformational coupling between $FliM_M$ and either $FliG_C$ or $FliG_N$. The atomic details of the rotor–stator interactions or the linkage between $FliG_M$ and $FliF_{Cper}$ are poorly understood.

Charged residues on a single $FliG_C$ α-helix (torque helix) are the primary determinant of rotor–stator interactions. $FliG_N$ co-folds with a FliF C-terminal fragment ($FliG_N.FliF_{C\text{-}tail}$) to form a domain with a similar architecture to $FliG_M$; $FliG_N$ full-frame or deletion fusions can be compensated by engineered modifications in the $FliG_M$ domain. The co-folded domain connects to $FliF_C{}^{per}$, the assembly scaffold for the C ring. The matching 33–34 symmetries of $FliF_C{}^{per}$ and the C ring support the 1:1 stoichiometry for $FliG_N$,$FliF_{C\text{-}tail}$ seen in the crystal structures. A chained $FliG_C$–$FliG_M$ stack is the working model for rotor ring organization, but its assembly is probably modulated by other C ring components whose genes encode *mot* alleles [8,9]. There is not a straightforward match of the FliG subunit symmetry with the 26 steps per revolution resolved in rotation assays. Other stator–rotor contacts may determine step periodicity. The in-situ conformation and dynamics of the linkage between the chained $FliG_C$–$FliG_M$, the co-folded $FliG_N.FliF_{C\text{-}tail}$ domain and $FliF_C{}^{per}$ remain to be determined.

The connectivity of the molecular linkage between the $FliM_N$ CheY binding site and the $FliG_C$ torque helix is now understood in broad outlines. $FliM_N$ tethered the CheY associates with secondary binding sites on the basal C ring that may differ with species. $FliM_M$ monomers fluctuate between bi-stable conformations that likely reflect the CCW and CW operational states. These states propagate across the C ring via inter-domain contacts and across a conserved interface to $FliG_M$. $FliG_M$ and its stacking contact with $FliG_C$ ARM-C are dynamic. $Helix_{MC}$ melting within the flexible inter-domain GG linker in conjunction with adjacent α-helices orchestrates these dynamics. The FliGc intra-domain MFXF hinge amplifies $FliG_M$ motions to cause a large re-orientation of the $FliG_C$ $\alpha C_{3\text{-}6}$ torque helix subdomain. These reorientations may reflect CCW <-> CW transitions. CW and CCW-locked motors have different torque velocity relations. Substitutions that alter the fold of $FliG_C$ $\alpha C_{3\text{-}6}$ affect MFXF hinge dynamics and selectively impair CCW rotor-stator interactions. Residue substitutions that enhance CW or CCW bias have different $FliG_M$ dynamics. Rotation state is influenced by changes in C ring size due to subunit number or packing variations. C ring morphology is also altered by the presence and activity of the stator complexes. The molecular basis of the coupling between $FliG_M$ dynamics and CCW <-> CW transitions is not defined, let alone the coupling between the C ring dynamics and the torque–velocity relation. Adaptive subunit exchange of GFP-tagged basal-ring components may influence switch ultra-sensitivity to CheY activity. The C-ring is disrupted by single residue perturbations [141] and might also be impaired by GFP, which has a similar size to the proteins tagged. However, the CCW adaptive increase in basal body fluorescence is difficult to explain by GFP perturbation of C ring assembly. Nevertheless, the correlation between adaptation and subunit turnover is qualitative and determination of whether it is determinative or incidental must await structural elucidation of the turnover mechanism. Finally, the integrated picture is based on studies of a few species. The complete diversity of the species under current study is substantially more, as glimpsed in motor output, CheY function, basal C ring composition and C ring architecture. The determination of the conformational transitions of the stator complexes during the work cycle is a likely prerequisite for the explanation of their long-range effects on C ring morphology. More generally, elucidation, as opposed to description, of switch diversity is a severe challenge that may be intractable is the absence of the fundamental signal and energy transduction mechanism.

Mathematical models have closely tracked the progress in motor physiology. The conformational spread model has been modified and extended, while a fundamentally different non-equilibrium motor model has been developed. These advances have led to the appreciation that switch operation cannot be understood in isolation from motor mechanics. However, the details of the chemical machinery that these models seek to explain has remained at the 2003 level. There is an urgent need for a top–down development of these models to discriminate between possible molecular mechanisms. MD simulations based on X-ray structures, NMR, ESR spectra and foot-printing techniques offer a bottoms-up approach to supply the kinetics required to correlate the molecular level descriptors known thus far with motor physiology. An atomic-level model of a rotor module from even one species will be a game-changer for a kinetic model with the necessary predictive power, although the scale-up of the simulations will be a logistic challenge with present-day resources. Nevertheless, research on this remarkable biomolecular machine has consistently advanced in unforeseen ways with fundamental implications for protein energetics and allostery. The innovation of its research community will ensure that it continues to do so.

Funding: This article received no external funding.

Acknowledgments: The author is grateful for the following colleagues for comments, discussion and permission to use material—Fan Bai, David Blair, Phillippe Cluzel, Brian Crane, Michio Homma, Lawrence Lee, Jun Liu, Chien Jung-Lo, Tohru Minamino, David DeRosier, Jonathon Scholey, Dennis Thomas, Yuhai Tu, and Junhua Yuan.

Conflicts of Interest: The author declares no conflict of interest.

References

1. Lane, N. The unseen world: Reflections on Leeuwenhoek (1677) 'Concerning little animals'. *Philos. Trans. R. Soc. Lond. Ser. B Biol. Sci.* **2015**, *370*. [CrossRef] [PubMed]
2. Berg, H.C.; Anderson, R.A. Bacteria swim by rotating their flagellar filaments. *Nature* **1973**, *245*, 380–382. [CrossRef] [PubMed]
3. Berg, H.C. The rotary motor of bacterial flagella. *Annu. Rev. Biochem.* **2003**, *72*, 19–54. [CrossRef] [PubMed]
4. Khan, S.; Scholey, J.M. Assembly, Functions and Evolution of Archaella, Flagella and Cilia. *Curr. Biol.* **2018**, *28*, R278–R292. [CrossRef] [PubMed]
5. Parkinson, J.S. cheA, cheB, and cheC genes of Escherichia coli and their role in chemotaxis. *J. Bacteriol.* **1976**, *126*, 758–770. [CrossRef] [PubMed]
6. Yamaguchi, S.; Aizawa, S.; Kihara, M.; Isomura, M.; Jones, C.J.; Macnab, R.M. Genetic evidence for a switching and energy-transducing complex in the flagellar motor of Salmonella typhimurium. *J. Bacteriol.* **1986**, *168*, 1172–1179.
7. Kihara, M.; Homma, M.; Kutsukake, K.; Macnab, R.M. Flagellar switch of Salmonella typhimurium: Gene sequences and deduced protein sequences. *J. Bacteriol.* **1989**, *171*, 3247–3257.
8. Sockett, H.; Yamaguchi, S.; Kihara, M.; Irikura, V.M.; Macnab, R.M. Molecular analysis of the flagellar switch protein FliM of Salmonella typhimurium. *J. Bacteriol.* **1992**, *174*, 793–806. [CrossRef]
9. Irikura, V.M.; Kihara, M.; Yamaguchi, S.; Sockett, H.; Macnab, R.M. Salmonella typhimurium fliG and fliN mutations causing defects in assembly, rotation, and switching of the flagellar motor. *J. Bacteriol.* **1993**, *175*, 802–810. [CrossRef]
10. Zhou, J.; Lloyd, S.A.; Blair, D.F. Electrostatic interactions between rotor and stator in the bacterial flagellar motor. *Proc. Natl. Acad. Sci. USA* **1998**, *95*, 6436–6441.
11. Lloyd, S.A.; Blair, D.F. Charged residues of the rotor protein FliG essential for torque generation in the flagellar motor of Escherichia coli. *J. Mol. Biol.* **1997**, *266*, 733–744. [CrossRef] [PubMed]
12. Blair, D.F.; Berg, H.C. Restoration of torque in defective flagellar motors. *Science* **1988**, *242*, 1678–1681. [CrossRef] [PubMed]
13. Zhou, J.; Sharp, L.L.; Tang, H.L.; Lloyd, S.A.; Billings, S.; Braun, T.F.; Blair, D.F. Function of protonatable residues in the flagellar motor of Escherichia coli: A critical role for Asp 32 of MotB. *J. Bacteriol.* **1998**, *180*, 2729–2735. [CrossRef]
14. Khan, I.H.; Reese, T.S.; Khan, S. The cytoplasmic component of the bacterial flagellar motor. *Proc. Natl. Acad. Sci. USA* **1992**, *89*, 5956–5960. [PubMed]
15. Francis, N.R.; Sosinsky, G.E.; Thomas, D.; DeRosier, D.J. Isolation, characterization and structure of bacterial flagellar motors containing the switch complex. *J. Mol. Biol.* **1994**, *235*, 1261–1270. [CrossRef]
16. Zhao, R.; Pathak, N.; Jaffe, H.; Reese, T.S.; Khan, S. FliN is a major structural protein of the C-ring in the Salmonella typhimurium flagellar basal body. *J. Mol. Biol.* **1996**, *261*, 195–208. [CrossRef]
17. Tang, H.; Braun, T.F.; Blair, D.F. Motility protein complexes in the bacterial flagellar motor. *J. Mol. Biol.* **1996**, *261*, 209–221. [CrossRef]
18. Lux, R.; Kar, N.; Khan, S. Overproduced Salmonella typhimurium flagellar motor switch complexes. *J. Mol. Biol.* **2000**, *298*, 577–583. [PubMed]
19. Taylor, K.A.; Glaeser, R.M. Retrospective on the early development of cryoelectron microscopy of macromolecules and a prospective on opportunities for the future. *J. Struct. Biol.* **2008**, *163*, 214–223. [CrossRef]

20. Shaikh, T.R.; Gao, H.; Baxter, W.T.; Asturias, F.J.; Boisset, N.; Leith, A.; Frank, J. SPIDER image processing for single-particle reconstruction of biological macromolecules from electron micrographs. *Nat. Protoc.* **2008**, *3*, 1941–1974. [CrossRef]
21. Thomas, D.R.; Morgan, D.G.; DeRosier, D.J. Rotational symmetry of the C ring and a mechanism for the flagellar rotary motor. *Proc. Natl. Acad. Sci. USA* **1999**, *96*, 10134–10139. [CrossRef] [PubMed]
22. Thomas, D.; Morgan, D.G.; DeRosier, D.J. Structures of bacterial flagellar motors from two FliF-FliG gene fusion mutants. *J. Bacteriol.* **2001**, *183*, 6404–6412. [PubMed]
23. Young, H.S.; Dang, H.; Lai, Y.; DeRosier, D.J.; Khan, S. Variable symmetry in Salmonella typhimurium flagellar motors. *Biophys. J.* **2003**, *84*, 571–577. [PubMed]
24. Kubori, T.; Shimamoto, N.; Yamaguchi, S.; Namba, K.; Aizawa, S. Morphological pathway of flagellar assembly in Salmonella typhimurium. *J. Mol. Biol.* **1992**, *226*, 433–446. [PubMed]
25. Ueno, T.; Oosawa, K.; Aizawa, S. M ring, S ring and proximal rod of the flagellar basal body of Salmonella typhimurium are composed of subunits of a single protein, FliF. *J. Mol. Biol.* **1992**, *227*, 672–677. [PubMed]
26. Meister, M.; Berg, H.C. The stall torque of the bacterial flagellar motor. *Biophys. J.* **1987**, *52*, 413–419. [CrossRef]
27. Chen, X.; Berg, H.C. Torque-speed relationship of the flagellar rotary motor of Escherichia coli. *Biophys. J.* **2000**, *78*, 1036–1041. [CrossRef]
28. Sowa, Y.; Hotta, H.; Homma, M.; Ishijima, A. Torque-speed relationship of the Na+-driven flagellar motor of Vibrio alginolyticus. *J. Mol. Biol.* **2003**, *327*, 1043–1051. [CrossRef]
29. Berry, R.M.; Berg, H.C. Absence of a barrier to backwards rotation of the bacterial flagellar motor demonstrated with optical tweezers. *Proc. Natl. Acad. Sci. USA* **1997**, *94*, 14433–14437. [CrossRef]
30. Berry, R.M.; Berg, H.C. Torque generated by the flagellar motor of Escherichia coli while driven backward. *Biophys. J.* **1999**, *76*, 580–587. [CrossRef]
31. Fahrner, K.A.; Ryu, W.S.; Berg, H.C. Biomechanics: Bacterial flagellar switching under load. *Nature* **2003**, *423*, 938. [CrossRef] [PubMed]
32. Kudo, S.; Magariyama, Y.; Aizawa, S. Abrupt changes in flagellar rotation observed by laser dark-field microscopy. *Nature* **1990**, *346*, 677–680. [CrossRef] [PubMed]
33. Lloyd, S.A.; Whitby, F.G.; Blair, D.F.; Hill, C.P. Structure of the C-terminal domain of FliG, a component of the rotor in the bacterial flagellar motor. *Nature* **1999**, *400*, 472–475. [CrossRef] [PubMed]
34. Brown, P.N.; Hill, C.P.; Blair, D.F. Crystal structure of the middle and C-terminal domains of the flagellar rotor protein FliG. *EMBO J.* **2002**, *21*, 3225–3234. [CrossRef] [PubMed]
35. Garza, A.G.; Biran, R.; Wohlschlegel, J.A.; Manson, M.D. Mutations in motB suppressible by changes in stator or rotor components of the bacterial flagellar motor. *J. Mol. Biol.* **1996**, *258*, 270–285. [CrossRef]
36. Stock, A.M.; Mottonen, J.M.; Stock, J.B.; Schutt, C.E. Three-dimensional structure of CheY, the response regulator of bacterial chemotaxis. *Nature* **1989**, *337*, 745–749. [CrossRef]
37. Silversmith, R.E.; Bourret, R.B. Throwing the switch in bacterial chemotaxis. *Trends Microbiol.* **1999**, *7*, 16–22. [CrossRef]
38. Yan, D.; Cho, H.S.; Hastings, C.A.; Igo, M.M.; Lee, S.Y.; Pelton, J.G.; Stewart, V.; Wemmer, D.E.; Kustu, S. Beryllofluoride mimics phosphorylation of NtrC and other bacterial response regulators. *Proc. Natl. Acad. Sci. USA* **1999**, *96*, 14789–14794. [CrossRef]
39. Welch, M.; Oosawa, K.; Aizawa, S.; Eisenbach, M. Phosphorylation-dependent binding of a signal molecule to the flagellar switch of bacteria. *Proc. Natl. Acad. Sci. USA* **1993**, *90*, 8787–8791. [CrossRef]
40. Eisenbach, M.; Caplan, S.R. Bacterial chemotaxis: Unsolved mystery of the flagellar switch. *Curr. Biol.* **1998**, *8*, R444–R446. [CrossRef]
41. Lee, S.Y.; Cho, H.S.; Pelton, J.G.; Yan, D.; Henderson, R.K.; King, D.S.; Huang, L.; Kustu, S.; Berry, E.A.; Wemmer, D.E. Crystal structure of an activated response regulator bound to its target. *Nat. Struct. Biol.* **2001**, *8*, 52–56. [CrossRef] [PubMed]
42. McEvoy, M.M.; Bren, A.; Eisenbach, M.; Dahlquist, F.W. Identification of the binding interfaces on CheY for two of its targets, the phosphatase CheZ and the flagellar switch protein fliM. *J. Mol. Biol.* **1999**, *289*, 1423–1433. [CrossRef] [PubMed]
43. Khan, S.; Pierce, D.; Vale, R.D. Interactions of the chemotaxis signal protein CheY with bacterial flagellar motors visualized by evanescent wave microscopy. *Curr. Biol.* **2000**, *10*, 927–930. [CrossRef]

44. Scharf, B.E.; Fahrner, K.A.; Turner, L.; Berg, H.C. Control of direction of flagellar rotation in bacterial chemotaxis. *Proc. Natl. Acad. Sci. USA* **1998**, *95*, 201–206. [CrossRef] [PubMed]
45. Cluzel, P.; Surette, M.; Leibler, S. An ultrasensitive bacterial motor revealed by monitoring signaling proteins in single cells. *Science* **2000**, *287*, 1652–1655. [PubMed]
46. Sagi, Y.; Khan, S.; Eisenbach, M. Binding of the chemotaxis response regulator CheY to the isolated, intact switch complex of the bacterial flagellar motor: Lack of cooperativity. *J. Biol. Chem.* **2003**, *278*, 25867–25871.
47. Khan, S.; Macnab, R.M. The steady-state counterclockwise/clockwise ratio of bacterial flagellar motors is regulated by protonmotive force. *J. Mol. Biol.* **1980**, *138*, 563–597. [CrossRef]
48. Duke, T.A.; Le Novere, N.; Bray, D. Conformational spread in a ring of proteins: A stochastic approach to allostery. *J. Mol. Biol.* **2001**, *308*, 541–553. [CrossRef]
49. Pervushin, K.; Riek, R.; Wider, G.; Wuthrich, K. Attenuated T2 relaxation by mutual cancellation of dipole-dipole coupling and chemical shift anisotropy indicates an avenue to NMR structures of very large biological macromolecules in solution. *Proc. Natl. Acad. Sci. USA* **1997**, *94*, 12366–12371. [CrossRef]
50. Thomas, D.R.; Francis, N.R.; Xu, C.; DeRosier, D.J. The three-dimensional structure of the flagellar rotor from a clockwise-locked mutant of Salmonella enterica serovar Typhimurium. *J. Bacteriol.* **2006**, *188*, 7039–7048. [CrossRef]
51. Chalfie, M.; Tu, Y.; Euskirchen, G.; Ward, W.W.; Prasher, D.C. Green fluorescent protein as a marker for gene expression. *Science* **1994**, *263*, 802–805. [CrossRef] [PubMed]
52. Nirody, J.A.; Sun, Y.-R.; Lo, C.-J. The biophysicist's guide to the bacterial flagellar motor. *Adv. Phys.* **2017**, *2*, 324–343. [CrossRef]
53. Bai, F.; Branch, R.W.; Nicolau, D.V., Jr.; Pilizota, T.; Steel, B.C.; Maini, P.K.; Berry, R.M. Conformational spread as a mechanism for cooperativity in the bacterial flagellar switch. *Science* **2010**, *327*, 685–689. [CrossRef] [PubMed]
54. Nakamura, S.; Hanaizumi, Y.; Morimoto, Y.V.; Inoue, Y.; Erhardt, M.; Minamino, T.; Namba, K. Direct observation of speed fluctuations of flagellar motor rotation at extremely low load close to zero. *Mol. Microbiol.* **2020**, *113*, 755–765. [CrossRef]
55. Sowa, Y.; Rowe, A.D.; Leake, M.C.; Yakushi, T.; Homma, M.; Ishijima, A.; Berry, R.M. Direct observation of steps in rotation of the bacterial flagellar motor. *Nature* **2005**, *437*, 916–919. [CrossRef]
56. Nakamura, S.; Kami-ike, N.; Yokota, J.P.; Minamino, T.; Namba, K. Evidence for symmetry in the elementary process of bidirectional torque generation by the bacterial flagellar motor. *Proc. Natl. Acad. Sci. USA* **2010**, *107*, 17616–17620. [CrossRef]
57. Pilizota, T.; Brown, M.T.; Leake, M.C.; Branch, R.W.; Berry, R.M.; Armitage, J.P. A molecular brake, not a clutch, stops the Rhodobacter sphaeroides flagellar motor. *Proc. Natl. Acad. Sci. USA* **2009**, *106*, 11582–11587. [CrossRef]
58. Mora, T.; Yu, H.; Sowa, Y.; Wingreen, N.S. Steps in the bacterial flagellar motor. *PLoS Comput. Biol.* **2009**, *5*, e1000540. [CrossRef]
59. Korobkova, E.A.; Emonet, T.; Park, H.; Cluzel, P. Hidden stochastic nature of a single bacterial motor. *Phys. Rev. Lett.* **2006**, *96*, 058105. [CrossRef]
60. Wang, F.; Yuan, J.; Berg, H.C. Switching dynamics of the bacterial flagellar motor near zero load. *Proc. Natl. Acad. Sci. USA* **2014**, *111*, 15752–15755. [CrossRef]
61. Turner, L.; Ryu, W.S.; Berg, H.C. Real-time imaging of fluorescent flagellar filaments. *J. Bacteriol.* **2000**, *182*, 2793–2801. [CrossRef] [PubMed]
62. Yuan, J.; Fahrner, K.A.; Berg, H.C. Switching of the bacterial flagellar motor near zero load. *J. Mol. Biol.* **2009**, *390*, 394–400. [CrossRef] [PubMed]
63. Yuan, J.; Fahrner, K.A.; Turner, L.; Berg, H.C. Asymmetry in the clockwise and counterclockwise rotation of the bacterial flagellar motor. *Proc. Natl. Acad. Sci. USA* **2010**, *107*, 12846–12849. [CrossRef] [PubMed]
64. Wang, F.; Shi, H.; He, R.; Wang, R.; Zhang, R.; Yuan, J. Non-equilibrium effects in the allosteric regulation of the bacterial flagellar switch. *Nat. Phys.* **2017**, *13*, 710–714. [CrossRef]
65. Bai, F.; Che, Y.S.; Kami-ike, N.; Ma, Q.; Minamino, T.; Sowa, Y.; Namba, K. Populational heterogeneity vs. temporal fluctuation in Escherichia coli flagellar motor switching. *Biophys. J.* **2013**, *105*, 2123–2129. [CrossRef]
66. Mora, T.; Bai, F.; Che, Y.S.; Minamino, T.; Namba, K.; Wingreen, N.S. Non-genetic individuality in Escherichia coli motor switching. *Phys. Biol.* **2011**, *8*, 024001. [CrossRef]

67. Fukuoka, H.; Sagawa, T.; Inoue, Y.; Takahashi, H.; Ishijima, A. Direct imaging of intracellular signaling components that regulate bacterial chemotaxis. *Sci. Signal.* **2014**, *7*, ra32. [CrossRef]
68. Fukuoka, H.; Inoue, Y.; Terasawa, S.; Takahashi, H.; Ishijima, A. Exchange of rotor components in functioning bacterial flagellar motor. *Biochem. Biophys. Res. Commun.* **2010**, *394*, 130–135. [CrossRef]
69. Delalez, N.J.; Berry, R.M.; Armitage, J.P. Stoichiometry and turnover of the bacterial flagellar switch protein FliN. *mBio* **2014**, *5*, e01216-14. [CrossRef]
70. Yuan, J.; Berg, H.C. Ultrasensitivity of an adaptive bacterial motor. *J. Mol. Biol.* **2013**, *425*, 1760–1764. [CrossRef]
71. Tu, Y. The nonequilibrium mechanism for ultrasensitivity in a biological switch: Sensing by Maxwell's demons. *Proc. Natl. Acad. Sci. USA* **2008**, *105*, 11737–11741. [CrossRef] [PubMed]
72. Bai, F.; Minamino, T.; Wu, Z.; Namba, K.; Xing, J. Coupling between switching regulation and torque generation in bacterial flagellar motor. *Phys. Rev. Lett.* **2012**, *108*, 178105. [CrossRef] [PubMed]
73. Ma, Q.; Nicolau, D.V., Jr.; Maini, P.K.; Berry, R.M.; Bai, F. Conformational spread in the flagellar motor switch: A model study. *PLoS Comput. Biol.* **2012**, *8*, e1002523. [CrossRef]
74. Ma, Q.; Sowa, Y.; Baker, M.A.; Bai, F. Bacterial Flagellar Motor Switch in Response to CheY-P Regulation and Motor Structural Alterations. *Biophys. J.* **2016**, *110*, 1411–1420. [CrossRef] [PubMed]
75. Tu, Y. Driven to peak. *Nat. Phys.* **2017**, *13*, 1–2. [CrossRef]
76. Park, S.Y.; Lowder, B.; Bilwes, A.M.; Blair, D.F.; Crane, B.R. Structure of FliM provides insight into assembly of the switch complex in the bacterial flagella motor. *Proc. Natl. Acad. Sci. USA* **2006**, *103*, 11886–11891. [CrossRef]
77. Lowder, B.J.; Duyvesteyn, M.D.; Blair, D.F. FliG subunit arrangement in the flagellar rotor probed by targeted cross-linking. *J. Bacteriol.* **2005**, *187*, 5640–5647. [CrossRef]
78. Vartanian, A.S.; Paz, A.; Fortgang, E.A.; Abramson, J.; Dahlquist, F.W. Structure of flagellar motor proteins in complex allows for insights into motor structure and switching. *J. Biol. Chem.* **2012**, *287*, 35779–35783. [CrossRef]
79. Paul, K.; Gonzalez-Bonet, G.; Bilwes, A.M.; Crane, B.R.; Blair, D. Architecture of the flagellar rotor. *EMBO J.* **2011**, *30*, 2962–2971. [CrossRef]
80. Lam, K.H.; Lam, W.W.; Wong, J.Y.; Chan, L.C.; Kotaka, M.; Ling, T.K.; Jin, D.Y.; Ottemann, K.M.; Au, S.W. Structural basis of FliG-FliM interaction in Helicobacter pylori. *Mol. Microbiol.* **2013**, *88*, 798–812. [CrossRef]
81. Pandini, A.; Morcos, F.; Khan, S. The Gearbox of the Bacterial Flagellar Motor Switch. *Structure* **2016**, *24*, 1209–1220. [CrossRef] [PubMed]
82. Brown, P.N.; Terrazas, M.; Paul, K.; Blair, D.F. Mutational analysis of the flagellar protein FliG: Sites of interaction with FliM and implications for organization of the switch complex. *J. Bacteriol.* **2007**, *189*, 305–312. [CrossRef] [PubMed]
83. Nishikino, T.; Hijikata, A.; Miyanoiri, Y.; Onoue, Y.; Kojima, S.; Shirai, T.; Homma, M. Rotational direction of flagellar motor from the conformation of FliG middle domain in marine Vibrio. *Sci. Rep.* **2018**, *8*, 17793. [CrossRef] [PubMed]
84. Zhang, H.; Lam, K.H.; Lam, W.W.L.; Wong, S.Y.Y.; Chan, V.S.F.; Au, S.W.N. A putative spermidine synthase interacts with flagellar switch protein FliM and regulates motility in Helicobacter pylori. *Mol. Microbiol.* **2017**, *106*, 690–703. [CrossRef] [PubMed]
85. Dyer, C.M.; Quillin, M.L.; Campos, A.; Lu, J.; McEvoy, M.M.; Hausrath, A.C.; Westbrook, E.M.; Matsumura, P.; Matthews, B.W.; Dahlquist, F.W. Structure of the constitutively active double mutant CheYD13K Y106W alone and in complex with a FliM peptide. *J. Mol. Biol.* **2004**, *342*, 1325–1335. [CrossRef] [PubMed]
86. Dyer, C.M.; Dahlquist, F.W. Switched or not?: The structure of unphosphorylated CheY bound to the N terminus of FliM. *J. Bacteriol.* **2006**, *188*, 7354–7363. [CrossRef]
87. Pandini, A.; Fornili, A.; Fraternali, F.; Kleinjung, J. Detection of allosteric signal transmission by information-theoretic analysis of protein dynamics. *FASEB J. Off. Publ. Fed. Am. Soc. Exp. Biol.* **2012**, *26*, 868–881. [CrossRef]
88. Fraiberg, M.; Afanzar, O.; Cassidy, C.K.; Gabashvili, A.; Schulten, K.; Levin, Y.; Eisenbach, M. CheY's acetylation sites responsible for generating clockwise flagellar rotation in Escherichia coli. *Mol. Microbiol.* **2015**, *95*, 231–244. [CrossRef]
89. Wheatley, P.; Gupta, S.; Chen, Y.; Petzold, C.J.; Ralston, C.R.; Blair, D.F.; Khan, S. Allosteric priming of *E. coli* CheY by the flagellar motor protein FliM. *BioRxiv* **2020**. [CrossRef]

90. Dyer, C.M.; Vartanian, A.S.; Zhou, H.; Dahlquist, F.W. A molecular mechanism of bacterial flagellar motor switching. *J. Mol. Biol.* **2009**, *388*, 71–84. [CrossRef]
91. Sarkar, M.K.; Paul, K.; Blair, D. Chemotaxis signaling protein CheY binds to the rotor protein FliN to control the direction of flagellar rotation in Escherichia coli. *Proc. Natl. Acad. Sci. USA* **2010**, *107*, 9370–9375. [CrossRef] [PubMed]
92. Brown, P.N.; Mathews, M.A.; Joss, L.A.; Hill, C.P.; Blair, D.F. Crystal structure of the flagellar rotor protein FliN from Thermotoga maritima. *J. Bacteriol.* **2005**, *187*, 2890–2902. [CrossRef]
93. Sarkar, M.K.; Paul, K.; Blair, D.F. Subunit organization and reversal-associated movements in the flagellar switch of Escherichia coli. *J. Biol. Chem.* **2010**, *285*, 675–684. [CrossRef] [PubMed]
94. Notti, R.Q.; Bhattacharya, S.; Lilic, M.; Stebbins, C.E. A common assembly module in injectisome and flagellar type III secretion sorting platforms. *Nat. Commun.* **2015**, *6*, 7125. [CrossRef]
95. McDowell, M.A.; Marcoux, J.; McVicker, G.; Johnson, S.; Fong, Y.H.; Stevens, R.; Bowman, L.A.; Degiacomi, M.T.; Yan, J.; Wise, A.; et al. Characterisation of Shigella Spa33 and Thermotoga FliM/N reveals a new model for C-ring assembly in T3SS. *Mol. Microbiol.* **2016**, *99*, 749–766. [CrossRef]
96. Gonzalez-Pedrajo, B.; Minamino, T.; Kihara, M.; Namba, K. Interactions between C ring proteins and export apparatus components: A possible mechanism for facilitating type III protein export. *Mol. Microbiol.* **2006**, *60*, 984–998. [CrossRef]
97. Abrusci, P.; Vergara-Irigaray, M.; Johnson, S.; Beeby, M.D.; Hendrixson, D.R.; Roversi, P.; Friede, M.E.; Deane, J.E.; Jensen, G.J.; Tang, C.M.; et al. Architecture of the major component of the type III secretion system export apparatus. *Nat. Struct. Mol. Biol.* **2013**, *20*, 99–104. [CrossRef]
98. Sircar, R.; Greenswag, A.R.; Bilwes, A.M.; Gonzalez-Bonet, G.; Crane, B.R. Structure and activity of the flagellar rotor protein FliY: A member of the CheC phosphatase family. *J. Biol. Chem.* **2013**, *288*, 13493–13502. [CrossRef]
99. Bischoff, D.S.; Ordal, G.W. Identification and characterization of FliY, a novel component of the Bacillus subtilis flagellar switch complex. *Mol. Microbiol.* **1992**, *6*, 2715–2723. [CrossRef]
100. Szurmant, H.; Ordal, G.W. Diversity in chemotaxis mechanisms among the bacteria and archaea. *Microbiol. Mol. Biol. Rev. MMBR* **2004**, *68*, 301–319. [CrossRef]
101. Ward, E.; Kim, E.A.; Panushka, J.; Botelho, T.; Meyer, T.; Kearns, D.B.; Ordal, G.; Blair, D.F. Organization of the Flagellar Switch Complex of Bacillus subtilis. *J. Bacteriol.* **2019**, *201*. [CrossRef] [PubMed]
102. Lam, K.H.; Ip, W.S.; Lam, Y.W.; Chan, S.O.; Ling, T.K.; Au, S.W. Multiple conformations of the FliG C-terminal domain provide insight into flagellar motor switching. *Structure* **2012**, *20*, 315–325. [CrossRef] [PubMed]
103. Kinoshita, M.; Namba, K.; Minamino, T. Effect of a clockwise-locked deletion in FliG on the FliG ring structure of the bacterial flagellar motor. *Genes Cells* **2018**, *23*, 241–247. [CrossRef]
104. Van Way, S.M.; Millas, S.G.; Lee, A.H.; Manson, M.D. Rusty, jammed, and well-oiled hinges: Mutations affecting the interdomain region of FliG, a rotor element of the Escherichia coli flagellar motor. *J. Bacteriol.* **2004**, *186*, 3173–3181.
105. Lee, L.K.; Ginsburg, M.A.; Crovace, C.; Donohoe, M.; Stock, D. Structure of the torque ring of the flagellar motor and the molecular basis for rotational switching. *Nature* **2010**, *466*, 996–1000. [CrossRef]
106. Zhulin, I.B. By Staying Together, Two Genes Keep the Motor Running. *Structure* **2017**, *25*, 214–215. [CrossRef]
107. Minamino, T.; Imada, K.; Kinoshita, M.; Nakamura, S.; Morimoto, Y.V.; Namba, K. Structural insight into the rotational switching mechanism of the bacterial flagellar motor. *PLoS Biol.* **2011**, *9*, e1000616.
108. Sircar, R.; Borbat, P.P.; Lynch, M.J.; Bhatnagar, J.; Beyersdorf, M.S.; Halkides, C.J.; Freed, J.H.; Crane, B.R. Assembly states of FliM and FliG within the flagellar switch complex. *J. Mol. Biol.* **2015**, *427*, 867–886. [CrossRef]
109. Baker, M.A.; Hynson, R.M.; Ganuelas, L.A.; Mohammadi, N.S.; Liew, C.W.; Rey, A.A.; Duff, A.P.; Whitten, A.E.; Jeffries, C.M.; Delalez, N.J.; et al. Domain-swap polymerization drives the self-assembly of the bacterial flagellar motor. *Nat. Struct. Mol. Biol.* **2016**, *23*, 197–203. [CrossRef]
110. Pandini, A.; Kleinjung, J.; Rasool, S.; Khan, S. Coevolved Mutations Reveal Distinct Architectures for Two Core Proteins in the Bacterial Flagellar Motor. *PLoS ONE* **2015**, *10*, e0142407. [CrossRef]
111. Dos Santos, R.N.; Khan, S.; Morcos, F. Characterization of C-ring component assembly in flagellar motors from amino acid coevolution. *R. Soc. Open Sci.* **2018**, *5*, 171854. [CrossRef] [PubMed]
112. Khan, S.; Guo, T.W.; Misra, S. A coevolution-guided model for the rotor of the bacterial flagellar motor. *Sci. Rep.* **2018**, *8*, 11754. [CrossRef] [PubMed]

113. Miyanoiri, Y.; Hijikata, A.; Nishino, Y.; Gohara, M.; Onoue, Y.; Kojima, S.; Kojima, C.; Shirai, T.; Kainosho, M.; Homma, M. Structural and Functional Analysis of the C-Terminal Region of FliG, an Essential Motor Component of Vibrio Na(+)-Driven Flagella. *Structure* **2017**, *25*, 1540–1548.e1543. [CrossRef] [PubMed]
114. Yakushi, T.; Yang, J.; Fukuoka, H.; Homma, M.; Blair, D.F. Roles of charged residues of rotor and stator in flagellar rotation: Comparative study using H+-driven and Na+-driven motors in Escherichia coli. *J. Bacteriol.* **2006**, *188*, 1466–1472. [CrossRef] [PubMed]
115. Onoue, Y.; Takekawa, N.; Nishikino, T.; Kojima, S.; Homma, M. The role of conserved charged residues in the bidirectional rotation of the bacterial flagellar motor. *Microbiologyopen* **2018**, *7*, e00587. [CrossRef] [PubMed]
116. Takekawa, N.; Kojima, S.; Homma, M. Contribution of many charged residues at the stator-rotor interface of the Na+-driven flagellar motor to torque generation in Vibrio alginolyticus. *J. Bacteriol.* **2014**, *196*, 1377–1385. [CrossRef]
117. Kojima, S.; Nonoyama, N.; Takekawa, N.; Fukuoka, H.; Homma, M. Mutations targeting the C-terminal domain of FliG can disrupt motor assembly in the Na(+)-driven flagella of Vibrio alginolyticus. *J. Mol. Biol.* **2011**, *414*, 62–74. [CrossRef]
118. Morimoto, Y.V.; Nakamura, S.; Hiraoka, K.D.; Namba, K.; Minamino, T. Distinct roles of highly conserved charged residues at the MotA-FliG interface in bacterial flagellar motor rotation. *J. Bacteriol.* **2013**, *195*, 474–481. [CrossRef]
119. Chang, Y.; Moon, K.H.; Zhao, X.; Norris, S.J.; Motaleb, M.A.; Liu, J. Structural insights into flagellar stator-rotor interactions. *eLife* **2019**, *8*, e48979. [CrossRef]
120. Chilcott, G.S.; Hughes, K.T. Coupling of flagellar gene expression to flagellar assembly in Salmonella enterica serovar typhimurium and Escherichia coli. *Microbiol. Mol. Biol. Rev. MMBR* **2000**, *64*, 694–708. [CrossRef]
121. Kim, E.A.; Panushka, J.; Meyer, T.; Carlisle, R.; Baker, S.; Ide, N.; Lynch, M.; Crane, B.R.; Blair, D.F. Architecture of the Flagellar Switch Complex of Escherichia coli: Conformational Plasticity of FliG and Implications for Adaptive Remodeling. *J. Mol. Biol.* **2017**, *429*, 1305–1320. [CrossRef] [PubMed]
122. Sakai, T.; Miyata, T.; Terahara, N.; Mori, K.; Inoue, Y.; Morimoto, Y.V.; Kato, T.; Namba, K.; Minamino, T. Novel Insights into Conformational Rearrangements of the Bacterial Flagellar Switch Complex. *mBio* **2019**, *10*, e00079-19. [CrossRef] [PubMed]
123. Levenson, R.; Zhou, H.; Dahlquist, F.W. Structural insights into the interaction between the bacterial flagellar motor proteins FliF and FliG. *Biochemistry* **2012**, *51*, 5052–5060. [CrossRef] [PubMed]
124. Ogawa, R.; Abe-Yoshizumi, R.; Kishi, T.; Homma, M.; Kojima, S. Interaction of the C-Terminal Tail of FliF with FliG from the Na+-Driven Flagellar Motor of Vibrio alginolyticus. *J. Bacteriol.* **2015**, *197*, 63–72. [CrossRef] [PubMed]
125. Lynch, M.J.; Levenson, R.; Kim, E.A.; Sircar, R.; Blair, D.F.; Dahlquist, F.W.; Crane, B.R. Co-Folding of a FliF-FliG Split Domain Forms the Basis of the MS:C Ring Interface within the Bacterial Flagellar Motor. *Structure* **2017**, *25*, 317–328. [CrossRef]
126. Xue, C.; Lam, K.H.; Zhang, H.; Sun, K.; Lee, S.H.; Chen, X.; Au, S.W.N. Crystal structure of the FliF-FliG complex from Helicobacter pylori yields insight into the assembly of the motor MS-C ring in the bacterial flagellum. *J. Biol. Chem.* **2018**, *293*, 2066–2078. [CrossRef]
127. Johnson, S.; Fong, Y.H.; Deme, J.C.; Furlong, E.J.; Kuhlen, L.; Lea, S.M. Symmetry mismatch in the MS-ring of the bacterial flagellar rotor explains the structural coordination of secretion and rotation. *Nat. Microbiol.* **2020**. [CrossRef]
128. Takekawa, N.; Nishiyama, M.; Kaneseki, T.; Kanai, T.; Atomi, H.; Kojima, S.; Homma, M. Sodium-driven energy conversion for flagellar rotation of the earliest divergent hyperthermophilic bacterium. *Sci. Rep.* **2015**, *5*, 12711. [CrossRef]
129. Zhao, X.; Norris, S.J.; Liu, J. Molecular architecture of the bacterial flagellar motor in cells. *Biochemistry* **2014**, *53*, 4323–4333. [CrossRef]
130. Terashima, H.; Kawamoto, A.; Morimoto, Y.V.; Imada, K.; Minamino, T. Structural differences in the bacterial flagellar motor among bacterial species. *Biophys. Phys.* **2017**, *14*, 191–198. [CrossRef]
131. Murphy, G.E.; Leadbetter, J.R.; Jensen, G.J. In situ structure of the complete Treponema primitia flagellar motor. *Nature* **2006**, *442*, 1062–1064. [CrossRef] [PubMed]
132. Liu, J.; Lin, T.; Botkin, D.J.; McCrum, E.; Winkler, H.; Norris, S.J. Intact flagellar motor of Borrelia burgdorferi revealed by cryo-electron tomography: Evidence for stator ring curvature and rotor/C-ring assembly flexion. *J. Bacteriol.* **2009**, *191*, 5026–5036. [CrossRef] [PubMed]

133. Henderson, L.D.; Matthews-Palmer, T.R.S.; Gulbronson, C.J.; Ribardo, D.A.; Beeby, M.; Hendrixson, D.R. Diversification of Campylobacter jejuni Flagellar C-Ring Composition Impacts Its Structure and Function in Motility, Flagellar Assembly, and Cellular Processes. *mBio* **2020**, *11*, e02286-19. [CrossRef] [PubMed]
134. Beeby, M.; Ribardo, D.A.; Brennan, C.A.; Ruby, E.G.; Jensen, G.J.; Hendrixson, D.R. Diverse high-torque bacterial flagellar motors assemble wider stator rings using a conserved protein scaffold. *Proc. Natl. Acad. Sci. USA* **2016**, *113*, E1917–E1926. [CrossRef]
135. wwPDB Consortium. Protein Data Bank: The single global archive for 3D macromolecular structure data. *Nucleic Acids Res.* **2019**, *47*, D520–D528. [CrossRef]
136. Seeliger, D.; Haas, J.; de Groot, B.L. Geometry-based sampling of conformational transitions in proteins. *Structure* **2007**, *15*, 1482–1492. [CrossRef]
137. Gao, R.; Bouillet, S.; Stock, A.M. Structural Basis of Response Regulator Function. *Annu. Rev. Microbiol.* **2019**, *73*, 175–197. [CrossRef]
138. Hyakutake, A.; Homma, M.; Austin, M.J.; Boin, M.A.; Hase, C.C.; Kawagishi, I. Only one of the five CheY homologs in Vibrio cholerae directly switches flagellar rotation. *J. Bacteriol.* **2005**, *187*, 8403–8410. [CrossRef]
139. Diepold, A.; Armitage, J.P. Type III secretion systems: The bacterial flagellum and the injectisome. *Philos. Trans. R. Soc. Lond. Ser. B Biol. Sci.* **2015**, *370*. [CrossRef]
140. Pallen, M.J.; Gophna, U. Bacterial flagella and Type III secretion: Case studies in the evolution of complexity. *Genome Dyn.* **2007**, *3*, 30–47.
141. Zhao, R.; Schuster, S.C.; Khan, S. Structural effects of mutations in Salmonella typhimurium flagellar switch complex. *J. Mol. Biol.* **1995**, *251*, 400–412. [CrossRef] [PubMed]

Review

Phylogenetic Distribution, Ultrastructure, and Function of Bacterial Flagellar Sheaths

Joshua Chu [1], Jun Liu [2] and Timothy R. Hoover [3,*]

1 Department of Microbiology, Cornell University, Ithaca, NY 14853, USA; jc2568@cornell.edu
2 Microbial Sciences Institute, Department of Microbial Pathogenesis, Yale University, West Haven, CT 06516, USA; jliu@yale.edu
3 Department of Microbiology, University of Georgia, Athens, GA 30602, USA
* Correspondence: trhoover@uga.edu; Tel.: +1-706-542-2675

Received: 30 January 2020; Accepted: 26 February 2020; Published: 27 February 2020

Abstract: A number of Gram-negative bacteria have a membrane surrounding their flagella, referred to as the flagellar sheath, which is continuous with the outer membrane. The flagellar sheath was initially described in *Vibrio metschnikovii* in the early 1950s as an extension of the outer cell wall layer that completely surrounded the flagellar filament. Subsequent studies identified other bacteria that possess flagellar sheaths, most of which are restricted to a few genera of the phylum Proteobacteria. Biochemical analysis of the flagellar sheaths from a few bacterial species revealed the presence of lipopolysaccharide, phospholipids, and outer membrane proteins in the sheath. Some proteins localize preferentially to the flagellar sheath, indicating mechanisms exist for protein partitioning to the sheath. Recent cryo-electron tomography studies have yielded high resolution images of the flagellar sheath and other structures closely associated with the sheath, which has generated insights and new hypotheses for how the flagellar sheath is synthesized. Various functions have been proposed for the flagellar sheath, including preventing disassociation of the flagellin subunits in the presence of gastric acid, avoiding activation of the host innate immune response by flagellin, activating the host immune response, adherence to host cells, and protecting the bacterium from bacteriophages.

Keywords: flagellum; flagellar sheath; *Helicobacter*; *Vibrio*; cardiolipin

---

1. Introduction

The bacterial flagellum is a complex organelle used for motility and is organized into three basic structures referred to as the basal body, hook and filament. Of these structures, the filament is the most prominent, forming a thin, helical structure that is typically 5–10 µm in length and is several times longer than the body of the bacterial cell. The filament is composed of tens of thousands of copies of a single flagellin protein or of multiple closely related flagellin proteins that self-assemble to form a hollow, tubular structure. In most bacterial species, the flagellar filament is exposed directly to the surrounding medium. The filament in several genera of Gram-negative bacteria, however, is surrounded by a membranous sheath that is contiguous with the outer membrane. Flagella in these bacteria are located almost exclusively at the cell pole, and occur as a single flagellum at one cell pole (polar flagellum), as a single flagellum at each cell pole (amphitrichous or bipolar flagella) or as multiple flagella at one cell pole (lophotrichous flagella). In bacteria with a lophotrichous arrangement of flagella, each flagellum is enclosed within a separate sheath. Other types of flagellar sheaths have been described, such as one found in the marine magnetotactic bacterium MO-1, which has a flagellar sheath composed of glycoprotein [1]. The MO-1 flagellar sheath differs further from membranous flagellar sheaths in that it surrounds a flagellar bundle consisting of multiple flagella rather than surrounding each flagellum [1]. Spirochetes enclose their flagella within the periplasmic space, which is somewhat analogous to the flagellar sheath in that the flagella of spirochetes are separated from the surrounding medium by a membrane. For additional information on the periplasmic flagella of spirochetes, we refer the reader to a recent review by Wolgemuth [2]. For the purposes of our review, we focus on bacteria that possess membranous flagellar sheaths. It has been nearly four decades since the last comprehensive review of the bacterial flagellar sheath [3], which was a major impetus for this review.

2. Phylogenetic Distribution of Flagellar Sheaths

Accurately assessing how widely distributed flagellar sheaths are among bacterial species is not a trivial task since reports on novel species often fail to indicate the presence or absence of a flagellar sheath. Moreover, when reports of novel bacterial species do indicate the presence of a flagellar sheath, they often omit a description of the ultrastructure of the sheath or do not include electron micrographs that clearly show the ultrastructure of the sheath. With these caveats in mind, membranous flagellar sheaths are found primarily in a handful of genera that are scattered throughout the phylum Proteobacteria. Within the Proteobacteria, representatives from five classes (Alphaproteobacteria, Betaproteobacteria, Gammaproteobacteria, Deltaproteobacteria and Epsilonproteobacteria) are reported to possess a membranous flagellar sheath. In addition to the Proteobacteria, some members of the phylum Planctomyces appear to have membranous flagellar sheaths, including *Pirella marina* and several members of the genus *Planctomycetes* [4,5].

Members of Alphaproteobacteria that have flagellar sheaths include *Seliberia stellate, Azospirillum brasilense, Rhodospirillum centenum,* and *Brucella melitensis,* all of which possess a single polar flagellum [6–9]. Within the Betaproteobacteria, the soil bacterium and plant pathogen *Robbsia andropogonis* (formerly known as *Burkholderia andropogonis, Pseudomonas andropogonis, Pseudomonas stizolobii,* as well as other names) has a single polar sheathed flagellum [10]. Busse and Auling indicate that members of the genus *Achromobacter*, which belong to the class Betaproteobacteria, have a peritrichous arrangement of sheathed flagella [11]. They do not report whether the *Achromobacter* flagellar sheaths are membranous, although the ultrastructure of *Achromobacter xylosoxidans* flagella appears consistent with that of a membranous flagellar sheath [12]. If the flagella of *Achromobacter* are indeed surrounded by a membranous sheath, this would be the only example of a bacterium with peritrichous sheathed flagella of which we are aware. Members of Gammaproteobacteria that have flagellar sheaths include *Halorhodospira adbelmalekii* [13], several species of *Pseudoalteromonas* [14–17], and most or all *Vibrio* species [18]. All of these members of Gammaproteobacteria possess either a single or multiple polar flagella that they use for swimming. In addition to producing a polar sheathed flagellum for swimming, various marine *Vibrio* species, including *Vibrio parahaemolyticus, Vibrio alginolyticus, Vibrio harveyi,* and *Vibrio shilonii,* elaborate lateral flagella that are used for rapid movement on surfaces [19–21]. The lateral flagella of the marine *Vibrio* species, as well as those of *A. brasilense* and *R. centenum,* lack a sheath [8,19–22], indicating that sheath biosynthesis is not inherently linked with flagellum biogenesis in these bacteria. *Bdellovibrio bacteriovorus, Bacteriovorax stolpii,* and *Bacteriovorax starrii,* which are predators of other Gram-negative bacteria, are members of the class Deltaproteobacteria that possess a single polar sheathed flagellum [23–25]. Within the class Epsilonproteobacteria, only members of the genus *Helicobacter* are reported to have flagellar sheaths. Most *Helicobacter* species possess a single sheathed polar flagellum or bipolar sheathed flagella; although the most extensively studied species, *Helicobacter pylori,* has lophotrichous sheathed flagella, and several *Helicobacter* species have unsheathed polar flagella. Interestingly, *Helicobacter* species that possess flagellar sheaths and *Helicobacter* species that have unsheathed flagella appear to segregate into distinct phylogenetic groups [26].

3. Composition of Flagellar Sheaths

Early ultrastructural studies of flagellar sheaths from various bacteria revealed two electron dense layers separated by a region of less electron density, consistent with the flagellar sheath being a unit membrane [3]. For some bacteria, including *B. bacteriovorus, B. melitensis, H. pylori,* and *V. fischeri,* a bulb-like structure is typically observed at the distal end of the flagellar sheath [7,27–29]. The detailed electron micrographs from the early studies suggested the flagellar sheath was contiguous with the outer membrane, and recent cryo-electron tomography (cryo-ET) studies of sheathed flagella from various bacteria have yielded detailed images that confirm the structural continuity between the outer membrane and flagellar sheath [30–34]. Figure 1 shows tomograms of *V. cholerae* and *H. pylori* sheathed flagella (authors' data).

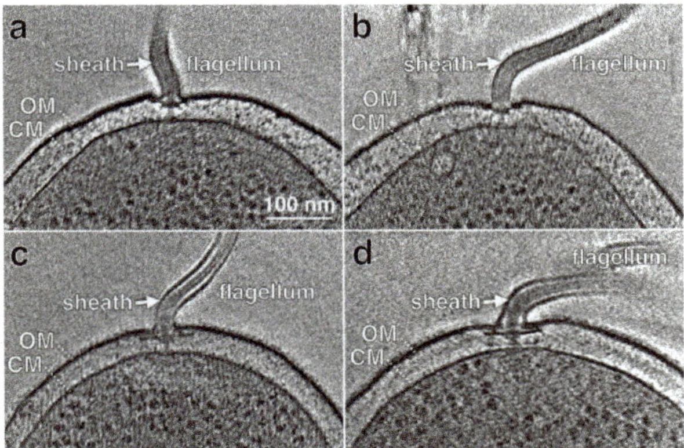

Figure 1. Cryot-ET reconstructions of intact cells show sheathed flagella. (**a**,**b**) Two representative sections from cryo-ET reconstructions of *V. cholerae* cells. (**c**,**d**) Two representative sections from cryo-ET reconstructions of *H. pylori* cells. The arrows indicate the flagellar sheath. For each flagellum, note the central core that consists of the hook and filament. The outer (OM) and cytoplasmic membranes (CM) are indicated.

Early studies on the flagellar sheath sought to determine if the composition of the sheath was similar to that of the outer membrane by using antibodies directed against lipopolysaccharide (LPS) or other surface antigens in the outer membrane. Hranitzky and co-workers reported for *V. cholerae* that antibodies directed against a crude flagellar sheath preparation reacted strongly with a component in both the sheath and surface of the cell body [35]. Conversely, Yang and co-workers found that antibodies directed against a purified antigen from the cell body of *V. cholerae* cross-reacted with and immobilized the flagellum [36]. Although Hranitzky and co-workers reported that antibodies directed against *V. cholerae* LPS did not bind to the flagellar sheath [35], a subsequent study with *V. cholerae* found that anti-LPS did indeed recognize the flagellar sheath [37]. While these studies indicated that at least some components are shared between the outer membrane and flagellar sheath, determining whether specific macromolecules localize to the flagellar sheath required further biochemical characterization of isolated flagellar sheaths.

Little information is available on the lipid composition of the flagellar sheath for any bacterium, but the data that are available are intriguing. Thomashow and Rittenberg reported that the LPS of the flagellar sheath of *B. bacteriovorous* was moderately enriched (~2.7-fold) for the fatty acid nonadecenoic acid ($C_{19:1}$) and depleted greatly (~17-fold) in β-hydroxymyristic acid (3-OH $C_{14:0}$) compared to LPS from bdellovibrios grown on *Escherichia coli* as a host [29]. This observation indicates *B. bacteriovorous* partitions specific LPS species into the flagellar sheath that differ from those in the outer membrane. Interestingly, the total LPS (i.e., LPS from both the outer membrane and flagellar sheath) from bdellovibrios grown axenically (i.e., on medium instead of host cells) was similar to the flagellar sheath LPS in that it was enriched for nonadecenoic acid and depleted for β-hydroxymyristic acid [29]. Using immunogold labelling, Norqvist and Wolf-Watz identified a surface antigen in the fish pathogen *Vibrio anguillarum* that localized specifically to the flagellar sheath [38]. The *V. anguillarum* surface antigen was resistant to proteinase K, but sensitive to periodic acid treatment. In addition, the antigen was absent in mutants that had transposon insertions in *virA* and *virB*, which encode enzymes involved in LPS O1 antigen biosynthesis. Collectively, these findings indicate the *V. anguillarum* surface antigen is LPS [38], and further suggest that like *B. bacteriovorus*, *V. anguillarum* partitions specific LPS species to the flagellar sheath. The mechanisms these bacteria use to segregate specific LPS species to the flagellar sheath are unknown.

In an examination of the fatty acid composition of *H. pylori* flagellar sheaths, Geis and co-workers found there were no fatty acids uniquely associated with the flagellar sheath, but the fatty acid composition profile of the sheath does differ from that of the whole cell membranes [39]. Cardiolipin is a phospholipid that accumulates in regions of membranes that have negative curvature, such as the cell pole and septal regions in rod-shaped bacteria [40–43]. Given the propensity of cardiolipin to accumulate in membranes with negative curvature, one might expect the flagellar sheath to contain significant amounts of cardiolipin. Consistent with this hypothesis, the *H. pylori* flagellar sheath appears to contain high amounts of cardiolipin [44]. In further support of the hypothesis, the two most abundant fatty acids in *H. pylori* flagellar sheaths, myristic acid ($C_{14:0}$) and cyclopropane nonadecanoic acid ($C_{19:0}$ cyc) [39], are also the two most common fatty acids in cardiolipin species from *H. pylori* [45–47].

A number of studies have investigated the proteins associated with bacterial flagellar sheaths, although the information on flagellar sheath proteins is still scant. Knowledge of the proteins that localize to the flagellar sheath is critical for understanding the function of the sheath. Early studies on characterizing flagellar sheath proteins relied on serological approaches or SDS-polyacrylamide gel electrophoresis to identify proteins that appeared to be enriched in the flagellar sheath [29,35,39,48–51]. These studies typically reported the sizes of the putative flagellar sheath proteins, but did not identify or further characterize the proteins. The *B. bacteriovorus* flagellar sheath was reported to have substantially less total protein (23%–28% dry weight) than that of outer membranes from other bacteria, which typically ranges from 40 to 70% [29]. Using monoclonal antibodies to an outer membrane fraction from *H. pylori* NCTC 11637, Doig and Trust identified six protein antigens that either localized within or were associated with the outer membrane, but did not recognize the flagellar sheath, which suggested the proteomes of the outer membrane and flagellar sheath of *H. pylori* differ from each other [48]. Bari and co-workers identified three outer membrane proteins associated with the *V. cholerae* flagellum-OmpU and OmpT, which are porins, and VC1894, which is a predicted collagen-binding surface adhesion [52]. Disrupting *ompU* or *ompT* in *V. cholerae* resulted in several flagellum-associated defects, including reduced motility, thinner flagella, increased proportion of non-flagellated cells, and increased release of flagellin into the growth medium [52]. These findings suggest OmpU and OmpT help to stabilize the *V. cholerae* flagellar sheath.

An important question is whether there are proteins that localize preferentially to the flagellar sheath. One protein reported to localize specifically to the *H. pylori* flagellar sheath is the *H. pylori* adhesion A (HpaA), although there are conflicting reports on the localization of this protein. HpaA was first described as a hemagglutinin that was shown by immunogold labelling to be located on the cell surface, but was not detectable on the flagellar sheath [53]. A subsequent study indicated HpaA occurred predominantly in the cytoplasmic fraction of Sarkosyl-solubilized cells, with only trace amounts of HpaA in the inner membrane fraction and no detectable HpaA in the outer membrane fraction [54]. Other immunogold labelling studies showed HpaA was specifically localized to the flagellar sheath [51,55]; while a later immunogold labeling study examined localization of HpaA in five *H. pylori* strains and detected HpaA on both the flagellar sheath and the bacterial surface in all of the strains [56]. The case for HpaA being a flagellar sheath protein is very compelling, and some of the discrepancies between the reports on the surface location of the protein may be attributed to differences in growth phases of the *H. pylori* cultures, media used for growing the bacteria, strain variability or choice of antibody [55,56].

In a study of *H. pylori* genes predicted to encode proteins secreted by the type V (autotransporter) pathway, Radin and co-workers found one of these proteins localized specifically to the flagellar sheath, which was determined by immunogold electron microscopy and fluorescence microscopy using an antibody to a c-myc tag introduced within the autotransporter [57]. Given the association of the autotransporter with the flagellar sheath, the researchers designated the protein as FaaA (flagellar-associated autotransporter A). Disrupting *faaA* resulted in reduced motility and a variety of defects in flagellum biosynthesis or stability, which included increased number of non-flagellated cells, reduced number of flagella per cell, increased frequency of broken flagella, and increased proportion of flagella that localized to nonpolar sites [57]. *faaA* is not part of any of the known flagellar gene regulons in *H. pylori* [58]. FaaA is synthesized and localized to the cell surface in a *H. pylori* 26695 strain that does not produce flagella [59]. Taken together, these findings suggest a role for FaaA in assembly of the flagellum and/or flagellar sheath, but expression of *faaA* is not tightly coupled with the expression of flagellar genes.

The physiological role of FaaA is not known, although it is required for optimal colonization in a mouse animal model by *H. pylori* during early stages of infection [57]. The colonization deficiency of the *faaA* mutant may result from the defects in motility and flagellum biosynthesis since *H. pylori* must be motile to penetrate the gastric mucus layer to colonize the gastric mucosa [60]. Proteins secreted by the autotransporter pathway have two domains, a secreted passenger domain and a β-barrel domain that inserts in the outer membrane and facilitates transport of the passenger domain across the outer membrane [61]. Passenger domains participate in a variety of cellular functions, including enzymatic activities (e.g., proteases, lipases/esterases), contact-dependent growth inhibition, immune evasion, cytotoxicity, cyto-/hemolysis, adherence, biofilm formation, auto-agglutination, and activation of actin polymerases for intracellular motility [61]. Depending on the autotransporter, the passenger domain is cleaved and released outside the cell after it is transported across the outer membrane or it remains linked to the β-barrel domain and is exposed on the cell surface. Since the c-myc tag used to examine the localization of FaaA was introduced into the passenger domain [57], FaaA belongs to this later class of autotransporters.

4. Rotation of the Sheathed Flagellum

Researchers who initially studied flagellar sheaths raised questions about how to apply the rotary model for the bacterial flagellum mechanism. Fuerst proposed two models for how the flagellar filament and sheath cooperated in motility [62]. The first model proposed the filament and sheath rotate together. This model requires the flagellar sheath to be rigid and the intersection of the base of the flagellar sheath and outer membrane to be discontinuous and fluid to enable the sheath to rotate with the filament. The sheath would also interact with the filament, perhaps through hydrophobic interactions, to generate a rigid membrane structure. In the second model, which makes fewer assumptions about the nature of the outer membrane and is seemingly more plausible, the filament rotates freely within a flexible wave-propagating sheath [62]. In this second model, the membrane of the flagellar sheath must be flexible enough to allow distortion by the rotational forces induced by the filament, but robust enough to remain associated with the cell body. Fuerst proposed experiments to examine the movement of polystyrene latex beads attached to the flagellar sheath through anti-sheath antibodies as a possible way to distinguish between his two models [62]; however, to the best of our knowledge, there are no reports that address the behavior of the flagellar sheath as the filament rotates.

Rotation of the sheathed flagellum of various *Vibrio* species is a major source for outer membrane vesicles (OMVs) that are released from the bacterial cell [63,64]. Aschtgen and co-workers demonstrated that the amount of OMVs released is proportional to the number of sheathed flagella per cell [63]. Specifically, the researchers showed that *V. cholerae*, which has a single polar sheathed flagellum, released fewer OMVs than *V. parahaemolyticus* or *V. fischeri*, which have multiple polar sheathed flagella. The researchers also showed that *E. coli*, which has unsheathed peritrichous flagella, released the fewest amount of OMVs, and a non-flagellated *E. coli* strain released the same amount of OMVs as its parental strain [63]. It is not known how flagellar rotation results in release of OMVs, but it seems likely that they are shed from the flagellar sheath as the flagellum rotates. Membrane blebs have been observed at the tip and shaft of *Vibrio* flagellar sheaths [65,66], and these blebs may be the source of OMVs that are released during rotation of the flagellum [63].

The outer membrane of Gram-negative bacteria is an asymmetrical lipid bilayer with LPS at the outer leaflet and phospholipids at the inner leaflet, and serves as an effective barrier to antibiotics, detergents and other toxic compounds. Exposure of bacterial cells to antimicrobial peptides or metal chelating agents such as EDTA leads to shedding of LPS and allows phospholipids from the inner leaflet to move into the outer leaflet, thereby compromising the outer membrane as a barrier [67]. Loss of LPS and other macromolecules associated with the flagellar sheath as the flagellum rotates could similarly lead to the migration of phospholipids from the inner leaflet to the outer leaflet, which could be deleterious to the bacterium. Consistent with this hypothesis, the *B. bacteriovorus* flagellar sheath was reported to have a higher proportion of phospholipids than typical outer membranes, and the authors of this study speculated that this might account for the unusual sensitivity of bdellovibrios to detergents [29]. Gram-negative bacteria have mechanisms for removing phospholipids from the outer leaflet of the outer member to maintain lipid asymmetry [68,69]. If shedding of sheath material does indeed compromise the flagellar sheath as a barrier, one might expect some bacteria that possess flagellar sheaths to have robust mechanisms for maintaining the lipid asymmetry of the outer membrane and sheath.

5. Biogenesis of the Flagellar Sheath

Little is known regarding the biosynthesis of the flagellar sheath in any bacterial species, which makes any attempt to attribute the phylogenetic distribution of the flagellar sheath to horizontal gene transfer or convergent evolution highly speculative. Regardless of the evolutionary history of the flagellar sheath, it likely is not coincidental that almost all bacteria reported to have a flagellar sheath possess polar flagella. The bacterial cell pole has unique physiochemical properties that may have facilitated the evolution of a membranous sheath at this location. For example, the accumulation of cardiolipin at the bacterial cell pole is attributed to its ability to form clusters or microdomains, which exhibit a high intrinsic curvature and therefore have a lower energy when localized to regions of the membrane with negative curvature [70,71]. In addition to forming microdomains, cardiolipin induces other changes in the physical properties of membranes that may be critical for assembly of the flagellar sheath, such as the ability to form nonbilayer structures [72–74] and decreasing lateral interactions within the monolayer leaflet, which lowers the energy needed to stretch membranes [75]. Localization of specific proteins to the cell pole may also contribute to formation of the flagellar sheath. Cardiolipin interacts strongly with many proteins [76] and, in some cases, cardiolipin is required for recruitment of specific proteins to the cell pole [43,77–79]. Additional mechanisms for localizing specific proteins to the cell pole that do not involve cardiolipin exist in rod-shaped bacteria [80], and such mechanisms could also facilitate flagellar sheath biosynthesis.

An important matter regarding flagellar sheath biogenesis is ascertaining the degree to which it is coupled to assembly of the flagellar filament. Richardson and co-workers generated non-motile mutants of *V. cholerae* following transposon mutagenesis, and identified five mutants that produced sheath-like structures that lacked the flagellar core [81]. In the coreless sheath mutants, a sheath-like structure was observed in about half the cells, and in contrast to the wild-type flagellum, the sheath-like structures were located almost always (>99%) at non-polar sites [81]. The coreless sheaths were elongated like normal sheaths, but in contrast to normal sheaths, the diameters of the coreless sheaths were irregular. These findings suggest that sheath biogenesis and flagellar assembly in *V. cholerae* can be uncoupled. This uncoupling, however, may be strain specific as the researchers were only able to isolate coreless sheath mutants from the classical strain of *V. cholerae*, and not the El Tor strain [81]. Unfortunately, the genes that were disrupted in the coreless sheath mutants were never identified, which would have provided clues that might explain the molecular basis for the unusual phenotype of these mutants. Ferooz and Letesson reported that in mutants of *B. melitensis* where *fliF* (encodes MS-ring protein), *flgE* (encodes hook protein), *fliC* (encodes flagellin) or *ftcR* (encodes flagellar gene master regulator) were deleted, coreless sheaths could be observed on some of the cells [7]. These findings are intriguing since the MS-ring is one of the earliest flagellar structures to be assembled, which suggests sheath biogenesis in *B melitensis* can initiate and proceed in the absence of any flagellar structure.

In contrast to the studies with *V. cholerae* and *B. melitensis* flagellar mutants [7,81], studies with wild-type *H. pylori* suggest assembly of the flagellum and sheath biogenesis are tightly coupled. In a high-throughput cryo-ET approach, Qin and co-workers visualized over 300 *H. pylori* flagella, which allowed them to image intermediate structures during flagellum assembly [32]. Figure 2 shows the intermediate structures of the flagellar sheath that are observed during flagellum assembly (author's data). Based on the series of *H. pylori* flagellum assembly intermediates, the growing rod assembly seems to push against the outer membrane and deform it. As the hook is assembled and grows, the outer membrane is deformed further and eventually forms a bubble that surrounds the hook. During filament assembly, the flagellar sheath and filament appear to elongate simultaneously, and the bulb-like structure seen in the mature flagellum is present in the nascent flagellum [32].

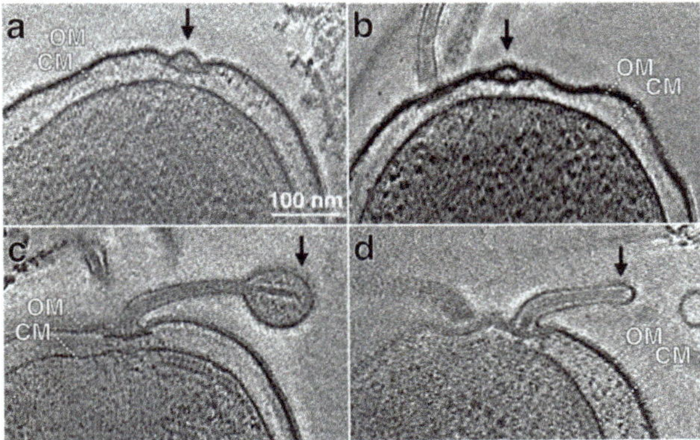

Figure 2. Cryo-ET reconstructions of intact cells show early stages of flagellar assembly and sheath formation. (**a**,**b**) Two representative sections from cryo-ET reconstructions of *H. pylori* cells show flagellar basal bodies without hook and filament. (**c**,**d**) Two representative sections from cryo-ET reconstructions of *H. pylori* cells show short flagellum. The outer (OM) and cytoplasmic membranes (CM) are indicated.

Virtually nothing is known about proteins that have roles in flagellar sheath biogenesis. The only studies that have shed any light on proteins with potential roles in sheath biosynthesis have been done with *V. alginolyticus*. In a cryo-ET analysis of the *V. alginolyticus* sheathed flagellum, Zhu and co-workers observed a ring-like structure associated with the base of the flagellar sheath [33]. The structure, designated as the O-ring, was located on the exterior side of the outer membrane, which displayed a striking 90° bend at the site of the O-ring [33]. The location of the O-ring and apparent deformation in the outer membrane that it elicits suggests a critical role for the O-ring in formation or function of the flagellar sheath. Figure 3 presents a model for assembly of the flagellum and flagellar sheath in *V. alginolyticus*. The genes encoding the O-ring protein(s) have yet to be identified, which has prevented researchers from confirming a role for the O-ring in flagellar sheath biogenesis. Structures that are analogous to the O-ring have not been identified in any other bacteria with sheathed flagella, indicating that any role for the O-ring in flagellar sheath biogenesis in *V. alginolyticus* is not universal among bacteria that possess flagellar sheaths.

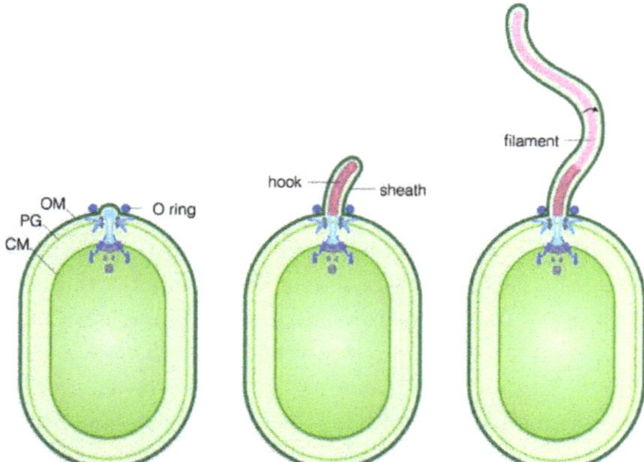

Figure 3. A schematic of assembly and sheath formation of a sheathed flagellum of *Vibrio alginolyticus*. The O-ring is assembled on the exterior side of the outer membrane at the point where the nascent flagellar sheath emerges. As the hook and filament are assembled, the flagellar sheath extends to encase these structures. The O-ring remains positioned at the base of the flagellar sheath where it stabilizes or induces a sharp bend in the outer membrane as it transitions into the flagellar sheath. The outer (OM), cytoplasmic membranes (CM), and peptidoglycan (PG) are indicated.

The flagellar motors of many bacteria have embellishments that are absent in the archetypical flagellar motors of *E. coli* and *S. enterica* serovar Typhimurium [30–33,82]. One such embellishment is the H-ring, which is closely associated with the L-ring/P-ring complex and in close proximity to the outer membrane in *Vibrio* species [30,33]. The H-ring is located on the periplasmic side of the outer membrane, and the proteins that comprise the H-ring (FlgO and FlgT) have been identified [30,83,84]. Zhu and co-workers demonstrated that deletion of *flgO* or *flgT* in *V. alginolyticus* disrupted formation of the H-ring and resulted in a many of the flagella being located in the periplasm [84]. About 80% of the flagella in the *flgT* mutant were located in the periplasm, compared with about 10% of the flagella in the *flgO* mutant and none of the flagella in the parental strain having a periplasmic location [84]. Some of the filaments of the periplasmic flagella protruded through the outer membrane at sites that were far from the cell pole. Some of the protruding filaments were encased in a flagellar sheath, while others lacked a sheath [84]. Taken together, these observations suggest the H-ring assists the flagellum in penetrating the outer membrane and forming the flagellar sheath.

6. Proposed Functions for Flagellar Sheaths

A variety of functions have been proposed for bacterial flagellar sheaths, however, the lack of sheath-less mutants for any bacterial species makes it difficult to confirm proposed functions for the flagellar sheath. The flagellar sheath may have multiple functions within a given bacterial species, and functions of the sheath may vary between species. For *H. pylori*, one of the original proposed functions for the flagellar sheath was to protect the filament subunits from dissociation in the presence of gastric acid. The *H. pylori* flagellar sheath has also been proposed to be involved in adherence. In support of the proposed role of the *H. pylori* flagellar sheath in adherence, the putative adhesion HpaA is reported to be located in the flagellar sheath. While HpaA was originally described as a sialic acid binding adhesion [53,85], supporting evidence for this activity is still lacking [54,55], and so it is unclear if HpaA does indeed have a role in adherence. Nevertheless, HpaA is required for colonization of the mouse model [86].

Another proposed function for the flagellar sheath is escaping detection of the flagellins by the host innate immune system. Toll-like receptor 5 (TLR5) is a surface exposed host receptor that recognizes flagellin [87]. Binding of flagellin to TLR5 stimulates proinflammatory cytokine production, which induces an inflammatory response that can lead to active clearance of the invading bacterium and an enhancement of the adaptive immune response [88]. Yoon and Mekalanos demonstrated that compared to the unsheathed flagella of *Salmonella enterica* serovar Typhimurium, the sheathed flagella of *V. cholerae* were significantly reduced in their relative potency to trigger the host innate response [89]. The *V. cholerae* flagellins and *S. enterica* serovar Typhimurium flagellin were similar in their potencies to trigger the host innate response, indicating that the *V. cholerae* flagellar sheath is effective in hiding immunogenic flagellins [89]. TLR5 recognizes a highly conserved region of flagellin that is required for flagellum assembly in members of the Gammaproteobacteria [90]. Flagellins of *H. pylori* and other members of the Epsilonproteobacteria lack the conserved region that interacts with TLR5 [91], which suggests the flagellar sheath does not play a major role in avoiding triggering the host innate immune response by flagellin in these bacteria.

The flagellins of several bacterial species are glycosylated (i.e., post-translationally modified by the covalent attachment of carbohydrates to specific amino acids), and some of these bacteria possess flagellar sheaths [92,93]. Flagellar glycans have roles in a variety of processes, including flagellar filament assembly, motility, autoaggultination, adherence to and invasion of host cells, virulence, and evasion of the host innate immune system [94–100]. *H. pylori* flagellins FlaA and FlaB are modified with a single type of glycan, pseudaminic acid, and flagellin glycosylation is required for assembly of the flagellar filament [100]. Flagellin glycosylation in *Campylobacter jejuni,* which is closely related to *H. pylori*, is also required for filament assembly, but the glycans of the *C. jejuni* flagellins are much more heterogeneous, and include various derivatives of pseudaminic acid and a derivative of legionaminic acid [101,102]. Logan proposed that the *H. pylori* flagellar sheath prevents recognition of the flagellin glycan by the host immune system, which may have decreased the evolutionary pressure for glycan heterogeneity in this bacterium [92].

As discussed previously, rotation of the sheathed flagellum of *Vibrio* species releases OMVs, which are known to have important roles in host signaling in symbiosis and pathogenesis. In the symbiosis between *V. fischeri* and the Hawaiian bobtail squid, *Euprymna scolopes*, LPS associated with the OMVs induces apoptotic cell death within the surface epithelium of the squid light organ that is required for its normal development [63,64]. Vanhove and co-workers found that OMVs released from *V. tasmaniensis*, a facultative intracellular pathogen of oyster haemocytes, contained several virulence factors that could be delivered to host cells either extracellularly or intracellularly [66]. The presence of several flagellar proteins in the OMVs and the occurrence of membrane blebs on flagellar sheaths suggested that some of the OMVs originated from the flagellar sheath [66]. It is not known, however, if OMVs derived from the flagellar sheath contain virulence factors.

OMVs also have a potential role in innate bacterial defense, as Manning and Kuehn showed OMVs protected enterotoxigenic *E. coli* from certain outer membrane-acting stressors, such as antimicrobial peptides and T4 bacteriophage [103]. OMVs interacted with antimicrobial peptides in a dose-dependent manner; and irreversibly bound phage, as well as reduced the ability of phage to infect once attached to the OMV [103]. In a somewhat related study, Zhang and co-workers demonstrated that rotation of the polar sheathed flagellum reduced absorption of phage OWB to *V. parahaemolyticus* [104]. Mutations that prevented either the synthesis or rotation of the polar flagellum enhanced the ability of the phage to lyse the bacterium [104]. The authors of this study suggested rotation of the sheathed flagellum of *V. parahaemolyticus* protects the bacterium from phage by releasing OMVs that bound the phage [104]. Alternatively, rotation of the polar flagellum is a mechanosensory mechanism that regulates gene expression [105], and mutations that prevent rotation of the polar flagellum may alter the cell surface to enhance phage absorption.

Another potential function of flagellar sheaths is to protect bacteria from flagellotropic phages, a group of phages that use the flagellar filament as a host receptor for attachment. The infection mechanism of flagellotropic phages is poorly understood, but flagellar rotation is required for infection and is thought to facilitate translocation of the phage along the filament to the cell surface [106]. It is possible flagellar sheaths evolved as a mechanism to hide the flagellar filament from flagellotropic phages. Consistent with this hypothesis, we are unaware of any reports on flagellotropic phages of bacteria that possess flagellar sheaths; although flagellotropic phages of bacterial species closely related to bacteria that have flagellar sheaths have been reported, such as phage F342 of *C. jejuni* [107].

7. Conclusions and Future Directions

The 1983 review of bacterial flagellar sheaths by Sjoblad and colleagues begins with the statement, "Although bacterial flagellar sheaths were observed over 30 years ago, they may still be characterized as structures in search of a function" [3]. Some of the assumed roles for flagellar sheaths in 1983, such as adherence, are still considered as possible roles for flagellar sheaths today. And although additional roles have been postulated for bacterial flagellar sheaths over the last 37 years, limitations in our knowledge of flagellar sheath biosynthesis and the lack of mutants that synthesize sheath-less flagella thwart efforts to confirm proposed functions for flagellar sheaths.

The limited number of studies that have examined the composition of flagellar sheaths have indicted the sheath is both similar to and different from the outer membrane. Differences in the LPS composition of flagellar sheaths and the outer membrane indicated by some studies [29,38] require an unknown mechanism to segregate specific LPS species within what appears to otherwise be a contiguous membrane. Mechanisms for localizing specific proteins to the sheath are easier to envision. For example, bacterial proteins are localized to the cell pole through a diffusion-capture mechanism in which proteins are inserted into the membrane where they can diffuse until encountering a geometrical cue (e.g., membrane curvature) or biochemical cue (e.g., specific phospholipids or other proteins already localized to the site) [80]. Given the unique physiochemical properties of flagellar sheaths (e.g., shape, phospholipid composition), such a diffusion-capture mechanism is likely to be responsible for the localization of proteins to the flagellar sheath. Studies in *V. cholerae* and *H. pylori* have identified proteins that appear to localize to the flagellar sheath [51,52,55,57], and future studies in these bacteria, as well as other bacterial species, will most certainly lead to the identification of additional flagellar sheath proteins. Dissecting the lipid and protein composition of bacterial flagellar sheaths is critical for understanding the function and biogenesis of these unique structures.

One of the most fascinating areas for future investigations into flagellar sheaths is understanding how these structures are assembled. Making headway in understanding the molecular mechanisms that control flagellar sheath biogenesis will require a combination of genetic, biochemical, and structural approaches. High-throughput cryo-ET studies, like that done by Qin and co-workers with *H. pylori* [32], will need to be done with other bacterial species. Identifying and disrupting genes that encode structural features intimately associated with flagellar sheaths, such as the O-ring of *V. alginolyticus*, will be required to ascertain the roles these genes play in flagellar sheath biosynthesis. Creative genetic screens will be needed to identify genes that are required for flagellar sheath biosynthesis and, hopefully, lead to the generation of mutant strains that produce sheath-less flagella, which can be used to test the requirement of the flagellar sheath in host colonization and pathogenesis. Biochemical studies will be needed to examine the composition of flagellar sheaths, as well as confirm the predicted activities of the products of candidate genes for sheath biosynthesis identified through genetic and genomic approaches.

Author Contributions: Conceptualization, J.C. and T.R.H.; writing—original draft preparation, J.C.; writing—review and editing, T.R.H.; visualization, J.L.; funding acquisition, T.R.H. and J.L. All authors have read and agreed to the published version of the manuscript.

Funding: This research was funded by NIH grant AI140444 to T.H., and NIH grants GM107629 and AI087946 to J.L.

Acknowledgments: We thank Katherine Gibson and Anna Karls for their thoughtful feedback on the manuscript.

Conflicts of Interest: The authors declare no conflict of interest.

References

1. Lefevre, C.T.; Santini, C.L.; Bernadac, A.; Zhang, W.J.; Li, Y.; Wu, L.F. Calcium ion-mediated assembly and function of glycosylated flagellar sheath of marine magnetotactic bacterium. *Mol. Microbiol* **2010**, *78*, 1304–1312. [CrossRef] [PubMed]
2. Wolgemuth, C.W. Flagellar motility of the pathogenic spirochetes. *Semin Cell Dev. Biol.* **2015**, *46*, 104–112. [CrossRef] [PubMed]
3. Sjoblad, R.D.; Emala, C.W.; Doetsch, R.N. Invited review: Bacterial flagellar sheaths: Structures in search of a function. *Cell Motil.* **1983**, *3*, 93–103. [CrossRef] [PubMed]
4. Fuerst, J.A. The planctomycetes: Emerging models for microbial ecology, evolution and cell biology. *Microbiology* **1995**, *141*, 1493–1506. [CrossRef] [PubMed]
5. Schlesner, H. *Pirella marina* sp. nov., a budding, peptidoglycan-less bacterium from brackish water. *Syst. Appl. Microbiol.* **1986**, *8*, 177–180. [CrossRef]
6. Burygin, G.L.; Shirokov, A.A.; Shelud'ko, A.V.; Katsy, E.I.; Shchygolev, S.Y.; Matora, L.Y. Detection of a sheath on *Azospirillum brasilense* polar flagellum. *Microbiology* **2007**, *76*, 728–734. [CrossRef]
7. Ferooz, J.; Letesson, J.J. Morphological analysis of the sheathed flagellum of *Brucella melitensis*. *BMC Res. Notes* **2010**, *3*, 333. [CrossRef]
8. Ragatz, L.; Jiang, Z.Y.; Bauer, C.E.; Gest, H. Macroscopic phototactic behavior of the purple photosynthetic bacterium *Rhodospirillum centenum*. *Arch. Microbiol.* **1995**, *163*, 1–6. [CrossRef]
9. Schmidt, J.M.; Starr, M.P. Unidirectional polar growth of cells of *Seliberia stellata* and aquatic seliberia-like bacteria revealed by immunoferritin labeling. *Arch. Microbiol.* **1984**, *138*, 89–95. [CrossRef]
10. Fuerst, J.A.; Hayward, A.C. The sheathed flagellum of *Pseudomonas stizolobii*. *J. Gen. Microbiol.* **1969**, *58*, 239–245. [CrossRef]
11. Busse, H.-J.; Auling, G. Achromobacter. In *Bergy's Manual of Systematics of Archaea and Bacteria*; Whitman, W., Rainey, F., Kampfer, P., Trujillo, M., Chun, J., DeVos, P., Eds.; John Wiley & Sons: Hoboken, NJ, USA, 2015.
12. Yabuuchi, E.; Yano, I.; Goto, S.; Tanimura, E.; Ito, T.; Ohyama, A. Description of *Achromobacter xylosoxidans* Yabuuchi and Ohyama 1971. *Int. J. Syst. Bacteriol.* **1974**, *24*, 470–477. [CrossRef]
13. Imhoff, J.F.; Truper, H.G. *Ectothiorhodospira abdelmalekii* sp. nov., a new halophilic and alkaliphilic phototrophic bacterium. *Zbl Bakt Hyg I Abt Orig C* **1981**, *2*, 228–234. [CrossRef]
14. Enger, O.; Nygaard, H.; Solberg, M.; Schei, G.; Nielsen, J.; Dundas, I. Characterization of *Alteromonas denitrificans* sp. nov. *Int. J. Syst. Bacteriol.* **1987**, *37*, 416–421. [CrossRef]
15. Hansen, A.J.; Ingebritsen, A.; Weeks, O.B. Flagellation of *Flavabacterium piscicida*. *J. Bacteriol.* **1963**, *86*, 602–603. [CrossRef] [PubMed]
16. Holmstrom, C.; James, S.; Neilan, B.A.; White, D.C.; Kjelleberg, S. *Pseudoalteromonas tunicata* sp. nov., a bacterium that produces antifouling agents. *Int. J. Syst. Bacteriol.* **1998**, *48*, 1205–1212. [CrossRef] [PubMed]
17. Novick, J.J.; Tyler, M.E. Isolation and characterization of *Alteromonas luteoviolacea* strains with sheathed flagella. *Int. J. Syst. Bacteriol.* **1985**, *35*, 111–113. [CrossRef]
18. Farmer, J.J., III; Janda, J.M.; Brenner, F.W.; Cameron, D.N.; Birkhead, K.M. Genus I. *Vibrio* Pacini 1854, 411AL. In *Bergey's Manual of Systematic Bacteriology*, 2nd ed.; Brenner, D.J., Krieg, N.R., Staley, J.R., Garrity, G., Eds.; Springer: New York, NY, USA, 2005; Volume 2, Part B; pp. 494–546.
19. Gonzalez, Y.; Venegas, D.; Mendoza-Hernandez, G.; Camarena, L.; Dreyfus, G. Na^+- and H^+-dependent motility in the coral pathogen *Vibrio shilonii*. *Fems. Microbiol. Lett.* **2010**, *312*, 142–150. [CrossRef]
20. McCarter, L.L. Polar flagellar motility of the Vibrionaceae. *Microbiol. Mol. Biol. Rev.* **2001**, *65*, 445–462, table of contents. [CrossRef]
21. McCarter, L.L. Dual flagellar systems enable motility under different circumstances. *J. Mol. Microbiol. Biotechnol* **2004**, *7*, 18–29. [CrossRef]
22. Tarrand, J.J.; Krieg, N.R.; Dobereiner, J. A taxonomic study of the *Spirillum lipoferum* group, with descriptions of a new genus, *Azospirillum* gen. nov. and two species, *Azospirillum lipoferum* (Beijerinck) comb. nov. and *Azospirillum brasilense* sp. nov. *Can. J. Microbiol.* **1978**, *24*, 967–980. [CrossRef]
23. Burnham, J.C.; Hashimoto, T.; Conti, S.F. Electron microscopic observations on the penetration of *Bdellovibrio bacteriovorus* into gram-negative bacterial hosts. *J. Bacteriol.* **1968**, *96*, 1366–1381. [CrossRef] [PubMed]
24. Seidler, R.J.; Starr, M.P. Structure of the flagellum of *Bdellovibrio bacteriovorus*. *J. Bacteriol.* **1968**, *95*, 1952–1955. [CrossRef] [PubMed]

25. Williams, H.; Baer, M. Genus II. *Bacteriovorax* Baer, Ravel, Chun, Hill and Williams 2000, 222^{VP}. In *Bergey's Manual of Systematic Bacteriology*, 2nd ed.; Brenner, D.J., Krieg, N.R., Staley, J.R., Garrity, G., Eds.; Springer: New York, NY, USA, 2005; Volume 2, Part C; pp. 1053–1057.
26. Berthenet, E.; Benejat, L.; Menard, A.; Varon, C.; Lacomme, S.; Gontier, E.; Raymond, J.; Boussaba, O.; Toulza, O.; Ducournau, A.; et al. Whole-genome sequencing and bioinformatics as pertinent tools to support Helicobacteracae taxonomy, based on three strains suspected to belong to novel *Helicobacter* species. *Front. Microbiol.* **2019**, *10*, 2820. [CrossRef] [PubMed]
27. Allen, R.D.; Baumann, P. Structure and arrangement of flagella in species of the genus *Beneckea* and *Photobacterium fischeri*. *J. Bacteriol.* **1971**, *107*, 295–302. [CrossRef] [PubMed]
28. Geis, G.; Leying, H.; Suerbaum, S.; Mai, U.; Opferkuch, W. Ultrastructure and chemical analysis of *Campylobacter pylori* flagella. *J. Clin. Microbiol.* **1989**, *27*, 436–441. [CrossRef]
29. Thomashow, L.S.; Rittenberg, S.C. Isolation and composition of sheathed flagella from *Bdellovibrio bacteriovorus* 109J. *J. Bacteriol.* **1985**, *163*, 1047–1054. [CrossRef]
30. Beeby, M.; Ribardo, D.A.; Brennan, C.A.; Ruby, E.G.; Jensen, G.J.; Hendrixson, D.R. Diverse high-torque bacterial flagellar motors assemble wider stator rings using a conserved protein scaffold. *Proc. Natl. Acad. Sci. USA* **2016**, *113*, E1917–E1926. [CrossRef]
31. Chen, S.; Beeby, M.; Murphy, G.E.; Leadbetter, J.R.; Hendrixson, D.R.; Briegel, A.; Li, Z.; Shi, J.; Tocheva, E.I.; Muller, A.; et al. Structural diversity of bacterial flagellar motors. *EMBO J.* **2011**, *30*, 2972–2981. [CrossRef]
32. Qin, Z.; Lin, W.T.; Zhu, S.; Franco, A.T.; Liu, J. Imaging the motility and chemotaxis machineries in *Helicobacter pylori* by cryo-electron tomography. *J. Bacteriol.* **2017**, *199*, e00695-16. [CrossRef]
33. Zhu, S.; Nishikino, T.; Hu, B.; Kojima, S.; Homma, M.; Liu, J. Molecular architecture of the sheathed polar flagellum in *Vibrio alginolyticus*. *Proc. Natl. Acad. Sci. USA* **2017**, *114*, 10966–10971. [CrossRef]
34. Zhu, S.; Nishikino, T.; Takekawa, N.; Terashima, H.; Kojima, S.; Imada, K.; Homma, M.; Liu, J. In situ structure of the *Vibrio* polar flagellum reveals distinct outer membrane complex and its specific interaction with the stator. *J. Bacteriol.* **2019**. [CrossRef] [PubMed]
35. Hranitzky, K.W.; Mulholland, A.; Larson, A.D.; Eubanks, E.R.; Hart, L.T. Characterization of a flagellar sheath protein of *Vibrio cholerae*. *Infect. Immun.* **1980**, *27*, 597–603. [CrossRef] [PubMed]
36. Yang, G.C.; Schrank, G.D.; Freeman, B.A. Purification of flagellar cores of *Vibrio cholerae*. *J. Bacteriol.* **1977**, *129*, 1121–1128. [CrossRef] [PubMed]
37. Fuerst, J.A.; Perry, J.W. Demonstration of lipopolysaccharide on sheathed flagella of *Vibrio cholerae* O:1 by protein A-gold immunoelectron microscopy. *J. Bacteriol.* **1988**, *170*, 1488–1494. [CrossRef]
38. Norqvist, A.; Wolf-Watz, H. Characterization of a novel chromosomal virulence locus involved in expression of a major surface flagellar sheath antigen of the fish pathogen *Vibrio anguillarum*. *Infect. Immun.* **1993**, *61*, 2434–2444. [CrossRef]
39. Geis, G.; Suerbaum, S.; Forsthoff, B.; Leying, H.; Opferkuch, W. Ultrastructure and biochemical studies of the flagellar sheath of *Helicobacter pylori*. *J. Med. Microbiol.* **1993**, *38*, 371–377. [CrossRef]
40. Bernal, P.; Munoz-Rojas, J.; Hurtado, A.; Ramos, J.L.; Segura, A. A *Pseudomonas putida* cardiolipin synthesis mutant exhibits increased sensitivity to drugs related to transport functionality. *Environ. Microbiol.* **2007**, *9*, 1135–1145. [CrossRef]
41. Kawai, F.; Shoda, M.; Harashima, R.; Sadaie, Y.; Hara, H.; Matsumoto, K. Cardiolipin domains in *Bacillus subtilis* Marburg membranes. *J. Bacteriol.* **2004**, *186*, 1475–1483. [CrossRef]
42. Mileykovskaya, E.; Dowhan, W. Visualization of phospholipid domains in *Escherichia coli* by using the cardiolipin-specific fluorescent dye 10-N-nonyl acridine orange. *J. Bacteriol.* **2000**, *182*, 1172–1175. [CrossRef]
43. Romantsov, T.; Helbig, S.; Culham, D.E.; Gill, C.; Stalker, L.; Wood, J.M. Cardiolipin promotes polar localization of osmosensory transporter ProP in *Escherichia coli*. *Mol. Microbiol.* **2007**, *64*, 1455–1465. [CrossRef]
44. Chu, J. Understanding the Role of Cardiolipin in Helicobacter Pylori Flagellar Synthesis. Ph.D. Thesis, University of Georgia, Athens, Georgia, 2019.
45. Chu, J.K.; Zhu, S.; Herrera, C.M.; Henderson, J.C.; Liu, J.; Trent, M.S.; Hoover, T.R. Loss of a cardiolipin synthase in *Helicobacter pylori* G27 blocks flagellum assembly. *J. Bacteriol.* **2019**, *201*. [CrossRef] [PubMed]
46. Hirai, Y.; Haque, M.; Yoshida, T.; Yokota, K.; Yasuda, T.; Oguma, K. Unique cholesteryl glucosides in *Helicobacter pylori*: Composition and structural analysis. *J. Bacteriol.* **1995**, *177*, 5327–5333. [CrossRef] [PubMed]

47. Zhou, P.; Hu, R.; Chandan, V.; Kuolee, R.; Liu, X.; Chen, W.; Liu, B.; Altman, E.; Li, J. Simultaneous analysis of cardiolipin and lipid A from *Helicobacter pylori* by matrix-assisted laser desorption/ionization time-of-flight mass spectrometry. *Mol. Biosyst* **2012**, *8*, 720–725. [CrossRef]
48. Doig, P.; Trust, T.J. Identification of surface-exposed outer membrane antigens of *Helicobacter pylori*. *Infect. Immun.* **1994**, *62*, 4526–4533. [CrossRef] [PubMed]
49. Furuno, M.; Sato, K.; Kawagishi, I.; Homma, M. Characterization of a flagellar sheath component, PF60, and its structural gene in marine *Vibrio*. *J. Biochem.* **2000**, *127*, 29–36. [CrossRef] [PubMed]
50. Luke, C.J.; Kubiak, E.; Cockayne, A.; Elliott, T.S.; Penn, C.W. Identification of flagellar and associated polypeptides of *Helicobacter* (formerly *Campylobacter*) *pylori*. *Fems. Microbiol. Lett.* **1990**, *59*, 225–230. [CrossRef]
51. Luke, C.J.; Penn, C.W. Identification of a 29 kDa flagellar sheath protein in *Helicobacter pylori* using a murine monoclonal antibody. *Microbiology* **1995**, *141*, 597–604. [CrossRef]
52. Bari, W.; Lee, K.M.; Yoon, S.S. Structural and functional importance of outer membrane proteins in *Vibrio cholerae* flagellum. *J. Microbiol.* **2012**, *50*, 631–637. [CrossRef]
53. Evans, D.G.; Evans, D.J., Jr.; Moulds, J.J.; Graham, D.Y. N-acetylneuraminyllactose-binding fibrillar hemagglutinin of *Campylobacter pylori*: A putative colonization factor antigen. *Infect. Immun.* **1988**, *56*, 2896–2906. [CrossRef]
54. O'Toole, P.W.; Janzon, L.; Doig, P.; Huang, J.; Kostrzynska, M.; Trust, T.J. The putative neuraminyllactose-binding hemagglutinin HpaA of *Helicobacter pylori* CCUG 17874 is a lipoprotein. *J. Bacteriol.* **1995**, *177*, 6049–6057. [CrossRef]
55. Jones, A.C.; Logan, R.P.H.; Foynes, S.; Cackayne, A.; Wren, B.W.; Penn, C.W. A flagellar sheath protein of *Helicobacter pylori* is identical to HpaA, a putative *N*-acetylneuraminyllactose-binding hemagglutinin, but is not an adhesin for AGS cells. *J. Bacteriol.* **1997**, *179*, 5643–5647. [CrossRef] [PubMed]
56. Lundstrom, A.M.; Blom, K.; Sundaeus, V.; Bolin, I. HpaA shows variable surface localization but the gene expression is similar in different *Helicobacter pylori* strains. *Microb Pathog* **2001**, *31*, 243–253. [CrossRef] [PubMed]
57. Radin, J.N.; Gaddy, J.A.; Gonzalez-Rivera, C.; Loh, J.T.; Algood, H.M.; Cover, T.L. Flagellar localization of a *Helicobacter pylori* autotransporter protein. *MBio* **2013**, *4*, e00613-12. [CrossRef] [PubMed]
58. Niehus, E.; Gressmann, H.; Ye, F.; Schlapbach, R.; Dehio, M.; Dehio, C.; Stack, A.; Meyer, T.F.; Suerbaum, S.; Josenhans, C. Genome-wide analysis of transcriptional hierarchy and feedback regulation in the flagellar system of *Helicobacter pylori*. *Mol. Microbiol.* **2004**, *52*, 947–961. [CrossRef] [PubMed]
59. Voss, B.J.; Gaddy, J.A.; McDonald, W.H.; Cover, T.L. Analysis of surface-exposed outer membrane proteins in *Helicobacter pylori*. *J. Bacteriol.* **2014**, *196*, 2455–2471. [CrossRef]
60. Ottemann, K.M.; Lowenthal, A.C. *Helicobacter pylori* uses motility for initial colonization and to attain robust infection. *Infect. Immun.* **2002**, *70*, 1984–1990. [CrossRef]
61. Meuskens, I.; Saragliadis, A.; Leo, J.C.; Linke, D. Type V secretion systems: An overview of passenger domain functions. *Front. Microbiol.* **2019**, *10*, 1163. [CrossRef]
62. Fuerst, J.A. Bacterial sheathed flagella and the rotary motor model for the mechanism of bacterial motility. *J. Biol.* **1980**, *84*, 761–774. [CrossRef]
63. Aschtgen, M.S.; Lynch, J.B.; Koch, E.; Schwartzman, J.; McFall-Ngai, M.; Ruby, E. Rotation of *Vibrio fischeri* flagella produces outer membrane vesicles that induce host development. *J. Bacteriol.* **2016**, *198*, 2156–2165. [CrossRef]
64. Brennan, C.A.; Hunt, J.R.; Kremer, N.; Krasity, B.C.; Apicella, M.A.; McFall-Ngai, M.J.; Ruby, E.G. A model symbiosis reveals a role for sheathed-flagellum rotation in the release of immunogenic lipopolysaccharide. *Elife* **2014**, *3*, e01579. [CrossRef]
65. Millikan, D.S.; Ruby, E.G. *Vibrio fischeri* flagellin A is essential for normal motility and for symbiotic competence during initial squid light organ colonization. *J. Bacteriol.* **2004**, *186*, 4315–4325. [CrossRef] [PubMed]
66. Vanhove, A.S.; Duperthuy, M.; Charriere, G.M.; Le Roux, F.; Goudenege, D.; Gourbal, B.; Kieffer-Jaquinod, S.; Coute, Y.; Wai, S.N.; Destoumieux-Garzon, D. Outer membrane vesicles are vehicles for the delivery of *Vibrio tasmaniensis* virulence factors to oyster immune cells. *Environ. Microbiol.* **2015**, *17*, 1152–1165. [CrossRef]
67. Nikaido, H. Molecular basis of bacterial outer membrane permeability revisited. *Microbiol Mol. Biol. Rev.* **2003**, *67*, 593–656. [CrossRef] [PubMed]

68. Dekker, N. Outer-membrane phospholipase A: Known structure, unknown biological function. *Mol. Microbiol.* **2000**, *35*, 711–717. [CrossRef] [PubMed]
69. Malinverni, J.C.; Silhavy, T.J. An ABC transport system that maintains lipid asymmetry in the gram-negative outer membrane. *Proc. Natl. Acad. Sci. USA* **2009**, *106*, 8009–8014. [CrossRef] [PubMed]
70. Huang, K.C.; Mukhopadhyay, R.; Wingreen, N.S. A curvature-mediated mechanism for localization of lipids to bacterial poles. *PLoS Comput. Biol.* **2006**, *2*, e151. [CrossRef] [PubMed]
71. Renner, L.D.; Weibel, D.B. Cardiolipin microdomains localize to negatively curved regions of *Escherichia coli* membranes. *Proc. Natl. Acad. Sci USA* **2011**, *108*, 6264–6269. [CrossRef]
72. Dahlberg, M. Polymorphic phase behavior of cardiolipin derivatives studied by coarse-grained molecular dynamics. *J. Phys. Chem. B* **2007**, *111*, 7194–7200. [CrossRef]
73. Schlame, M. Cardiolipin synthesis for the assembly of bacterial and mitochondrial membranes. *J. Lipid. Res.* **2008**, *49*, 1607–1620. [CrossRef]
74. Schlame, M.; Rua, D.; Greenberg, M.L. The biosynthesis and functional role of cardiolipin. *Prog. Lipid. Res.* **2000**, *39*, 257–288. [CrossRef]
75. Nichols-Smith, S.; Teh, S.Y.; Kuhl, T.L. Thermodynamic and mechanical properties of model mitochondrial membranes. *Biochim Biophys. Acta* **2004**, *1663*, 82–88. [CrossRef] [PubMed]
76. Planas-Iglesias, J.; Dwarakanath, H.; Mohammadyani, D.; Yanamala, N.; Kagan, V.E.; Klein-Seetharaman, J. Cardiolipin interactions with proteins. *Biophys. J.* **2015**, *109*, 1282–1294. [CrossRef] [PubMed]
77. Buskirk, S.W.; Lafontaine, E.R. *Moraxella catarrhalis* expresses a cardiolipin synthase that impacts adherence to human epithelial cells. *J. Bacteriol.* **2014**, *196*, 107–120. [CrossRef] [PubMed]
78. Romantsov, T.; Gonzalez, K.; Sahtout, N.; Culham, D.E.; Coumoundouros, C.; Garner, J.; Kerr, C.H.; Chang, L.; Turner, R.J.; Wood, J.M. Cardiolipin synthase A colocalizes with cardiolipin and osmosensing transporter ProP at the poles of *Escherichia coli* cells. *Mol. Microbiol.* **2018**, *107*, 623–638. [CrossRef] [PubMed]
79. Rossi, R.M.; Yum, L.; Agaisse, H.; Payne, S.M. Cardiolipin synthesis and outer membrane localization are required for *Shigella flexneri* virulence. *MBio* **2017**, *8*. [CrossRef] [PubMed]
80. Treuner-Lange, A.; Sogaard-Andersen, L. Regulation of cell polarity in bacteria. *J. Cell Biol.* **2014**, *206*, 7–17. [CrossRef]
81. Richardson, K.; Nixon, L.; Mostow, P.; Kaper, J.B.; Michalski, J. Transposon-induced non-motile mutants of *Vibrio cholerae*. *J. Gen. Microbiol.* **1990**, *136*, 717–725. [CrossRef]
82. Chaban, B.; Coleman, I.; Beeby, M. Evolution of higher torque in *Campylobacter*-type bacterial flagellar motors. *Sci. Rep.* **2018**, *8*, 97. [CrossRef]
83. Terashima, H.; Koike, M.; Kojima, S.; Homma, M. The flagellar basal body-associated protein FlgT is essential for a novel ring structure in the sodium-driven *Vibrio* motor. *J. Bacteriol.* **2010**, *192*, 5609–5615. [CrossRef]
84. Zhu, S.; Nishikino, T.; Kojima, S.; Homma, M.; Liu, J. The *Vibrio* H-ring facilitates the outer membrane penetration of the polar sheathed flagellum. *J. Bacteriol.* **2018**, *200*. [CrossRef]
85. Evans, D.G.; Karjalainen, T.K.; Evans, D.J., Jr.; Graham, D.Y.; Lee, C.H. Cloning, nucleotide sequence, and expression of a gene encoding an adhesin subunit protein of *Helicobacter pylori*. *J. Bacteriol.* **1993**, *175*, 674–683. [CrossRef] [PubMed]
86. Carlsohn, E.; Nystrom, J.; Bolin, I.; Nilsson, C.L.; Svennerholm, A.M. HpaA is essential for *Helicobacter pylori* colonization in mice. *Infect. Immun.* **2006**, *74*, 920–926. [CrossRef] [PubMed]
87. Hayashi, F.; Smith, K.D.; Ozinsky, A.; Hawn, T.R.; Yi, E.C.; Goodlett, D.R.; Eng, J.K.; Akira, S.; Underhill, D.M.; Aderem, A. The innate immune response to bacterial flagellin is mediated by Toll-like receptor 5. *Nature* **2001**, *410*, 1099–1103. [CrossRef]
88. Kawasaki, T.; Kawai, T. Toll-like receptor signaling pathways. *Front. Immunol.* **2014**, *5*, 461. [CrossRef] [PubMed]
89. Yoon, S.S.; Mekalanos, J.J. Decreased potency of the *Vibrio cholerae* sheathed flagellum to trigger host innate immunity. *Infect. Immun.* **2008**, *76*, 1282–1288. [CrossRef] [PubMed]
90. Smith, K.D.; Andersen-Nissen, E.; Hayashi, F.; Strobe, K.; Bergman, M.A.; Barrett, S.L.; Cookson, B.T.; Aderem, A. Toll-like receptor 5 recognizes a conserved site on flagellin required for protofilament formation and bacterial motility. *Nat. Immunol.* **2003**, *4*, 1247–1253. [CrossRef] [PubMed]
91. Lee, S.K.; Stack, A.; Katzowitsch, E.; Aizawa, S.I.; Suerbaum, S.; Josenhans, C. *Helicobacter pylori* flagellins have very low intrinsic activity to stimulate human gastric epithelial cells via TLR5. *Microbes Infect.* **2003**, *5*, 1345–1356. [CrossRef]

92. Logan, S.M. Flagellar glycosylation-a new component of the motility repertoire? *Microbiology* **2006**, *152*, 1249–1262. [CrossRef]
93. Merino, S.; Tomas, J.M. Gram-negative flagella glycosylation. *Int. J. Mol. Sci.* **2014**, *15*, 2840–2857. [CrossRef]
94. Arora, S.K.; Neely, A.N.; Blair, B.; Lory, S.; Ramphal, R. Role of motility and flagellin glycosylation in the pathogenesis of *Pseudomonas aeruginosa* burn wound infections. *Infect. Immun.* **2005**, *73*, 4395–4398. [CrossRef]
95. Ewing, C.P.; Andreishcheva, E.; Guerry, P. Functional characterization of flagellin glycosylation in *Campylobacter jejuni* 81-176. *J. Bacteriol.* **2009**, *191*, 7086–7093. [CrossRef]
96. Guerry, P.; Ewing, C.P.; Schirm, M.; Lorenzo, M.; Kelly, J.; Pattarini, D.; Majam, G.; Thibault, P.; Logan, S. Changes in flagellin glycosylation affect *Campylobacter* autoagglutination and virulence. *Mol. Microbiol.* **2006**, *60*, 299–311. [CrossRef] [PubMed]
97. Hanuszkiewicz, A.; Pittock, P.; Humphries, F.; Moll, H.; Rosales, A.R.; Molinaro, A.; Moynagh, P.N.; Lajoie, G.A.; Valvano, M.A. Identification of the flagellin glycosylation system in *Burkholderia cenocepacia* and the contribution of glycosylated flagellin to evasion of human innate immune responses. *J. Biol. Chem.* **2014**, *289*, 19231–19244. [CrossRef] [PubMed]
98. Howard, S.L.; Jagannathan, A.; Soo, E.C.; Hui, J.P.; Aubry, A.J.; Ahmed, I.; Karlyshev, A.; Kelly, J.F.; Jones, M.A.; Stevens, M.P.; et al. *Campylobacter jejuni* glycosylation island important in cell charge, legionaminic acid biosynthesis, and colonization of chickens. *Infect. Immun.* **2009**, *77*, 2544–2556. [CrossRef] [PubMed]
99. Ichinose, Y.; Taguchi, F.; Yamamoto, M.; Ohnishi-Kameyama, M.; Atsumi, T.; Iwaki, M.; Manabe, H.; Kumagai, M.; Nguyen, Q.T.; Nguyen, C.L.; et al. Flagellin glycosylation is ubiquitous in a broad range of phytopathogenic bacteria. *J. Gen. Plant. Path* **2013**, *79*, 359–365. [CrossRef]
100. Schirm, M.; Soo, E.C.; Aubry, A.J.; Austin, J.; Thibault, P.; Logan, S.M. Structural, genetic and functional characterization of the flagellin glycosylation process in *Helicobacter pylori*. *Mol. Microbiol.* **2003**, *48*, 1579–1592. [CrossRef]
101. Logan, S.M.; Hui, J.P.; Vinogradov, E.; Aubry, A.J.; Melanson, J.E.; Kelly, J.F.; Nothaft, H.; Soo, E.C. Identification of novel carbohydrate modifications on *Campylobacter jejuni* 11168 flagellin using metabolomics-based approaches. *FEBS J.* **2009**, *276*, 1014–1023. [CrossRef]
102. Schirm, M.; Schoenhofen, I.C.; Logan, S.M.; Waldron, K.C.; Thibault, P. Identification of unusual bacterial glycosylation by tandem mass spectrometry analyses of intact proteins. *Anal. Chem.* **2005**, *77*, 7774–7782. [CrossRef]
103. Manning, A.J.; Kuehn, M.J. Contribution of bacterial outer membrane vesicles to innate bacterial defense. *BMC Microbiol.* **2011**, *11*, 258. [CrossRef]
104. Zhang, H.; Li, L.; Zhao, Z.; Peng, D.; Zhou, X. Polar flagella rotation in *Vibrio parahaemolyticus* confers resistance to bacteriophage infection. *Sci. Rep.* **2016**, *6*, 26147. [CrossRef]
105. Belas, R. Biofilms, flagella, and mechanosensing of surfaces by bacteria. *Trends Microbiol.* **2014**, *22*, 517–527. [CrossRef] [PubMed]
106. Guerrero-Ferreira, R.C.; Viollier, P.H.; Ely, B.; Poindexter, J.S.; Georgieva, M.; Jensen, G.J.; Wright, E.R. Alternative mechanism for bacteriophage adsorption to the motile bacterium *Caulobacter crescentus*. *Proc. Natl. Acad. Sci. USA* **2011**, *108*, 9963–9968. [CrossRef] [PubMed]
107. Baldvinsson, S.B.; Sorensen, M.C.; Vegge, C.S.; Clokie, M.R.; Brondsted, L. *Campylobacter jejuni* motility is required for infection of the flagellotropic bacteriophage F341. *Appl. Environ. Microbiol.* **2014**, *80*, 7096–7106. [CrossRef] [PubMed]

Review

Flagella and Swimming Behavior of Marine Magnetotactic Bacteria

Wei-Jia Zhang [1,2] and Long-Fei Wu [2,3,*]

1 Laboratory of Deep-Sea Microbial Cell Biology, Institute of Deep-sea Science and Engineering, Chinese Academy of Sciences, Sanya 572000, China; wzhang@idsse.ac.cn
2 International Associated Laboratory of Evolution and Development of Magnetotactic Multicellular Organisms, F-13402 CNRS-Marseille, France/CAS-Sanya 572000, China
3 Aix Marseille Univ, CNRS, LCB, IMM, IM2B, CENTURI, F-13402 Marseille, France
* Correspondence: wu@imm.cnrs.fr; Tel.: +33-4-9116-4157

Received: 25 February 2020; Accepted: 15 March 2020; Published: 16 March 2020

Abstract: Marine environments are generally characterized by low bulk concentrations of nutrients that are susceptible to steady or intermittent motion driven by currents and local turbulence. Marine bacteria have therefore developed strategies, such as very fast-swimming and the exploitation of multiple directional sensing–response systems in order to efficiently migrate towards favorable places in nutrient gradients. The magnetotactic bacteria (MTB) even utilize Earth's magnetic field to facilitate downward swimming into the oxic–anoxic interface, which is the most favorable place for their persistence and proliferation, in chemically stratified sediments or water columns. To ensure the desired flagella-propelled motility, marine MTBs have evolved an exquisite flagellar apparatus, and an extremely high number (tens of thousands) of flagella can be found on a single entity, displaying a complex polar, axial, bounce, and photosensitive magnetotactic behavior. In this review, we describe gene clusters, the flagellar apparatus architecture, and the swimming behavior of marine unicellular and multicellular magnetotactic bacteria. The physiological significance and mechanisms that govern these motions are discussed.

Keywords: flagellar number and position; north-seeking and south-seeking; magnetic and photo-response

---

1. Introduction

Magnetotactic bacteria (MTB) are a group of phylogenetically, morphologically, and physiologically diverse Gram-negative bacteria [1,2]. They share the common capability of synthesizing unique intracellular organelles, the magnetosomes, i.e., single-domain magnetic crystals of magnetite or greigite, which are enveloped by membranes (Figure 1). Cytoskeleton MamK filaments enable the magnetosomes to be organized into chains [3–5]. Magnetosome chains impart a net magnetic dipole moment to the cell, which allows cells to align and swim along geomagnetic field lines [6]. This behavior, referred to as magnetotaxis, is believed to facilitate microaerophilic or anaerobic MTB to locate at the preferable oxic–anoxic interface in chemically stratified sediments or water columns [1].

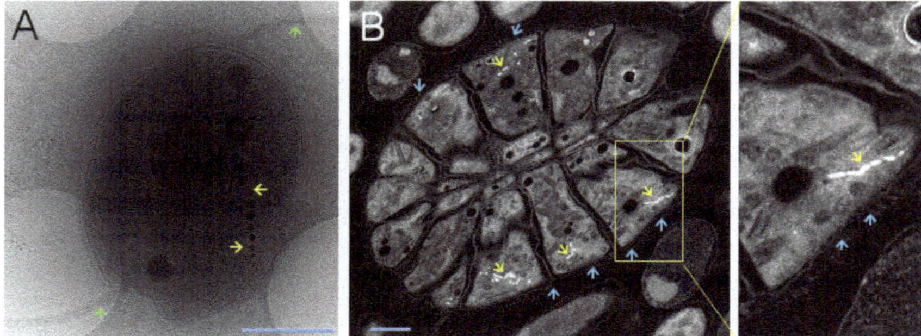

Figure 1. Magnetosomes and flagella of magnetotactic bacteria. (**A**) Bilophotrichously flagellated MO-1 cells possess two sheathed flagellar bundles (green arrow) and one magnetosome chain (yellow arrow). (**B**) Peritrichously flagellated ellipsoidal magnetoglobule with flagella (blue arrows) and magnetosomes (yellow arrows). Only the portion of flagellar filaments in the surface matrix was preserved during sample preparation. Scale bar is equal to 0.5 μm. Courtesy of the electron cryotomography micrograph (**A**) from Dr. J. Ruan and Professor K. Namba, and of the Scan-TEM high-angle annular dark-field (STEM-HAADF) mode micrograph (**B**) of ultrathin sections of high-pressure freezing/freeze substitution fixation (HPF/FS) fixed ellipsoidal magnetoglobule from Professor N. Menguy and Dr. A. Kosta.

Phylogenetically, magnetotactic bacteria are members of several classes of the Proteobacteria phylum including the Alpha-, Gamma-, Delta-, Zeta-, *Candidatus* Lambda-, *Candidatus* Eta-classes, the Nitrospirae phylum, the *Candidatus* Omnitrophica phylum, the *Candidatus* Latescibacteria phylum, and the Planctomycetes phylum [7]. They present various morphotypes including cocci, spirilla, rod-shaped, vibrio, and more complex multicellular magnetotactic prokaryotes that are also called magnetoglobules (MMP) [1,8].

Magnetotactic bacteria are found worldwide in aquatic environments from freshwater to marine ecosystems. Here, we will discuss mainly three types of marine magnetotactic bacteria because of their complex flagellar architecture and peculiar motile behavior. The first is the spirillum *Magnetospira* sp. strain QH-2 isolated from the intertidal sediments of the China Sea [9]. Phylogenetically, QH-2 belongs to Rhodospirillaceae and is closely related to two freshwater magnetotactic spirilla, *Magnetospirillum magneticum* AMB-1 and *Magnetospirillum gryphiswaldense* MSR-1. Yet, certain traits such as the synthesis of osmoprotectant, Na^+-dependent NADH-quinone oxidoreductase, and Na^+-motive force driven flagellar motors, make QH-2 better suited to a marine sedimentary lifestyle than its freshwater counterparts [10]. The second is the ovoid-coccoid *Magnetococcus massalia* strain MO-1 isolated from sediments of the Mediterranean Sea (Figure 1A) [11]. MO-1 belongs to the newly established class *Candidatus* Etatproteobacteria and possesses the most exquisite flagellar apparatus [12]. The third group is the magnetoglobules that have developed both multicellular and magnetotactic properties during their evolution. To date, magnetotactic multicellular prokaryotes are found only in marine environments [8]. They exhibit peculiar patterns of motility by coordinatively rotating tens of thousands of peritrichous flagella (Figure 1B), including both polar and axis magneto-aerotaxis, ping-pong motion, and photophobic and photokinesis swimming patterns.

2. Flagellar Apparatus of Marine Magnetotactic Bacteria

Flagella provide one of the most highly efficient means of bacterial locomotion and play a pivotal role in adhesion, biofilm formation, and host invasion [13–16]. Bacterial flagella share a basic tripartite structure; the basal body, the hook, and the filament [17]. The basal body contains a reversible rotary motor made of a rotor, a drive shaft, a bushing, and about a dozen stators. The stator forms the proton or sodium ion pathway and converts ion flow across the cytoplasmic membrane into the mechanical work required for flagellar motor rotation. The basal body also contains the flagellar protein export apparatus, which recognizes, unfolds, and translocates flagellar components into the central channel and to the distal, growing end of the flagellum [15,17]. Flagellar filaments have a helical structure and function as a screw, where rotation pushes or pulls the cell. Despite structural similarities, bacterial flagella exhibit extensive variations in both number and placement between species, and this criterion had been used in bacterial taxonomy in the past. Bacteria may have a single flagellum (monotrichous) at one end of the cell (polar flagellum), or a single flagellum at both ends (amphitrichous), numerous flagella in a tuft (lophotrichous), or flagella distributed all over the cell (peritrichous). The three model magnetotactic bacteria reviewed here possess amphitrichous, bilophotrichous, and peritrichous flagella that underpin complex magnetotactic motion.

2.1. Flagellar Apparatus of Amphitrichously Flagellated Magnetospira sp. Strain QH-2

The spirillum *Magnetospira* sp. strain QH-2 was isolated from the intertidal sediments of the China Sea [9]. The cells are amphitrichously flagellated with a single flagellum at each pole, their composition and structure are probably the simplest when compared to the bilophotrichous flagella of MO-1 and the peritrichous flagella of the multicellular magnetoglobules. Genomic analysis identified flagellum synthesis genes at 10 locations (Figure 2) [10]. Intriguingly, multiple genes coding for either proton-driven or sodium ion-driven motors were identified, including a single *pomA*, a single *motB*, and two complete sets (*pomAB*/*pomA-motB*), although the paralogs share limited similarity (below 45%). In addition to the flagellar biosynthesis genes, well conserved in most prokaryotes, two genes annotated as O-b-N-acetylglucosaminyltransferase were identified in flagellar gene clusters. They contain a flagellin and several flagellar biosynthesis regulatory genes, demonstrating their function in flagellin glycosylation.

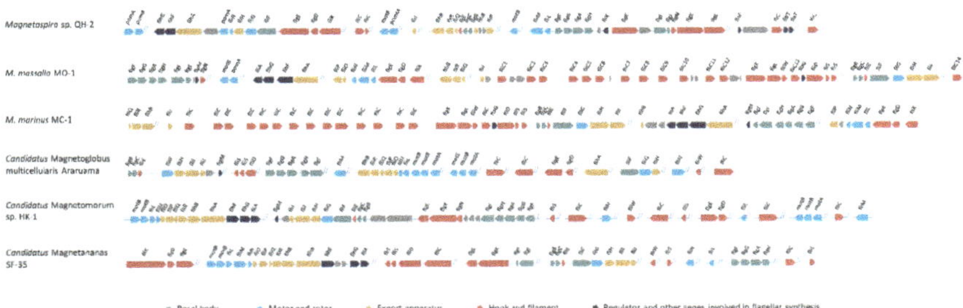

Figure 2. Organization of flagella genes in model magnetotactic bacteria. The data are derived from genomic data of amphitrichously flagellated *Magnetospira* sp. QH-2 [10], bilophotrichously flagellated *M. massalia* strain MO-1 [18] and *M. marinus* strain MC-1 [19], peritrichous flagellated spherical magnetoglobules Ca. M. multicellularis Araruama [20], Ca. Magnetomorum strain HK-1 [21], and ellipsoidal magnetoglobules Ca. Magnetananas updated from the incomplete genome sequence [22]. Separated localization of the gene clusters is marked by double slashes. Arrows show the genes and their transcriptional direction; their lengths are proportional to the size of the genes.

2.2. Flagellar Apparatus of Bilophotrichously Flagellated M. massalia Strain MO-1

M. massalia strain MO-1 synthesizes two sheathed flagellar bundles on the long axis side of its ovoid body (Figure 1A). Each bundle is composed of 7 flagella and 24 fibrils. The flagella are organized in a 2:3:2 array, and each of them is surrounded by 6 fibrils; altogether they constitute seven intertwined hexagonal arrays [12]. It has been hypothesized that the 24 fibrils might counter rotate between the 7 flagellar filaments to minimize the friction that would be generated if the flagella were directly packed together in a tight bundle [12]. The closely related *M. marinus* strain MC-1 and several marine bilophotrichously flagellated magnetotactic cocci seem to possess a flagellar apparatus with a similar architecture [11,23–26]. Recently, an even more complex flagellar apparatus consisting of 19 flagella arranged in a 3:4:5:4:3 array within the flagellar bundle has been observed in a magnetotactic cocci found in the biogenic sediments of a Mariana–Yap seamount [27]. Questions inevitably arose about these exquisite flagellar apparatuses, such as: What is the factor that determines the accurate localization of flagella and why are they constrained within a sheath structure?

Sheath or pseudo-sheaths, each enclosing a single flagellum, have been reported for *Caulobacter crescentus* [28], *Pseudomonas rhodos* [29], *Vibrio* spp. [16], *Helicobacter pylori* [30], and *Bdellovibrio bacteriovorus* [31]. These flagellar sheath structures are believed to be an extension of the outer membrane. In contrast, the sheath of MO-1 is assembled from a large (>350 kDa) glycoprotein and in a calcium ion-dependent manner made into a left-handed helical structure [12,32]. The sheathed bundle of seven flagella produces a thrust force, which is nine times greater than an unsheathed one, and this is indispensable for the smooth swimming motion of MO-1 cells [12,32]. In addition, as all strains possessing similar flagellar structures reside in marine sediments, the presence of a sheath could possibly protect the filaments from breaking whilst moving through sands, and this implies that there is an evolutionary adaptation to this habitat.

The complexity of this bilophotrichous sheathed flagellar is further demonstrated by using genome sequences. Genomic analysis revealed that the genetic structures of flagellar synthesis genes in strains MO-1 and MC-1 are well conserved (Figure 2). Most intriguingly, they possess the highest number of flagellin paralogs (14 flagellin genes in strain MO-1 and 15 in strain MC-1) found in bacterial genomes to date [19,33]. In both strains, most flagellin genes are spread in a tandem array, while a single *fliC* resides in a more compact flagellar gene cluster consisting of *flgKL* (hook-associated proteins), *fliW* (antagonist of general regulator CsrA), a putative *flaG* gene (function unknown), *fliD* (filament cap), and two *fliS* (chaperon) (Figure 2). There is no obvious element, such as insertion-sequence (IS) elements or duplicated flanking sequences, which could explain the mechanism of duplication of these *fliC* paralogs. As indicated by quantitative PCR (q-PCR) and mass spectrometry analyses, all 14 flagellins in MO-1 are expressed, highly glycosylated, and present in the flagellar filaments, although they differ significantly in quantity [33]. The biological significance of highly redundant flagellins and the way they make up the filament, i.e., whether each flagellin forms an individual simple filament or whether multiple flagellins form complex segmented or mosaic filaments, requires in-depth research. Nevertheless, the flagella of MO-1 cells show unprecedented complexity in spatial organization and flagellin redundancy in unicellular microorganisms.

2.3. Peritrichous Flagella of Multicellular Magnetoglobules

There are two kinds of magnetoglobules. In 1983, Farina et al. discovered spherical or mulberry-like magnetoglobules in the Rodrigo de Freitas lagoon in Brazil [34]. Typically, 15–45 bacterial cells arrange themselves with a helical geometry in a multicellular entity [35]. Since then, these types of magnetoglobules have been observed worldwide [36–42]. The second morphotype, the ellipsoidal or pineapple-like magnetoglobules, were observed in the Mediterranean Sea [8,43,44], the China Sea, and the Pacific Ocean [45–49]. Approximately 60 cells axisymmetrically assemble along the longitudinal axis to achieve a one-layer hollow entity that is held by a lattice at the surface [8]. Phylogenetic studies have identified eleven species belonging to six genera of spherical magnetoglobules and nine species belonging to six genera of ellipsoidal magnetoglobules. They formed branches of a magnetoglobule clade, which are affiliated with Deltaproteobacteria, but are distinguished from another multicellular Deltaproteobacteria, the myxobacteria [8]. Both morphotypes exhibit a conspicuous periphery–core architecture. Juxtaposed membranes adhere together cells surrounding the core lumen where material and information exchange may occur among the cells. Magnetoglobules possess multiple magnetosome chains arranged along their long axis at the cell periphery. The surface of magnetoglobules is covered by approximately tens of thousands of flagella (Figure 1) [8,50].

Genomic analysis revealed the following salient features of the genes required for flagella synthesis in magnetoglobules. First, they possess well-conserved gene clusters containing (*motA*)-2*motB*-*fliRQPONL*-*flhB*-(*flhAFG*-*fliA*) (Figure 2). Second, they have multiple copies of several genes involved in motor rotation, such as *motAB* that code for proton–ion driven motors and *fliN* codes for a part of the motor switch complex, which modulates the motor activity. It is noticeable that the second copy of *fliN* is twice the size of the copy in the conserved cluster. The long *fliN* of the switch complex component might be involved in the coordination of flagellar rotation. Third, they have 2–3 copies of flagellin *fliC* genes, of which one copy is longer than the usual *fliC* genes. Finally, the *flhF* and *flhG* genes controlling the flagellar number and position are highly conserved in magnetoglobules. They may bear intrinsic characteristics for the regular implantation of thousands of flagella at the outer surface of magnetoglobule cells.

3. Magnetotaxis Behavior of Marine Magnetotactic Bacteria

Magnetotactic bacteria are capable of aligning and swimming along the geomagnetic field lines. The efficiency of magnetic orientation depends on the local redox gradient and latitude of the habitats where the MTB dwell, as well as on the flagellar apparatus of MTB cells.

3.1. Polar and Axial Magnetotaxis

Magnetotaxis and aerotaxis work together in MTB to perform a so-called "magneto-aerotaxis". Two different magneto-aerotactic mechanisms, termed polar and axial magnetotaxis, are found in different bacterial species [1,24]. In droplets of samples on a microscope slide or cover, there is an oxygen gradient that is created due to the diffusion of oxygen from the peripheric edge toward the center. When inspected with the optical microscope under oxic conditions, polar magnetotactic bacteria swim persistently in one direction, either the north or the south, in the magnetic field. In contrast, axial magnetotactic cells swim in either direction along the magnetic field lines with frequent, spontaneous reversals of swimming direction without turning around.

The bilophotrichously flagellated *M. massalia* strain MO-1 exhibits a polar magnetotactic behavior, swimming northwards along the geomagnetic field lines by means of two sheathed flagellar bundles, at speeds of up to 300 μm/s, with frequent changes from a right to a left hand helical trajectory [11]. Freshwater amphitrichously flagellated *M. magneticum* AMB-1 shares a similar morphology with the marine *Magnetospira* sp. strain QH-2, and its swimming behavior has been the most extensively studied. Asymmetric rotation of the flagella (counterclockwise at the lagging pole and clockwise at the leading pole) enables the cell to "run" while symmetric rotation triggers cell tumbling [51]. AMB-1 cells frequently tumble and change swimming direction, displaying the typical axial magnetotactic behavior. Peritrichous magnetoglobules collected from the Mediterranean Sea swim preferentially northward, a polar magnetotaxis. However, at times, some of them randomly change swimming direction southward and subsequently change back to a north-seeking swim [8]. This is a typical behavior of axial magnetotaxis. The stochastic backward motion may play a similar physiological function to the tumbling of *Escherichia coli* that allows bacteria to randomly explore the favorable direction in which to go. Therefore, a given MTB may perform both polar and axial magnetotactic motilities that are not reciprocally exclusive, and the alternative usage is part of the adaptation strategy.

3.2. Bounce Motion

Magnetoglobules display a canonical escape or ping-pong motion. It is composed of a sudden accelerated excursion from the droplet edge towards the center opposing the direction of magnetotaxis. At variable distances, they decelerate, stop, and swim with acceleration back to the droplet edge [8, 34–36,41,42,44,45,48,52–57]. In fact, the ping-pong motion is not restricted to magnetoglobules; other morphotypes of MTB also display this kind of motility. The small cell sizes make observations difficult. Some of the big rod-shaped MTB exhibit obvious escape motion as shown in Figure 3A and Vdieo S1 ping-pong motion of big rod-shaped MTB.

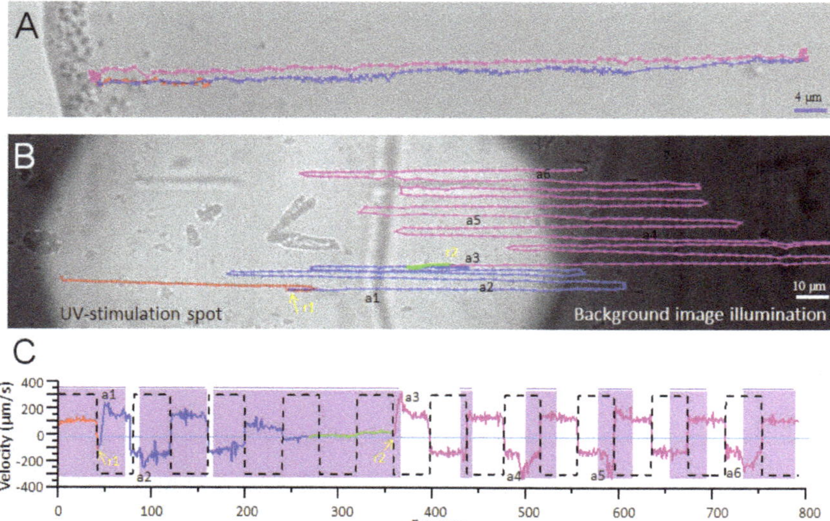

Figure 3. Ping-pong motion and photo-sensitive motility. (**A**) a magnetotactic bacillus of ~ 4 µm swims northward (red track) until the edge of the droplet. Then, it swims southward, opposite to the north-seeking swimming direction to the center of the droplet (blue track), which is followed by a returning north-seeking (magenta track). (**B**) is a representative photosensitive swimming behavior of magnetoglobules and (**C**) is an ImageJ analysis of the data [8]. The dot-line square curve indicates the north direction of the alternating magnetic field. Positive velocity means that the magnetoglobule swims from left to right on the image whilst the negative values are opposite. The velocity curve colors in (**C**) correspond to the same colors of the swim tracks in (**B**). When the velocity curve is on the same side as the field curve using zero velocity line as a reference, the magnetoglobule exhibits a north-seeking magnetotaxis (e.g., red and green tracks), otherwise it displays a south-seeking magnetotaxis (blue and magenta tracks). Violet areas show the swimming of the magnetoglobule in the UV spot. Yellow arrows with r1 and r2 indicate the sudden change of swimming direction to south-seeking; a1 to a6 show the accelerations.

The ping-pong motion can be observed when cells hit the edge of the droplets or other kinds of obstacles, such as the wall of microchannels [8]. In addition, both the unicellular *M. massalia* strain MO-1 [58] and multicellular magnetoglobules [8] exhibit a conspicuous backward motion when they encounter particles. In all conditions, cells are prevented from swimming in a magnetotaxis direction, and exhibit a bounce motion. The mechanism involved in the mechanical sensing of microchannel walls and particles might be different from that of the surface/border of the droplets.

3.3. Photo-Sensitive Magnetotaxis: Photophobic Response and Photokinesis

Sunlight consists of electromagnetic waves, of which high energetic radiation is harmful for living organisms. Fortunately, the geomagnetic field protects living beings from the deleterious effect of radiation. In addition, the geomagnetic field provides a pervasive and reliable source of directional and positional information for various organisms to use as an orientation cue, which maps migrating or homing routes. Magnetotactic bacteria have developed means of sensing not only the geomagnetic field, but also certain wavelengths of sunlight.

Microbes react to light illumination in different ways depending on their physiological properties. Phototaxis refers to cells swimming along the direction of a light beam towards (positive) or away from (negative) a light source [59]. In reaction to a sudden change of light intensity, photophobic microbes will swim to lower intensity whereas scotophobic microbes will move to higher intensity regions. Photokinesis describes the change in velocity (speed and direction) in response to light. The freshwater *M. magneticum* AMB-1 exhibits phototaxis behavior that is independent of the wavelength and magnetotaxis [60]. Photophobic swimming has been reported for unicellular *Magnetospira* sp. QH-2 [9] and multicellular magnetoglobules [20,37,41,42,44,45,48]. In this case, magnetotaxis drives cells to the edge of droplets. In reaction to illumination with blue (450–480 nm), violet (400–410 nm), and ultraviolet light (330–385 nm), the bacteria swim towards the center, in the opposite direction of magnetotaxis, with increased acceleration, which is similar to ping-pong motion, but is distinct due to the absence of a return swim. The reaction time is proportional to the wavelength: the shorter the wavelength, the quicker the reaction.

Interestingly, ellipsoidal magnetoglobules show a photokinesis behavior. Reverse fluorescence microscopes generally have two light-sources. One is a transmission background visible light (tungsten halogen lamp) for observation and imaging, and the other is an epi-illumination for fluorescence excitation. The second light source can be used to analyze the photo effect on swimming behavior by illumination at a given wavelength on a defined area [8]. The swimming of ellipsoidal magnetoglobules collected from the Mediterranean Sea was maintained within the illumination spots via the application of an alternate uniform magnetic field in order to periodically reverse the swimming direction of magnetoglobules (Figure 3B). At times, magnetoglobules suddenly changed their swimming direction being opposite to the initial magnetotaxis direction with increased acceleration, when stimulated with UV light (385 nm). Variable proportions of magnetoglobules reacted to violet (430 nm) [8]. This is a typical photokinesis behavior, i.e., changing the swim speed and direction. The dependence of the wavelength and intensity of the light stimulus remains to be characterized.

3.4. Physiological Function of Magnetotaxis

Magnetotactic bacteria live at, or just below, the oxic–anoxic interface or redoxocline in aquatic habitats. Interestingly, magnetotactic bacteria collected from the Northern Hemisphere swim preferentially northward, in parallel with the geomagnetic field lines (north-seeking (NS)) [23], and those from the Southern Hemisphere swim preferentially antiparallel to the geomagnetic field lines to the magnetic south pole (south-seeking (SS)) [61]. The geomagnetic field is inclined downward from horizontal in the Northern Hemisphere, and upward in the Southern Hemisphere, with the inclination magnitude increasing from the equator to the poles. Therefore, the hypothetical physiological function of magnetotaxis can be that magnetotaxis guides the cells in each hemisphere downward to the less oxygenated regions of the aquatic habitat [1].

Marine sediments are characterized by opposing oxygen and reductant (e.g., sulfide) gradients within the upper millimeters of the sediments, which are covered by air-saturated seawater. The pattern of the gradients constantly changes due to the convective water currents at the sediment surface, dynamic metabolism of microbe populations, or periodic exposure to the air during low tide. In order to adapt to these ever-changing environmental parameters, magnetotactic bacteria have to combine magnetotaxis with aerotaxis. Moreover, penetration downward from the water phase into the sediments and swimming in the water pockets requires robust flagellar propellers. As a consequence of environmental selection, the *M. massalia* strain MO-1 synthesizes sheath-protected, well-organized, and highly coordinated flagellar apparatus that ensure a high swimming velocity [11]. When they encounter the bulk of an obstacle, MO-1 cells can squeeze through them or change direction using the bounce motion, thereby circumventing the obstacles [58]. Backward swimming occurs using various angles between the translation and field axes, which provides a large range of swimming directions in order to circumvent the obstacle. The robust flagellar apparatus and versatile swimming capacity give MO-1 cells a competitive fitness in marine sediments. In addition, studies with the axenic culture of MO-1 provide compelling evidence to support the physiological significance of magnetotaxis.

The cultures in polystyrene plastic tubes exhibit a vertical downward oxidation–reduction potential (ORP, or redox) gradient, and a radiate gradient with an oxygen concentration decrease from the peripheral zone to the center, due to the diffusion of oxygen across the tube wall. During growth, MO-1 generates an oxic–anoxic–oxic oxycline pattern and forms two bacterial swarm bands. Remarkably, the upper band, where the magnetic field is parallel to the direction of the redox potential decrease, consists of >95% north-seeking (NS) cells, while the lower bacterial band, where the downward magnetic field lines are opposite to the upward direction of redox potential decrease, is composed of >90% south-seeking (SS) cells [62]. In both loci, cells with the 'correct' magnetotaxis polarity that are directed to swim towards the direction of the redox potential decrease are selected. Therefore, these observations are consistent with the hypothesis of magnetotaxis function and indicate the configuration of the ORP and magnetic field direction on a magnetotactic direction [62]. Further analysis by incubating MO-1 cells in a shielded, hypo-magnetic field (2 nT) showed that bacterial growth produces irregular forms of oxycline. Most importantly, the biomass of the cultures incubated in a hypo-magnetic environment are two orders of magnitude lower than those in the geomagnetic field, and could not grow at all when inoculated with a low quantity of cells [58]. This clearly demonstrates that magnetotaxis does present an advantage for the growth of MO-1 in the oxycline, and is even essential at low cell densities.

Magnetoglobules swim faster than most unicellular bacteria and are large in size, which has an advantage in mitigating the risk of predation. Magnetoglobules dwell in intertidal sediments as deep as 30 cm and undergo seasonal vertical movement in response to nutrient distribution changes [63]. Shapiro et al. have suggested that photophobic behavior enables magnetoglobules to optimize their location to adapt to circadian variations in chemical gradients and light intensity [37]. Indeed, genes involved in controlling the circadian rhythm have been found in the genomes of magnetoglobules [20,22]. Therefore, magnetoglobules seem to have adopted a multicellularity and photosensitive magnetotaxis in order to adapt to shallow marine environments.

4. Mechanism of Magnetotaxis

In seeking an environment optimal for their growth, bacteria change swimming direction frequently by changing the direction of flagellar rotation. Our current understanding of chemotaxis stems mainly, from the extensively studied, peritrichously flagellated enterobacteria *E. coli* and *Salmonella* spp. [15]. These alternate between periods of "run" and "tumble" and the swimming pattern is determined by the direction of the flagellar motor rotation. When the motor rotates in the counterclockwise (CCW) direction (as viewed from the distal end of the filament), several flagellar filaments form a loose bundle to propel the cell forward to run. When the motor reverses its rotation to clockwise (CW), the bundle falls apart and the cell tumbles [15]. Monopolar flagellum pushes marine vibrio forward by CCW rotation and pulls it backwards through CW rotation [64]. According to the prevailing hypothesis, magnetotactic bacteria align passively along the geomagnetic field lines, which guide swimming downward from the oxic zone to the oxic–anoxic interface, by rotating their flagella counterclockwise [1]. When located in the anoxic zone magnetotactic bacteria swim upward by reversing the direction of flagellar rotation from counterclockwise to clockwise. This is a simplified assumption, because it does not explain the coordinated rotation of bilophotrichous flagella of MO-1 and tens of thousands peritrichous flagella of magnetoglobules. MO-1 cells swim for very long distances without stopping, until they encounter an obstacle, which causes them to turn their bodies and swim against the magnetic field to circumvent the obstacle [58]. Such behavior is in contrast with the model of aligned forward–backward motion. The ovoid MO-1 cells possess two flagellar bundles on the long axis side of their body (Figure 1) [11,12]. These cells rotate around and translate along their short body axis [58]. It is consistent with the fact that the two flagellar bundles are placed on the long body axis, thus presumably generating the propulsion along the short body axis [65]. Electron cryotomography (ECT) analysis revealed that the magnetosome chains in MO-1 cells are roughly along the long body axis in >90% of cells, or with angles of less than 45° to the short body axis in 5% of cells [12]. Therefore, the direction of the magnetic moment is not parallel to the short body axis in most MO-1 cells, hence MO-1 cells are not perfectly aligned along the magnetic field lines while they swim. The direction of magnetic dipole moment exhibits a cyclical change perpendicular to their translation direction. The poor alignment of magnetic moment along the magnetic field lines enables backward swimming with a body reversal in bounce motion. In contrast, ellipsoidal magnetoglobules align well in magnetic fields and their bodies remain in the same direction when swimming backwards [8]. It is noteworthy that the backward swimming in bounce motion and the axial magnetotaxis of magnetoglobules start with the highest acceleration and have higher instantaneous velocity than the forward swimming. Hence, magnetoglobules seem to steer their flagella according to the magnetic direction of their swimming. Greenberg et al. have analyzed the kinematics of ping-pong motility in magnetic fields and proposed a receptor-mediated mechanism for sensing the magnetic field by spherical magnetoglobules [56].

Bacteria can sense a wide range of environmental signals that steer bacterial locomotion through the extensively studied chemotaxis mechanism [64,66]. The chemoreceptors, methyl-accepting chemotaxis proteins (MCPs), detect the stimuli, and control, through histidine protein kinase CheA, the phosphorylation state of the response regulator CheY. Phospho-CheY interacts with the flagellar motor and switches the rotation direction. Rotation in one direction results in smooth swimming, whilst switching the rotation direction may lead to backward motion, tumbling, or stopping swimming, depending on the bacterial species [64]. We have proposed a chemotaxis-like magnetotaxis mechanism for the freshwater *M. magneticum* strain AMB-1. We have shown that the Amb0994, an MCP-like protein, lacks the periplasmic signal molecule-binding domain, and interacts with cytoskeleton MamK filaments, on which the magnetosome chain is connected [67]. Our hypothesis is that poor alignment of magnetosome chains in the magnetic field would generate a magnetic torque that applies a mechanical strength on the MamK filament. Interaction between the MamK filament and Amb0994 converts the mechanical signal to a biochemical signal, i.e., phosphorylation of CheA. Subsequent phosphorylation of CheY and its binding onto the flagellar motor would slow down or stop the rotation of flagella, to avoid them from swimming in the wrong direction. Two results are consistent with this hypothesis. Overexpression of Amb0994 interferes with the AMB-1 response to the reversal of the magnetic field [67]. Deletion of the *amb0994* gene resulted in the failure of AMB-1 cells to align with the magnetic field lines in a weak biologically relevant magnetic field, and this dysfunction was recovered by in trans complementation of the mutant [68]. These results support the chemotaxis-like magnetotaxis mechanism. Considering the morphological and physiological diversity of magnetotactic bacteria, various magnetotactic mechanisms might be used.

Coordinated swimming behavior is a fundamental feature that emerged during the evolution of multicellularity. Magnetoglobules exhibit a highly complex motion: polar and axial magnetotaxis, bounce motion, photophobic response, and magneto-photokinesis [8]. Bacterial photo-sensing might rely directly on dedicated photoreceptors, or indirectly on the products of photosynthesis or other illumination by-products, i.e., reactive oxygen species, ATP, change of intracellular redox, or force proton motif. Six types of photosensory proteins using four kinds of chromophores are well characterized [69,70]. Among them, two groups, cryptochromes and sensory rhodopsins, are involved in photo-responsive motion.

Flavin-based cryptochrome serves as magnetoreceptor for migratory birds to exploit the geomagnetic field for direction and mapping [71]. Blue-light excitation of cryptochrome proteins in the retina creates a radical–pair consisting of molecules with a single unpaired electron. The spins of the two unpaired electrons are either antiparallel to one another (singlet state) or parallel (triplet state). As with a compass, the spin of one unpaired electron is primarily influenced by the magnetism of a nearby atomic nucleus, and the other is further away from the nucleus and influenced only by Earth's magnetic field [71,72]. The difference in the field shifts the radical pair between two quantum states with differing chemical reactivity. Therefore, a change in surrounding magnetic field affects the interconversion and the reaction direction, which results in an output signal being transferred to the neural system in animals [71,73,74]. The radical pair compass is light-dependent, involves quantum entanglement, and is thus considered as a representative example of quantum biology [72].

Prokaryotic rhodopsins (proteorhodopsins) are involved in photomotility at two levels [75]. They function as photo-driven ion pumps, where proteorhodopsins translocate ions across cytoplasmic membrane and establish ion gradients upon capture of light. In turn, the gradients drive the flagellar motors for motility [76]. Sensory rhodopsins are directly involved in phototaxis of archaea *Halobacterium halobium*, which are attracted to long wavelength visible light (red-light attraction), and repelled by shorter wavelength light (blue-light repellence). Together, two phototaxis receptors, sensory rhodopsin I (SRI) and sensory rhodopsin II (SRII), and two transducers, haloarchaeal transducer for SRI (HtrI) and haloarchaeal transducer for SRII (HtrII) form two phototaxis reception complexes [77]. Retinal-containing SRI or SRII are transmembrane proteins that encircle cognate HtrI or HtrII. The transducers HtrI and HtrII are structurally and functionally similar to MCP proteins. The SR-Htr complexes modulate the CheA kinase activity and steer flagellar rotation through integration with a switch regulator CheY. Orange-light activates SRI that interacts with HtrI and transiently inhibits CheA kinase activity. Reduced concentrations of phosphorylated CheY decreases the probability of switching motor rotation. As a consequence, the cell continues swimming towards the orange light, displaying the red-light attraction behavior [77]. In contrast, blue-light activation of SRII excites transient activation of CheA, an increase of phospho-CheY concentration, and has the probability of switching flagellar motor rotation direction, which leads to the blue-repellence. In addition to these two simple, direct reaction processes, sequential activation of SRI by orange followed by near-UV results in a strong repellent response. Sensory proteorhodopsin has been found in marine bacteria but their physiological function has not been demonstrated yet [77].

We have not identified genes that encode for either cryptochrome or proteorhodopsin in the genomes of *Mangetospira* sp. QH-2 [10], *M. massalia* MO-1 [18], spherical magnetoglobules [21,78], or incomplete genomes of ellipsoidal magnetoglobules. Therefore, the photo-sensing observed in these magnetotactic bacteria might be performed with other kinds of photoreceptors, or indirectly through chemical and physical reactions. Short wavelength light induces photoreaction and creates active oxygen species, which modify physiological conditions and triggers cellular reaction. In ellipsoidal magnetoglobules, we observed the fence-like structure, which looks like photosynthetic membrane lamellae and could be an appropriate candidate for accommodating the photoreceptors involved in photo-sensing [8]. It might also function as a grating to relay and convert light signals. Multicellular magnetotactic prokaryotes displayed a helical trajectory of swimming and reacted to illumination with UV-light perpendicular to the translation direction. They changed the magnetotaxis direction and velocity suddenly within the illumination area. Therefore, the magneto-photokinesis is unlikely to be a result of the detection of an intracellular spatial light gradient. The sudden change of swim direction under constant illumination would suggest the cumulating effect of periodical exposure of photoreceptive structures to UV-light, or the production of harmful by-products. Therefore, multicellular magnetotactic prokaryotes reversed their swimming direction to escape from the deleterious light. It remains an enigma how thousands of flagella of 60–80 cells coordinate their rotations to propel the swimming direction away from the default magnetotaxis orientation.

5. Conclusions

Magnetotaxis is an obvious magnetic field reactive swimming behavior, and little is known about the mechanism of magnetoreception. Despite extensive studies of magnetotactic bacteria over the last two decades, it remains a question of debate whether bacteria steer their flagellar motors in response to the state of their alignment in magnetic fields. Light is electromagnetic radiation and it affects magnetotaxis. What might be the connection between magnetic and optical stimuli? Photoreceptors known to be responsible for photomotion have not been identified in magnetotactic bacteria, in spite of the advances in metagenomics. The scarcity of axenic marine bacterial cultures makes the study of photo-sensitive magnetotaxis mechanisms even more complicated. The paradigm of bacterial chemotaxis is underpinned by intracellular diffusion of phosphorylated proteins and their binding to flagellar motors, in order to steer the swimming behavior in response to environmental stimuli [66]. How is a signal transmitted across multiple membranes to reach tens of thousands of flagellar motors at the surface of approximately 60 cells? Considering the particle and wave duality of photons, the application of the quantum concept might provide a solution and shed some light onto the complex study of magnetic photokinesis.

Supplementary Materials: The following are available online at http://www.mdpi.com/2218-273X/10/3/460/s1, Video S1: ping-pong motion of big rod-shaped MTB.

Author Contributions: L.-F.W. has prepared the Figures 1 and 3 and W.-J.Z. has prepared the Figure 2. L.-F.W. and W.-J.Z. have written the manuscript. All authors have read and agreed to the published version of the manuscript.

Funding: This work was supported in part by a funding from the Excellence Initiative of Aix-Marseille University—A*Midex, a French "Investissements d'Avenir" programme, by grants 91751202 and 91751108 from the NSFC, grants 2018YFC0309904, 2016YFC0302502, and 2016YFC0304905 from the National Key R&D Program of China, grants Y950071 and Y9719105 from the Deep-sea Technology Innovation institute, grants 2018YD01 and 2018YD02 from Sanya City, and grants from the CNRS for LIA-MagMC.

Acknowledgments: We acknowledge Juanfang Ruan, Keiichi Namba, Nicolas Menguy, and Artemis Kosta for the courtesy of the micrographs used in Figure 1, and Claire-Lise Santini for continuous valuable assistance.

Conflicts of Interest: The authors declare no conflicts of interest.

References

1. Bazylinski, D.A.; Frankel, R.B. Magnetosome formation in prokaryotes. *Nat. Rev. Microbiol.* **2004**, *2*, 217–230. [CrossRef] [PubMed]
2. Schüler, D. Genetics and cell biology of magnetosome formation in magnetotactic bacteria. *FEMS Microbiol. Rev.* **2008**, *32*, 654–672. [CrossRef] [PubMed]
3. Komeili, A.; Li, Z.; Newman, D.K.; Jensen, G.J. Magnetosomes are cell membrane invaginations organized by the actin-like protein MamK. *Science* **2006**, *311*, 242–245. [CrossRef]
4. Scheffel, A.; Gruska, M.; Faivre, D.; Linaroudis, A.; Plitzko, J.M.; Schuler, D. An acidic protein aligns magnetosomes along a filamentous structure in magnetotactic bacteria. *Nature* **2006**, *440*, 110–114. [CrossRef]
5. Pradel, N.; Santini, C.L.; Bernadac, A.; Fukumori, Y.; Wu, L.-F. Biogenesis of actin-like bacterial cytoskeletal filaments destined for positioning prokaryotic magnetic organelles. *Proc. Natl. Acad. Sci. USA* **2006**, *103*, 17485–17489. [CrossRef]
6. Blakemore, R.P. Magnetotactic bacteria. *Annu. Rev. Microbiol.* **1982**, *36*, 217–238. [CrossRef]
7. Lin, W.; Zhang, W.; Zhao, X.; Roberts, A.P.; Paterson, G.A.; Bazylinski, D.A.; Pan, Y. Genomic expansion of magnetotactic bacteria reveals an early common origin of magnetotaxis with lineage-specific evolution. *ISME J.* **2018**, *12*, 1508–1519. [CrossRef]
8. Qian, X.X.; Santini, C.L.; Kosta, A.; Menguy, N.; Le Guenno, H.; Zhang, W.; Li, J.; Chen, Y.R.; Liu, J.; Alberto, F.; et al. Juxtaposed membranes underpin cellular adhesion and display unilateral cell division of multicellular magnetotactic prokaryotes. *Environ. Microbiol.* **2019**. [CrossRef]
9. Zhu, K.; Pan, H.; Li, J.; Yu-Zhang, K.; Zhang, S.-D.; Zhang, W.-Y.; Zhou, K.; Yue, H.; Pan, Y.; Xiao, T.; et al. Isolation and characterization of a marine magnetotactic spirillum axenic culture QH-2 from an intertidal zone of the China Sea. *Res. Microbiol.* **2010**, *161*, 276–283. [CrossRef] [PubMed]
10. Ji, B.; Zhang, S.D.; Arnoux, P.; Rouy, Z.; Alberto, F.; Philippe, N.; Murat, D.; Zhang, W.J.; Rioux, J.B.; Ginet, N.; et al. Comparative genomic analysis provides insights into the evolution and niche adaptation of marine Magnetospira sp. QH-2 strain. *Environ. Microbiol.* **2013**. [CrossRef]
11. Lefèvre, C.T.; Bernadac, A.; Yu-Zhang, K.; Pradel, N.; Wu, L.-F. Isolation and characterization of a magnetotactic bacterial culture from the Mediterranean Sea. *Environ. Microbiol.* **2009**, *11*, 1646–1657. [CrossRef] [PubMed]
12. Ruan, J.; Kato, T.; Santini, C.-L.; Miyata, T.; Kawamoto, A.; Zhang, W.-J.; Bernadac, A.; Wu, L.-F.; Namba, K. Architecture of a flagellar apparatus in the fast-swimming magnetotactic bacterium MO-1. *Proc. Natl. Acad. Sci. USA* **2012**, *109*, 20643–20648. [CrossRef] [PubMed]
13. Terashima, H.; Kojima, S.; Homma, M. Flagellar motility in bacteria structure and function of flagellar motor. *Int. Rev. Cell Mol. Biol.* **2008**, *270*, 39–85. [PubMed]
14. Macnab, R.M. How bacteria assemble flagella. *Annu. Rev. Microbiol.* **2003**, *57*, 77–100. [CrossRef] [PubMed]
15. Berg, H.C. The rotary motor of bacterial flagella. *Annu. Rev. Biochem.* **2003**, *72*, 19–54. [CrossRef]
16. McCarter, L.L. Polar flagellar motility of the *Vibrionaceae*. *Microbiol. Mol. Biol. Rev.* **2001**, *65*, 445–462. [CrossRef]

17. Nakamura, S.; Minamino, T. Flagella-Driven Motility of Bacteria. *Biomolecules* **2019**, *9*, 279. [CrossRef]
18. Ji, B.; Zhang, S.D.; Zhang, W.J.; Rouy, Z.; Alberto, F.; Santini, C.L.; Mangenot, S.; Gagnot, S.; Philippe, N.; Pradel, N.; et al. The chimeric nature of the genomes of marine magnetotactic coccoid-ovoid bacteria defines a novel group of Proteobacteria. *Environ. Microbiol.* **2017**, *19*, 1103–1119. [CrossRef]
19. Schübbe, S.; Williams, T.J.; Xie, G.; Kiss, H.E.; Brettin, T.S.; Martinez, D.; Ross, C.A.; Schuler, D.; Cox, B.L.; Nealson, K.H.; et al. Complete genome sequence of the chemolithoautotrophic marine magnetotactic coccus strain MC-1. *Appl. Environ. Microbiol.* **2009**, *75*, 4835–4852.
20. Abreu, F.; Morillo, V.; Nascimento, F.F.; Werneck, C.; Cantao, M.E.; Ciapina, L.P.; de Almeida, L.G.; Lefèvre, C.T.; Bazylinski, D.A.; de Vasconcelos, A.T.; et al. Deciphering unusual uncultured magnetotactic multicellular prokaryotes through genomics. *ISME J.* **2013**, *8*, 1055–1068. [CrossRef]
21. Kolinko, S.; Richter, M.; Glockner, F.O.; Brachmann, A.; Schuler, D. Single-cell genomics reveals potential for magnetite and greigite biomineralization in an uncultivated multicellular magnetotactic prokaryote. *Environ. Microbiol. Rep.* **2014**, *6*, 524–531. [CrossRef] [PubMed]
22. Leao, P.; Chen, Y.R.; Abreu, F.; Wang, M.; Zhang, W.J.; Zhou, K.; Xiao, T.; Wu, L.F.; Lins, U. Ultrastructure of ellipsoidal magnetotactic multicellular prokaryotes depicts their complex assemblage and cellular polarity in the context of magnetotaxis. *Environ. Microbiol.* **2017**, *19*, 2151–2163. [CrossRef] [PubMed]
23. Blakemore, R. Magnetotactic bacteria. *Science* **1975**, *190*, 377–379. [CrossRef] [PubMed]
24. Frankel, R.B.; Bazylinski, D.A.; Johnson, M.S.; Taylor, B.L. Magneto-aerotaxis in marine coccoid bacteria. *Biophys. J.* **1997**, *73*, 994–1000. [CrossRef]
25. Zhou, K.; Pan, H.; Yue, H.; Xiao, T.; Wu, L.-F. Architecture of flagellar apparatus of marine magnetotactic cocci from Qingdao. *Marine Sci.* **2010**, *34*, 88–92.
26. Bazylinski, D.A.; Williams, T.J.; Lefèvre, C.T.; Berg, R.J.; Zhang, C.L.; Bowser, S.S.; Dean, A.J.; Beveridge, T.J. *Magnetococcus marinus* gen. nov., sp. nov., a marine, magnetotactic bacterium that represents a novel lineage (*Magnetococcaceae* fam. nov.; *Magnetococcales* ord. nov.) at the base of the *Alphaproteobacteria*. *Int. J. Syst. Evol. Microbiol.* **2013**, *63*, 801–808. [CrossRef]
27. Liu, J.; Zhang, W.; Li, X.; Li, X.; Chen, X.; Li, J.-H.; Teng, Z.; Xu, C.; Santini, C.-L.; Zhao, L.; et al. Bacterial community structure and novel species of magnetotactic bacteria in sediments from a seamount in the Mariana volcanic arc. *Sci. Rep.* **2017**, *7*, 17964. [CrossRef]
28. Trachtenberg, S.; DeRosier, D.J. A three-start helical sheath on the flagellar filament of *Caulobacter crescentus*. *J. Bacteriol.* **1992**, *174*, 6198–6206. [CrossRef]
29. Schmitt, R.; Raska, I.; Mayer, F. Plain and complex flagella of *Pseudomonas rhodos*: Analysis of fine structure and composition. *J. Bacteriol.* **1974**, *117*, 844–857. [CrossRef]
30. Geis, G.; Leying, H.; Suerbaum, S.; Mai, U.; Opferkuch, W. Ultrastructure and chemical analysis of *Campylobacter pylori* flagella. *J. Clin. Microbiol.* **1989**, *27*, 436–441. [CrossRef]
31. Thomashow, L.S.; Rittenberg, S.C. Isolation and composition of sheathed flagella from *Bdellovibrio bacteriovorus* 109J. *J. Bacteriol.* **1985**, *163*, 1047–1054. [CrossRef] [PubMed]
32. Lefèvre, C.T.; Santini, C.L.; Bernadac, A.; Zhang, W.J.; Li, Y.; Wu, L.F. Calcium ion-mediated assembly and function of glycosylated flagellar sheath of marine magnetotactic bacterium. *Mol. Microb.* **2010**, *78*, 1304–1312. [CrossRef] [PubMed]
33. Zhang, W.J.; Santini, C.L.; Bernadac, A.; Ruan, J.; Zhang, S.D.; Kato, T.; Li, Y.; Namba, K.; Wu, L.F. Complex spatial organization and flagellin composition of flagellar propeller from marine magnetotactic ovoid strain MO-1. *J. Mol. Biol.* **2012**, *416*, 558–570. [CrossRef] [PubMed]
34. Farina, M.; Lins de Barros, H.; Motta de Esquivel, D.; Danon, J. Ultrastructure of a magnetotactic microorganism. *Biol. Cell.* **1983**, *48*, 85–88.
35. Keim, C.N.; Martines, J.L.; Lins de Barros, H.; Lins, U.; Farina, M. Structure, behavior, ecology and diversity of multicellular magnetotactic prokaryotes. In *Magnetoreception and Magnetosomes in Bacteria*; Schüler, D., Ed.; Springer: Berlin/Heidelberg, Germany, 2006; pp. 104–132.
36. Rodgers, F.G.; Blakemore, R.P.; Blakemore, N.A.; Frankel, R.B.; Bazylinski, D.A.; Maratea, D.; Rodgers, C. Intercellular structure in a many-celled magnetotactic prokaryote. *Arch. Microbiol.* **1990**, *154*, 18–22. [CrossRef]
37. Shapiro, O.H.; Hatzenpichler, R.; Buckley, D.H.; Zinder, S.H.; Orphan, V.J. Multicellular photo-magnetotactic bacteria. *Environ. Microbiol. Rep.* **2011**, *3*, 233–238. [CrossRef]

38. Simmons, S.L.; Edwards, K.J. Unexpected diversity in populations of the many-celled magnetotactic prokaryote. *Environ. Microbiol.* **2007**, *9*, 206–215. [CrossRef]
39. Winklhofer, M.; Abracado, L.G.; Davila, A.F.; Keim, C.N.; Lins de Barros, H.G. Magnetic optimization in a multicellular magnetotactic organism. *Biophys. J.* **2007**, *92*, 661–670. [CrossRef]
40. Wenter, R.; Wanner, G.; Schüler, D.; Overmann, J. Ultrastructure, tactic behaviour and potential for sulfate reduction of a novel multicellular magnetotactic prokaryote from North Sea sediments. *Environ. Microbiol.* **2009**, *11*, 1493–1505. [CrossRef]
41. Zhou, K.; Zhang, W.Y.; Pan, H.M.; Li, J.H.; Yue, H.D.; Xiao, T.; Wu, L.F. Adaptation of spherical multicellular magnetotactic prokaryotes to the geochemically variable habitat of an intertidal zone. *Environ. Microbiol.* **2013**, *15*, 1595–1605. [CrossRef]
42. Zhang, R.; Chen, Y.R.; Du, H.J.; Zhang, W.Y.; Pan, H.M.; Xiao, T.; Wu, L.F. Characterization and phylogenetic identification of a species of spherical multicellular magnetotactic prokaryotes that produces both magnetite and greigite crystals. *Res. Microbiol.* **2014**, *65*, 481–489. [CrossRef] [PubMed]
43. Lefèvre, C.; Bernadac, A.; Pradel, N.; Wu, L.-F.; Yu-Zhang, K.; Xiao, T.; Yonnet, J.-P.; Lebouc, A.; Song, T.; Fukumori, Y. Characterization of mediterranean magnetotactic bacteria. *J. Oce. Univ. China* **2007**, *6*, 5–9. [CrossRef]
44. Chen, Y.R.; Zhang, W.Y.; Zhou, K.; Pan, H.M.; Du, H.J.; Xu, C.; Xu, J.H.; Pradel, N.; Santini, C.L.; Li, J.H.; et al. Novel species and expanded distribution of ellipsoidal multicellular magnetotactic prokaryotes. *Environ. Microbiol. Rep.* **2016**, *8*, 218–226. [CrossRef] [PubMed]
45. Chen, Y.R.; Zhang, R.; Du, H.J.; Pan, H.M.; Zhang, W.Y.; Zhou, K.; Li, J.H.; Xiao, T.; Wu, L.F. A novel species of ellipsoidal multicellular magnetotactic prokaryotes from Lake Yuehu in China. *Environ. Microbiol.* **2015**, *17*, 637–647. [CrossRef] [PubMed]
46. Dong, Y.; Li, J.; Zhang, W.; Zhang, W.; Zhao, Y.; Xiao, T.; Wu, L.-F.; Pan, H. The detection of magnetotactic bacteria in deep sea sediments from the east Pacific Manganese Nodule Province. *Environ. Microb. Rep.* **2016**, *8*, 239–249. [CrossRef] [PubMed]
47. Du, H.-J.; Chen, Y.-R.; Zhang, R.; Pan, H.-M.; Zhang, W.-Y.; Zhou, K.; Wu, L.-F.; Xiao, T. Temporal distributions and environmental adaptations of two types of multicellular magnetotactic prokaryote in the sediments of Lake Yuehu, China. *Environ. Microb. Rep.* **2015**, *7*, 538–546. [CrossRef]
48. Zhou, K.; Zhang, W.Y.; Yu-Zhang, K.; Pan, H.M.; Zhang, S.D.; Zhang, W.J.; Yue, H.D.; Li, Y.; Xiao, T.; Wu, L.F. A novel genus of multicellular magnetotactic prokaryotes from the Yellow Sea. *Environ. Microbiol.* **2012**, *14*, 405–413. [CrossRef]
49. Teng, Z.; Zhang, Y.; Zhang, W.; Pan, H.; Xu, J.; Huang, H.; Xiao, T.; Wu, L.-F. Diversity and Characterization of Multicellular Magnetotactic Prokaryotes from Coral Reef Habitats of the Paracel Islands, South China Sea. *Front. Microbiol.* **2018**, *9*. [CrossRef]
50. Silva, K.T.; Abreu, F.; Almeida, F.P.; Keim, C.N.; Farina, M.; Lins, U. Flagellar apparatus of south-seeking many-celled magnetotactic prokaryotes. *Microsc. Res. Tech.* **2007**, *70*, 10–17. [CrossRef]
51. Murat, D.; Herisse, M.; Espinosa, L.; Bossa, A.; Alberto, F.; Wu, L.F. Opposite and Coordinated Rotation of Amphitrichous Flagella Governs Oriented Swimming and Reversals in a Magnetotactic Spirillum. *J. Bacteriol.* **2015**, *197*, 3275–3282. [CrossRef]
52. Lins de Barros, H.G.; Esquivel, D.M.; Farina, M. Magnetotaxis. *Sci. Prog.* **1990**, *74*, 347–359. [PubMed]
53. Lins, U.; Kachar, B.; Farina, M. Imaging faces of shadowed magnetite (Fe(3)O(4)) crystals from magnetotactic bacteria with energy-filtering transmission electron microscopy. *Microsc. Res. Tech.* **1999**, *46*, 319–324. [CrossRef]
54. Simmons, S.L.; Sievert, S.M.; Frankel, R.B.; Bazylinski, D.A.; Edwards, K.J. Spatiotemporal distribution of marine magnetotactic bacteria in a seasonally stratified coastal salt pond. *Appl. Environ. Microbiol.* **2004**, *70*, 6230–6239. [CrossRef] [PubMed]
55. Keim, C.N.; Abreu, F.; Lins, U.; Lins de Barros, H.; Farina, M. Cell organization and ultrastructure of a magnetotactic multicellular organism. *J. Struct. Biol.* **2004**, *145*, 254–262. [CrossRef] [PubMed]
56. Greenberg, M.; Canter, K.; Mahler, I.; Tornheim, A. Observation of magnetoreceptive behavior in a multicellular magnetotactic prokaryote in higher than geomagnetic fields. *Biophys. J.* **2005**, *88*, 1496–1499. [CrossRef] [PubMed]

57. Kolinko, I.; Lohsse, A.; Borg, S.; Raschdorf, O.; Jogler, C.; Tu, Q.; Posfai, M.; Tompa, E.; Plitzko, J.M.; Brachmann, A.; et al. Biosynthesis of magnetic nanostructures in a foreign organism by transfer of bacterial magnetosome gene clusters. *Nat. Nanotechnol.* **2014**, *9*, 193–197. [CrossRef] [PubMed]
58. Zhang, S.D.; Petersen, N.; Zhang, W.J.; Cargou, S.; Ruan, J.; Murat, D.; Santini, C.L.; Song, T.; Kato, T.; Notareschi, P.; et al. Swimming behaviour and magnetotaxis function of the marine bacterium strain MO-1. *Environ. Microbiol. Rep.* **2014**, *6*, 14–20. [CrossRef]
59. Wilde, A.; Mullineaux, C.W. Light-controlled motility in prokaryotes and the problem of directional light perception. *FEMS Microbiol. Rev.* **2017**, *41*, 900–922. [CrossRef]
60. Chen, C.; Ma, Q.; Jiang, W.; Song, T. Phototaxis in the magnetotactic bacterium Magnetospirillum magneticum strain AMB-1 is independent of magnetic fields. *Appl. Microbiol. Biotechnol.* **2010**. [CrossRef]
61. Blakemore, R.P.; Frankel, R.B.; Kalmijn, A.J. South-seeking magnetotactic bacteria in the Southern Hemisphere. *Nature* **1980**, *286*, 384–385. [CrossRef]
62. Zhang, W.-J.; Chen, C.; Li, Y.; Song, T.; Wu, L.-F. Configuration of redox gradient determines magnetotactic polarity of the marine bacteria MO-1. *Environ. Microbiol. Rep.* **2010**, *2*, 646–650. [CrossRef] [PubMed]
63. Liu, J.; Zhang, W.; Du, H.; Leng, X.; Li, J.-H.; Pan, H.; Xu, J.; Wu, L.-F.; Xiao, T. Seasonal changes in the vertical distribution of two types of multicellular magnetotactic prokaryotes in the sediment of Lake Yuehu, China. *Environ. Microb. Rep.* **2018**, *10*, 475–484. [CrossRef] [PubMed]
64. Hazelbauer, G.L.; Falke, J.J.; Parkinson, J.S. Bacterial chemoreceptors: High-performance signaling in networked arrays. *Trends Biochem. Sci.* **2008**, *33*, 9–19. [CrossRef] [PubMed]
65. Yang, C.; Chen, C.; Ma, Q.; Wu, L.-F.; Song, T. Dynamic model and motion mechanism of magnetotactic bacteria with two lateral flagellar bundles. *J. Bio. Eng.* **2012**, *9*, 200–210. [CrossRef]
66. Wadhams, G.H.; Armitage, J.P. Making sense of it all: Bacterial chemotaxis. *Nat. Rev. Mol. Cell Biol.* **2004**, *5*, 1024–1037. [CrossRef] [PubMed]
67. Philippe, N.; Wu, L.-F. An MCP-like protein interacts with the MamK cytoskeleton and is involved in magnetotaxis in *Magnetospirillum magneticum* AMB-1. *J. Mol. Biol.* **2010**, *400*, 309–322. [CrossRef]
68. Zhu, X.; Ge, X.; Li, N.; Wu, L.F.; Luo, C.; Ouyang, Q.; Tu, Y.; Chen, G. Angle sensing in magnetotaxis of Magnetospirillum magneticum AMB-1. *Integr. Biol.* **2014**, *6*, 706–713. [CrossRef]
69. Gomelsky, M.; Hoff, W.D. Light helps bacteria make important lifestyle decisions. *Trends Microbiol.* **2011**, *19*, 441–448. [CrossRef]
70. Kottke, T.; Xie, A.; Larsen, D.S.; Hoff, W.D. Photoreceptors Take Charge: Emerging Principles for Light Sensing. *Annu. Rev. Biophys.* **2018**. [CrossRef]
71. Hore, P.J.; Mouritsen, H. The Radical-Pair Mechanism of Magnetoreception. *Annu. Rev. Biophys.* **2016**, *45*, 299–344. [CrossRef]
72. Ball, P. Physics of life: The dawn of quantum biology. *Nature* **2011**, *474*, 272–274. [CrossRef] [PubMed]
73. Ritz, T.; Adem, S.; Schulten, K. A model for photoreceptor-based magnetoreception in birds. *Biophys. J.* **2000**, *78*, 707–718. [CrossRef]
74. Liedvogel, M.; Mouritsen, H. Cryptochromes–a potential magnetoreceptor: What do we know and what do we want to know? *J. R. Soc. Interface* **2010**, *7* Suppl 2, S147–S162. [CrossRef]
75. Govorunova, E.G.; Sineshchekov, O.A.; Li, H.; Spudich, J.L. Microbial Rhodopsins: Diversity, Mechanisms, and Optogenetic Applications. *Annu. Rev. Biochem.* **2017**, *86*, 845–872. [CrossRef] [PubMed]
76. Fuhrman, J.A.; Schwalbach, M.S.; Stingl, U. Proteorhodopsins: An array of physiological roles? *Nat. Rev. Microbiol.* **2008**, *6*, 488–494. [CrossRef] [PubMed]
77. Spudich, J.L. The multitalented microbial sensory rhodopsins. *Trends Microbiol.* **2006**, *14*, 480–487. [CrossRef] [PubMed]
78. Abreu, F.; Martins, J.L.; Silveira, T.S.; Keim, C.N.; de Barros, H.G.; Filho, F.J.; Lins, U. 'Candidatus Magnetoglobus multicellularis', a multicellular, magnetotactic prokaryote from a hypersaline environment. *Int. J. Syst. Evol. Microbiol.* **2007**, *57*, 1318–1322. [CrossRef] [PubMed]

MDPI

Review

Spirochete Flagella and Motility

Shuichi Nakamura

Department of Applied Physics, Graduate School of Engineering, Tohoku University, 6-6-05 Aoba, Aoba-ku, Sendai, Miyagi 980-8579, Japan; naka@bp.apph.tohoku.ac.jp; Tel.: +81-22-795-5849

Received: 11 March 2020; Accepted: 3 April 2020; Published: 4 April 2020

Abstract: Spirochetes can be distinguished from other flagellated bacteria by their long, thin, spiral (or wavy) cell bodies and endoflagella that reside within the periplasmic space, designated as periplasmic flagella (PFs). Some members of the spirochetes are pathogenic, including the causative agents of syphilis, Lyme disease, swine dysentery, and leptospirosis. Furthermore, their unique morphologies have attracted attention of structural biologists; however, the underlying physics of viscoelasticity-dependent spirochetal motility is a longstanding mystery. Elucidating the molecular basis of spirochetal invasion and interaction with hosts, resulting in the appearance of symptoms or the generation of asymptomatic reservoirs, will lead to a deeper understanding of host–pathogen relationships and the development of antimicrobials. Moreover, the mechanism of propulsion in fluids or on surfaces by the rotation of PFs within the narrow periplasmic space could be a designing base for an autonomously driving micro-robot with high efficiency. This review describes diverse morphology and motility observed among the spirochetes and further summarizes the current knowledge on their mechanisms and relations to pathogenicity, mainly from the standpoint of experimental biophysics.

Keywords: spirochetes; periplasmic flagella; motility; chemotaxis; molecular motor

1. Introduction

Motility systems of living organisms are currently classified into 18 types [1]. Even when focusing on bacteria only, the motility is diverse when bacterial species are concerned [2]. A major motility form would be the flagella-dependent swimming well observed and described in *Escherichia coli* and *Salmonella enterica*, and these species have helical flagella extending to the cell exterior. Spirochetes, which are members of a group of gram-negative bacteria with a spiral or flat-wave cell body, also show flagella-dependent motility, but their flagella are hidden within the periplasmic space and are thus called periplasmic flagella (PFs). Externally flagellated bacteria are propelled by direct interaction of flagella and fluid, whereas spirochetes swim by rolling or undulation of a cell body driven by PFs rotation beneath the outer membrane. Physics difference results in an invalidation of applying the canonical model obtained from external flagella to spirochetal periplasmic flagella.

This review article describes the motility of spirochetes while connecting it with the unique structures of their cell bodies and PFs. Taxonomically, the phylum *Spirochaetae* is classified into *Leptospiraceae*, *Brachyspiraceae*, *Spirochaetaceae*, and *Brevinemataceae* families, containing pathogenic species, for example, *Leptospira interrogans* (leptospirosis), *Brachyspira hyodysenteriae* (swine dysentery), *Borrelia burgdorferi* (Lyme disease), and *Treponema pallidum* (syphilis). As observed with other motile pathogens, spirochete motility is an essential virulence factor. Thus, the last part of this review discusses the involvement of motility in spirochetal pathogenicity.

2. Cell Structure

A schematic of the basic structure shared among spirochete species is shown in Figure 1a. The protoplasmic cylinder consists of a cytoplasm, a cytoplasmic membrane, and a peptidoglycan layer, which is covered by the outer membrane. Each PF filament connects with a basal motor called the flagellar motor that is embedded in the cytoplasmic membrane and the peptidoglycan layer via a short, bent structure corresponding to the universal joint hook in the *E. coli* flagellar motor (details are described below) [3]. The morphologies of the cell body and the PF as well as the number of PFs greatly differ among species, and those of three representative species are summarized in Table 1. The cell body of *Borrelia* spp. exhibits a flat-wave shape and contains 7~11 PFs long enough to overlap with those extending from the other end at the center of the cell body [4–7]. *Brachyspira* spp. appear to have a flat-wave body because of their non-spiral, almost straight configuration observed in swimming cells [8], but no explicit evidence has been reported. *Brachyspira* PFs overlap at the cell center, and so do those of *Borrelia* [9]. The cell morphology of *Leptospira* spp. is distinguished from the other two spirochetes by a small cell width and short wavelength [4,10]. The protoplasmic cylinder of *Leptospira* (Figure 1b,c) is relatively rigid, maintaining the helix parameters even during swimming, whereas both ends of the cell body are frequently transformed, as described later [11–14]. Unlike *Borrelia* and *Brachyspira*, PFs of *Leptospira* are too short to overlap [15].

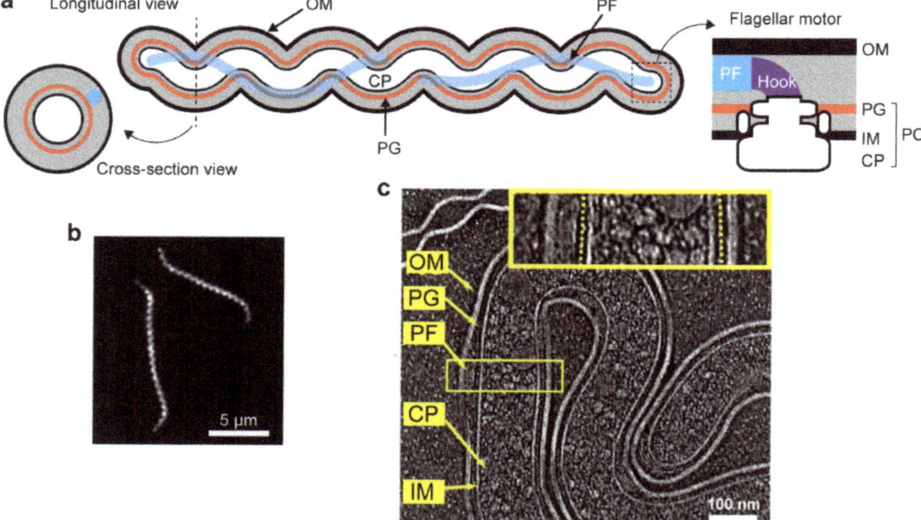

Figure 1. Spirochetal cell structure. (**a**) Schematics of longitudinal and zoom-in cross-section views of the cell structure and the flagellar motor shared by spirochete species; outer membrane (OM), periplasmic flagellum (PF), peptidoglycan layer (PG), inner membrane (IM), cytoplasm (CP), and protoplasmic cylinder (PC) are shown. If readers view from the hook to the motor, the flagellar motor rotates in a counterclockwise (CCW) direction at one pole of a single cell, whereas the motor at another cell pole rotates in a clockwise (CW) direction. (**b**) Dark-field micrograph of *Leptospira biflexa*. (**c**) Longitudinal slice image obtained by cryo-electron tomography of *L. biflexa* (adapted from [14] with permission from the publisher). OM, IM, and PF are clearly visible, and PGs observed in the yellow square are indicated by yellow dashed lines in the enlarged view (inset).

Table 1. Comparison of the cell structure and the periplasmic flagella (PFs) among three spirochete species.

Species (Disease)	Cell Morphology		Cell Body Parameters			PF				Ref.
			Length	Width	Wavelength	Number	Shape	Overlap	Proteins	
Borrelia burgdorferi (Lyme disease)		**Flat wave**	~20 μm	~0.3 μm	~2.8 μm	14~22	Left-handed helix	Yes	FlaA FlaB	[4–7]
Brachyspira hyodysenteriae (Swine dysentery)		Flat wave?	~10 μm	~0.3 μm	~4 μm	16~18	Left-handed helix	Yes	FlaA FlaB1,2,3	[8,16–18]
Leptospira interrogans (Leptospirosis)		Right-handed helix	~20 μm	~0.15 μm	~0.7 μm	2	Coiled shape	No	FlaA1,2 FlaB1,2 FcpA, FcpB	[4,10,15, 19–23]

3. Periplasmic Flagella

3.1. Physical Properties of the PF Filament

The flagellar filament of *E. coli* functions as a screw propeller through interaction with fluid [24]. In contrast, spirochete PFs are thought to rotate or transform the cell body by intimate contact with cell membranes, although direct observation of the PF rotation has not been successful. Another important role of the PF is to establish a wavy morphology, similar to a cytoskeleton, and the PF dependence of spirochete morphology has been observed in the periodontal disease-associated spirochetes *Treponema denticola* [25], *B. burgdorferi* [26,27], and *Leptospira* spp. [15,19–22]. For example, the loss of the PF in *B. burgdorferi* straightens the entire cell body [26]. In contrast, *Leptospira* PF depletion affects only the bent morphology of the cell ends, and the short-pitch helix in the protoplasmic cylinder is believed to be maintained by a bacterial actin homolog, MreB [28]. Both the cell body and the PF can be considered elastic materials, and the observed PF-dependent spirochete morphology is a consequence of the mechanical interaction between these two elastic bodies of different stiffness [29,30]. This difference in stiffness between the cell body and the PF can be evaluated by calculating the ratio of bending moduli (A), that is, (A_{Cell}/A_{PF}), based on which a theoretical study predicted an A_{Cell}/A_{PF} ratio of ~0.15 for *Leptospira* [29]; the PF is stiffer than the cell body. Another model showed an A_{Cell}/A_{PF} ratio of ~5 for *Borrelia*, which was consistent with the experimental value obtained by stiffness measurements of the borrelial cell body and the PF using optical tweezers [30]; in this case, the PF is stiffer than the cell body. The elastic properties of the cell body and the PF are crucial determinants of species-specific morphology and are thought to be related to the swimming mechanism described later [31].

The filament is connected to the flagellar motor via a hook structure. The hook in *E. coli* consists of the flagellar hook protein (FlgE) and is flexible enough to function as a universal joint to transmit the torque generated by the basal motor to the filament, regardless of the direction [24]. Although the spirochetal hook is also formed by FlgE, *T. denticola* FlgE features self-catalytic intersubunit crosslinking between conserved lysine and cysteine residues, thereby conferring structural stability [32]. The proper stiffness of the hook could be important for the interaction between the PF and the cell body.

3.2. Structure of the PF Filament

The *E. coli* flagellar filament is formed by tens of thousands of copies of a single flagellin protein, FliC [24]. Species with more complicated flagella are composed of multiple flagellins, for example, *Campylobacter jejuni* (FlaA and FlaB) and *Caulobacter crescentus* (FljJ, FljK, FljL, FljM, FljN, and FljO) [24]. All spirochete PFs known also consist of more than two proteins, and they generally contain FlaA and FlaB. In *B. burgdorferi*, FlaB forms the entire PF filament, and FlaA is believed to be localized around the base of the filament near the basal motor [27]. The PFs of *B. hyodysenteriae* and *Leptospira* spp. comprise a core filament and sheath [16]. In *B. hyodysenteriae*, three FlaB proteins (FlaB1-3) assemble to form a helical core filament (2.4 μm in wavelength and 0.6 μm in helix diameter), and an FlaA protein assembles to form a straight sheath; association of the FlaB core with the FlaA sheath determines the morphology of the fully assembled PFs (2.8 μm in wavelength and 0.9 μm in helix diameter) [17,18]. Synthesis of the PF and swimming motility in *B. hyodysenteriae* are affected by double knockout of *flaB1-flaB2* but not by double knockout of *flaB1-flaB3* or single knockout of *flaB3*, highlighting the importance of FlaB1 and FlaB2 in the *Brachyspira* core filament and the possibility of functional compensation between these two proteins [18]. In *Leptospira* spp., PF also consists of the core and the sheath, and six proteins have been identified as PF components: FlaA1, FlaA2, FlaB1, FlaB2, FcpA, and FcpB. PFs isolated from leptospiral cells exhibit a coiled shape [15], but the core filament is straight in the absence of a sheath, indicating that the sheath is indispensable for bending the leptospiral PF [19,21]. The PF core filament of the non-pathogenic species *Leptospira biflexa* is formed by FlaB1 and FlaB2 [19]. The remaining four proteins are involved in synthesizing the sheath or in coiling the PF through core–sheath interactions; however, their roles are not fully elucidated. Deletion of *flaA1* and *flaA2* does not affect the synthesis of the sheath [20], whereas *fcpA* knockout mutants lack a sheath [19,26]. Immunoprecipitations showed the interaction of FcpA with FlaB1 and FlaA2 [19]. These results suggest that FcpA is a major sheath component and plays a central role in coiling via its interaction with the core filament. Recently, cryo-electron microscopy revealed that FcpB is a sheath protein that is localized along the outer curve of the PF, suggesting a contribution to PF coiling [22,23].

3.3. Flagellar Motor

Spirochetes and externally flagellated species share fundamental motor parts for rotation, a rotor and a dozen stator units (torque generators) [24], but spirochetes flagellar motor has some spirochete-specific structures, resulting in a unique performance. Motor torque is generated by interaction between the rotor and the stator [33]. Assuming that the force generated by a single stator unit (F_S) is the same among species, the produced motor torque (M) depends on the radius of the rotor ring ($r_R \approx$ the distance between the motor axis and the rotor-stator contact point) and the number of stator units assembled to the motor (N_S): $M = F_S \times r_R \times N_S$ [34]. Cryo-electron tomography showed that the rotor ring in spirochete motor is larger than that in other external flagellar motors: ~31 nm for *B. burgdorferi*, ~20 nm for *S. enterica*, ~22 nm for *Vibrio fischeri*, and ~27 nm for *C. jejuni* [34]. Thus, the flagellar motor with a larger rotor ring allows more stators to surround the rotor. In addition to the geometrical advantage, the number of assembled stators of externally flagellated species is dynamically altered by changes in load against the motor and the input energy for rotation (e.g., N_S is decreased up to one near zero load) [24,35–38], whereas the maximum number of stator units could be incorporated into motors under any conditions in spirochetes [3,39–41]. Such stable assembly of the spirochete stators is thought to involve a spirochete-specific motor component called "P-collar" conserved in *T. primitia* [39], *T. pallidum* [41], *B. burgdorferi* [3], *L interrogans*, and *L. biflexa* [40]; perhaps the part plays a key role in stator assembly [34]. This knowledge predicts that the spirochetal motor can produce higher torque, which is supported by motility measurements showing that *Leptospira* spp. produce a stall torque of ~4000 pN nm [10], whereas the stall torque of *E. coli* is ~2000 pN nm [42].

4. Swimming Motility

4.1. PF-Dependent Swimming

In externally flagellated bacteria, when viewed from behind a swimming cell, a left-handed helical flagellum rotates counterclockwise (CCW), which is balanced by the clockwise (CW) rotation of the cell body (Figure 2a) [43]. In the case of spirochetes, the protoplasmic cylinder is believed to be rotated in the opposite direction of the PF rotation (Figure 2b) [14]. Rotation of the PFs of *Borrelia* and *Brachyspira* drives wave propagation along the cell body, thus providing thrust for swimming [44]. In contrast, the swimming form of *Leptospira* is more complex. When viewing a swimming *Leptospira* cell from its posterior side, the PF transforms both ends of the cell body into a left-handed spiral or a hook shape and gyrates the bent ends in a CCW fashion; concurrently, the PF rotates the right-handed protoplasmic cylinder in a CW manner (Figure 2c) [11,12]. The majority of thrust for *Leptospira* swimming is given by gyration of the spiral end and rolling of the protoplasmic cylinder [10]. However, correlative speed variation between the protoplasmic cylinder and the hook end was observed [14], suggesting that *Leptospira* swimming depends on mechanical communication among the three rotating parts.

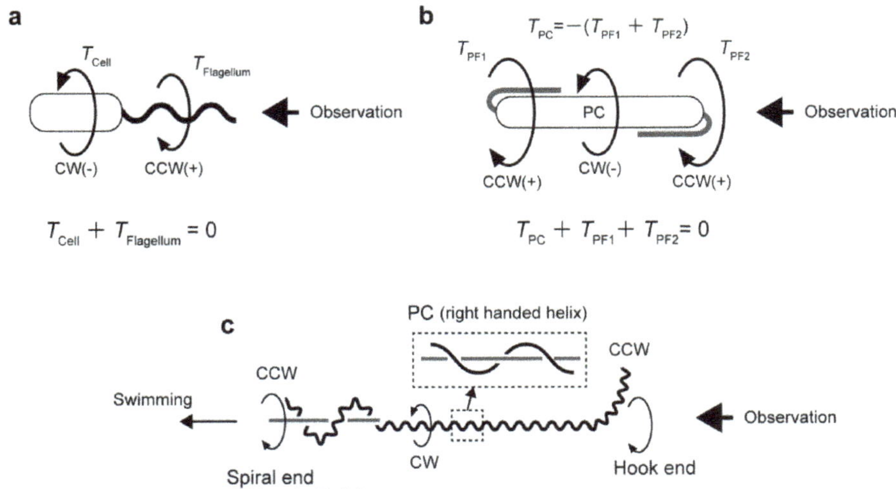

Figure 2. Mechanical models for bacterial swimming. (**a**) Steady-state swimming of an externally flagellated bacterium. Torques of the cell body (T_{Cell}) and flagellum ($T_{Flagellum}$) are balanced, that is, their sum is zero. (**b**) Schematic of spirochetal swimming, where the outer membrane is ignored. The protoplasmic cylinder (PC) is rotated by the counter torque of the periplasmic flagella (PFs) rotating at both ends of the cell body. (**c**) Swimming model for *Leptospira*. Rotational directions are indicated by large arrows.

4.2. Energy Input for Spirochete Motility

The bacterial flagellar motor is fueled by the ion motive force (*IMF*), which is the sum of the membrane voltage ($\Delta\psi$) and the ion concentration gap between the cell exterior and interior (ΔpI). *E. coli* and *S. enterica* use the proton motive force ($PMF = \Delta\psi + \Delta pH$) for flagellar rotation, whereas *Vibrio cholerae* uses the sodium motive force ($SMF = \Delta\psi + \Delta pNa$) [24]. The coupling ion used in torque generation by the flagellar motor depends on the type of stator units [45]. The MotA/MotB complex present in *E. coli* and *S. enterica* is an H^+-type stator, and the PomA/PomB complex of *Vibrio* spp. is a Na^+-type stator. *Vibrio alginolyticus* uses MotA/MotB and PomA/PomB stators for the lateral flagella and polar flagellum, respectively [46,47]. *Bacillus subtilis* also possesses both H^+-type MotA/MotB and Na^+-type MotP/MotS complexes [48,49]. Such hybrid stator systems can exchange stator units in response to changes in environmental conditions, such as pH and viscosity [50]. The coupling ion for spirochete motility was investigated in some species by using ionophores and Na^+ inhibitors, showing that *B. burgdorferi* [51] and *Spirochaeta aurantia* [52] utilize H^+ for swimming, because they are completely paralyzed by the protonophore carbonyl cyanide m-chlorophenylhydrazone (CCCP). Swimming of *L. biflexa* is also inhibited by CCCP in acidic to neutral pH, while some residual motility is observed under alkaline conditions, even in the presence of CCCP [53]. Moreover, addition of Na^+ to the medium enhances leptospiral motility [53]. These results suggest the possibility that the major coupling ion for *Leptospira* swimming is H^+, and that Na is used secondarily in alkaline conditions.

4.3. Coordinated Rotation of PFs

The flagellar motor rotates both CCW and CW, and a reversal of the direction of motor rotation results in a change in the swimming direction. In *E. coli*, a rotational switch from CCW to CW unravels the flagellar bundle and thus causes an instant tumbling motion, which is followed by swimming in a randomly determined direction upon returning to CCW rotation [24,33]. Motor reversal from CCW to CW rotation in the polarly flagellated bacterium *V. alginolyticus* changes the swimming direction from forward to backward, whereas the reversal from CW to CCW causes "buckling" of the flagellum at the hook, resulting in a 90 degree change in swimming direction [54]. These motor reversal-based changes in swimming direction are related to bacterial chemotaxis, which may be stimulated by chemicals, temperature, light, and other trigger mechanisms [55]. In spirochetes, rotational directions of PFs are important for directed swimming [6,44]. According to the schematic structure shown in Figure 1a, the flagellar motors residing at both cell ends have to rotate in opposite directions to each other; if they rotate in the same direction, the cell body will not be rotated due to the counterbalance of torques generated by the two motors or the inability to swim due to a twist of the cell body. This mechanical model suggests that asymmetric rotation and synchronized motor reversal between PFs are required for the cells to swim smoothly and change swimming direction [44].

Coordinated rotation of *E. coli* flagellar motors can be observed when they reside close to each other, which was explained by diffusion of the phosphorylated chemotaxis response regulator CheY (CheY-P) within the cytoplasm. CheY-P molecules generated in response to methylation of the methyl-accepting chemotaxis protein (MCP) bind to a rotor protein FliM and induce a conformational change of the rotor. As a result, the rotor switch rotation direction from CCW to CW. The delay time of reversal observed between the two motors is consistent with the diffusion time of CheY-P (~100 ms) [56]. CheY is also involved in spirochete chemotaxis [57–60], but whether its diffusion can manage signal transduction between motors depends on the distance. CheY-P diffusion could be effective in *E. coli* cells that are 1–2 μm in length [56] but not for rapid coordination [61] of spirochete motors that are more than 10 μm apart from each other. Using the equation giving time t for diffusing x with the diffusion constant D, $t = x^2/2D$, CheY with a diffusion coefficient of $D \approx 10\ \mu m^2/s$ [56,62] can be estimated to take 5 s for diffusing 10 μm. This estimation suggests that a CheY-independent mechanism could control the rapid swimming reversal observed in spirochetes. Furthermore, a chemotaxis-deficient *B. burgdorferi* mutant (*cheA* knockout strain) swims straight without reversal, indicating that asymmetric rotation of PFs at different poles of a single cell during steady-state swimming is not related to the chemotaxis system [44]. *B. burgdorferi* possesses two *fliG* homologs, *fliG1* and *fliG2*. FliG1 plays a central role for torque generation through interaction with stator units. FliG2 is essential for PF synthesis in *B. burgdorferi* [63]. Knockout of *fliG1* does not affect PF synthesis, but subcellular localization studies on FliG1 tagged with green fluorescent protein (GFP) revealed that the localization of FliG1 is asymmetric [63]. This suggests the possibility that asymmetric PF rotation observed for *B. burgdorferi* can be attributed to structural differences in flagellar motors residing at both cell ends. Furthermore, a mathematical model predicted the importance of the interaction between PFs at the cell center. In a borrelial model with a single PF, free swimming of the spirochete was reproduced by assuming that both ends of the PF are anchored to the cell body (intimate interaction between PFs) but not by assuming that only one end of the PF is anchored (no interaction between PFs). In the case of *Leptospira* with short PFs, given that the leptospiral cell body is stiffer than PFs [29], torque transmission from one end to the other may occur along the cell body instead of being mediated by direct contact between PFs.

4.4. Translation Versus Rotation

Swimming speeds differ significantly among species (Figure 3a). *E. coli* and *Salmonella* spp. swim at 20–30 μm/s [64,65], while *C. crescentus* (~60 μm/s) [66], *V. cholerae* (~100 μm/s) [67], and the magnetotactic marine bacterium MO-1 (~300 μm/s) [68] are examples of faster swimmers. In comparison with externally flagellated bacteria, the swimming speed of spirochetes in liquid media is much slower. The fastest swimmer is *Leptospira* spp. (~15 μm/s) [10,69], which is followed by *B. burgdorferi* (~7 μm/s) [70], *Brachyspira pilosicoli* (~5 μm/s) [8], and *Treponema pallidum* (~2 μm/s) [71]. Swimming speeds are correlated with cell body rotation rates or wave frequencies (Figure 3b). Dividing the swimming speed v by the rotation rate or the wave frequency f gives the migration distance achieved by one revolution of the helical body, that is, v/f. The ratio of v/f to helix pitch p, $(v/f)/p$, is similar to motion efficiency; for example, equal values of v/f and p, that is, $(v/f)/p = 1$, indicate swimming without slip [72]. The $(v/f)/p$ ratios of *S. enterica* and *V. alginolyticus* are ~0.1 [64] and ~0.07 [72], respectively, meaning that these bacteria move by less than 10% of the helix pitch of their flagella by one flagellar revolution. *B pilosicoli* and *L. biflexa* show $(v/f)/p$ values of ~0.17 [8] and ~0.27 [73], respectively, showing slightly more efficient swimming than external flagella-driven motility. Spirochetal $(v/f)/p$ values increase with viscosity, leading to increased swimming speeds at high viscosity (described below).

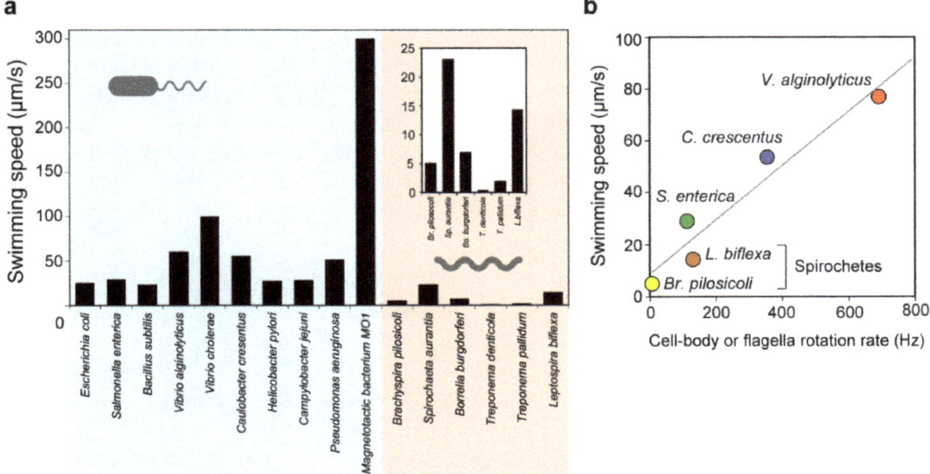

Figure 3. Speeds of bacterial motility. (**a**) Swimming or gliding speeds of various bacterial species. Spirochete-derived data are enlarged in the inset. Refer to the following literature for the corresponding swimming measurements: *E. coli* [65], *S. enterica* [74], *B. subtilis* [49], *V. alginolyticus* [75], *V. cholerae* [67], *C. crescentus* [66], *Helicobacter pylori* [76], *C. jejuni* [77], *Pseudomonas aeruginosa* [78], magnetotactic bacterium MO-1 [68], *B. pilosicoli* [8], *S. aurantia* [79], *B. burgdorferi* [70], *T. denticola* [80], *T. pallidum* [71], and *L. biflexa* [10]. (**b**) Relationships between rotation rates and swimming speeds: *S. enterica* [64], *V. alginolyticus* [72], *C. crescentus* [81], *B. pilosicoli* [8], and *L. biflexa* [10].

4.5. Effect of Viscosity on Swimming Motility

Although the swimming ability of spirochetes seems to be inferior to that of other flagellated bacteria (Figure 3), spirochete swimming is known to be improved by increased viscosity. Kaiser and Doetsch reported that the swimming speed of *L. biflexa* monotonically increased with viscosity in methylcellulose solutions [82]. Similar phenomena have been observed in *B. burgdorferi* [83], *T. denticola* [80], and *B. pilosicoli* [8]. *T. denticola* cannot swim at all in medium without polymers, but smooth translation is allowed by the addition of methylcellulose to the medium (~6 µm/s in 1% methylcellulose 4000 solution) [80]. However, swimming motilities of these spirochetes cannot be improved by all types of viscous fluids but only by gel-like, heterogeneous polymer solutions, for example those containing methylcellulose, polyvinylpyrrolidone (PVP), or mucin [8,69,83,84]. These linear polymers form a quasi-rigid network and are thus treated as viscoelastic fluids [85]. In contrast, the swimming speeds of *B. pilosicoli* [8], *L. biflexa* [10], and *B. burgdorferi* slow down in the presence of the branched polymer Ficoll that does not form a network [71]. Measurements in *B. pilosicoli* highlighted that the v/f value of this spirochete was improved by addition of PVP but not Ficoll [8]. Although the mechanisms by which spirochete motilities are influenced by the differences in microscopic polymer structure are not fully understood, viscoelasticity is believed to be related to this unique phenomenon.

Leptospira are known to be attracted to higher viscosity, and the mechanism of this so-called "viscotaxis" was explained by the viscosity-dependent increment of swimming speed [86]. However, a recent motility study using *Leptospira* proposed another plausible model of taxis-like behavior, which was based on the result that a change in viscosity affects the reversal frequency in swimming direction [13]. When a leptospiral cell swims with the anterior spiral (S) end and the posterior hook

(H) end (SH form), the transformation into symmetric cell morphology (SS or HH form) interrupts swimming transiently, although the cell keeps rotating (Figure 4a). Leptospiral swimming is restarted by transformation from symmetric to asymmetric forms, and the swimming direction after exhibiting symmetric morphologies is determined by the cell forming SH or HS. The transformation process of SH-SS/HH-SH causes a pause of swimming but does not change the swimming direction (stepping movement), whereas SH-SS/HH-HS turns the swimming direction by 180 degrees (reversal movement) (Figure 4b) [13]. Takabe et al. measured the stepping and the reversal events of individual leptospiral cells in various viscous solutions containing methylcellulose, Ficoll, or the major viscous agent for tissue mucin, showing that the reversal frequency increased with viscosity (Figure 4c) [13]. The reversal movement returns the cell to its original position, indicating that there is no net migration. Thus, viscosity-dependent impairment of net migration occurs due to the increment of the reversal event that results in trapping leptospires in areas with higher viscosity, which could assist the accumulation of spirochetes in the mucus layer in vivo (Figure 4d).

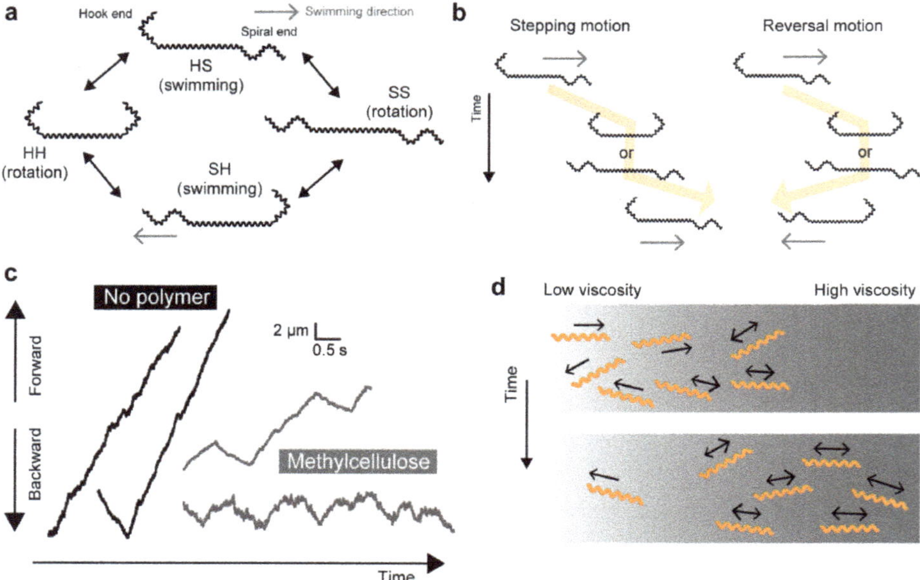

Figure 4. Effect of viscosity on *Leptospira* swimming. (**a**) Association of cell morphology and swimming in *Leptospira*. The spirochete can swim while displaying asymmetric morphologies (SH or HS), with the front end pointing towards the swimming direction and usually displaying a spiral shape. (**b**) Definition of stepping and reversal motions. (**c**) Reversal movements are enhanced by the addition of methylcellulose to the medium. (**d**) A plausible explanation of "viscotaxis" in *Leptospira*. Enhanced swimming reversal with elevated viscosity suppresses net migration of *Leptospira* cells, facilitating an accumulation of spirochetes in high viscosity areas.

5. Chemotaxis

Early studies on chemotaxis using *E. coli* and *S. enterica* showed that these are attracted to nutritious substrates, such as sugars and amino acids, but are repelled by harmful ones, such as alcohols. Notably, not all of the attractants and repellants are related to metabolism [87,88]. In spirochetes, *S. aurantia* shows an attraction response to many sugars, such as glucose, xylose, galactose, and fructose [79], whereas *B. hyodysenteriae* is attracted to serine, fucose, and lactose [89]. *B. burgdorferi* does not respond to common chemicals, such as sugars and amino acids, but is attracted to rabbit serum and is repelled by ethanol and butanol [51]. Both pathogenic and saprophyte *Leptospira* spp. are attracted not only to their sole carbon sources, i.e., long-chain fatty acids, but also to sugars (e.g., glucose) that cannot be metabolized in *Leptospira* [90–92]. Chemotaxis to hemoglobin was observed in the pathogenic species *L. interrogans* but not in saprophytes [93].

Chemotaxis is closely related to the reversal of flagellar rotation, as described in Section 4.3. Motor reversal in peritrichous bacteria results in an exploration of the environment by repeated run-and-tumble movements [24,33] and causes back-and-forth movements with ~90 degree changes in swimming direction by buckling in the case of polarly flagellated bacteria [54]. The swimming pattern of spirochetes involves back-and-forth motions, and attractants increase the persistency of their directed runs [91]. However, when swimming freely in liquid medium, the spirochetal back-and-forth movement cannot result in changes in direction as large as *Vibrio*, because the spirochete cell body is elastic but not too flexible to be buckled by mechanical stress. A physical study on *Leptospira* showed that such a long and spiral body has a larger diffusion coefficient than a simple rod, suggesting that the exploration of spirochetes involves passive Brownian motion in addition to active swimming [94].

6. Movement on Solid Surfaces

Pseudomonas aeruginosa not only swim with a polar flagellum but can also move on a solid surface using pili in a process called twitching motility [2,95]. To that effect, ambivalent motility of *P. aeruginosa* is realized by two distinct machineries specialized for movement in liquid and on solid media, respectively. A major motility form of spirochetes is swimming, but *Leptospira* spp. can move both in liquid and on solid surfaces. Cox and Twigg first reported leptospiral snake-like movement on a smooth surface, which was called "crawling" [96]. For moving while attached to surfaces, *Mycoplasma mobile* uses abundant leg-like protein complexes that are expressed on the cell surface; these legs successively catch and release sialylated oligosaccharides on surfaces, thereby propelling the cell [97]. Another gliding bacterium, *Myxococcus xanthus*, has a machinery that is composed of intracellular motor proteins and an external adhesive complex (Agl-Glt) [98]. Leptospiral swimming is a result of flagella-dependent motility, but a machinery specialized in crawling has yet to be identified. Charon et al. observed that microbeads attached to the leptospiral cell surface via anti-whole cell antibody freely move along the cell body, suggesting that unspecialized antigens residing on the outer sheath are involved in crawling motility by functioning as mobile adhesins [99]. A recent study by Tahara et al. showed that crawling is completely inhibited by CCCP, indicating that PMF-dependent PF rotation drives crawling (Figure 5a) [73]. Furthermore, it was revealed that modification of glass surfaces with anti-lipopolysaccharide (LPS) antibody affects the crawling speed and that anti-LPS antibody-coated microbeads move on the outer bacterial membrane. These results suggest that LPS is responsible for crawling, serving as one of the adhesins anchoring the cell to the surface (Figure 5b–d) [73]. Electron microscopic observation of a hamster liver infected by pathogenic leptospires showed entry of leptospiral cells into the intercellular junction of hepatocytes [100], implying that leptospiral pathogenicity could involve adherence of spirochetes to host cells, followed by crawling (discussed in Section 7).

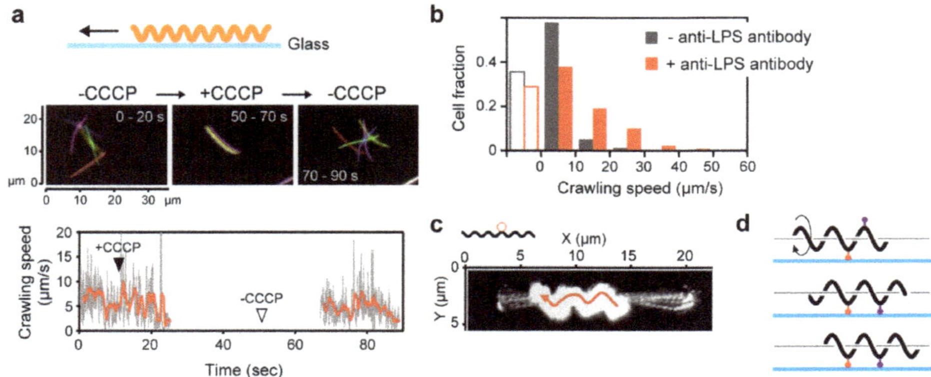

Figure 5. Crawling motility of *Leptospira*. (**a**) Effect of carbonyl cyanide m-chlorophenylhydrazone (CCCP) on *Leptospira* crawling on a glass surface. (**b**) Effect of anti- lipopolysaccharide (LPS) antibody on crawling speed. Open bars indicate the fractions of cells adhered to the glass without crawling. (**c**) Movement of a microbead coated with anti-LPS antibody on the leptospiral cell surface. Sequential frames of a movie were superimposed to show the bead trajectory. (**d**) Schematic explanation of crawling. Adhesive molecules (red and purple symbols), such as LPS, anchor the cell to a surface, and PF-dependent rolling of the protoplasmic cylinder propels the cell.

7. Motility as A Virulence Factor

In general, bacterial flagella and motility are related to virulence, such as invasion, adhesion, and others [101,102]. Motility is an essential virulence factor for pathogenic spirochetes, and loss of motility due to a lack of flagellar genes attenuates infections with *B. burgdorferi* [63], *B. hyodysenteriae* [103], and *L. interrogans* [20,21]. Invasion of *B. burgdorferi* via a tick bite induces a hallmark rash, called erythema migrans, at the initial stage of Lyme disease. Motility analyses of *B. burgdorferi* using the mouse dermis showed three distinct motilities of the spirochete, which were termed translocating, wriggling, and lunging [70]. The translocating state is similar to swimming in solutions, whereas the wriggling (the entire cell body is fixed in place but keeps undulation) and the lunging (the cell body is partially fixed on the surface) states are observed only in the dermis or the gelatin resembling the mouse dermis. The translocation is essential for dissemination within the host, and transient adhesion by wriggling and lunging is thought to be involved in changing the moving direction and evading host immune system [70]. *Brachyspira* spp. penetrate the epithelial mucosa with one end of the cell body moving in the same direction, and this well-aligned colonization is called "false-brush-border", which could involve directed motility of spirochetes [104]. In *Leptospira* spp., pathogenic strains are classified into ~300 serovars based on the structural difference in LPS, and the severity of the infection outcome depends on the combination of host species and leptospiral serovars [105]. Although the details on the relationship between motility of *Leptospira* serovars and their host-dependent pathogenicity remain unknown, the crawling motility mediated by leptospiral LPS and other adhesion molecules is a potential key factor [73,106]. Recently, we measured adhesivity and crawling of some leptospiral serovars on kidney cells derived from various mammalian hosts, including humans, showing close correlation of the measured parameters with the symptom severity of the host–serovar pairs; pairs causing more severe symptoms, such as hemorrhage, jaundice, and nephritis, show high adhesivity and persistent crawling of leptospires on the host cells [106]. This knowledge is an important step toward understanding the host–pathogen relationship to develop novel antimicrobials for targeting pathogen dynamics.

8. Conclusions and Perspectives

Members of the spirochetes share a basic cell structure, but their configurations, PF compositions, and motility forms are extremely diverse. Remarkable advancements in cryo-electron microscopy/tomography have unveiled many spirochete-specific structures, such as the motor scaffold P-collar, fully assembled stator units, and a combination of multiple proteins for establishing the unique morphology of PFs. These are important clues to discuss high torque generation by the spirochetal flagellar motor. Motility measurements by optical microscopy showed improved efficiency of swimming motility in gel-like fluids and viscosity-dependent enhancement of swimming reversal, probably facilitating an accumulation of spirochetes in viscous milieus that exist abundantly within a host body. A recent study showed the close relationship of the spirochetal movements over host cell surfaces and the severity of the symptoms caused, giving crucial insight into the practical role of bacterial motility as a virulence factor.

Although the knowledge summarized in this review deepened the understanding of the mechanics of spirochete motility and its biological significance, there are still many issues remaining, such as the interaction between spirochetes and viscoelastic fluids, signal transduction for the coordinated rotation of PFs between both cell ends, and the molecular basis of crawling motility on the host cells. Further studies on these subjects will advance biomimetic technology and prompt the development of novel prevention/medication strategies.

Funding: This work was supported in part by Grants-in-Aid for Scientific Research from the Japan Society for the Promotion of Science (JSPS KAKENHI Grant Numbers 18K07100) and Grant-in-Aid for Scientific Research on Innovative Areas "Harmonized Supramolecular Motility Machinery and Its Diversity" (Grant Numbers 15H01307).

Acknowledgments: The author thank K. Takabe, Md. S. Islam, J. Xu, A. Kawamoto, N. Koizumi, and S. Kudo for critical discussion related to research referred in this review.

Conflicts of Interest: The authors declare no conflict of interest.

References

1. Miyata, M.; Robinson, R.C.; Uyeda, T.Q.P.; Fukumori, Y.; Fukushima, S.; Haruta, S.; Homma, M.; Inaba, K.; Ito, M.; Kaito, C.; et al. Tree of motility – A proposed history of motility systems in the tree of life. *Genes Cells* **2020**, *25*, 6–21. [CrossRef] [PubMed]
2. Jarrell, K.F.; McBride, M.J. The surprisingly diverse ways that prokaryotes move. *Nat. Rev. Micro.* **2008**, *6*, 466–476. [CrossRef] [PubMed]
3. Zhao, X.; Zhang, K.; Boquoi, T.; Hu, B.; Motaleb, M.A.; Miller, K.A.; James, M.E.; Charon, N.W.; Manson, M.D.; Norris, S.J.; et al. Cryoelectron tomography reveals the sequential assembly of bacterial flagella in *Borrelia Burgdorferi*. *PNAS* **2013**, *110*, 14390–14395. [CrossRef] [PubMed]
4. Goldstein, S.F.; Buttle, K.F.; Charon, N.W. Structural analysis of the *Leptospiraceae* and *Borrelia burgdorferi* by high-voltage electron microscopy. *J. Bacteriol.* **1996**, *178*, 6539–6545. [CrossRef] [PubMed]
5. Charon, N.W.; Goldstein, S.F.; Marko, M.; Hsieh, C.; Gebhardt, L.L.; Motaleb, M.A.; Wolgemuth, C.W.; Limberger, R.J.; Rowe, N. The flat-ribbon configuration of the periplasmic flagella of *Borrelia burgdorferi* and its relationship to motility and morphology. *J. Bacteriol.* **2009**, *191*, 600–607. [CrossRef] [PubMed]
6. Charon, N.W.; Cockburn, A.; Li, C.; Liu, J.; Miller, K.A.; Miller, M.R.; Motaleb, M.A.; Wolgemuth, C.W. The unique paradigm of spirochete motility and chemotaxis. *Annu. Rev. Microbiol.* **2012**, *66*, 349–370. [CrossRef]
7. Goldstein, S.F.; Charon, N.W.; Kreiling, J.A. *Borrelia burgdorferi* swims with a planar waveform similar to that of eukaryotic flagella. *Proc. Natl. Acad. Sci. USA* **1994**, *91*, 3433–3437. [CrossRef]
8. Nakamura, S.; Adachi, Y.; Goto, T.; Magariyama, Y. Improvement in motion efficiency of the spirochete *Brachyspira pilosicoli* in viscous environments. *Biophys. J.* **2006**, *90*, 3019–3026. [CrossRef]
9. Tasu, C.; Nakamura, S.; Tazawa, H.; Hara, H.; Adachi, Y. Morphological properties of a human intestinal spirochete first isolated from a patient with diarrhea in Japan. *Microbiol. Immunol.* **2003**, *47*, 989–996. [CrossRef]

10. Nakamura, S.; Leshansky, A.; Magariyama, Y.; Namba, K.; Kudo, S. Direct measurement of helical cell motion of the spirochete leptospira. *Biophys. J.* **2014**, *106*, 47–54. [CrossRef] [PubMed]
11. Berg, H.C.; Bromley, D.B.; Charon, N.W. Leptospiral motility. *Symp. Soc. Gen. Microbiol.* **1978**, *28*, 285–294.
12. Goldstein, S.F.; Charon, N.W. Multiple-exposure photographic analysis of a motile spirochete. *Proc. Natl. Acad. Sci. USA* **1990**, *87*, 4895–4899. [CrossRef] [PubMed]
13. Takabe, K.; Tahara, H.; Islam, M.S.; Affroze, S.; Kudo, S.; Nakamura, S. Viscosity-dependent variations in the cell shape and swimming manner of *Leptospira*. *Microbiology* **2017**, *163*, 153–160. [CrossRef] [PubMed]
14. Takabe, K.; Kawamoto, A.; Tahara, H.; Kudo, S.; Nakamura, S. Implications of coordinated cell-body rotations for *Leptospira* motility. *Biochem. Biophys. Res. Commun.* **2017**, *491*, 1040–1046. [CrossRef]
15. Bromley, D.B.; Charon, N.W. Axial filament involvement in the motility of *Leptospira interrogans*. *J. Bacteriol.* **1979**, *137*, 1406–1412. [CrossRef] [PubMed]
16. Li, C.; Motaleb, A.; Sal, M.; Goldstein, S.F.; Charon, N.W. Spirochete periplasmic flagella and motility. *J. Mol. Microbiol. Biotechnol.* **2000**, *2*, 345–354.
17. Li, C.; Corum, L.; Morgan, D.; Rosey, E.L.; Stanton, T.B.; Charon, N.W. The spirochete FlaA periplasmic flagellar sheath protein impacts flagellar helicity. *J. Bacteriol.* **2000**, *182*, 6698–6706. [CrossRef] [PubMed]
18. Li, C.; Wolgemuth, C.W.; Marko, M.; Morgan, D.G.; Charon, N.W. Genetic analysis of spirochete flagellin proteins and their involvement in motility, filament assembly, and flagellar morphology. *J. Bacteriol.* **2008**, *190*, 5607–5615. [CrossRef]
19. Sasaki, Y.; Kawamoto, A.; Tahara, H.; Kasuga, K.; Sato, R.; Ohnishi, M.; Nakamura, S.; Koizumi, N. Leptospiral flagellar sheath protein FcpA interacts with FlaA2 and FlaB1 in Leptospira biflexa. *PLoS ONE* **2018**, *13*, e0194923. [CrossRef]
20. Lambert, A.; Picardeau, M.; Haake, D.A.; Sermswan, R.W.; Srikram, A.; Adler, B.; Murray, G.A. FlaA proteins in *Leptospira interrogans* are essential for motility and virulence but are not required for formation of the flagellum sheath. *Infect. Immun.* **2012**, *80*, 2019–2025. [CrossRef]
21. Wunder, E.A.; Figueira, C.P.; Benaroudj, N.; Hu, B.; Tong, B.A.; Trajtenberg, F.; Liu, J.; Reis, M.G.; Charon, N.W.; Buschiazzo, A.; et al. A novel flagellar sheath protein, FcpA, determines filament coiling, translational motility and virulence for the *Leptospira* spirochete. *Mol. Microbiol.* **2016**, *101*, 457–470. [CrossRef] [PubMed]
22. Wunder, E.A.J.; Slamti, L.; Suwondo, D.N.; Gibson, K.H.; Shang, Z.; Sindelar, C.V.; Trajtenberg, F.; Buschiazzo, A.; Ko, A.I.; Picardeau, M. FcpB Is a surface filament protein of the endoflagellum required for the motility of the spirochete *Leptospira*. *Front. Cell. Infect. Microbiol.* **2018**, *8*. [CrossRef] [PubMed]
23. Gibson, K.H.; Trajtenberg, F.; Wunder, E.A.; Brady, M.R.; San Martin, F.; Mechaly, A.; Shang, Z.; Liu, J.; Picardeau, M.; Ko, A.; et al. An asymmetric sheath controls flagellar supercoiling and motility in the leptospira spirochete. *eLife* **2020**, *9*, e53672. [CrossRef] [PubMed]
24. Nakamura, S.; Minamino, T. Flagella-driven motility of bacteria. *Biomolecules* **2019**, *9*, 279. [CrossRef] [PubMed]
25. Ruby, J.D.; Li, H.; Kuramitsu, H.; Norris, S.J.; Goldstein, S.F.; Buttle, K.F.; Charon, N.W. Relationship of *Treponema denticola* periplasmic flagella to irregular cell morphology. *J. Bacteriol.* **1997**, *179*, 1628–1635. [CrossRef]
26. Motaleb, M.A.; Corum, L.; Bono, J.L.; Elias, A.F.; Rosa, P.; Samuels, D.S.; Charon, N.W. Borrelia burgdorferi periplasmic flagella have both skeletal and motility functions. *PNAS* **2000**, *97*, 10899–10904. [CrossRef]
27. Sal, M.S.; Li, C.; Motalab, M.A.; Shibata, S.; Aizawa, S.; Charon, N.W. *Borrelia burgdorferi* uniquely regulates its motility genes and has an intricate flagellar hook-basal body structure. *J. Bacteriol.* **2008**, *190*, 1912–1921. [CrossRef]
28. Slamti, L.; de Pedro, M.A.; Guichet, E.; Picardeau, M. Deciphering morphological determinants of the helix-shaped *Leptospira*. *J. Bacteriol.* **2011**, *193*, 6266–6275. [CrossRef]
29. Kan, W.; Wolgemuth, C.W. The shape and dynamics of the *Leptospiraceae*. *Biophys. J.* **2007**, *93*, 54–61. [CrossRef]
30. Dombrowski, C.; Kan, W.; Motaleb, M.A.; Charon, N.W.; Goldstein, R.E.; Wolgemuth, C.W. The elastic basis for the shape of *Borrelia burgdorferi*. *Biophys. J.* **2009**, *96*, 4409–4417. [CrossRef]
31. Vig, D.K.; Wolgemuth, C.W. Swimming dynamics of the lyme disease spirochete. *Phys. Rev. Lett.* **2012**, *109*, 218104. [CrossRef] [PubMed]

32. Miller, M.R.; Miller, K.A.; Bian, J.; James, M.E.; Zhang, S.; Lynch, M.J.; Callery, P.S.; Hettick, J.M.; Cockburn, A.; Liu, J.; et al. Spirochaete flagella hook proteins self-catalyse a lysinoalanine covalent crosslink for motility. *Nat. Microbiol.* **2016**, *1*, 1–8. [CrossRef]
33. Berg, H.C. The rotary motor of bacterial flagella. *Annu. Rev. Biochem.* **2003**, *72*, 19–54. [CrossRef] [PubMed]
34. Beeby, M.; Ribardo, D.A.; Brennan, C.A.; Ruby, E.G.; Jensen, G.J.; Hendrixson, D.R. Diverse high-torque bacterial flagellar motors assemble wider stator rings using a conserved protein scaffold. *PNAS* **2016**, *113*, E1917–E1926. [CrossRef] [PubMed]
35. Lele, P.P.; Hosu, B.G.; Berg, H.C. Dynamics of mechanosensing in the bacterial flagellar motor. *PNAS* **2013**, *110*, 11839–11844. [CrossRef] [PubMed]
36. Tipping, M.J.; Delalez, N.J.; Lim, R.; Berry, R.M.; Armitage, J.P. Load-dependent assembly of the bacterial flagellar motor. *MBio* **2013**, *4*, e00551-13. [CrossRef] [PubMed]
37. Nord, A.L.; Gachon, E.; Perez-Carrasco, R.; Nirody, J.A.; Barducci, A.; Berry, R.M.; Pedaci, F. Catch bond drives stator mechanosensitivity in the bacterial flagellar motor. *Proc. Natl. Acad. Sci. USA* **2017**, *114*, 12952–12957. [CrossRef]
38. Castillo, D.J.; Nakamura, S.; Morimoto, Y.V.; Che, Y.-S.; Kami-ike, N.; Kudo, S.; Minamino, T.; Namba, K. The C-terminal periplasmic domain of MotB is responsible for load-dependent control of the number of stators of the bacterial flagellar motor. *Biophysics* **2013**, *9*, 173–181. [CrossRef]
39. Murphy, G.E.; Leadbetter, J.R.; Jensen, G.J. In situ structure of the complete *Treponema primitia* flagellar motor. *Nature* **2006**, *442*, 1062–1064. [CrossRef]
40. Raddi, G.; Morado, D.R.; Yan, J.; Haake, D.A.; Yang, X.F.; Liu, J. Three-dimensional structures of pathogenic and saprophytic *Leptospira* species revealed by cryo-electron tomography. *J. Bacteriol.* **2012**, *194*, 1299–1306. [CrossRef]
41. Liu, J.; Howell, J.K.; Bradley, S.D.; Zheng, Y.; Zhou, Z.H.; Norris, S.J. Cellular architecture of *Treponema pallidum*: Novel flagellum, periplasmic cone, and cell envelope as revealed by cryo electron tomography. *J. Mol. Biol.* **2010**, *403*, 546–561. [CrossRef] [PubMed]
42. Sowa, Y.; Berry, R.M. Bacterial flagellar motor. *Q. Rev. Biophys.* **2008**, *41*, 103–132. [CrossRef] [PubMed]
43. DiLuzio, W.R.; Turner, L.; Mayer, M.; Garstecki, P.; Weibel, D.B.; Berg, H.C.; Whitesides, G.M. *Escherichia coli* swim on the right-hand side. *Nature* **2005**, *435*, 1271–1274. [CrossRef] [PubMed]
44. Li, C.; Bakker, R.G.; Motaleb, M.A.; Sartakova, M.L.; Cabello, F.C.; Charon, N.W. Asymmetrical flagellar rotation in *Borrelia burgdorferi* nonchemotactic mutants. *Proc. Natl. Acad. Sci. USA* **2002**, *99*, 6169–6174. [CrossRef] [PubMed]
45. Thormann, K.M.; Paulick, A. Tuning the flagellar motor. *Microbiology* **2010**, *156*, 1275–1283. [CrossRef]
46. Kawagishi, I.; Maekawa, Y.; Atsumi, T.; Homma, M.; Imae, Y. Isolation of the polar and lateral flagellum-defective mutants in *Vibrio alginolyticus* and identification of their flagellar driving energy sources. *J. Bacteriol.* **1995**, *177*, 5158–5160. [CrossRef]
47. Kawagishi, I.; Imagawa, M.; Imae, Y.; McCarter, L.; Homma, M. The sodium-driven polar flagellar motor of marine *Vibrio* as the mechanosensor that regulates lateral flagellar expression. *Mol. Microbiol.* **1996**, *20*, 693–699. [CrossRef]
48. Ito, M.; Hicks, D.B.; Henkin, T.M.; Guffanti, A.A.; Powers, B.D.; Zvi, L.; Uematsu, K.; Krulwich, T.A. MotPS is the stator-force generator for motility of alkaliphilic *Bacillus*, and its homologue is a second functional Mot in *Bacillus subtilis*: Alkaliphile MotPS and its *B. subtilis* homologue. *Mol. Microbiol.* **2004**, *53*, 1035–1049. [CrossRef]
49. Ito, M.; Terahara, N.; Fujinami, S.; Krulwich, T.A. Properties of motility in *Bacillus subtilis* powered by the H^+-coupled MotAB flagellar stator, Na^+-coupled MotPS or hybrid stators MotAS or MotPB. *J. Mol. Biol.* **2005**, *352*, 396–408. [CrossRef]
50. Minamino, T.; Terahara, N.; Kojima, S.; Namba, K. Autonomous control mechanism of stator assembly in the bacterial flagellar motor in response to changes in the environment. *Mol. Microbiol.* **2018**, *109*, 723–734. [CrossRef]
51. Shi, W.; Yang, Z.; Geng, Y.; Wolinsky, L.E.; Lovett, M.A. Chemotaxis in *Borrelia burgdorferi*. *J. Bacteriol.* **1998**, *180*, 231–235. [CrossRef] [PubMed]
52. Goulbourne, E.A.; Greenberg, E.P. Relationship between proton motive force and motility in *Spirochaeta aurantia*. *J. Bacteriol.* **1980**, *143*, 1450–1457. [CrossRef] [PubMed]

53. Islam, M.S.; Morimoto, Y.V.; Kudo, S.; Nakamura, S. H^+ and Na^+ are involved in flagellar rotation of the spirochete *Leptospira*. *Biochem. Biophys. Res. Commun.* **2015**, *466*, 196–200. [CrossRef] [PubMed]
54. Son, K.; Guasto, J.S.; Stocker, R. Bacteria can exploit a flagellar buckling instability to change direction. *Nat. Phys.* **2013**, *9*, 494–498. [CrossRef]
55. Bray, D. *Cell Movements: From Molecules to Motility*; Garland Science: New York, NY, USA, 2000; ISBN 978-0-203-83358-2.
56. Terasawa, S.; Fukuoka, H.; Inoue, Y.; Sagawa, T.; Takahashi, H.; Ishijima, A. Coordinated reversal of flagellar motors on a single *Escherichia coli* cell. *Biophys. J.* **2011**, *100*, 2193–2200. [CrossRef]
57. Li, Z.-H.; Dong, K.; Yuan, J.-P.; Hu, B.-Y.; Liu, J.-X.; Zhao, G.-P.; Guo, X.-K. Characterization of the *cheY* genes from *Leptospira interrogans* and their effects on the behavior of *Escherichia coli*. *Biochem. Biophys. Res. Commun.* **2006**, *345*, 858–866. [CrossRef]
58. Novak, E.A.; Sekar, P.; Xu, H.; Moon, K.H.; Manne, A.; Wooten, R.M.; Motaleb, M.A. The *Borrelia burgdorferi* CheY3 response regulator is essential for chemotaxis and completion of its natural infection cycle. *Cell. Microbiol.* **2016**, *18*, 1782–1799. [CrossRef]
59. Motaleb, M.A.; Sultan, S.Z.; Miller, M.R.; Li, C.; Charon, N.W. CheY3 of *Borrelia burgdorferi* is the key response regulator essential for chemotaxis and forms a long-lived phosphorylated intermediate. *J. Bacteriol.* **2011**, *193*, 3332–3341. [CrossRef]
60. Bellgard, M.I.; Wanchanthuek, P.; La, T.; Ryan, K.; Moolhuijzen, P.; Albertyn, Z.; Shaban, B.; Motro, Y.; Dunn, D.S.; Schibeci, D.; et al. Genome sequence of the pathogenic intestinal spirochete *Brachyspira hyodysenteriae* reveals adaptations to its lifestyle in the porcine large intestine. *PLoS ONE* **2009**, *4*, e4641. [CrossRef]
61. Charon, N.W.; Daughtry, G.R.; McCuskey, R.S.; Franz, G.N. Microcinematographic analysis of tethered Leptospira illini. *J. Bacteriol.* **1984**, *160*, 1067–1073. [CrossRef]
62. Segall, J.E.; Ishihara, A.; Berg, H.C. Chemotactic signaling in filamentous cells of *Escherichia coli*. *J. Bacteriol.* **1985**, *161*, 51–59. [CrossRef]
63. Li, C.; Xu, H.; Zhang, K.; Liang, F.T. Inactivation of a putative flagellar motor switch protein FliG1 prevents *Borrelia burgdorferi* from swimming in highly viscous media and blocks its infectivity. *Mol. Microbiol.* **2010**, *75*, 1563–1576. [CrossRef]
64. Magariyama, Y.; Sugiyama, S.; Kudo, S. Bacterial swimming speed and rotation rate of bundled flagella. *FEMS Microbiol. Lett.* **2001**, *199*, 125–129. [CrossRef]
65. Minamino, T.; Imae, Y.; Oosawa, F.; Kobayashi, Y.; Oosawa, K. Effect of intracellular pH on rotational speed of bacterial flagellar motors. *J. Bacteriol.* **2003**, *185*, 1190–1194. [CrossRef]
66. Faulds-Pain, A.; Birchall, C.; Aldridge, C.; Smith, W.D.; Grimaldi, G.; Nakamura, S.; Miyata, T.; Gray, J.; Li, G.; Tang, J.X.; et al. Flagellin redundancy in *Caulobacter crescentus* and its implications for flagellar filament assembly. *J. Bacteriol.* **2011**, *193*, 2695–2707. [CrossRef]
67. Kojima, S.; Yamamoto, K.; Kawagishi, I.; Homma, M. The polar flagellar motor of *Vibrio cholerae* is driven by an Na^+ motive force. *J. Bacteriol.* **1999**, *181*, 1927–1930. [CrossRef]
68. Ruan, J.; Kato, T.; Santini, C.-L.; Miyata, T.; Kawamoto, A.; Zhang, W.-J.; Bernadac, A.; Wu, L.-F.; Namba, K. Architecture of a flagellar apparatus in the fast-swimming magnetotactic bacterium MO-1. *Proc. Natl. Acad. Sci. USA* **2012**, *109*, 20643–20648. [CrossRef]
69. Takabe, K.; Nakamura, S.; Ashihara, M.; Kudo, S. Effect of osmolarity and viscosity on the motility of pathogenic and saprophytic *Leptospira*. *Microbiol. Immunol.* **2013**, *57*, 236–239. [CrossRef]
70. Harman, M.W.; Dunham-Ems, S.M.; Caimano, M.J.; Belperron, A.A.; Bockenstedt, L.K.; Fu, H.C.; Radolf, J.D.; Wolgemuth, C.W. The heterogeneous motility of the Lyme disease spirochete in gelatin mimics dissemination through tissue. *PNAS* **2012**, *109*, 3059–3064. [CrossRef]
71. Harman, M.; Vig, D.K.; Radolf, J.D.; Wolgemuth, C.W. Viscous dynamics of Lyme disease and syphilis spirochetes reveal flagellar torque and drag. *Biophys. J.* **2013**, *105*, 2273–2280. [CrossRef]
72. Magariyama, Y.; Sugiyama, S.; Muramoto, K.; Kawagishi, I.; Imae, Y.; Kudo, S. Simultaneous measurement of bacterial flagellar rotation rate and swimming speed. *Biophys. J.* **1995**, *69*, 2154–2162. [CrossRef]
73. Tahara, H.; Takabe, K.; Sasaki, Y.; Kasuga, K.; Kawamoto, A.; Koizumi, N.; Nakamura, S. The mechanism of two-phase motility in the spirochete *Leptospira*: Swimming and crawling. *Sci. Adv.* **2018**, *4*, eaar7975. [CrossRef]

74. Nakamura, S.; Morimoto, Y.V.; Kami-ike, N.; Minamino, T.; Namba, K. Role of a conserved prolyl residue (Pro173) of MotA in the mechanochemical reaction cycle of the proton-driven flagellar motor of *Salmonella*. *J. Mol. Biol.* **2009**, *393*, 300–307. [CrossRef]
75. Atsumi, T.; Maekawa, Y.; Yamada, T.; Kawagishi, I.; Imae, Y.; Homma, M. Effect of viscosity on swimming by the lateral and polar flagella of *Vibrio alginolyticus*. *J. Bacteriol.* **1996**, *178*, 5024–5026. [CrossRef]
76. Celli, J.P.; Turner, B.S.; Afdhal, N.H.; Keates, S.; Ghiran, I.; Kelly, C.P.; Ewoldt, R.H.; McKinley, G.H.; So, P.; Erramilli, S.; et al. *Helicobacter pylori* moves through mucus by reducing mucin viscoelasticity. *Proc. Natl. Acad. Sci. USA* **2009**, *106*, 14321–14326. [CrossRef]
77. Apel, D.; Ellermeier, J.; Pryjma, M.; DiRita, V.J.; Gaynor, E.C. Characterization of *Campylobacter jejuni* RacRS reveals roles in the heat shock response, motility, and maintenance of cell length homogeneity. *J. Bacteriol.* **2012**, *194*, 2342–2354. [CrossRef]
78. Shigematsu, M.; Meno, Y.; Misumi, H.; Amako, K. The measurement of swimming velocity of *Vibrio cholerae* and *Pseudomonas aeruginosa* using the video tracking methods. *Microbiol. Immunol.* **1995**, *39*, 741–744. [CrossRef]
79. Greenberg, E.P.; Canale-Parola, E. Chemotaxis in *Spirochaeta aurantia*. *J. Bacteriol.* **1977**, *130*, 485–494. [CrossRef]
80. Ruby, J.D.; Charon, N.W. Effect of temperature and viscosity on the motility of the spirochete *Treponema denticola*. *FEMS Microbiol. Lett.* **1998**, *169*, 251–254. [CrossRef]
81. Li, G.; Tang, J.X. Low flagellar motor torque and high swimming efficiency of *Caulobacter crescentus* swarmer cells. *Biophys. J.* **2006**, *91*, 2726–2734. [CrossRef]
82. Kaiser, G.E.; Doetsch, R.N. Enhanced translational motion of *Leptospira* in viscous environments. *Nature* **1975**, *255*, 656–657. [CrossRef]
83. Kimsey, R.B.; Spielman, A. Motility of Lyme disease spirochetes in fluids as viscous as the extracellular matrix. *J. Infect. Dis.* **1990**, *162*, 1205–1208. [CrossRef]
84. Magariyama, Y.; Kudo, S. A Mathematical explanation of an increase in bacterial swimming speed with viscosity in linear-polymer solutions. *Biophys. J.* **2002**, *83*, 733–739. [CrossRef]
85. Berg, H.C.; Turner, L. Movement of microorganisms in viscous environments. *Nature* **1979**, *278*, 349–351. [CrossRef]
86. Petrino, M.G.; Doetsch, R.N. "Viscotaxis", a new behavioural response of *Leptospira interrogans* (*biflexa*) strain B16. *J. Gen. Microbiol.* **1978**, *109*, 113–117. [CrossRef]
87. Melton, T.; Hartman, P.E.; Stratis, J.P.; Lee, T.L.; Davis, A.T. Chemotaxis of *Salmonella typhimurium* to Amino Acids and Some Sugars. *J. Bacteriol.* **1978**, *133*, 708–716. [CrossRef]
88. Tso, W.-W.; Adler, J. Negative chemotaxis in *Escherichia coli*. *J. Bacteriol.* **1974**, *118*, 560–576. [CrossRef]
89. Kennedy, M.J.; Yancey, R.J. Motility and chemotaxis in *Serpulina hyodysenteriae*. *Vet. Microbiol.* **1996**, *49*, 21–30. [CrossRef]
90. Lambert, A.; Takahashi, N.; Charon, N.W.; Picardeau, M. Chemotactic behavior of pathogenic and nonpathogenic *Leptospira* species. *Appl. Environ. Microbiol.* **2012**, *78*, 8467–8469. [CrossRef]
91. Islam, M.S.; Takabe, K.; Kudo, S.; Nakamura, S. Analysis of the chemotactic behaviour of *Leptospira* using microscopic agar-drop assay. *FEMS Microbiol. Lett.* **2014**, *356*, 39–44. [CrossRef]
92. Affroze, S.; Islam, M.S.; Takabe, K.; Kudo, S.; Nakamura, S. Characterization of leptospiral chemoreceptors using a microscopic agar drop assay. *Curr. Microbiol.* **2016**, *73*, 202–205. [CrossRef]
93. Yuri, K.; Takamoto, Y.; Okada, M.; Hiramune, T.; Kikuchi, N.; Yanagawa, R. Chemotaxis of leptospires to hemoglobin in relation to virulence. *Infect. Immun.* **1993**, *61*, 2270–2272. [CrossRef]
94. Butenko, A.V.; Mogilko, E.; Amitai, L.; Pokroy, B.; Sloutskin, E. Coiled to diffuse: Brownian motion of a helical bacterium. *Langmuir* **2012**, *28*, 12941–12947. [CrossRef]
95. Wall, D.; Kaiser, D. Type IV pili and cell motility. *Mol. Microbiol.* **1999**, *32*, 1–10. [CrossRef]
96. Cox, P.J.; Twigg, G.I. Leptospiral motility. *Nature* **1974**, *250*, 260–261. [CrossRef]
97. Miyata, M. Unique centipede mechanism of *Mycoplasma* gliding. *Annu. Rev. Microbiol.* **2010**, *64*, 519–537. [CrossRef]
98. Faure, L.M.; Fiche, J.-B.; Espinosa, L.; Ducret, A.; Anantharaman, V.; Luciano, J.; Lhospice, S.; Islam, S.T.; Tréguier, J.; Sotes, M.; et al. The mechanism of force transmission at bacterial focal adhesion complexes. *Nature* **2016**, *539*, 530–535. [CrossRef]

99. Charon, N.W.; Lawrence, C.W.; O'Brien, S. Movement of antibody-coated latex beads attached to the spirochete *Leptospira interrogans*. *Proc. Natl. Acad. Sci. USA* **1981**, *78*, 7166–7170. [CrossRef]
100. Miyahara, S.; Saito, M.; Kanemaru, T.; Villanueva, S.Y.A.M.; Gloriani, N.G.; Yoshida, S. Destruction of the hepatocyte junction by intercellular invasion of *Leptospira* causes jaundice in a hamster model of Weil's disease. *Int. J. Exp. Path.* **2014**, *95*, 271–281. [CrossRef]
101. Haiko, J.; Westerlund-Wikström, B. The role of the bacterial flagellum in adhesion and virulence. *Biology* **2013**, *2*, 1242. [CrossRef]
102. Josenhans, C.; Suerbaum, S. The role of motility as a virulence factor in bacteria. *Int. J. Med. Microbiol.* **2002**, *291*, 605–614. [CrossRef] [PubMed]
103. Rosey, E.L.; Kennedy, M.J.; Yancey, R.J. Dual *flaA1 flaB1* mutant of *Serpulina hyodysenteriae* expressing periplasmic flagella is severely attenuated in a murine model of swine dysentery. *Infect. Immun.* **1996**, *64*, 4154–4162. [CrossRef] [PubMed]
104. Kraaz, W.; Pettersson, B.; Thunberg, U.; Engstrand, L.; Fellström, C. *Brachyspira aalborgi* infection diagnosed by culture and 16S ribosomal DNA sequencing using human colonic biopsy specimens. *J. Clin. Microbiol.* **2000**, *38*, 3555–3560. [CrossRef] [PubMed]
105. Picardeau, M. Virulence of the zoonotic agent of leptospirosis: Still terra incognita? *Nat. Rev. Micro.* **2017**, *15*, 297–307. [CrossRef] [PubMed]
106. Xu, J.; Koizumi, N.; Nakamura, S. Adhesivity and motility of a zoonotic spirochete: Implications in host-dependent pathogenicity. *bioRxiv* **2020**. [CrossRef]

Review

Living in a Foster Home: The Single Subpolar Flagellum Fla1 of *Rhodobacter sphaeroides*

Laura Camarena [1,*] and Georges Dreyfus [2,*]

1. Depto. Biología Molecular y Biotecnología, Instituto de Investigaciones Biomédicas, Universidad Nacional Autónoma de México, Ciudad Universitaria, CDMX 04510, Mexico
2. Depto. Genética Molecular, Instituto de Fisiología Celular, Universidad Nacional Autónoma de México, Ciudad Universitaria, CDMX 04510, Mexico

* Correspondence: rosal@unam.mx (L.C.); gdreyfus@ifc.unam.mx (G.D.); Tel.: +55-562-29222 (L.C.); +55-562-25618 (G.D.)

Received: 28 March 2020; Accepted: 13 May 2020; Published: 16 May 2020

Abstract: *Rhodobacter sphaeroides* is an α-proteobacterium that has the particularity of having two functional flagellar systems used for swimming. Under the growth conditions commonly used in the laboratory, a single subpolar flagellum that traverses the cell membrane, is assembled on the surface. This flagellum has been named Fla1. Phylogenetic analyses have suggested that this flagellar genetic system was acquired from an ancient γ-proteobacterium. It has been shown that this flagellum has components homologous to those present in other γ-proteobacteria such as the H-ring characteristic of the *Vibrio* species. Other features of this flagellum such as a straight hook, and a prominent HAP region have been studied and the molecular basis underlying these features has been revealed. It has also been shown that FliL, and the protein MotF, mainly found in several species of the family *Rhodobacteraceae*, contribute to remodel the amphipathic region of MotB, known as the plug, in order to allow flagellar rotation. In the absence of the plug region of MotB, FliL and MotF are dispensable. In this review we have covered the most relevant aspects of the Fla1 flagellum of this remarkable photosynthetic bacterium.

Keywords: bacterial flagellum; *Rhodobacter sphaeroides*; motility; FliL; FlgT; flagellar rod; flagellar hook; FlgP

---

1. Introduction

1.1. The Flagellar Structure

The bacterial flagellum is driven by a complex molecular motor. The flagellar basal body contains the rotor and the export apparatus, and is composed of numerous proteins arranged in several rings and a central rod (reviewed recently in [1]). The MS ring, embedded in the internal membrane, is the base platform for the assembly of the rod. At the center of the MS ring a flagellar-specific export system is responsible for the export of most of the axial proteins that form the basal body [2–5]. The flagellar rod traverses the cell envelope and in its proximal end is formed by FlgB, FlgC and FlgF, and the distal end by FlgG [6]. Around the distal rod, the P and L rings act as a bushing allowing rod penetration through the peptidoglycan and the outer membrane, respectively [7,8]. This process is favored by the action of the bifunctional protein FlgJ that acts as a scaffolding rod-capping protein and also possesses glucosaminidase activity to penetrate the cell wall [9–12]. Once the rod reaches the outer membrane, the hook is assembled outside the cell. This structure transmits torque to the flagellar filament [13]. The physical properties of these two axial structures are different given that the filament is a long rigid helix and the hook is a short flexible structure that acts as a universal joint [14–16]. The distal end of the hook is connected to the filament via the hook associated proteins, FlgK and FlgL that mediate the transition from the flexible hook to the rigid filament [6,17]. The filament is the most prominent component, typically 5–10 μm in length and is several times longer than the cell body. The flagellar motor contains a stator that is composed by multiple units of the MotA/MotB complexes (4:2 stoichiometry) that form an ion channel that conducts the ions (H^+ or Na^+) of the transmembrane electrochemical gradient and generates motor rotation that propels the bacterial cell [1,18–20] (Figure 1). Recruitment of the MotA/B complexes and activation of the proton channel are complex processes that have been extensively reviewed recently [1,18,19,21]. Briefly, it is important to mention that recruitment of these complexes to the basal body has been related to the interaction of the cytoplasmic loop of MotA with FliG, which is part of the C-ring (Figure 1) [22–24]. Besides, in *Vibrio* the proteins MotY and MotX form the periplasmic T-ring that interacts with PomB (equivalent to MotB in *Vibrio*) and stabilize the stator complexes [25,26]. Activation of the proton channel requires extensive remodeling of the periplasmic region of MotB [27–32], and it has been proposed that the flagellar protein FliL participates in this process [33–37].

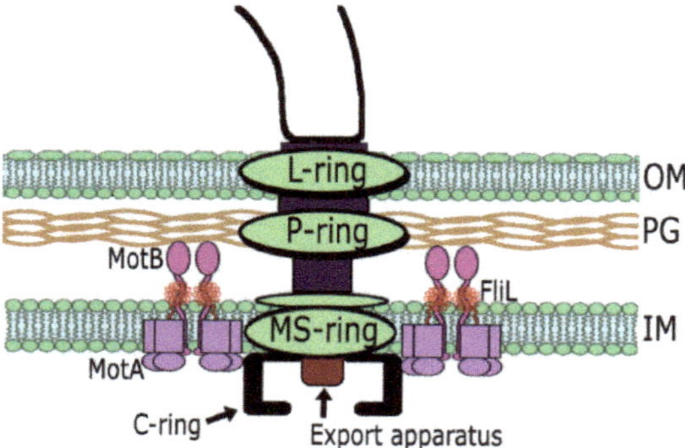

Figure 1. Scheme of the bacterial flagellum that shows the most relevant elements of the core structure that is common to several species of Gram-negative microorganisms.

The flagellum has been thoroughly studied in various bacterial species, and recently the advancement of cryo-electron tomography, a powerful non-invasive technique, has revealed a high complexity and variability of its ultrastructural components [38–44].

1.2. The Two Flagellar Systems of Rhodobacter sphaeroides

R. sphaeroides is an α-proteobacterium from the non-taxonomic group of the purple non-sulfur photosynthetic bacteria. This microorganism frequently found in lakes and stagnant water bodies has a versatile metabolism since it grows by aerobic or anaerobic respiration, photosynthesis or fermentation [45]. The genome of several strains of *R. sphaeroides* has been sequenced and it consists of two chromosomes and several plasmids [46–48] or by one chromosome, one chromide and several plasmids as it has been recently suggested [49,50]. This microorganism was described as motile [45]. The characterization of these motile cells revealed the presence of a single subpolar flagellum (later named as Fla1) (see Figure 2) that promotes a swimming pattern characterized by linear runs interrupted by short stop events. This bacterium swims in liquid medium at average velocities of 80 to 45 μm/s [51,52]. The flagellar motor is dependent on the H^+ gradient and it rotates unidirectionally interrupted by short stop periods [51,53]. During the stop events the flagellum is locked by an unknown mechanism [54]. The initial characterization revealed that most of the genes encoding for this structure were clustered and its organization in the genome was similar to that found in other well characterized bacteria, such as *Escherichia coli* and *Salmonella enterica* [55]. However, further studies on the molecular structure of this flagellum have shown that it has particular components that evoke those found in *Vibrio* [18,56,57].

When the genome sequence of *R. sphaeroides* was completed, the presence of a second flagellar gene cluster was evident [46]. The cluster was later named *fla2*, given that it could potentially form a complete functional flagellum. However, these genes were apparently not expressed according to microarray studies (data accessible at NCBI GEO database accession, GSE139, and GSE12269) [58,59]. Phylogenetic studies revealed that the *fla2* cluster is vertically inherited in this bacterium, whereas the *fla1* genes were probably acquired by a horizontal transfer event from an ancestral γ-proteobacterium [60]. Later on, we showed that the expression of the *fla2* genes was possible under specific conditions in the laboratory [61]. Nonetheless, the signals that triggers in nature the expression of these genes remain to be determined. The expression of the second gene cluster gives rise to several polar flagella that, like the Fla1 flagellum, allow *R. sphaeroides* to swim in a liquid medium [60,62] (Figure 2). The number of flagella per cell ranges from two to nine with an average of 4.5 [62]. The chemosensory response of the Fla2 flagella is controlled by a set of CheY proteins, i.e., CheY1, CheY2, CheY5 that, until the *fla2* cluster was expressed, lacked a motility phenotype when mutated [63–65].

The evolution of the Fla1 flagellum has allowed its adaptation to support efficient swimming of this bacterium. In this review we present the outstanding features of this flagellum and the main differences with the *fla2* genetic system.

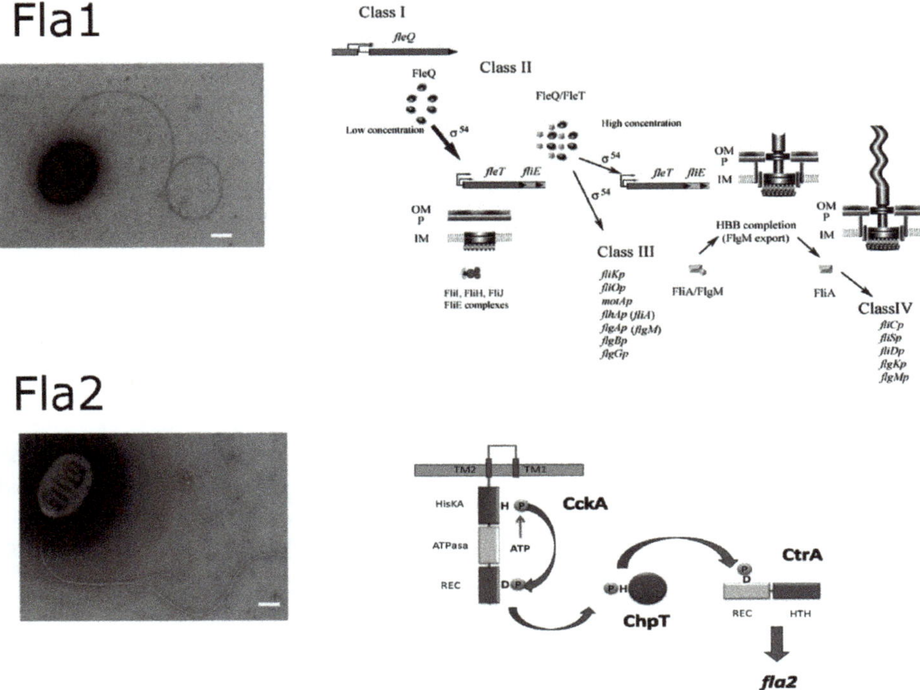

Figure 2. Electron micrograph showing *Rhodobacter sphaeroides* expressing either the Fla1 flagellum or Fla2 flagella. Cells were grown separately and under different growth conditions (for details see, [62]). Bar = 500 nm. The schemes showing the regulatory pathway of each flagellar system are shown at the right side of each micrograph [60,61].

2. Overview of the Flagellar Genetic System in *R. sphaeroides*

The *fla1* genes are mainly organized in a single locus that also includes several genes related to the chemotactic response of this flagellum, as well as several regulatory genes. This region is located in chromosome I and it is comprised of approximately 56.6 kb; other flagellar genes whose products are part of this flagellum are *motAB*, that are not located within this cluster [46,47,66].

The σ^{54} factor (RpoN) together with the RNA polymerase core (E) is responsible of the expression of the *fla1* genes. The gene encoding for this particular sigma factor, *rpoN2*, is located within the *fla1* flagellar locus [67,68]. It should be noted that *R. sphaeroides* has the peculiarity of having four different genes encoding for σ^{54} (*rpoN1* to *rpoN4*), and only $\sigma^{54\text{-}2}$ is responsible for the expression of the flagellar genes [68]. Phylogenetic analysis suggests that the different copies of the *rpoN* genes arose from duplication events followed by selection processes that allowed them to specialize [69]. $E\sigma^{54\text{-}2}$ also controls the expression of some chemotactic genes as is explained later.

It is known that the σ^{54} factor bound to the catalytic core of the RNA polymerase (E) is unable to form an open complex for transcription initiation. This step requires that an activator protein remodels the DNA-$E\sigma^{54}$ complex by hydrolyzing ATP [70,71]. The master activator protein for the expression of the class I flagellar genes is FleQ that together with $E\sigma^{54\text{-}2}$ promotes the expression of an operon that includes a second σ^{54} activator protein named FleT as well as the genes *fliEFGHIJ*. FleQ forms a heterodimeric complex with FleT and activates the expression of the class III flagellar genes. In this gene class the sigma factor σ^{28}, also called FliA, and its anti-sigma protein FlgM are expressed, as well as the genes encoding the components required to complete the basal body, the hook, and the stator proteins MotA and MotB. When the hook is completed, FlgM is exported out of the cell and FliA directs the RNA polymerase to express the class IV flagellar genes such as *fliC* and *fliD* encoding flagellin and the filament cap protein, respectively [55,72] (Figure 2).

In accordance with the expression of the *fla1* genes, FleQ and $E\sigma^{54\text{-}2}$ also activate the expression of the chemotactic genes located within the flagellar locus, achieving the expression of the cytoplasmic chemotactic receptor *tlpT*, and the chemotactic signal transduction system that includes *cheA4* and *cheA3, cheW4, cheR3, cheB2 and cheY6* [65,73,74]. FliA is responsible for the expression of the chemotactic operon that includes *cheY4* and the chemotactic receptor *mcpG*, which is localized in chromosome II [73]. Other chemotactic components that control rotation of Fla1 are encoded in the chemotactic operon *cheOp*2 that includes *cheY3*, *cheA2*, *cheW2* and *cheW3*, *cheR2*, *cheB1* and *tlpC*. This operon is expressed by the housekeeping σ^{70} factor and also from a promoter dependent on σ^{28} [73,75]. The control of the chemotactic response mediated by these proteins is complex and it has been reviewed elsewhere [74].

On the other hand, the expression of the *fla2* genes requires the absence of the Fla1 flagellum, and the activation of a two-component system, formed by the histidine kinase CckA, the phosphotransferase ChpT and the response regulator CtrA [61]. Details of the mechanisms that control CckA activation and the negative control of Fla1 over Fla2 are currently being studied by our group. Nevertheless, when CtrA is phosphorylated by CckA, the expression of the *fla2* genes is turned on. These genes include those within the *fla2* cluster (of approx. 32 kb), *fliM* and *fliG* that are located elsewhere in chromosome I, as well as *flaA* (flagellin), and its regulators *flaF* and *flbT* that are located in plasmid A [76,77]. Recently it was demonstrated that CtrA also activates the expression of the chemotactic operon *cheOp1* that includes three chemotactic receptors i.e., *mcpA*, *mcpB* and *tlpS*, as well as the chemotaxis genes *cheD*, *cheX*, *cheW1*, *cheR1*, *cheY1*, *cheY2* and *cheY5* [73]. It has also been shown that all these components specifically control the chemosensory response of the Fla2 flagella [64]; CtrA also activates other chemotactic receptors [77]; however, it remains to be tested if these receptors affect the chemotactic response mediated by this flagellum.

3. The Hook and Basal Body

Initial characterization of the Fla1 flagellum revealed two prominent features that contrasted from the canonical well-studied flagellum from *E. coli* and *S. enterica.* One of these features was that Fla1 has a straight hook and the second is that it shows a bulky hook-associated-protein (HAP) region [78–80] (Figure 3A). The bulky HAP region correlates with the large molecular mass of FlgK1 with 1363 residues, which is three times larger than its homologue in *S. enterica*. FlgK1 has well-conserved N and C-terminal regions with residues present in orthologous proteins, and a large central non-conserved region of 860 residues that accounts for the large molecular mass of this protein. Discrete deletions of 100 amino acids within this non-conserved region revealed that the complete protein is required for normal swimming since practically all these mutants showed a severe reduction in swimming velocity and jiggling trajectories. Importantly, cells expressing FlgK1 lacking residues 340–440 or 840–940 located in the non-conserved region, produced flagella indistinguishable from the wild-type; nevertheless, the mutant cells were unable to swim in liquid medium, revealing that these non-conserved regions are indeed relevant to handle the load exerted by motor rotation [81]. The presence of at least three flagellin-hook IN motifs (pFam07196) detected with the HMMER software package [82] and at least two internal repeats detected with SMART (Simple Modular Architecture Research Tool) [83,84], suggests that this central region could be the result of several processes of internal duplication. So far, few studies have addressed the relevance of the HAP region and its influence on the correct polymorphic shape of the filament when torque is applied [85].

As mentioned above, another characteristic feature of the Fla1 flagellum is the presence of a straight hook (Figure 3B). Purified flagella showed a straight hook in a wide range of pH values, from 4 to 9 [66]. This is in contrast with other bacteria such as *E. coli*, *S. enterica*, and the α proteobacterium *Magnetospirillum magnetotacticum* that have a curved hook [66]. The *R. sphaeroides* hook protein FlgE1 is 50% similar to FlgE from *S. enterica* ($FlgE_{Se}$), however it has twice as many proline residues than its counterpart $FlgE_{Se}$ (23/423 versus 12/403), and several of these residues are clustered in short regions not found in $FlgE_{Se}$ [86]. According to the structural model defined for $FlgE_{Se}$ [15,87–89], one of these insertions is located in the Dc domain and the other in the D1 domain. A deletion of six residues in one of these regions did not prevent hook assembly but the structure was conspicuously curved (Figure 3C). The swimming trajectories of these cells were wavy instead of the smooth trajectories commonly seen for wild type *R. sphaeroides* cells [86]. This mutation affects the D1 domain that participates in the axial interactions between subunits. Interestingly, it has been recently shown that a short insertion in the Dc domain of $FlgE_{Se}$ made the hook straight. From this study, it was suggested that the Dc domain acts as a structural switch to coordinate axial packing interactions of the D1 domain with the supercoiling of the hook structure [90]. Therefore, these studies concur on the role of the axial packing interactions of D1 domains of the FlgE protein to profoundly affect the final structure of the hook.

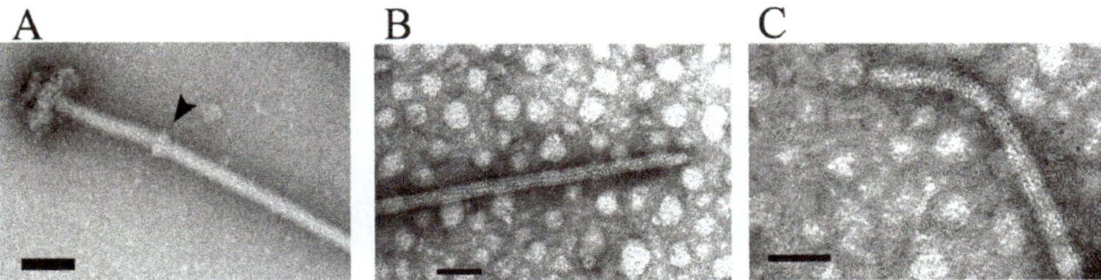

Figure 3. Electron micrographs showing (**A**) wild-type Fla1 filament-hook-basal body [62], arrow denotes the bulky HAP region; (**B**) sheared Fla1 wild-type filament-hook; (**C**) sheared Fla1 filament-hook from a mutant lacking residues 91–96 of FlgE [86]. Bar = 50 nm.

4. Rod Assembly and Opening of the Peptidoglycan Barrier

Another interesting aspect of the basal body is the order in which the different subunits that form the flagellar rod are assembled. Previously, work in *S. enterica*, showed that FliE and FlgB formed the proximal end of the rod; likewise, previous reports indicated that FlgG is the most distal component. However, the order of assembly was not known i.e., if FlgC or FlgF followed after FlgB. Using purified preparations of the five different rod proteins from *R. sphaeroides*, a possible assembly order was recently suggested. In this study, specific interactions between FliE and FlgB, FlgB and FlgF, and between FlgC and FlgG, were detected. From these results, it was proposed that the order of assembly of the rod proteins in *R. sphaeroides* is FliE, FlgB, FlgF, FlgC and FlgG [91]. This order is different to the one proposed for the Gram-positive bacterium *Bacillus subtilis* and the spirochete *Borrelia burgdorferi*, where it was suggested that the rod proteins are assembled in the following order: FliE, FlgB, FlgC, FlhO (FlgF), and FlgG [40,92]. The difference between the order of assembly proposed for *R. sphaeroides* and

B. subtilis or *B. burgdorferi* could be explained by the different experimental approaches used in these studies or possibly due to an actual difference between these organisms in the order of assembly of the rod structure. The limited amount of studies that addresses this issue prevent a comparison with species related to *R. sphaeroides*.

Additional proteins are required during the assembly process of this axial structure. In *Salmonella* a chaperone protein (FlgJ) has a dual function, as a scaffold and also as a muramidase that degrades the peptidoglycan layer to facilitate rod penetration [9]. In contrast, in *R. sphaeroides* FlgJ lacks the muramidase domain but it retains its ability to function as a scaffold for rod assembly [93]. It was also found that a gene in the *flgG* operon codes for a protein that has a signal sequence at its N-terminus followed by a soluble lytic transglycosylase domain, and could act as a muramidase to remodel the peptidoglycan wall [94]. The protein encoded by this gene is indeed a flagellar soluble lytic transglycosylase named SltF that specifically interacts with FlgJ through its C-terminus. SltF is exported to the periplasm by means of the SecA pathway where it encounters the scaffold protein that directs it to the specific site in the peptidoglycan layer that will be remodeled to allow the passage of the rod [94,95]. Given that SltF is exported by the general secretion pathway, it is possible that this protein must be distributed throughout the periplasmic space potentially causing widespread damage. However, it was recently shown that the enzymatic activity of SltF is modulated by the interaction of the different rod proteins. It is stimulated by the flagellar rod protein FlgB, and inhibited by FlgF [96].

5. The Flagellar Motor of *R. sphaeroides*

In the absence of a chemical gradient *R. sphaeroides* swims following a random pattern of runs and stops. During the run periods the Fla1 flagellum rotates unidirectionally in the clockwise direction and uses H^+ as the coupling ion. When rotation stops, the filament coils up against the cell body, and the swimming trajectory changes [51,97]. Biochemical and genetic studies of the flagellar motor have revealed that, apart from the core structure characterized in *E. coli* and *S. enterica*, other accessory components form part of this flagellum. In this regard it has been shown that in Fla1, proteins homologous to FlgT, FlgP, and MotF (a protein of restricted distribution in some species of the family *Rhodobacteraceae*), are part of this structure (Figure 4).

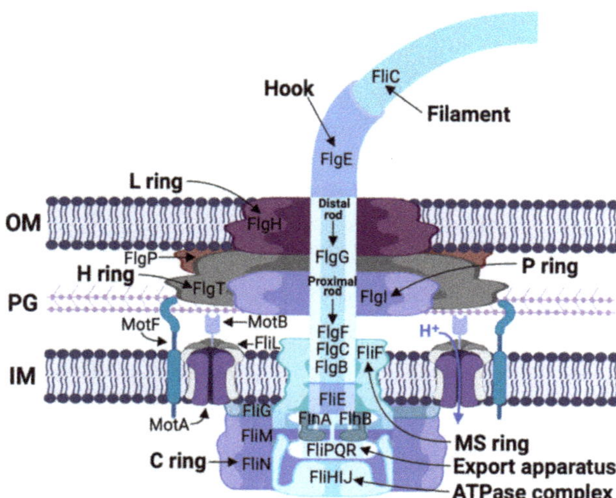

Figure 4. Schematic drawing of the *R. sphaeroides* flagellar motor. The model is based on the electron microscopic analysis of isolated flagella of *R. sphaeroides*, as well as inferences based on protein-protein interaction analysis and in situ visualization of the flagellar structure of *Vibrio*. The name of the different components and proteins that form this structure are indicated. This figure was created with BioRender.com (website: https://biorender.com/).

FlgT is a periplasmic protein exported by the general secretion pathway. It forms the H-ring that surrounds the PL-rings and it is widely distributed in several species of *Vibrio*, *Aeromonas*, *Pseudoalteromonas* and also several species of the family *Rhodobacteraceae* [98–100]. We have demonstrated that FlgT from *R. sphaeroides* forms a characteristic H-ring (Figure 5), and that this protein, apart from interacting with itself, interacts with FlgH that forms the L-ring of the flagellar core structure, where this interaction would assist to anchor the H-ring to the basal body [56,57]. However, in contrast with the situation observed in *V. alginolyticus*, and *V. cholerae* where the absence of FlgT affects flagellar assembly, as well as the penetration of the outer membrane [98,100,101]; in *R. sphaeroides* the absence of FlgT results in a Mot⁻ phenotype [56]. This indicates that in this bacterium the function of the H-ring is mainly associated with torque generation and motility and not with flagellar assembly. Although its role may not be direct, as discussed below, since FlgT interacts with other proteins that are directly related with torque generation.

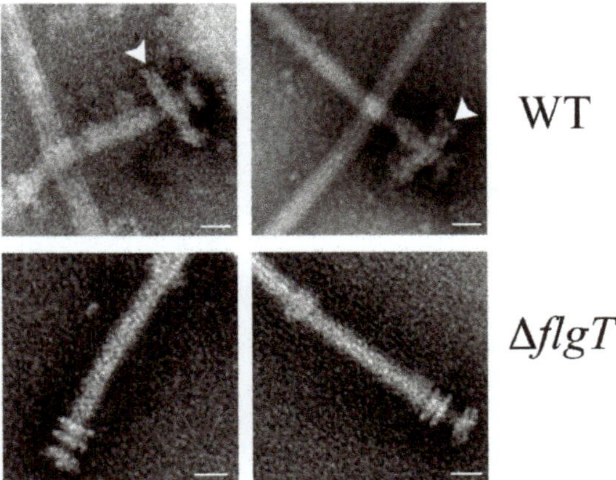

Figure 5. Electron micrographs showing isolated filament-hook-basal bodies from wild type cells expressing Fla1 and from a mutant lacking FlgT. White arrows denote the H-ring Bar = 20 nm [56].

The flagellar motor of *R. sphaeroides* also includes the protein FlgP that is an outer membrane lipoprotein essential for flagellum formation [57]. In this work, we observed that FlgP interacts with itself suggesting that it could form an oligomeric structure. It was also proposed that FlgP could form the basal ring that is located under the outer membrane, as it has been previously observed in *C. jejuni* and *V. fisheri* [102]. FlgP also interacts with FlgT and FlgH, these interactions would be an additional support for the formation of the basal disk [57]. Nevertheless, in the absence of FlgT, FlgP should be included given that the flagellar structure is formed; whereas in the absence of FlgP the flagellum is not assembled. More precisely, in the absence of FlgP the flagellar hook is not assembled, even though the hook protein FlgE is present in the cytoplasm. Hence, the anti-sigma factor FlgM is not exported from the cell, and the flagellar genes dependent on σ^{28}, such as those encoding for flagellin and other chemotactic proteins, are not expressed. In contrast, in Δ*flgP* mutants the flagellar rod is assembled; therefore, it was proposed that FlgP is required for a proper rod to hook transition [57]. Since, the L-ring assembly has also been related with this process it can be presumed that in *R. sphaeroides* the L-ring could be remodeled by the basal disk. It should be noted that in *V. alginolyticus* it has been proposed that FlgP forms the middle part of the H-ring [101]; however, given the different phenotypes associated with the loss of FlgT and FlgP, we chose to name the structures as H-ring and basal disk respectively, as it was proposed for *V. fischeri* [102].

A flagellar gene named *motF* was identified in *R. sphaeroides* and it is present in some species of the family *Rhodobacteraceae*. MotF is a 24 kDa protein that has a transmembrane region spanning from residue 54 to 74 and a large periplasmic C-terminal portion. It was shown that a proper localization of a fluorescent version of this protein is dependent on the presence of an activated proton channel, given that in the absence of MotA/B, or FliL, GFP-MotF forms several fluorescent foci per cell instead of the single one observed in wild-type cells. The presence of several different populations of GFP-MotF in these mutants could be caused by a weak association of GFP-MotF with the flagellar structure when the stator complexes are not present or activated [103].

Remarkably, Δ*motF*cells recovered the swimming ability by a secondary mutation in the amphipathic helix of MotB localized after the transmembrane segment of this protein [103]. This region known as the plug, has been proposed to prevent proton flow before the MotA/MotB complex associates with the flagellar structure [104]. In addition, eight extragenic suppressors of the Mot$^-$ phenotype caused by the absence of FliL also affected this specific region of MotB [34]. Surprisingly, all these *motB* mutant alleles were also able to suppress the Mot$^-$ phenotype of Δ*motF* [103] (Figure 6). Therefore, it is strongly suggested that FliL and MotF are implicated in remodeling the C-terminus of MotB and hence promote the activation of the proton channel. If the hydrophobicity of the amphipathic helix of the plug is reduced, as occurs in the suppressor mutants, the presence of FliL and MotF is dispensable for flagellar rotation.

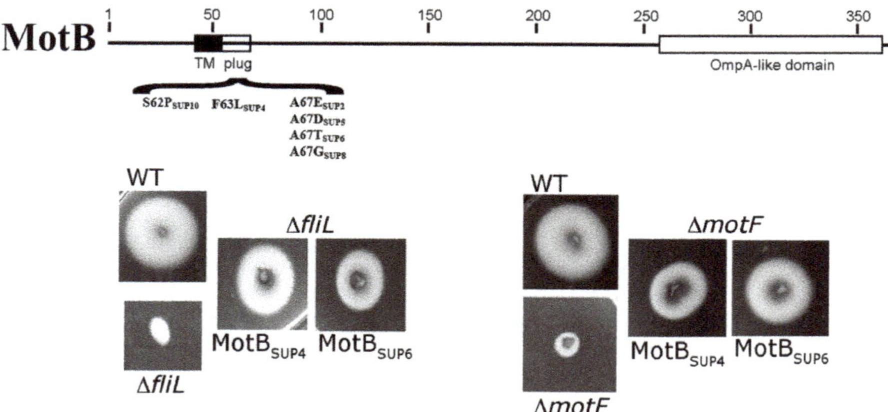

Figure 6. Scheme of the plug region of MotB showing the mutant alleles that suppress the Mot$^-$ phenotype of Δ*fliL* and Δ*motF*. Also shown are the swimming phenotypes on soft agar of the various strains and suppressor mutants [34,56,103].

It was observed that the H-ring (FlgT) is necessary for recruitment of GFP-MotF in the flagellar motor [56]; therefore, the role of FlgT on flagellar rotation could be indirect. In accordance with this possibility we detected that FlgT interacts with FliL and MotF, indicating that the H-ring could act as a hub to recruit or stabilize these proteins that are directly involved in the activation of the proton channel. However, FlgT also interacts with MotB and the mutants in MotB that act as secondary suppressors of Δ*fliL* and Δ*motF*, barely improve swimming of the Δ*flgT* strain, indicating that the H-ring could also participate in the recruitment of the stator complexes [56]. In this context it is important to mention that in *R. sphaeroides* there are no homologues of *motX* and *motY* whose products have been proposed to form the T-ring in *V. alginolyticus* that contributes to recruit the PomA/PomB (equivalent to MotA/B in *Vibrio*) complexes to the flagellar structure [25,101]. Therefore, FlgT could have possibly evolved in order to gain this role in *R. sphaeroides*.

6. Dominance of Fla1 over Fla2

An interesting question regarding the coexistence of the two flagellar systems in this free-living bacterium, is if there is cross-regulation between the two. Under the growth conditions commonly used in the laboratory, Fla2 flagella have never been observed, suggesting that something in the culture medium could favor the expression of the *fla1* genes and repress *fla2* expression. Routinely, the growth conditions used for the enrichment and isolation of phototrophic bacteria involve the use of organic acids as electron donors in the growth medium, under photoheterotrophic conditions. Therefore, these compounds were obvious candidates to be tested. We have recently demonstrated that the expression of the *cckA* and *ctrA* genes, that encode the histidine kinase and the response regulator of the two-component system that activates the expression of the *fla2* genes, is repressed when a high concentration (34 mM) of C4-dicarboxilic acids is present in the culture medium. However, the growth of the wild-type strain (WS8N) in the absence of C4-dicarboxilic acids is not sufficient to induce activation of the CckA/ChpT/CtrA two-component system and therefore the *fla2* genes are not transcribed. We have speculated that CckA should be activated by additional specific environmental conditions that remain to be understood. Nevertheless, it was observed that the cells carrying a mutation in the master regulator *fleQ* could acquire a gain of function mutation in CckA that allows the expression of the *fla2* genes. As a result, a homogeneous population of bacteria able to swim with the Fla2 flagella was obtained; however, when these cells were complemented with a plasmid expressing FleQ, most of the cells stopped synthesizing Fla2 flagella and the number of cells expressing Fla1 flagella increased sharply. Remarkably, cells carrying both types of flagella were never detected. This suggests that, under laboratory conditions, there is a clear dominance of the Fla1 system over Fla2 mediated by an unknown molecular mechanism [61]. The elucidation of this regulatory mechanism would shed light on how a complete set of foster genes were acquired and stably incorporated into the regulatory circuit that controls motility in this organism and also would reveal details of the evolutionary success of this resourceful bacterium.

7. Future Directions

Given that the *fla1* system was laterally acquired, it is important to elucidate the molecular mechanisms controlling biogenesis and rotation of this structure with particular emphasis on FlgP and FlgT, which are not present in the vertically inherited flagellar genes of several α-proteobacteria so far characterized. It is apparent that these proteins do not accomplish the same function that their homologues in *Vibrio*, suggesting that they may have evolved differently. In this context, it would be relevant to look deeper into the molecular role of the family *Rhodobacteraceae*-specific MotF protein and test its possible role as a stabilization element of the stator complexes (MotA/MotB).

Regarding flagellar biogenesis, it will be important to determine the molecular mechanisms underlying the localization and control of the activity of the soluble lytic transglycosylase (SltF) that is important, not only to understand, how a flagellar-specific lytic enzyme is controlled, but also to determine if these mechanisms are conserved in other type III secretion systems.

The in situ analysis of the structure of the Fla1 motor by combining cryo-EM and genetic studies is an important pending assignment to identify the hypothetic basal ring and the localization of MotF.

The elucidation of the genetic mechanisms that control the communication between the *fla1* and *fla2* genetic systems that results in a mutually exclusive expression, is of particular relevance.

Funding: This work was partially supported by DGAPA-UNAM (PAPIIT-IG200420).

Acknowledgments: We acknowledge Aurora Osorio, Teresa Ballado and Javier de la Mora, for technical support and David Rodriguez Méndez for the preparation of Figure 4.

Conflicts of Interest: The authors declare no conflicts of interest.

References

1. Nakamura, S.; Minamino, T. Flagella-Driven Motility of Bacteria. *Biomolecules* **2019**, *9*, 279. [CrossRef] [PubMed]
2. Macnab, R.M. Type III flagellar protein export and flagellar assembly. *Biochim. Biophys. Acta (BBA) Bioenerg.* **2004**, *1694*, 207–217. [CrossRef] [PubMed]

3. Kuhlen, L.; Abrusci, P.; Johnson, S.; Gault, J.; Deme, J.; Caesar, J.; Dietsche, T.; Mebrhatu, M.T.; Ganief, T.; Macek, B.; et al. Structure of the Core of the Type Three Secretion System Export Apparatus. *Nat. Struct. Mol. Biol.* **2018**, *25*, 583–590. [CrossRef] [PubMed]
4. Fukumura, T.; Makino, F.; Dietsche, T.; Kinoshita, M.; Kato, T.; Wagner, S.; Namba, K.; Imada, K.; Minamino, T. Assembly and stoichiometry of the core structure of the bacterial flagellar type III export gate complex. *PLoS Biol.* **2017**, *15*, e2002281. [CrossRef]
5. Zhao, X.; Norris, S.J.; Liu, J. Molecular Architecture of the Bacterial Flagellar Motor in Cells. *Biochemistry* **2014**, *53*, 4323–4333. [CrossRef]
6. Homma, M.; DeRosier, D.J.; Macnab, R.M. Flagellar hook and hook-associated proteins of *Salmonella typhimurium* and their relationship to other axial components of the flagellum. *J. Mol. Biol.* **1990**, *213*, 819–832. [CrossRef]
7. Jones, C.J.; Homma, M.; Macnab, R.M. Identification of proteins of the outer (L and P) rings of the flagellar basal body of *Escherichia coli*. *J. Bacteriol.* **1987**, *169*, 1489–1492. [CrossRef]
8. Akiba, T.; Yoshimura, H.; Namba, K. Monolayer crystallization of flagellar L-P rings by sequential addition and depletion of lipid. *Science* **1991**, *252*, 1544–1546. [CrossRef]
9. Hirano, T.; Minamino, T.; Macnab, R.M. The role in flagellar rod assembly of the N-terminal domain of *Salmonella* FlgJ, a flagellum-specific muramidase. *J. Mol. Biol.* **2001**, *312*, 359–369. [CrossRef]
10. Herlihey, F.A.; Moynihan, P.; Clarke, A.J. The Essential Protein for Bacterial Flagella Formation FlgJ Functions as a β-N-Acetylglucosaminidase. *J. Biol. Chem.* **2014**, *289*, 31029–31042. [CrossRef]
11. Zaloba, P.; Bailey-Elkin, B.A.; Derksen, M.; Mark, B.L. Structural and Biochemical Insights into the Peptidoglycan Hydrolase Domain of FlgJ from *Salmonella typhimurium*. *PLoS ONE* **2016**, *11*, e0149204. [CrossRef] [PubMed]
12. Nambu, T.; Minamino, T.; Macnab, R.M.; Kutsukake, K. Peptidoglycan-Hydrolyzing Activity of the FlgJ Protein, Essential for Flagellar Rod Formation in *Salmonella typhimurium*. *J. Bacteriol.* **1999**, *181*, 1555–1561. [CrossRef] [PubMed]
13. Matsunami, H.; Barker, C.S.; Yoon, Y.-H.; Wolf, M.; Samatey, F. Complete structure of the bacterial flagellar hook reveals extensive set of stabilizing interactions. *Nat. Commun.* **2016**, *7*, 13425. [CrossRef] [PubMed]
14. Shibata, S.; Matsunami, H.; Aizawa, S.-I.; Wolf, M. Torque transmission mechanism of the curved bacterial flagellar hook revealed by cryo-EM. *Nat. Struct. Mol. Biol.* **2019**, *26*, 941–945. [CrossRef]
15. Samatey, F.; Matsunami, H.; Imada, K.; Nagashima, S.; Shaikh, T.R.; Thomas, D.R.; Chen, J.Z.; DeRosier, D.J.; Kitao, A.; Namba, K. Structure of the bacterial flagellar hook and implication for the molecular universal joint mechanism. *Nature* **2004**, *431*, 1062–1068. [CrossRef]
16. Kato, T.; Makino, F.; Miyata, T.; Horváth, P.; Namba, K. Structure of the native supercoiled flagellar hook as a universal joint. *Nat. Commun.* **2019**, *10*, 5295–5298. [CrossRef]
17. Hong, H.J.; Kim, T.H.; Song, W.S.; Ko, H.-J.; Lee, G.-S.; Kang, S.G.; Kim, P.-H.; Yoon, S.-I. Crystal structure of FlgL and its implications for flagellar assembly. *Sci. Rep.* **2018**, *8*, 14307. [CrossRef]
18. Minamino, T.; Imada, K. The bacterial flagellar motor and its structural diversity. *Trends Microbiol.* **2015**, *23*, 267–274. [CrossRef]
19. Minamino, T.; Terahara, N.; Kojima, S.; Namba, K. Autonomous control mechanism of stator assembly in the bacterial flagellar motor in response to changes in the environment. *Mol. Microbiol.* **2018**, *109*, 723–734. [CrossRef]
20. Paul, K.; Gonzalez-Bonet, G.; Bilwes, A.M.; Crane, B.R.; Blair, D.F. Architecture of the flagellar rotor. *EMBO J.* **2011**, *30*, 2962–2971. [CrossRef]
21. Morimoto, Y.V.; Minamino, T. Structure and Function of the Bi-Directional Bacterial Flagellar Motor. *Biomolecules* **2014**, *4*, 217–234. [CrossRef] [PubMed]
22. Zhou, J.; Lloyd, S.A.; Blair, D.F. Electrostatic interactions between rotor and stator in the bacterial flagellar motor. *Proc. Natl. Acad. Sci. USA* **1998**, *95*, 6436–6441. [CrossRef] [PubMed]
23. Morimoto, Y.V.; Nakamura, S.; Kami-Ike, N.; Namba, K.; Minamino, T. Charged residues in the cytoplasmic loop of MotA are required for stator assembly into the bacterial flagellar motor. *Mol. Microbiol.* **2010**, *78*, 1117–1129. [CrossRef] [PubMed]
24. Morimoto, Y.V.; Nakamura, S.; Hiraoka, K.D.; Namba, K.; Minamino, T. Distinct Roles of Highly Conserved Charged Residues at the MotA-FliG Interface in Bacterial Flagellar Motor Rotation. *J. Bacteriol.* **2012**, *195*, 474–481. [CrossRef]

25. Terashima, H.; Fukuoka, H.; Yakushi, T.; Kojima, S.; Homma, M. The *Vibrio* motor proteins, MotX and MotY, are associated with the basal body of Na+-driven flagella and required for stator formation. *Mol. Microbiol.* **2006**, *62*, 1170–1180. [CrossRef]
26. Zhu, S.; Nishikino, T.; Takekawa, N.; Terashima, H.; Kojima, S.; Imada, K.; Homma, M.; Liu, J. In Situ Structure of the *Vibrio* Polar Flagellum Reveals a Distinct Outer Membrane Complex and Its Specific Interaction with the Stator. *J. Bacteriol.* **2020**, *202*. [CrossRef]
27. Kojima, S.; Takao, M.; Almira, G.; Kawahara, I.; Sakuma, M.; Homma, M.; Kojima, C.; Imada, K. The Helix Rearrangement in the Periplasmic Domain of the Flagellar Stator B Subunit Activates Peptidoglycan Binding and Ion Influx. *Structure* **2018**, *26*, 590–598.e5. [CrossRef]
28. Kojima, S.; Imada, K.; Sakuma, M.; Sudo, Y.; Kojima, C.; Minamino, T.; Homma, M.; Namba, K. Stator assembly and activation mechanism of the flagellar motor by the periplasmic region of MotB. *Mol. Microbiol.* **2009**, *73*, 710–718. [CrossRef]
29. Zhu, S.; Takao, M.; Li, N.; Sakuma, M.; Nishino, Y.; Homma, M.; Kojima, S.; Imada, K. Conformational change in the periplamic region of the flagellar stator coupled with the assembly around the rotor. *Proc. Natl. Acad. Sci. USA* **2014**, *111*, 13523–13528. [CrossRef]
30. O'Neill, J.; Xie, M.; Hijnen, M.; Roujeinikova, A. Role of the MotB linker in the assembly and activation of the bacterial flagellar motor. *Acta Crystallogr. Sect. D Biol. Crystallogr.* **2011**, *67*, 1009–1016. [CrossRef]
31. Reboul, C.F.; Andrews, D.A.; Nahar, M.F.; Buckle, A.M.; Roujeinikova, A. Crystallographic and Molecular Dynamics Analysis of Loop Motions Unmasking the Peptidoglycan-Binding Site in Stator Protein MotB of Flagellar Motor. *PLoS ONE* **2011**, *6*, e18981. [CrossRef] [PubMed]
32. Andrews, D.A.; Nesmelov, Y.E.; Wilce, J.A.; Roujeinikova, A. Structural analysis of variant of *Helicobacter pylori* MotB in its activated form, engineered as chimera of MotB and leucine zipper. *Sci. Rep.* **2017**, *7*, 13435. [CrossRef] [PubMed]
33. Partridge, J.D.; Nieto, V.; Harshey, R.M. A New Player at the Flagellar Motor: FliL Controls both Motor Output and Bias. *mBio* **2015**, *6*. [CrossRef] [PubMed]
34. Suaste-Olmos, F.; Domenzain, C.; Mireles-Rodríguez, J.C.; Poggio, S.; Osorio, A.; Dreyfus, G.; Camarena, L. The Flagellar Protein FliL Is Essential for Swimming in *Rhodobacter sphaeroides*. *J. Bacteriol.* **2010**, *192*, 6230–6239. [CrossRef] [PubMed]
35. Zhu, S.; Kumar, A.; Kojima, S.; Homma, M. FliL associates with the stator to support torque generation of the sodium-driven polar flagellar motor of *Vibrio*. *Mol. Microbiol.* **2015**, *98*, 101–110. [CrossRef]
36. Subramanian, S.; Kearns, D.B. Functional Regulators of Bacterial Flagella. *Annu. Rev. Microbiol.* **2019**, *73*, 225–246. [CrossRef]
37. Takekawa, N.; Isumi, M.; Terashima, H.; Zhu, S.; Nishino, Y.; Sakuma, M.; Kojima, S.; Homma, M.; Imada, K. Structure of *Vibrio* FliL, a New Stomatin-like Protein That Assists the Bacterial Flagellar Motor Function. *mBio* **2019**, *10*, e00292-19. [CrossRef]
38. Liu, J.; Howell, J.K.; Bradley, S.D.; Zheng, Y.; Zhou, Z.H.; Norris, S.J. Cellular Architecture of *Treponema pallidum*: Novel Flagellum, Periplasmic Cone, and Cell Envelope as Revealed by Cryo Electron Tomography. *J. Mol. Biol.* **2010**, *403*, 546–561. [CrossRef]
39. Chen, S.; Beeby, M.E.; Murphy, G.; Leadbetter, J.R.; Hendrixson, D.R.; Briegel, A.; Li, Z.; Shi, J.I.; Tocheva, E.; Müller, A.; et al. Structural diversity of bacterial flagellar motors. *EMBO J.* **2011**, *30*, 2972–2981. [CrossRef]
40. Zhao, X.; Zhang, K.; Boquoi, T.; Hu, B.; Motaleb, A.; Miller, K.A.; James, M.E.; Charon, N.W.; Manson, M.D.; Norris, S.J.; et al. Cryoelectron tomography reveals the sequential assembly of bacterial flagella in *Borrelia burgdorferi*. *Proc. Natl. Acad. Sci. USA* **2013**, *110*, 14390–14395. [CrossRef]
41. Zhu, S.; Nishikino, T.; Hu, B.; Kojima, S.; Homma, M.; Liu, J. Molecular architecture of the sheathed polar flagellum in *Vibrio alginolyticus*. *Proc. Natl. Acad. Sci. USA* **2017**, *114*, 10966–10971. [CrossRef] [PubMed]
42. Kaplan, M.; Ghosal, D.; Subramanian, P.; Oikonomou, C.M.; Kjær, A.; Pirbadian, S.; Ortega, D.R.; Briegel, A.; El-Naggar, M.Y.; Jensen, G.J. The presence and absence of periplasmic rings in bacterial flagellar motors correlates with stator type. *eLife* **2019**, *8*. [CrossRef] [PubMed]
43. Kaplan, M.; Subramanian, P.; Ghosal, D.; Oikonomou, C.M.; Pirbadian, S.; Starwalt-Lee, R.; Mageswaran, S.K.; Ortega, D.R.A.; Gralnick, J.; El-Naggar, M.Y.; et al. In situ imaging of the bacterial flagellar motor disassembly and assembly processes. *EMBO J.* **2019**, *38*, e100957. [CrossRef] [PubMed]
44. Zhu, S.; Schniederberend, M.; Zhitnitsky, D.; Jain, R.; Galán, J.E.; Kazmierczak, B.I.; Liu, J. In Situ Structures of Polar and Lateral Flagella Revealed by Cryo-Electron Tomography. *J. Bacteriol.* **2019**, *201*. [CrossRef] [PubMed]

45. Imhoff. *The Phototrophic Alpha Proteobacteria*, 3rd ed.; Dworkin, M.F.S., Rosenberg, E., Schleifer, K.-H., Stackebrandt, E., Eds.; Springer-Verlag: New York, NY, USA, 2006; Volume 5.
46. MacKenzie, C.; Choudhary, M.; Larimer, F.W.; Predki, P.F.; Stilwagen, S.; Armitage, J.P.; Barber, R.D.; Donohue, T.J.; Hosler, J.P.; Newman, J.E.; et al. The home stretch, a first analysis of the nearly completed genome of *Rhodobacter sphaeroides* 2.4.1. *Photosynth. Res.* **2001**, *70*, 19–41. [CrossRef]
47. Porter, S.L.; Wilkinson, D.A.; Byles, E.D.; Wadhams, G.H.; Taylor, S.; Saunders, N.; Armitage, J.P. Genome Sequence of *Rhodobacter sphaeroides* Strain WS8N. *J. Bacteriol.* **2011**, *193*, 4027–4028. [CrossRef]
48. Lim, S.-K.; Kim, S.J.; Cha, S.H.; Oh, Y.-K.; Rhee, H.-J.; Kim, M.-S.; Lee, J.K. Complete Genome Sequence of *Rhodobacter sphaeroides* KD131. *J. Bacteriol.* **2008**, *191*, 1118–1119. [CrossRef]
49. Harrison, P.W.; Lower, R.P.; Kim, N.K.; Young, J.P.W. Introducing the bacterial 'chromid': Not a chromosome, not a plasmid. *Trends Microbiol.* **2010**, *18*, 141–148. [CrossRef]
50. DiCenzo, G.C.; Finan, T.M. The Divided Bacterial Genome: Structure, Function, and Evolution. *Microbiol. Mol. Biol. Rev.* **2017**, *81*, 3. [CrossRef]
51. Armitage, J.P.; Macnab, R.M. Unidirectional, intermittent rotation of the flagellum of *Rhodobacter sphaeroides*. *J. Bacteriol.* **1987**, *169*, 514–518. [CrossRef]
52. Youle, M.; Rohwer, F.; Stacy, A.; Whiteley, M.; Steel, B.C.; Delalez, N.J.; Nord, A.; Berry, R.M.; Armitage, J.P.; Kamoun, S.; et al. The Microbial Olympics. *Nat. Rev. Microbiol.* **2012**, *10*, 583–588. [CrossRef] [PubMed]
53. Packer, H.; Harrison, D.; Dixon, R.; Armitage, J. The effect of pH on the growth and motility of *Rhodobacter sphaeroides* WS8 and the nature of the driving force of the flagellar motor. *Biochim. Biophys. Acta (BBA) Bioenerg.* **1994**, *1188*, 101–107. [CrossRef]
54. Pilizota, T.; Brown, M.T.; Leake, M.C.; Branch, R.W.; Berry, R.M.; Armitage, J.P. A molecular brake, not a clutch, stops the *Rhodobacter sphaeroides* flagellar motor. *Proc. Natl. Acad. Sci. USA* **2009**, *106*, 11582–11587. [CrossRef] [PubMed]
55. Poggio, S.; Osorio, A.; Dreyfus, G.; Camarena, L. The flagellar hierarchy of *Rhodobacter sphaeroides* is controlled by the concerted action of two enhancer-binding proteins. *Mol. Microbiol.* **2005**, *58*, 969–983. [CrossRef] [PubMed]
56. Fabela, S.; Domenzain, C.; De La Mora, J.; Osorio, A.; Ramirez-Cabrera, V.; Poggio, S.; Dreyfus, G.; Camarena, L. A Distant Homologue of the FlgT Protein Interacts with MotB and FliL and Is Essential for Flagellar Rotation in *Rhodobacter sphaeroides*. *J. Bacteriol.* **2013**, *195*, 5285–5296. [CrossRef] [PubMed]
57. Pérez-González, C.; Domenzain, C.; Poggio, S.; Gonzalez-Halphen, D.; Dreyfus, G.; Camarena, L. Characterization of FlgP, an Essential Protein for Flagellar Assembly in *Rhodobacter sphaeroides*. *J. Bacteriol.* **2018**, *201*. [CrossRef]
58. Arai, H.; Roh, J.H.; Kaplan, S. Transcriptome Dynamics during the Transition from Anaerobic Photosynthesis to Aerobic Respiration in *Rhodobacter sphaeroides* 2.4.1. *J. Bacteriol.* **2007**, *190*, 286–299. [CrossRef]
59. Roh, J.H.; Smith, W.E.; Kaplan, S. Effects of Oxygen and Light Intensity on Transcriptome Expression in *Rhodobacter sphaeroides* 2.4.1. *J. Biol. Chem.* **2003**, *279*, 9146–9155. [CrossRef]
60. Poggio, S.; Abreu-Goodger, C.; Fabela, S.; Osorio, A.; Dreyfus, G.; Vinuesa, P.; Camarena, L. A Complete Set of Flagellar Genes Acquired by Horizontal Transfer Coexists with the Endogenous Flagellar System in *Rhodobacter sphaeroides*. *J. Bacteriol.* **2007**, *189*, 3208–3216. [CrossRef]
61. Vega-Baray, B.; Domenzain, C.; Rivera, A.; Alfaro-López, R.; Gómez-César, E.; Poggio, S.; Dreyfus, G.; Camarena, L. The Flagellar Set Fla2 in *Rhodobacter sphaeroides* Is Controlled by the CckA Pathway and Is Repressed by Organic Acids and the Expression of Fla1. *J. Bacteriol.* **2015**, *197*, 833–847. [CrossRef]
62. De La Mora, J.; Uchida, K.; Del Campo, A.M.; Camarena, L.; Aizawa, S.-I.; Dreyfus, G. Structural Characterization of the Fla2 Flagellum of *Rhodobacter sphaeroides*. *J. Bacteriol.* **2015**, *197*, 2859–2866. [CrossRef] [PubMed]
63. Del Campo, A.M.; Ballado, T.; De La Mora, J.; Poggio, S.; Camarena, L.; Dreyfus, G. Chemotactic Control of the Two Flagellar Systems of *Rhodobacter sphaeroides* Is Mediated by Different Sets of CheY and FliM Proteins. *J. Bacteriol.* **2007**, *189*, 8397–8401. [CrossRef] [PubMed]
64. Campo, A.M.-D.; Ballado, T.; Camarena, L.; Dreyfus, G. In *Rhodobacter sphaeroides*, Chemotactic Operon 1 Regulates Rotation of the Flagellar System 2. *J. Bacteriol.* **2011**, *193*, 6781–6786. [CrossRef] [PubMed]
65. Shah, D.; Porter, S.L.; Martin, A.C.; Hamblin, P.A.; Armitage, J.P. Fine tuning bacterial chemotaxis: Analysis of *Rhodobacter sphaeroides* behaviour under aerobic and anaerobic conditions by mutation of the major chemotaxis operons and *cheY* genes. *EMBO J.* **2000**, *19*, 4601–4613. [CrossRef]
66. Aizawa, S.I. *The Flagellar World: Electron Microscopic Images of Bacterial flagella and Related Surface Structures from More Than 30 Species*; Elsevier Inc.: Amsterdam, The Netherlands, 2014.

67. Poggio, S.; Aguilar, C.; Osorio, A.; Gonzalez-Pedrajo, B.; Dreyfus, G.; Camarena, L. Sigma(54) promoters control expression of genes encoding the hook and basal body complex in *Rhodobacter sphaeroides*. *J. Bacteriol.* **2000**, *182*, 5787–5792. [CrossRef]
68. Poggio, S.; Osorio, A.; Dreyfus, G.; Camarena, L. The four different σ^{54} factors of *Rhodobacter sphaeroides* are not functionally interchangeable. *Mol. Microbiol.* **2002**, *46*, 75–85. [CrossRef]
69. Domenzain, C.; Camarena, L.; Osorio, A.; Dreyfus, G.; Poggio, S. Evolutionary origin of the *Rhodobacter sphaeroides* specialized RpoN sigma factors. *FEMS Microbiol. Lett.* **2011**, *327*, 93–102. [CrossRef]
70. Gao, F.; Danson, A.E.; Ye, F.; Jovanovic, M.; Buck, M.; Zhang, X. Bacterial Enhancer Binding Proteins—AAA+ Proteins in Transcription Activation. *Biomolecules* **2020**, *10*, 351. [CrossRef]
71. Zhang, N.; Buck, M. A Perspective on the Enhancer Dependent Bacterial RNA Polymerase. *Biomolecules* **2015**, *5*, 1012–1019. [CrossRef]
72. Peña-Sánchez, J.; Poggio, S.; Flores-Pérez, Ú.; Osorio, A.; Domenzain, C.; Dreyfus, G.; Camarena, L. Identification of the binding site of the σ 54 hetero-oligomeric FleQ/FleT activator in the flagellar promoters of *Rhodobacter sphaeroides*. *Microbiology* **2009**, *155*, 1669–1679. [CrossRef]
73. Hernandez-Valle, J.; Domenzain, C.; De La Mora, J.; Poggio, S.; Dreyfus, G.; Camarena, L. The Master Regulators of the Fla1 and Fla2 Flagella of *Rhodobacter sphaeroides* Control the Expression of Their Cognate CheY Proteins. *J. Bacteriol.* **2016**, *199*, e00670-16. [CrossRef] [PubMed]
74. Porter, S.L.; Wadhams, G.H.; Armitage, J.P. Signal processing in complex chemotaxis pathways. *Nat. Rev. Microbiol.* **2011**, *9*, 153–165. [CrossRef] [PubMed]
75. Martin, A.C.; Gould, M.; Byles, E.; Roberts, M.; Armitage, J.P. Two Chemosensory Operons of *Rhodobacter sphaeroides* Are Regulated Independently by Sigma 28 and Sigma 54. *J. Bacteriol.* **2006**, *188*, 7932–7940. [CrossRef] [PubMed]
76. Rivera-Osorio, A.; Osorio, A.; Poggio, S.; Dreyfus, G.; Camarena, L. Architecture of divergent flagellar promoters controlled by CtrA in *Rhodobacter sphaeroides*. *BMC Microbiol.* **2018**, *18*, 129. [CrossRef] [PubMed]
77. Hernández-Valle, J.; Sanchez-Flores, A.; Poggio, S.; Dreyfus, G.; Camarena, L. The CtrA Regulon of *Rhodobacter sphaeroides* Favors Adaptation to a Particular Lifestyle. *J. Bacteriol.* **2020**, *202*, JB.00678-19. [CrossRef]
78. Sockett, R.E.; Foster, J.C.A.; Armitage, J.P. Molecular Biology of the *Rhodobacter Sphaeroides* Flagellum. *Mol. Biol. Membr. Bound Complexes Phototrophic Bact.* **1990**, *53*, 473–478. [CrossRef]
79. González-Pedrajo, B.; Ballado, T.; Campos, A.; E Sockett, R.; Camarena, L.; Dreyfus, G. Structural and genetic analysis of a mutant of *Rhodobacter sphaeroides* WS8 deficient in hook length control. *J. Bacteriol.* **1997**, *179*, 6581–6588. [CrossRef]
80. West, M.A.; Dreyfus, G. Isolation and Ultrastructural Study of the Flagellar Basal Body Complex from *Rhodobacter sphaeroides*WS8 (Wild Type) and a Polyhook Mutant PG. *Biochem. Biophys. Res. Commun.* **1997**, *238*, 733–737. [CrossRef]
81. Castillo, D.J.; Ballado, T.; Camarena, L.; Dreyfus, G. Functional analysis of a large non-conserved region of FlgK (HAP1) from *Rhodobacter sphaeroides*. *Antonie Van Leeuwenhoek* **2009**, *95*, 77–90. [CrossRef]
82. Eddy, S.R. Multiple alignment using hidden Markov models. In Proceedings of the Third International Conference on Intelligent Systems for Molecular Biology, Cambridge, UK, 16–19 July 1995; Volume 3, pp. 114–120, ISBN 0-929280-83-0.
83. Schultz, J.; Milpetz, F.; Bork, P.; Ponting, C.P. SMART, a simple modular architecture research tool: Identification of signaling domains. *Proc. Natl. Acad. Sci. USA* **1998**, *95*, 5857–5864. [CrossRef]
84. Letunic, I.; Doerks, T.; Bork, P. SMART: Recent updates, new developments and status in 2015. *Nucleic Acids Res.* **2014**, *43*, D257–D260. [CrossRef]
85. Fahrner, K.A.; Block, S.M.; Krishnaswamy, S.; Parkinson, J.S.; Berg, H.C. A Mutant Hook-associated Protein (HAP3) Facilitates Torsionally Induced Transformations of the Flagellar Filament of *Escherichia coli*. *J. Mol. Biol.* **1994**, *238*, 173–186. [CrossRef] [PubMed]
86. Ballado, T.; Camarena, L.; González-Pedrajo, B.; Silva-Herzog, E.; Dreyfus, G. The Hook Gene (*flgE*) Is Expressed from the *flgBCDEF* Operon in *Rhodobacter sphaeroides*: Study of an *flgE* Mutant. *J. Bacteriol.* **2001**, *183*, 1680–1687. [CrossRef] [PubMed]
87. Fujii, T.; Kato, T.; Namba, K. Specific Arrangement of α-Helical Coiled Coils in the Core Domain of the Bacterial Flagellar Hook for the Universal Joint Function. *Structure* **2009**, *17*, 1485–1493. [CrossRef] [PubMed]

88. Shaikh, T.R.; Thomas, D.R.; Chen, J.Z.; Samatey, F.; Matsunami, H.; Imada, K.; Namba, K.; DeRosier, D.J. A partial atomic structure for the flagellar hook of *Salmonella typhimurium*. *Proc. Natl. Acad. Sci. USA* **2005**, *102*, 1023–1028. [CrossRef] [PubMed]
89. Horváth, P.; Kato, T.; Miyata, T.; Namba, K. Structure of *Salmonella* Flagellar Hook Reveals Intermolecular Domain Interactions for the Universal Joint Function. *Biomolecules* **2019**, *9*, 462. [CrossRef]
90. Hiraoka, K.D.; Morimoto, Y.V.; Inoue, Y.; Fujii, T.; Miyata, T.; Makino, F.; Minamino, T.; Namba, K. Straight and rigid flagellar hook made by insertion of the FlgG specific sequence into FlgE. *Sci. Rep.* **2017**, *7*, 46723. [CrossRef]
91. Osorio-Valeriano, M.; De La Mora, J.; Camarena, L.; Dreyfus, G. Biochemical Characterization of the Flagellar Rod Components of *Rhodobacter sphaeroides*: Properties and Interactions. *J. Bacteriol.* **2016**, *198*, 544–552. [CrossRef]
92. Burrage, A.M.; Vanderpool, E.; Kearns, D.B. Assembly Order of Flagellar Rod Subunits in *Bacillus subtilis*. *J. Bacteriol.* **2018**, *200*. [CrossRef]
93. González-Pedrajo, B.; De La Mora, J.; Ballado, T.; Camarena, L.; Dreyfus, G. Characterization of the *flgG* operon of *Rhodobacter sphaeroides* WS8 and its role in flagellum biosynthesis. *Biochim. Biophys. Acta (BBA) Gene Struct. Expr.* **2002**, *1579*, 55–63. [CrossRef]
94. De La Mora, J.; Ballado, T.; González-Pedrajo, B.; Camarena, L.; Dreyfus, G. The Flagellar Muramidase from the Photosynthetic Bacterium *Rhodobacter sphaeroides*. *J. Bacteriol.* **2007**, *189*, 7998–8004. [CrossRef]
95. De La Mora, J.; Osorio-Valeriano, M.; González-Pedrajo, B.; Ballado, T.; Camarena, L.; Dreyfus, G. The C Terminus of the Flagellar Muramidase SltF Modulates the Interaction with FlgJ in *Rhodobacter sphaeroides*. *J. Bacteriol.* **2012**, *194*, 4513–4520. [CrossRef] [PubMed]
96. Herlihey, F.A.; Osorio-Valeriano, M.; Dreyfus, G.; Clarke, A.J. Modulation of the Lytic Activity of the Dedicated Autolysin for Flagellum Formation SltF by Flagellar Rod Proteins FlgB and FlgF. *J. Bacteriol.* **2016**, *198*, 1847–1856. [CrossRef] [PubMed]
97. Armitage, J.P.; Pitta, T.P.; Vigeant, M.; Packer, H.L.; Ford, R.M. Transformations in Flagellar Structure of *Rhodobacter sphaeroides* and Possible Relationship to Changes in Swimming Speed. *J. Bacteriol.* **1999**, *181*, 4825–4833. [CrossRef] [PubMed]
98. Terashima, H.; Koike, M.; Kojima, S.; Homma, M. The Flagellar Basal Body-Associated Protein FlgT Is Essential for a Novel Ring Structure in the Sodium-Driven *Vibrio* Motor. *J. Bacteriol.* **2010**, *192*, 5609–5615. [CrossRef] [PubMed]
99. Merino, S.; Tomás, J.M. The FlgT Protein Is Involved in Aeromonas hydrophila Polar Flagella Stability and Not Affects Anchorage of Lateral Flagella. *Front. Microbiol.* **2016**, *7*, 7141. [CrossRef] [PubMed]
100. Cameron, D.E.; Urbach, J.M.; Mekalanos, J.J. A defined transposon mutant library and its use in identifying motility genes in *Vibrio* cholerae. *Proc. Natl. Acad. Sci. USA* **2008**, *105*, 8736–8741. [CrossRef]
101. Zhu, S.; Nishikino, T.; Kojima, S.; Homma, M.; Liu, J. The *Vibrio* H-Ring Facilitates the Outer Membrane Penetration of the Polar Sheathed Flagellum. *J. Bacteriol.* **2018**, *200*, JB-00387. [CrossRef]
102. Beeby, M.; Ribardo, D.A.; Brennan, C.A.; Ruby, E.G.; Jensen, G.J.; Hendrixson, D.R. Diverse high-torque bacterial flagellar motors assemble wider stator rings using a conserved protein scaffold. *Proc. Natl. Acad. Sci. USA* **2016**, *113*, E1917–E1926. [CrossRef]
103. Ramírez-Cabrera, V.; Poggio, S.; Domenzain, C.; Osorio, A.; Dreyfus, G.; Camarena, L. A Novel Component of the *Rhodobacter sphaeroides* Fla1 Flagellum Is Essential for Motor Rotation. *J. Bacteriol.* **2012**, *194*, 6174–6183. [CrossRef]
104. Hosking, E.R.; Vogt, C.; Bakker, E.P.; Manson, M.D. The *Escherichia coli* MotAB Proton Channel Unplugged. *J. Mol. Biol.* **2006**, *364*, 921–937. [CrossRef]

Article

Architecture of the Bacterial Flagellar Distal Rod and Hook of *Salmonella*

Yumiko Saijo-Hamano [1,†], Hideyuki Matsunami [2], Keiichi Namba [1,3] and Katsumi Imada [4,*]

1 Graduate School of Frontier Biosciences, Osaka University, 1-3 Yamadaoka, Suita, Osaka 565-0871, Japan
2 Molecular Cryo-Electron Microscopy Unit, Okinawa Institute of Science and Technology Graduate University, 1919-1 Tancha, Onna, Kunigami 904-0495, Japan
3 RIKEN Center for Biosystems Dynamics Research and SPring-8 Center, 1-3 Yamadoaka, Suita, Osaka 565-0871, Japan
4 Department of Macromolecular Science, Graduate School of Science, Osaka University, 1-1 Machikaneyama, Toyonaka, Osaka 560-0043, Japan
* Correspondence: kimada@chem.sci.osaka-u.ac.jp; Tel.: +81-06-6850-5455
† Current affiliation: Division of Structural Medicine and Anatomy, Department of Physiology and Cell Biology, Kobe University Graduate School of Medicine, Kobe, 650-0017 Japan.

Received: 17 June 2019; Accepted: 4 July 2019; Published: 7 July 2019

Abstract: The bacterial flagellum is a large molecular complex composed of thousands of protein subunits for motility. The filamentous part of the flagellum, which is called the axial structure, consists of the filament, the hook, and the rods, with other minor components—the cap protein and the hook associated proteins. They share a common basic architecture of subunit arrangement, but each part shows quite distinct mechanical properties to achieve its specific function. The distal rod and the hook are helical assemblies of a single protein, FlgG and FlgE, respectively. They show a significant sequence similarity but have distinct mechanical characteristics. The rod is a rigid, straight cylinder, whereas the hook is a curved tube with high bending flexibility. Here, we report a structural model of the rod constructed by using the crystal structure of a core fragment of FlgG with a density map obtained previously by electron cryomicroscopy. Our structural model suggests that a segment called L-stretch plays a key role in achieving the distinct mechanical properties of the rod using a structurally similar component protein to that of the hook.

Keywords: bacterial flagellum; crystal structure; electron cryomicroscopy; flagellar rod; hook

---

1. Introduction

Many motile bacteria move in liquid environments by rotating a helical filamentous organelle called the flagellum. The flagellum is a large molecular assembly of about 20–30 thousand of protein subunits of more than 20 types of proteins. The filamentous part of the flagellum, termed the axial structure, is rotated by a motor embedded in the cell membrane. The axial structure consists of three morphologically distinct regions, the rod, the hook, and the filament from the proximal to the distal end, with a few minor components [1,2]. The filament is a long helical propeller with a diameter of ~20 nm and a typical length of around 15 μm. The filament is a helical assembly of about 30,000 flagellin (FliC) subunits. The hook is a short, curved segment with an approximate length of 55 nm. The hook is a helical assembly of about 120 subunits of FlgE [3,4]. The function of the hook is a universal joint that transmits motor torque to the filament in any orientation relative to the motor axis. The rod is a straight structure with a length of about 30 nm. The rod is a helical cylinder that comprises four proteins: FlgB, FlgC, and FlgF in the proximal part and FlgG in the distal part [5–7]. The rod is a drive shaft that penetrates the peptidoglycan (PG) layer and the outer membrane and connects the hook with the rotor of the motor.

The axial components share a common architecture in the subunit arrangement and the domain arrangement. The axial subunit proteins are arranged in a helical array of 11 subunits in two turns of the 1-start helix. This arrangement produces 11 protofilaments, which are strands of the component proteins aligned nearly parallel to the filament axis [8–10]. The subunits form a concentric multi-layer tube in the axial structure. The inner tube is composed of α-helical coiled coils constructed by the N- and C- terminal regions of each component protein [11–15]. These terminal regions are disordered in monomeric state in solution but are folded into the coiled coils only when the subunits are incorporated into the flagellum [16,17].

Despite the common architecture, the mechanical properties of the axial structures are quite distinct. The filament forms a stiff, super helical structure as a propeller for efficient propulsion of the cell. The hook is flexible in bending but rigid against twisting to achieve a universal joint function, whereas the rod is a rigid straight cylinder to work as a drive shaft. Since the inner most tube shows a common structural feature, the differences in the mechanical and functional properties are ascribed to the structural difference of the outer regions. In fact, the amino acid sequence of the outer region of FliC (UniProtKB ID: P06179) differs completely from that of FlgE (UniProtKB ID: P0A1J1) or FlgG (UniProtKB ID: P0A1J3). However, FlgE and FlgG show a high sequence similarity to each other, even in the first outer region D1, although FlgE has an additional domain outside D1. Therefore, their structures have been investigated to understand the molecular basis of the specific properties of these structures.

The hook structure of the *Salmonella typhimurium* (St) has been studied by X-ray crystallography and electron cryomicroscopy (cryoEM) image analysis. St-FlgE is composed of three domains, D0, D1, and D2, and a region connecting D0 and D1 termed Dc. The D1 and D2 domains are composed of β-structures and are connected by a short stretch of an anti-parallel β-strands [18]. The D1 and D2 domains are loosely packed along the protofilament in the hook, which allows the sliding motion in the axial subunit interface. Therefore, the intersubunit distance can be compressed or extended up to ~2 nm. This property is thought to be a key factor for flexibility in bending [18–20]. On the other hand, the D2 domains are closely arranged along the 6-start direction on the outer surface of the hook. This structure greatly contributes to rigidity against twisting [18,21]. However, the structure of the Dc region is unclear because of the low resolution of the cryoEM density. Recently, the complete hook structure of *Campylobacter jejuni* (Cj) revealed that D0 and D1 domains are linked by an L-shaped extended structure termed L-stretch, which is composed of 50 residues following the N-terminal helix [15]. The L-stretch of the Cj-hook interacts with the neighboring three protofilaments, thereby stabilizing the hook structure [15]. Therefore, the Dc region of St-hook would be expected to adopt a structure similar to the L-stretch of Cj-hook.

A partial structural model of the distal rod has recently been constructed on the basis of a cryoEM density map of the distal rod at a 7 Å resolution obtained from a polyrod mutant of St-FlgG, a mutant that produces an unusually long distal rod, and a homology model of FlgG constructed based on the crystal structure of St-FlgE [22]. The helical symmetry of the distal rod is almost the same as that of the hook. Therefore, the subunit arrangement of the rod is very similar to that of the hook, but the orientation of each domain is significantly different. The D1 domain of FlgG stands upright to tightly interact with the neighboring subunits in the protofilament of the distal rod, whereas that of FlgE is tilted about 7° to make a gap along the protofilament. Moreover, the N-terminal helix of FlgG is longer than that of FlgE, so the N-terminal helix directly contacts with the axially neighboring subunit in the rod. These tight axial subunit interactions are thought to provide the bending rigidity to the distal rod. However, both structural models do not include the Dc region due to the limited resolution of the density map. The Dc region contains a FlgG specific segment not present in FlgE, and this segment is important for the rod to form a rigid, straight structure. A FlgE mutant with the insertion of the FlgG specific segment forms a straight and rigid hook [23]. Therefore, a complete and more precise structural model of the rod including the Dc region is needed for further understanding of the molecular basis of its mechanical and functional properties.

Here, we have constructed an atomic model of the St-rod structure on the basis of the crystal structure of a core fragment of St-FlgG (FlgG20) solved at 2.0 Å resolution in combination with the previous cryoEM map. FlgG20 more closely resembles Cj-FlgE than the core fragment of St-FlgE. Cj-FlgE also contains the FlgG specific segment that is missing in St-FlgE. Therefore, the complete distal rod model was constructed by using the similarities between FlgG20 and Cj-FlgE. This model revealed that the Dc region adopts the L-stretch structure and that the intersubunit interactions mediated by the L-stretch greatly contribute to stabilizing and rigidifying the distal rod structure.

2. Materials and Methods

2.1. Preparation and Crystallization of FlgG20

Details of expression, purification, and crystallization of FlgG20 (residues 47–227) from *Salmonella enterica* serovar Typhimurium and its selenomethionine derivative have been previously reported [24]. Key resources are summarized in Table S1. Briefly, FlgG20 was purified by affinity chromatography using a HisPrep FF 16/10 column (GE Healthcare Life Sciences, Little Chalfont, Buckinghamshire, UK) and treated with thrombin-protease (GE Healthcare Life Sciences) to remove His-tag, followed by anion-exchange chromatography using a RESOURCE Q (6 mL) (GE Healthcare Life Sciences). The best crystals were obtained from sitting drops with a reservoir solution containing polyethylene glycol monomethyl ether 2000 (Hampton Research, Aliso Viejo, CA, USA), ammonium sulfate (WAKO) and sodium acetate (pH 4.6) (Hampton Research).

2.2. Diffraction Data Collection and Structure Determination

X-ray diffraction data were collected at synchrotron beamlines BL38B1 and BL41XU in SPring-8 (Harima, Japan) with the approval of the Japan Synchrotron Radiation Research Institute (JASRI) (Proposal No. 2010B1013 and 2010B1901), as previously described [24]. The diffraction data were processed using iMOSFLM [25] and were scaled using SCALA [26] from the CCP4 program suite [27,28]. Data collection statistics are summarized in Table 1. The initial phase was determined using the SAD data of the Se-Met derivative and the initial model was automatically constructed with a program Phenix [29]. The model was manually modified with Coot [30] and refined with Phenix. The refinement converged to an R value of 20.0% and an R free value of 22.9% for the SAD data at a resolution of 2.0 Å. The final model contains 304 residues and 181 water molecules. A Ramachandran plot showed 98.0% and 2.0% residues located in the most favorable and allowed regions, respectively. Refinement statistics are summarized in Table 1.

Table 1. Data collection and refinement statistics.

Space Group	$P2_12_12_1$		
Cell dimensions (Å)	a = 47.5, b = 67.0, c = 110.3		
Wavelength (Å)	0.9790		
Resolution (Å)	36.8–2.0 (2.11–2.00)		
R_{merge}	0.106 (0.389)		
$I/\sigma I$	10.8 (4.0)		
Completeness (%)	98.1 (97.5)		
Redundancy	6.9 (6.6)		
	FlgG20	Rod (EMD-6683)	Hook (EMD-1647)
Resolution range (Å)	36.6–2.0 (2.08–2.00)	7.4	7.1
No. of reflections working	22,747 (2448)		
No. of reflections test	1217 (124)		
R_w (%)	20.0 (24.8)		
R_{free} (%)	22.9 (31.8)		
No. of protein atoms	2264		
No. of solvent atoms	192		
B-factors			
Protein atoms	37.0		
Solvent atoms	38.5		
Map correlation coefficient		0.729	0.713
Root mean square deviation bond length (Å)	0.002	0.03	0.00
Root mean square deviation bond angle (°)	0.537	0.73	0.59
Ramachandran plot (%)			
favored	98.0	93.0	94.3
allowed	2.0	7.0	5.7
outliers	0	0	0
Rotamer outliers (%)	0.4	9.27	7.14
All atom clash score		7.99	4.80

Values in parentheses are for the highest resolution shell. $R_w = \sum \| Fo | - | Fc \| / \sum | Fo |$, $R_{free} = \sum \| Fo | - | Fc \| / \sum | Fo |$.

2.3. Model Building of the Rod and Hook

The crystal structure of FlgG20 is roughly fitted in the EM density map of the rod (EMDataBank ID: EMD-6683) using UCSF Chimera [31]. The D0–D1 region of Cj-FlgE (PDB ID: 5JXL) was superimposed on FlgG20 by fitting the D1 domains of both molecules. Then, D0 and the L-stretch of Cj-FlgE were added to FlgG20 and replaced their sidechains with those of St-FlgG to construct the full-length FlgG model. The model was modified by fitting in the EM density using Coot. The residues 53–64 were removed due to the poor density of this region. Then, a rod segment model composed of 22 subunits (four turns along the 1-start helix) was produced using the helical symmetry of the rod (1-start: $\theta = 64.75°$, $z = 4.13$ Å) and was refined with real space refinement using Phenix under noncrystallographic symmetry (NCS) constraints. The NCS operator was not refined to preserve helical symmetry.

The hook model was constructed basically the same way as the rod model. The D0–D1 region of Cj-FlgE (PDB ID: 5JXL) was superimposed on FlgE (PDB ID: 3A69) by fitting the D1 domains of both molecules. D0 and the L-stretch of Cj-FlgE were fused to the D1–D2 domain of St-FlgE and replaced their sidechains with those of St-FlgE. The model was modified by fitting in the EM density of the hook (EMDataBank ID: EMD-1647) using Coot. A rod segment model composed of 22 subunits (four turns along the 1-start helix) was produced using the helical symmetry of St-hook (1-start: $\theta = 64.78°$, $z = 4.12$ Å) and was refined with real space refinement under NCS constraints using Phenix. The refinement statistics are summarized in Table 1.

Coordinates and structure factors have been deposited in the Protein Data Bank under accession codes 6JF2 (FlgG20), 6JZR (St-rod) and 6JZT (St-hook).

3. Results

3.1. Structure of FlgG20

The crystal asymmetric unit contains two FlgG20 molecules, mol A (80–224) and mol B (72–227), whose structures are almost identical to each other with a root-mean-square deviation (rmsd) of 0.60 Å for Cα atoms. The N-terminal 33 and C-terminal 3 residues of mol A and 25 residues of mol B were not traced because of poor electron density. FlgG20 shows a single domain structure composed of 15 β-strands and the loops connecting the β-strands (Figure 1A), and is similar to the D1 domain of FlgE except for the absence of the triangular loop (Figure S1A,B) and the conformation of the N- and C-terminal regions (Figure 1B). Interestingly, the N- and C-terminal regions of FlgG20 are more similar to *Campylobacter* FlgE (Cj-FlgE, PDB ID: 5JXL) [15] than the core fragment of *Salmonella* FlgE (St-FlgE31, PDB ID: 1WLG) [18] (Figure 1 and Figure S1). The β-stretch composed of the N- and C- terminal chains of FlgG20 (N85–T89 and Y218–T221, respectively) fits nicely on the corresponding regions of Cj-FlgE (Figure 1D).

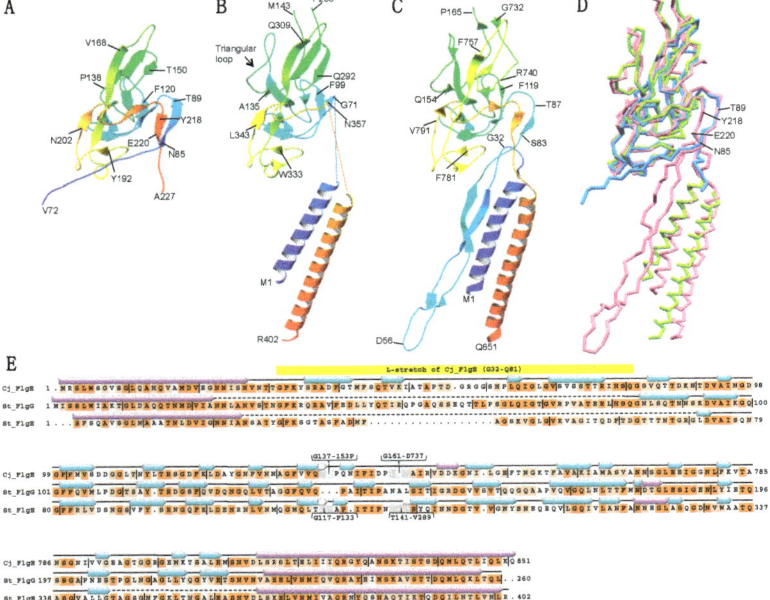

Figure 1. Structure of a core fragment of St-FlgG (FlgG20) from *Salmonella typhimurium* (St), FlgE from *Salmonella typhimurium* and FlgE from *Campylobacter jejuni* (Cj). Ribbon representation of the crystal structure of FlgG20 (**A**), the cryoEM structure model of D0 and D1 domains of St-FlgE (PDB ID: 3A69) (**B**), and the cryoEM structure of D0, L-stretch and D1 of Cj-FlgE (PDB ID: 5JXL) (**C**). The Dc region of the St-FlgE structure is not modeled due to the low resolution of the cryoEM map in (**B**). The models are color-coded from blue to red through the rainbow spectrum from the N to C terminus. (**D**) The superimposition of backbone models of FlgG20 (blue), St-FlgE (green) and Cj-FlgE (pink). The positions of N85, T89, Y218 and T221 of FlgG20 are indicated. (**E**) Structure-based sequence alignment of Cj-FlgG, St-FlgG and St-FlgE. Conserved residues are highlighted in dark orange (identical residues) or light orange (similar residues). Purple and cyan arrow indicates α-helix and β-strand, respectively. Molecular figures were drawn using MolFeat (Ver 3.6, FiatLux Corporation).

3.2. Atomic Model of the Distal Rod

The previous atomic model of the distal rod (PDB ID: 5WRH) was built by fitting a homology model of a fragment of FlgG (92–216) into a cryoEM map of the polyrod [22]. Although the homology model is similar to the D1 core of the FlgG20 crystal structure, the crystal structure includes a part of the region connecting D0 and D1 that was not present in the homology model (Figure S1). We, therefore, conducted rigid body fitting of the FlgG20 structure into the cryoEM map (EMDataBank ID: EMD-6683) using UCSF Chimera, and compared it with the previous model [22]. The C-terminal region of FlgG20 comes close to the N-terminal end of the C-terminal D0 helix of FlgG in the previous rod model. The N-terminal region of FlgG20 goes into the edge of the elongated density that was not assigned in the previous model. We found here that the elongated density can be assigned as the L-stretch, as is seen in Cj-FlgE, considering the structural similarity between FlgG20 (Figure 1A) and Cj-FlgE (Figure 1C) and the amino acid sequence similarity between St-FlgG and Cj-FlgE (Figure 1E). We superimposed the D0-L-stretch-D1 structure model of Cj-FlgE on the FlgG20 model fitted in the cryoEM map and realized that it is possible to fit the D0-L-stretch structure of Cj-FlgE into the elongated density with slight modification. Then, we constructed a new subunit model of the distal rod by connecting the FlgG20 model with the D0-L-stretch model built on the basis of the Cj-FlgE structure.

The new model revealed that the L-stretch extensively contacts with the D1 domains of other FlgG subunits and thereby reinforces the rod structure (Figure 2). The L-stretch extends along the D1 domain of FlgG in the −5 position toward that in the −10 position (Figure 2B–D). The location of the tip of the L-stretch (P52–S64) is unclear due to the poor density of this region. The L-stretch also interacts with the D1 domains of FlgG in the −11 and −16 positions and thereby mediates the 6-start interactions of the subunits in the −5 and −11 positions and those in the −10 and −16 positions (Figure 2C,D). The D1 domain interacts with the D1 domains of nearest-neighbor subunits in all directions, but only a few direct contacts are observed in each direction (Figure S2). Thus, the interactions mediated by the L-stretch greatly contribute to stabilizing and rigidifying the distal rod.

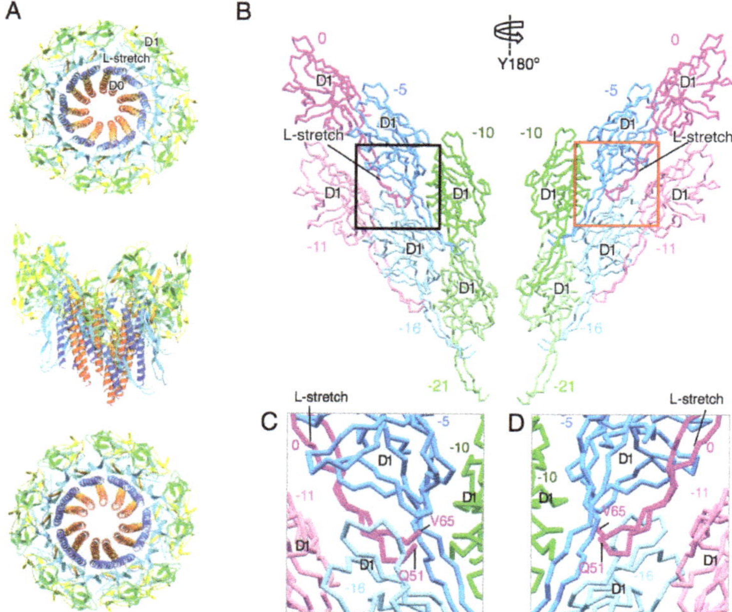

Figure 2. Structure model of the rod. (**A**) Two turns of the distal rod model. Eleven FlgG subunits along the one start helix are shown. Each subunit is colored from blue to red through the rainbow spectrum from the N to C terminus. Upper panel, viewed from the distal end; middle panel, side view; lower panel, viewed from the proximal end. (**B**) Subunit interaction of the D1 domain and the L-stretch in the rod. Six subunits in three adjacent protofilaments (two subunits from each protofilament) are shown. The number counted from the subunit 0 along the 1-start helical line is shown above or below each subunit. Left panel, view from the outside; right panel, view from the inside of the rod. (**C**) (**D**) Close up view of the L-stretch of the rod. The region in the black and red boxes in (**B**) are shown in (**C**) and (**D**), respectively. The tip of the L-stretch (P52-S64) is invisible due to poor density. Molecular figures were drawn using MolFeat (Ver 3.6, FiatLux Corporation).

3.3. Atomic Model of the Hook

The previous cryoEM map of the St-hook also showed an unassigned elongated density between the D1 domain and the D0 helices [21]. The shape and position of the density is very similar to the density corresponding to the L-stretch of the Cj-hook or that of the St-rod. Therefore, we superimposed the D0-L-stretch-D1 structure of Cj-FlgE (PDB ID: 5JXL) on the St-FlgE model (PDB ID: 3A69) in the cryoEM density of the St-hook (EMDataBank ID: EMD-1647). The D0 helices and the L-stretch of Cj-FlgE were well fitted into the EM density with slight modification (Figure 3). The amino acid sequence alignment indicates that the L-stretch of St-FlgE is 18 residues shorter than that of Cj-FlgE (Figure 1E) [32]. Consistent with this, the density of St-FlgE corresponding to the L-stretch is shorter than the other two and terminates at the position expected from the sequence. Thus, we built the model of the Dc region based on the Cj-FlgE structure and constructed a full-length St-hook model. The subunit interaction mediated by the L-stretch is not as extensive as in the distal rod (Figures 2 and 3). The L-stretch extends to the −5 direction and interacts with the inner surface of the D1 domains of the FlgE subunits in the −5 and −11 positions but not with the subunits in −10 and −16 positions (Figure 3B–D).

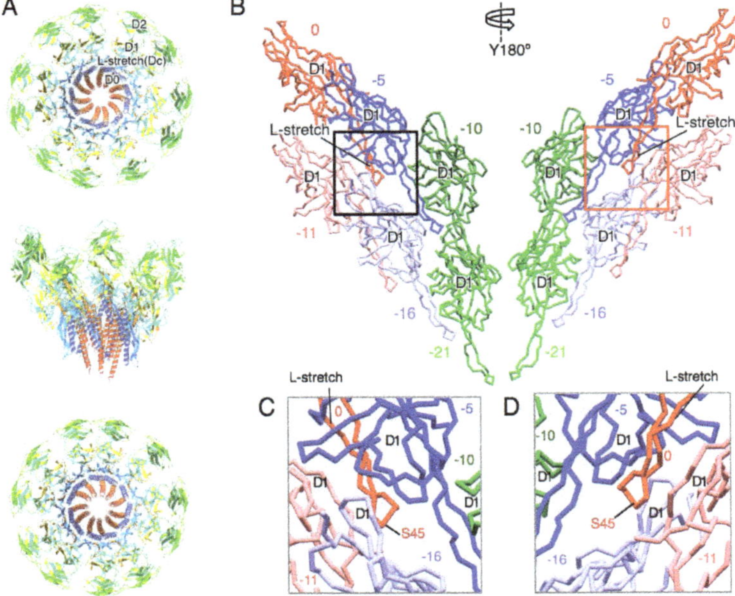

Figure 3. Structure model of the hook. (**A**) Two turns of the hook model. Eleven FlgE subunits along the one start helix are shown. Each subunit is colored from blue to red through the rainbow spectrum from the N to C terminus. Upper panel, viewed from the distal end; middle panel, side view; lower panel, viewed from the proximal end. (**B**) Subunit interaction of the D1 and the L-stretch in the hook. Six subunits in three adjacent protofilaments (two subunits from each protofilament) are shown. The number counted from the subunit 0 along the 1-start helical line is shown above or below each subunit. Left panel, view from the outside; right panel, view from the inside of the hook. (**C**) (**D**) Close up view of the L-stretch in the hook. The region in the black and red boxes in (**B**) are shown in (**C**) and (**D**), respectively. Molecular figures were drawn using MolFeat (Ver 3.6, FiatLux Corporation).

4. Discussion

The distal rod and hook show distinct mechanical properties, although they share the same helical symmetry and repeat distance and are composed of similar subunit proteins (FlgG and FlgE) with 39%

sequence identity. The rod is straight and rigid whereas the hook is curved and flexible and is easy to bend. Previous cryoEM studies revealed that the D1 domain of FlgE in the hook is tilted about 7° relative to that of FlgG in the distal rod. this difference resulted in a loose axial subunit packing of FlgE, making the hook flexible in bending unlike the rod. Moreover, a small gap between the D0 helices of the axially contacting FlgE subunits allows compression and extension of each protofilament [22].

The new models of the rod and hook have revealed marked contributions of the L-stretch to their mechanical properties. Interacting with the −5 and −10 subunits, the long L-stretch of FlgG fastens three adjacent protofilaments in the rod. In addition, the L-stretch mediates the 6-start interaction of −5 and −11 subunits and 0 and −11 subunits. These 5-start and 6-start interaction networks make the distal rod rigid in its straight form. Moreover, the tight interaction of L-stretch with the D1 domain in the −5 position may stabilize the upright orientation of FlgG, forming tight subunit packing in the distal rod. In contrast, the L-stretch of FlgE is shorter than that of FlgG and only interacts with the inner surface of the −5 and −10 subunits. These interactions probably reinforce the hook structure while allowing the axial compression and extension of each hook protofilament for the bending flexibility of the hook.

The rod length is regulated to around 25 nm in wild-type cells, but some FlgG mutants form unusually long rod structures, termed polyrods. More than 20 mutants, including a few residue deletion mutants, were isolated, and over half of the mutation sites were localized in the segment of residues 52–66 [32–34], which corresponds to the distal part of the L-stretch. Chevance et al. mapped the mutation sites on a homology model of FlgG and suggested that this region might interact with residues G183 and S197 in the neighboring FlgG subunit in the rod and that the other mutation sites are rather widely distributed [32]. We have now mapped these mutation sites on the new rod model and found that all of them are localized within the plausible space where the distal part (residues 52–66) of the L-stretch would be located, albeit most of them were not visible in the EM map (Figure 4). The absence of the EM density for the distal part of the L-stretch may be because the resolution of the EM density map was not high enough to clearly resolve this part, which is in close contact with neighboring subunits, as described above. The distal part of the L-stretch in the 0 position presumably interacts with the middle part of the L-stretch of the subunit in the −5 position and the D1 domain of the subunit in the −10 position. Therefore, the intermolecular interactions around the distal part of the L-stretch are likely to be the key factor for the stiffness of the rod. To confirm this idea, we need a cryoEM structure of the polyrod mutant, as well as a wild-type distal rod at a higher, near-atomic resolution.

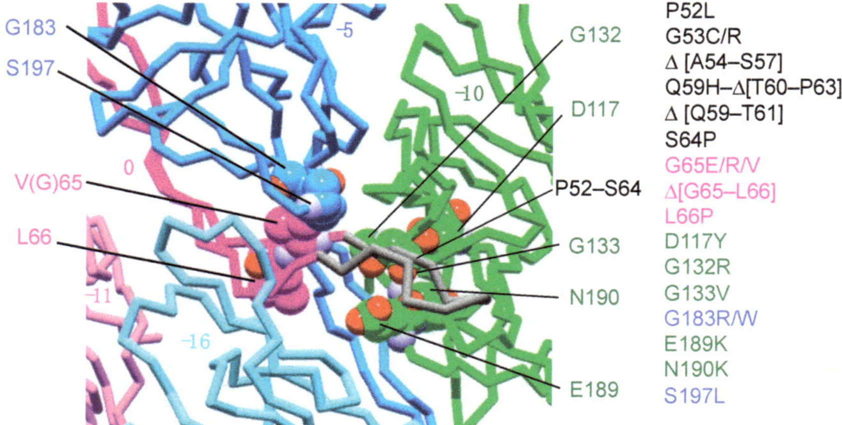

Figure 4. Polyrod mutation sites are localized around the tip of the L-stretch. The known polyrod mutation sites are mapped on the rod structure. Subunit at 0, −5, −10 positions are colored in magenta, blue, and green, respectively. The putative structure of the L-stretch tip of subunit 0 is shown in gray.

The polyrod mutation sites except for those in the L-stretch tip are indicated by ball models. Oxygen and nitrogen atoms are colored in red and purple, respectively. The carbon atoms are painted by the same color as used for the subunits. The polyrod mutations [32] are listed to the right of the figure. Molecular figures were drawn using MolFeat (Ver 3.6, FiatLux Corporation).

5. Conclusions

We constructed a structural model of the flagellar distal rod by fitting the crystal structure of FlgG20 into the previous cryoEM map of a polyrod and a new hook model using the previous cryoEM map of the hook. These structural models suggest that the L-stretch stabilizes the distal rod and plays a key role in achieving the distinct mechanical properties of the rod using a structurally similar protein to that of the hook.

Supplementary Materials: The following are available online, Figure S1: Structural comparison of the D1 domain, Figure S2: The rod model docked in the cryoEM map, Table S1: Key Resources.

Author Contributions: Conceptualization, Y.S.-H., K.N., and K.I.; methodology, Y.S.-H., K.N., and K.I.; validation, Y.S.-H. and K.I.; formal analysis, Y.S.-H. and K.I.; investigation, Y.S.-H., H.M. and K.I.; resources Y.S.-H., H.M., and K.I.; data curation, Y.S.-H. and K.I.; writing—original draft preparation, Y.S.-H. and K.I.; writing—review and editing, Y.S.-H., K.N., and K.I.; visualization, Y.S.-H. and K.I.; project administration, K.N. and K.I.; funding acquisition, Y.S.-H., H.M., K.N., and K.I.

Funding: This work was supported in part by MEXT/JSPS KAKENHI Grant Number JP24570132 (to Y.S-H.), JP17K07318 (to H.M.), JP21227006 (to K.N.), JP25000013 (to K.N.), JP15H02386 (to K.I.) and JP23115008 (to K.I.).

Acknowledgments: We thank the beamline staff at SPring-8 for technical help in the use of beamlines.

Conflicts of Interest: The authors declare no conflict of interest.

References

1. Macnab, R.M. How bacteria assemble flagella. *Annu. Rev. Microbiol.* **2003**, *57*, 77–100. [CrossRef] [PubMed]
2. Imada, K. Bacterial flagellar axial structure and its construction. *Biophys. Rev.* **2018**, *10*, 559–570. [CrossRef] [PubMed]
3. Kutsukake, K.; Suzuki, T.; Yamaguchi, S.; Iino, T. Role of gene flaFV on flagellar hook formation in Salmonella typhimurium. *J. Bacteriol.* **1979**, *140*, 267–275. [PubMed]
4. Jones, C.J.; Macnab, R.M.; Okino, H.; Aizawa, S. Stoichiometric analysis of the flagellar hook-(basal-body) complex of Salmonella typhimurium. *J. Mol. Biol.* **1990**, *212*, 377–387. [CrossRef]
5. Okino, H.; Isomura, M.; Yamaguchi, S.; Magariyama, Y.; Kudo, S.; Aizawa, S.I. Release of flagellar filament-hook-rod complex by a Salmonella typhimurium mutant defective in the M ring of the basal body. *J. Bacteriol.* **1989**, *171*, 2075–2082. [CrossRef] [PubMed]
6. Homma, M.; Kutsukake, K.; Hasebe, M.; Iino, T.; Macnab, R.M. FlgB, FlgC, FlgF and FlgG. A family of structurally related proteins in the flagellar basal body of Salmonella typhimurium. *J. Mol. Biol.* **1990**, *211*, 465–477. [CrossRef]
7. Kubori, T.; Shimamoto, N.; Yamaguchi, S.; Namba, K.; Aizawa, S. Morphological pathway of flagellar assembly in Salmonella typhimurium. *J. Mol. Biol.* **1992**, *226*, 433–446. [CrossRef]
8. O'Brien, E.J.; Bennett, P.M. Structure of straight flagella from a mutant Salmonella. *J. Mol. Biol.* **1972**, *70*, 133–152. [CrossRef]
9. Wagenknecht, T.; DeRosier, D.; Shapiro, L.; Weissborn, A. Three-dimensional reconstruction of the flagellar hook from Caulobacter crescentus. *J. Mol. Biol.* **1981**, *151*, 439–465. [CrossRef]
10. Wagenknecht, T.; DeRosier, D.J.; Aizawa, S.-I.; Macnab, R.M. Flagellar hook structures of Caulobacter and Salmonella and their relationship to filament structure. *J. Mol. Biol.* **1982**, *162*, 69–87. [CrossRef]
11. Mimori, Y.; Yamashita, I.; Murata, K.; Fujiyoshi, Y.; Yonekura, K.; Toyoshima, C.; Namba, K. The structure of the R-type straight flagellar filament of Salmonella at 9 A resolution by electron cryomicroscopy. *J. Mol. Biol.* **1995**, *249*, 69–87. [CrossRef] [PubMed]

12. Mimori-Kiyosue, Y.; Vonderviszt, F.; Yamashita, I.; Fujiyoshi, Y.; Namba, K. Direct interaction of flagellin termini essential for polymorphic ability of flagellar filament. *Proc. Natl. Acad. Sci. USA* **1996**, *93*, 15108–15113. [CrossRef] [PubMed]
13. Mimori-Kiyosue, Y.; Vonderviszt, F.; Namba, K. Locations of terminal segments of flagellin in the filament structure and their roles in polymerization and polymorphism. *J. Mol. Biol.* **1997**, *270*, 222–237. [CrossRef] [PubMed]
14. Yonekura, K.; Maki-Yonekura, S.; Namba, K. Complete atomic model of the bacterial flagellar filament by electron cryomicroscopy. *Nature* **2003**, *424*, 643–650. [CrossRef] [PubMed]
15. Matsunami, H.; Barker, C.S.; Yoon, Y.-H.; Wolf, M.; Samatey, F.A. Complete structure of the bacterial flagellar hook reveals extensive set of stabilizing interactions. *Nat. Commun.* **2016**, *7*, 13425. [CrossRef]
16. Vonderviszt, F.; Ishima, R.; Akasaka, K.; Aizawa, S.-I. Terminal disorder: A common structural feature of the axial proteins of bacterial flagellum? *J. Mol. Biol.* **1992**, *226*, 575–579. [CrossRef]
17. Saijo-Hamano, Y.; Uchida, N.; Namba, K.; Oosawa, K. In vitro characterization of FlgB, FlgC, FlgF, FlgG, and FliE, flagellar basal body proteins of Salmonella. *J. Mol. Biol.* **2004**, *339*, 423–435. [CrossRef] [PubMed]
18. Samatey, F.A.; Matsunami, H.; Imada, K.; Nagashima, S.; Shaikh, T.R.; Thomas, D.R.; Chen, J.Z.; Derosier, D.J.; Kitao, A.; Namba, K. Structure of the bacterial flagellar hook and implication for the molecular universal joint mechanism. *Nature* **2004**, *431*, 1062–1068. [CrossRef]
19. Shaikh, T.R.; Thomas, D.R.; Chen, J.Z.; Samatey, F.A.; Matsunami, H.; Imada, K.; Namba, K.; Derosier, D.J. A partial atomic structure for the flagellar hook of Salmonella typhimurium. *Proc. Natl. Acad. Sci. USA* **2005**, *102*, 1023–1028. [CrossRef]
20. Furuta, T.; Samatey, F.A.; Matsunami, H.; Imada, K.; Namba, K.; Kitao, A. Gap compression/extension mechanism of bacterial flagellar hook as the molecular universal joint. *J. Struct. Biol.* **2007**, *157*, 481–490. [CrossRef]
21. Fujii, T.; Kato, T.; Namba, K. Specific arrangement of alpha-helical coiled coils in the core domain of the bacterial flagellar hook for the universal joint function. *Struct. Lond. Engl. 1993* **2009**, *17*, 1485–1493. [CrossRef]
22. Fujii, T.; Kato, T.; Hiraoka, K.D.; Miyata, T.; Minamino, T.; Chevance, F.F.V.; Hughes, K.T.; Namba, K. Identical folds used for distinct mechanical functions of the bacterial flagellar rod and hook. *Nat. Commun.* **2017**, *8*, 14276. [CrossRef] [PubMed]
23. Hiraoka, K.D.; Morimoto, Y.V.; Inoue, Y.; Fujii, T.; Miyata, T.; Makino, F.; Minamino, T.; Namba, K. Straight and rigid flagellar hook made by insertion of the FlgG specific sequence into FlgE. *Sci. Rep.* **2017**, *7*, 46723. [CrossRef] [PubMed]
24. Saijo-Hamano, Y.; Matsunami, H.; Namba, K.; Imada, K. Expression, purification, crystallization and preliminary X-ray diffraction analysis of a core fragment of FlgG, a bacterial flagellar rod protein. *Acta Crystallograph. Sect. F Struct. Biol. Cryst. Commun.* **2013**, *69*, 547–550. [CrossRef] [PubMed]
25. Battye, T.G.G.; Kontogiannis, L.; Johnson, O.; Powell, H.R.; Leslie, A.G.W. iMOSFLM: A new graphical interface for diffraction-image processing with MOSFLM. *Acta Crystallogr. D Biol. Crystallogr.* **2011**, *67*, 271–281. [CrossRef] [PubMed]
26. Evans, P. Scaling and assessment of data quality. *Acta Crystallogr. D Biol. Crystallogr.* **2006**, *62*, 72–82. [CrossRef] [PubMed]
27. Collaborative Computational Project, Number 4 The CCP4 suite: Programs for protein crystallography. *Acta Crystallogr. D Biol. Crystallogr.* **1994**, *50*, 760–763. [CrossRef] [PubMed]
28. Winn, M.D.; Ballard, C.C.; Cowtan, K.D.; Dodson, E.J.; Emsley, P.; Evans, P.R.; Keegan, R.M.; Krissinel, E.B.; Leslie, A.G.W.; McCoy, A.; et al. Overview of the CCP4 suite and current developments. *Acta Crystallogr. D Biol. Crystallogr.* **2011**, *67*, 235–242. [CrossRef]
29. Adams, P.D.; Afonine, P.V.; Bunkóczi, G.; Chen, V.B.; Davis, I.W.; Echols, N.; Headd, J.J.; Hung, L.-W.; Kapral, G.J.; Grosse-Kunstleve, R.W.; et al. PHENIX: A comprehensive Python-based system for macromolecular structure solution. *Acta Crystallogr. D Biol. Crystallogr.* **2010**, *66*, 213–221. [CrossRef]
30. Emsley, P.; Lohkamp, B.; Scott, W.G.; Cowtan, K. Features and development of Coot. *Acta Crystallogr. D Biol. Crystallogr.* **2010**, *66*, 486–501. [CrossRef]
31. Pettersen, E.F.; Goddard, T.D.; Huang, C.C.; Couch, G.S.; Greenblatt, D.M.; Meng, E.C.; Ferrin, T.E. UCSF Chimera–a visualization system for exploratory research and analysis. *J. Comput. Chem.* **2004**, *25*, 1605–1612. [CrossRef] [PubMed]

32. Chevance, F.F.V.; Takahashi, N.; Karlinsey, J.E.; Gnerer, J.; Hirano, T.; Samudrala, R.; Aizawa, S.-I.; Hughes, K.T. The mechanism of outer membrane penetration by the eubacterial flagellum and implications for spirochete evolution. *Genes Dev.* **2007**, *21*, 2326–2335. [CrossRef] [PubMed]
33. Hirano, T.; Mizuno, S.; Aizawa, S.-I.; Hughes, K.T. Mutations in Flk, FlgG, FlhA, and FlhE That Affect the Flagellar Type III Secretion Specificity Switch in Salmonella enterica. *J. Bacteriol.* **2009**, *191*, 3938–3949. [CrossRef] [PubMed]
34. Cohen, E.J.; Hughes, K.T. Rod-to-Hook Transition for Extracellular Flagellum Assembly Is Catalyzed by the L-Ring-Dependent Rod Scaffold Removal. *J. Bacteriol.* **2014**, *196*, 2387–2395. [CrossRef] [PubMed]

Article

Structure of *Salmonella* Flagellar Hook Reveals Intermolecular Domain Interactions for the Universal Joint Function

Péter Horváth [1,†,‡], Takayuki Kato [1,*,‡], Tomoko Miyata [1] and Keiichi Namba [1,2,3,*]

1 Graduate School of Frontier Biosciences, Osaka University, 1-3 Yamadaoka, Suita, Osaka 565-0871, Japan
2 RIKEN SPring-8 Center and Center for Biosystems Dynamics Research, 1-3 Yamadaoka, Suita, Osaka 565-0871, Japan
3 JEOL YOKOGUSHI Research Alliance Laboratories, Osaka University, 1-3 Yamadaoka, Suita, Osaka 565-0871, Japan
* Correspondence: galy@fbs.osaka-u.ac.jp (T.K.); keiichi@fbs.osaka-u.ac.jp (K.N.); Tel.: +81-6-6879-4625 (T.K. & K.N.)
† Current address: National Center of Biotechnology, C/Darwin 3, Universidad Autónoma de Madrid. Campus de Cantoblanco, 28049 Madrid, Spain.
‡ These authors contributed equally.

Received: 17 August 2019; Accepted: 4 September 2019; Published: 9 September 2019

Abstract: The bacterial flagellum is a motility organelle consisting of a rotary motor and a long helical filament as a propeller. The flagellar hook is a flexible universal joint that transmits motor torque to the filament in its various orientations that change dynamically between swimming and tumbling of the cell upon switching the motor rotation for chemotaxis. Although the structures of the hook and hook protein FlgE from different bacterial species have been studied, the structure of *Salmonella* hook, which has been studied most over the years, has not been solved at a high enough resolution to allow building an atomic model of entire FlgE for understanding the mechanisms of self-assembly, stability and the universal joint function. Here we report the structure of *Salmonella* polyhook at 4.1 Å resolution by electron cryomicroscopy and helical image analysis. The density map clearly revealed folding of the entire FlgE chain forming the three domains D0, D1 and D2 and allowed us to build an atomic model. The model includes domain Dc with a long β-hairpin structure that connects domains D0 and D1 and contributes to the structural stability of the hook while allowing the flexible bending of the hook as a molecular universal joint.

Keywords: cryoEM; *Salmonella* hook; universal joint; helical image analysis

---

1. Introduction

The bacterial flagellum is a motility organelle and is a complex nanomachine consisting of more than 20 different proteins in different copy numbers ranging from a few to a few tens of thousands [1]. Bacteria swim in viscous liquid toward more favorable environments from less favorable ones for their survival, proliferation and/or infection [2–4]. The bacterial flagellum can be divided into three main parts: the transmembrane basal body that acts as a rotary motor as well as the flagellar protein export machine to construct the flagellar axial structures; the long filament extending into the cell exterior to function as a helical propeller; and the hook connecting the motor and filament as a universal joint to transmit motor torque to the filament.

The rotary motor of *Salmonella* is powered by proton motive force across the cell membrane and rotates the filament at around 300 revolutions per second [5–7]. The filament is a helical assembly of a single protein, FliC (flagellin), and a few tens of thousands of FliC molecules form a 10–15 μm long supercoiled tubular structure in a gently curved form to be a helical propeller [1], which is rotated by the motor to produce thrust for bacterial swimming [8,9]. The filament is normally a left-handed supercoil, and its counter-clockwise rotation by the motor produces thrust for cell swimming. *Salmonella* has several peritrichous flagella that form a bundle behind the cell to produce strong thrust, but the switching of motor rotation to clockwise direction causes a polymorphic transition of the filament to a few different right-handed supercoils, thereby making the bundle to fall apart, causing cell tumbling for chemotaxis and thermotaxis [10–12]. The relatively rigid structure of the filament against bending assures these left- and right-handed supercoils function as a propeller.

The *Salmonella* hook is a short, highly curved tubular structure built by helical assembly of about 120 copies of a single protein, FlgE [13–16]. The length of the hook is regulated to 55 nm ± 7 nm [17], and its highly curved structure and bending flexibility make it work as a universal joint, transmitting motor torque to the filament regardless of the filament orientation off-axis of the motor during run and tumble of the cell for taxis [18–20].

The distal end of the hook is connected to the filament via two hook-associated proteins, FlgK and FlgL, while its proximal end is directly connected with the rod within the basal body. The rod is rigid as a drive shaft that transmits motor torque to the filament through the hook. The rod, hook and filament form the axial structure of the flagellum and show structural similarity to each other as tubular structures composed of 11 protofilaments. Nevertheless, their diameters, mechanical properties and functions are distinct from each other. The rod consists of four different proteins, FlgB, FlgC, FlgF, and FlgG [21] in different copy numbers: around five for FlgB, FlgC and FlgF; and about 26 for FlgG [22]. The distal, major part of the rod is formed by FlgG, which shows a high (32%) sequence similarity to FlgE, and the structure of the polyrod formed by a FlgG mutant shares the same helical symmetry and axial repeat distance with that of the hook, explaining why the hook is directly connected to the rod without any adaptor proteins between them [23].

The structure of the bacterial flagellum has been intensively studied for *Salmonella enterica* serovar Typhimurium by X-ray crystallography and electron microscopy (cryoEM) for many different parts [24]. The crystal structure of a core fragment of hook protein FlgE as well as the structure of polyhook, which is an abnormally long hook produced by deletion of the hook ruler protein FliK, are also available for the *Salmonella* flagellum [25–27], and these structures provided deep insights into the structural mechanism for the universal joint function. The axial packing interactions of the FlgE subunits along the protofilament of the hook have evolved in such a way for the protofilament to be compressible and extensible by mechanical force to make the tubular structure made of 11 protofilaments quite flexible in bending while their lateral packing interactions keep the tubular structure stable to make it rigid against twisting. The FlgE molecule consists of three major domains, D0, D1 and D2, that are arranged radially from the inner core to the outer surface of the tubular structure and axially from the proximal to the distal end of the hook. The structures of these domains are well resolved by combined use of X-ray crystallography and cryoEM helical image analysis [25–27]. The FlgE chain is folded to form these three domains by starting from D0, going through D1 and D2, and coming back to D0 through D1. In addition, a previous cryoEM study identified a small, elongated domain Dc that connects domains D0 and D1, and this domain appears to play an important role in stabilizing the hook structure by intervening the packing interactions of neighboring subunits, but it was difficult to trace the chain to build an atomic model due to the resolution of the map being limited to 7.1 Å (at a Fourier shell correlation of 0.5), even though most of the secondary structures in the other three domains were clearly visible [23]. A recent high-resolution cryoEM study on the hook structure of *Campylobacter jejuni* revealed the domain structures that are nearly identical to those of *Salmonella* hook FlgE and also resolved the structure of domain Dc clearly as an L-shaped, long β-hairpin named L-stretch [28]. This long β-hairpin structure appears to fit well into the previous density map of domain Dc of *Salmonella* hook, but the length of the *Salmonella* peptide chain assigned to this Dc domain is much shorter than that of *Campylobacter* by 17 residues. So a question still remains as to how domain Dc with the β-hairpin conformation interacts with neighboring FlgE subunits to stabilize the entire hook structure without reducing its bending flexibility.

We therefore used cryoEM image analysis to obtain a high-resolution structure of *Salmonella* hook to build its complete atomic model. With recent advances in cryoEM techniques with the use of a direct electron detector camera for high-resolution image data collection, we have obtained a 3D density map at 4.1 Å resolution and built an atomic model that reveals the conformation of domain Dc and its intimate interactions with neighboring subunits. We describe the nature of these intersubunit interactions and discuss the role of domain Dc in the structural stability and universal joint function of the hook.

2. Materials and Methods

2.1. Sample Preparation and Electron Microscopy

Flagellar polyhooks were isolated from the SJW880 strain of *Salmonella enterica* serovar Typhimurium (Δ*fliK* mutant) according to the protocol by Aizawa et al. with slight modifications [29]. After isolation and purification, the polyhook solution was kept at 4 °C in a cold room overnight to make polyhooks straight. A 3.0-μL sample solution was applied onto a Quantifoil holey carbon molybdenum grid (R0.6/1.0, Quantifoil Micro Tools GmbH, Jena, Germany), and the grid was plunge-frozen into liquid ethane using Vitrobot (Thermo Fisher Scientific, Waltham, MA, USA). The grid was imaged manually with a JEM-3200FSC electron cryomicroscope (JEOL, Tokyo, Japan), equipped with a field emission electron gun operated at 300 kV, an Ω-type in-column energy filter with the energy-selection slit width set to 10 eV for zero-loss imaging, and a K2 Summit Direct Electron Detection Camera (Gatan, Pleasanton, CA, USA). The grid temperature was kept at 77 K by liquid N_2, and cryoEM images were recorded in the Super-Resolution mode of K2 camera with an image size of 7676 × 7420 pixels, at a nominal magnification of 50,000×, a defocus range of 0.5–2.5 μm, and a dose rate of 2.2 electrons/pixel/sec. Images were recorded in a dose fractionation mode, with a total exposure time of 6 s at a frame rate of 0.2 s/frame, and so 30 frames were recorded for each movie image, with a total cumulative dose of 82 electrons/Å^2.

2.2. Image Processing

Beam-induced sample motions on dose-fractionated frames were corrected by GPU accelerated MotionCor2 software [30]. First global motion correction was carried out on individual frames, and subsequently the corrected frames were divided into 5 × 5 patches and local motion was corrected. Gain reference and dose weighting was then applied, the first three frames were discarded, and the rest were averaged. Accurate determination, refinement and correction of the contrast transfer function was carried out by Gctf, a GPU-accelerated software [31]. RELION 2.0 [32] was then used for image processing and 3D image reconstruction.

In total, 42,293 segment images were extracted from 486 micrographs, 2D classification was performed, and 41,260 segment images were used for further image processing. In the 3D classification step, local search of helical symmetry (the twist angle and axial rise per unit) was performed with initial parameters of 64.7° for the twist angle and 4.1 Å for the axial rise [23]. In the 3D reconstruction, auto-refine step local search was repeated with initial parameters of 64.78° for the twist angle and 4.05 Å for the axial rise and a step size of 0.1 Å. The helical parameters converged nearly to the initial parameters, confirming the accuracy of the helical parameters. In the mask creation step prior to binarization, the input 3D map was lowpass-filtered to 10 Å resolution. The initial binary mask was extended with a raised-cosine soft edge with 15-pixel width in every direction. In the post-processing step, map sharpening was performed with the given detector MTF and a B-factor of −216. Resolution was determined by calculating the Fourier shell correlation (FSC) coefficient between two halves of the data by the gold-standard approach to avoid overfitting. The final resolution was 4.06 Å at the FSC of 0.143 (Figure S1).

2.3. Atomic Model Building

The previous atomic model of *Salmonella* hook (PDB ID: 3A69) [27] with an FlgE subunit consisting of domain D0, D1 and D2 was used as the initial model to fit into the 3D map domain by domain. The atomic model of domain Dc was built by a homology model-building software, MODELLER [33]. The target sequence was that of *Salmonella* FlgE domain Dc, and the 3D model template was the corresponding domain Dc of *C. jejuni* FlgE (PDB ID: 5JXL) [28]. The sequence identity between these two Dc domains was 35%. Then this homology model was fit into the 3D map by Rosetta [34] and refined with its asymmetric refinement procedure. The Ramachandran plot by Coot software [35] showed that 94% of the residues were in the preferred regions and 5% were in the allowed regions.

A homology model of *Salmonella* FlgG including domain Dc was also built by MODELLER [33] using a recent cryoEM model (PDB ID:5WRH) [23] as a template. The target sequence was that of *Salmonella* FlgG, and the 3D model template was again the corresponding Dc domain of *C. jejuni* FlgE (PDB ID: 5JXL) [28] as well as domains D0 and D1 of *Salmonella* FlgG (PDB ID:5WRH) [23]. The Ramachandran validation in Coot software showed that 89% of the residues were in the preferred regions and 7% in the allowed regions. The relaxation of the structure by searching the local conformation space was carried out by Rosetta Relax [34] by a modified scope of the Relax program, restricting the conformations that Relax can sample. All atomic model and maps are rendered by UCSF Chimera [36].

The cryoEM 3D density map has been deposited to the Electron Microscopy Data Bank (https://www.ebi.ac.uk/pdbe/emdb/) under accession code EMD-9974, and the atomic model coordinates have been deposited to the Protein Data Bank (https://pdbj.org) under accession code 6KFK.

3. Results and Discussion

3.1. Structure Determination

The hook was purified from a *fliK*-deficient mutant strain of *Salmonella*, SJW880 [37], which produces polyhooks that are structurally identical to the native hook but can grow as long as 1 μm. The polyhook solution was incubated at 4 °C overnight to convert the supercoiled form into being straight, a 3 μL of the solution was applied onto a Quantifoil holey carbon grid, and then the grid was plunge-frozen into liquid ethane using a vitrification device (Vitrobot, TFS) [23]. The grid was then observed by a JEOL electron cryomicroscope, JEM-3200FSC, equipped with an Ω-type energy filter, a Schottky-type field emission electron gun operated at 300 kV, and a Gatan K2 Summit direct electron detector camera in movie mode. Movie frames were corrected with MotionCor2 software [30]. Image selection criteria were based on the Thon ring evaluation, and image analysis was carried out by RELION 2 [32]. The final resolution of the three-dimensional image reconstruction was 4.1 Å by the criteria of a Fourier shell correlation of 0.143. The density map was used to build an atomic model, starting from a previous model of domains D0, D1 and D2 of *Salmonella* hook [23] and a homology model of domain Dc generated by MODELLER software [33] based on the model of *Campylobacter* hook [28]. The atomic model was refined through iterative local rebuilding by ROSETTA software [34].

3.2. Structure of *Salmonella* Hook and FlgE in the Hook

The 3D density map and the model of *Salmonella* hook are presented in Figure 1 in different views and slices to visualize the structure and helical array of each of the four domains in different colors: domain D2 in magenta; D1 in blue; Dc in green; and D0 in red, from left to right panels. In the previous model built based on a cryoEM density map [23], the model of domain D0 was not so accurate and that of domain Dc was missing due to limited resolution of the density map. The model of *Salmonella* FlgE is now complete with the full-length chain connected from the N- to the C-terminus as shown in Figure 2 in two different views. Domain D0 consists of the N- and C-terminal α-helices (Ser 1–Ala 27 and Leu 367–Arg 402, respectively) forming an antiparallel coiled coil, with the C-terminal helix facing the central channel and tilted from the hook axis by about 17°. Domain Dc is a long β-hairpin (Lys

32–Phe 60) running parallel with the N-terminal helix and with its length nearly the same as that of the N-terminal helix. Thus, domains D0 and Dc together form a compact domain, which we hereafter call domain D0-Dc. A stretched chain after the β-hairpin of domain Dc (Thr 61–Gly 71) and another stretched chain before the C-terminal helix (Gly 358–Asp 366), together with a loop connecting the N-terminal helix and β-hairpin (Thr 28–Phe 31), forms the surface of domain D0-Dc closely interacting with domain D1. Domains D1 (Leu 72–Lys 146 and Tyr 283–Asn 357) and D2 (Ser 147–Gly 282) both form compact globular domains consisting of β-sheets, β-hairpins and loops, as was revealed by X-ray crystallography of a core fragment of FlgE [25]. The pair of chains connecting domains D0-Dc and D1 and those connecting domains D1 and D2 both appear to be flexible enough for each of the 11 stranded hook protofilaments to be curved and twisted to different extents and directions when the hook is bent to different curvatures for the universal joint function.

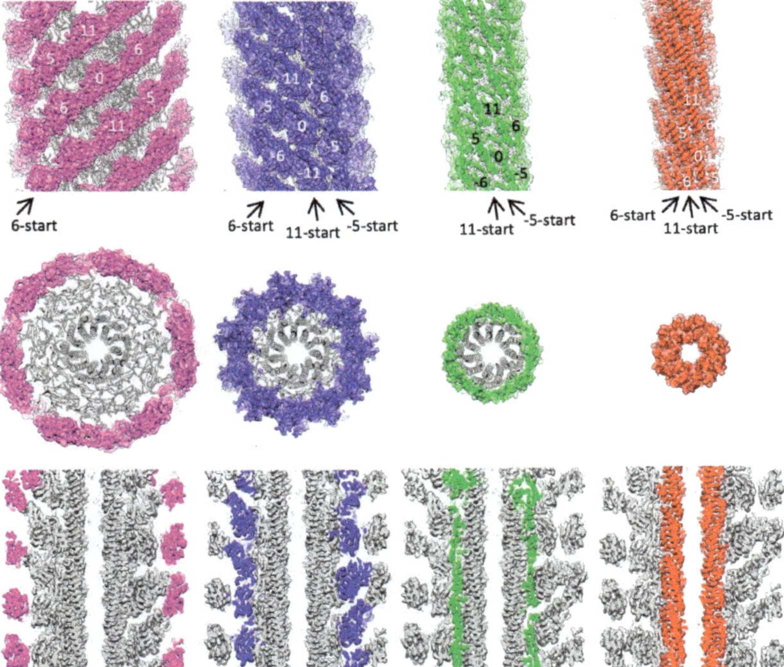

Figure 1. Structures and spatial arrangements of FlgE domains in the hook. The 3D density map and ribbon diagram of the atomic model are presented in different views and slices to visualize the structure and helical array of each of the four domains in different colors: D2, magenta; D1, blue; Dc, green; and D0, red, from left to right panels. The side views are shown in the upper panels, end-on views in the middle, and central slices along the hook axis in the lower. The diameters of these domain arrays are 172 Å, 142 Å, 85 Å, and 62 Å, respectively. The start numbers of the three major helical lines are labeled with arrows in the side views.

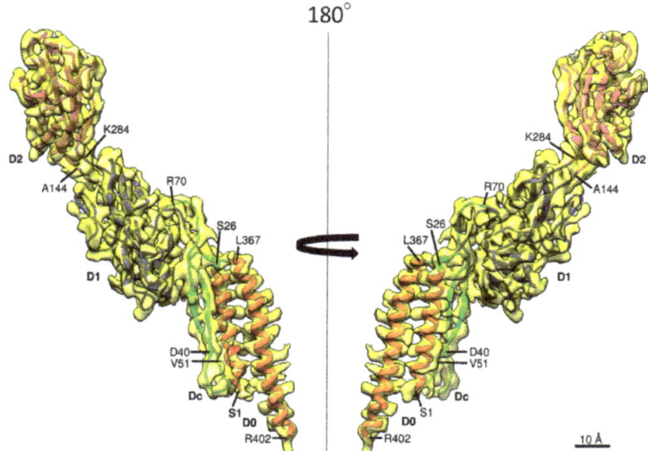

Figure 2. The 3D density map and model of a FlgE subunit. Two side views are shown, with the four domains colored as follows: D0, red; Dc, green; D1, gray; and D2, orange.

3.3. *Intersubunit Interactions*

Intersubunit interactions are shown for each of the four domains in the upper panel of Figure 1. FlgE subunits assemble into the tubular structure of the hook in a helical array, and the hook structure has three major helical lines: −5-start, 6-start and 11-start, where the number represents the number of helical strands, and the negative and positive numbers mean left- and right-handed helices, respectively. According to these helical lines, we call one of the subunits subunit 0 and its surrounding six subunits as subunit −5, 5, −6, 6, −11, and 11 (see upper panel of Figure 1).

The D2 domains (magenta) form continuous helical arrays along the 6-start helical lines on the hook surface but have no interactions along −5-start and 11-start helices, and therefore, these six stranded helical arrays of D2 domains are widely apart from each other. The D1 domains (blue) form a mesh-like array along all three major helical lines but their interactions are not very tight with a small gap in all the directions, suggesting that these interactions modestly contribute to the stability of the hook structure. The D1 domains are probably interconnected indirectly through one or two layers of water molecules between their domain surfaces, although water molecules are not visible at this resolution [38]. These large and small axial gaps between the D2 domains and between the D1 domains, respectively, are actually important for flexible bending of the hook for its universal joint function because the axial compression and extension of the protofilaments are essential for the bending motion.

In contrast, the Dc domains (green) have close interactions in the −5-start and 11-start helical lines, making a stable tubular structure, albeit the 6-start interactions are absent. The D0 domains (red) form intimate interactions in all the directions, with the N- and C-terminal helices forming a two-layer helical bundle where the C-terminal helices tile the inner surface of the central channel of the hook with polar and charged residues. The N-terminal helices, not facing the channel, have interactions only in the −5-start helical line, but together with Dc domains, they are involved in intricate intersubunit interactions. As described in the previous section, the N-terminal helix of domain D0 and the long β-hairpin of domain Dc form a tight interaction, strongly suggesting that domains D0 and Dc are likely to behave as one domain D0-Dc. A stretched loop of subunit −11 formed by residues 64–67 at the top of domain Dc connecting to domain D1 is involved in close tripartite interactions with the tip of the long β-hairpin of domain Dc (residues 40–52) and the N-terminal end (residues 1–7) of subunit 0 (Figure 3 upper panel). Thus, the tubular structure of the hook is stabilized and maintained mainly by the intersubunit interactions of D0-Dc domains alone in the inner core of the hook structure.

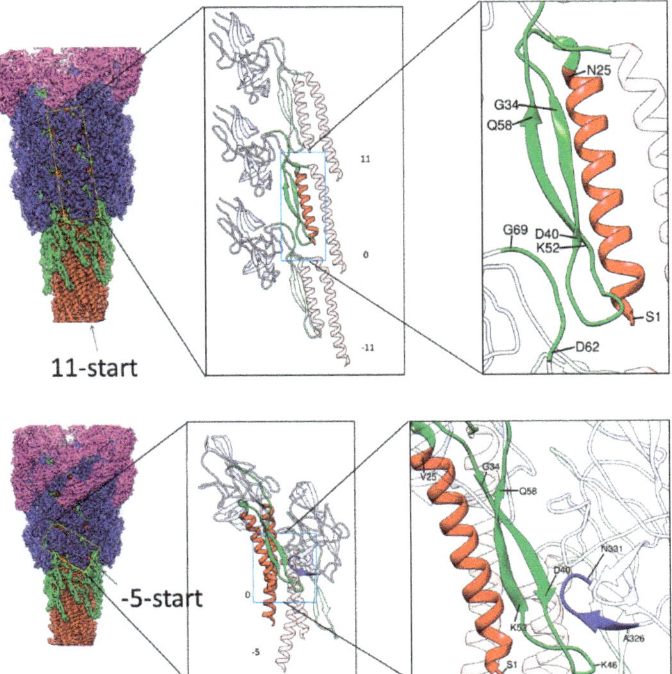

Figure 3. Intermolecular interactions along the 11-start and −5-start direction. Left panels are side views of the hook with the four domains of FlgE colored as in Figure 1, with outer domains removed towards the bottom. The yellow boxes in the left panel indicate the arrays of FlgE subunits magnified in the middle panel: Three subunits in the 11-start helical line in the upper row and two subunits in the −5-start in the lower row. Magnified images of some parts of FlgE are also colored as in Figure 1 in the middle and right panels. In the upper row, the N-terminal helix and the long β-hairpin of domain Dc of subunit 0 are colored red and green, respectively. In the lower row, the N- and C-terminal helices and the long β-hairpin of subunit 0 are colored red and green, respectively, and the tip loop of a short β-hairpin at the bottom of domain D1 of subunit −5 is colored blue. These colored parts in the middle panels are highlighted with light-blue boxes to indicate the portions further magnified in the right panels.

There is, however, an additional intersubunit interaction contributing to the structural stability that is found between different domains. Residues 38–40 in the middle of the β-hairpin of domain Dc of subunit 0 are closely interacting with the tip loop (residues 328–331) of a short β-hairpin of domain D1 of subunit −5 (Figure 3 lower panel). Thus, domain D1 is also involved in further stabilizing the hook structure through an intersubunit interaction with domain D0-Dc. Thus, these intersubunit interactions clearly indicate the importance of the long β-hairpin of domain Dc for the structural stability of the hook, as found by deletion mutation experiments [39].

Even in these tight interactions between D0-Dc domains stabilizing the hook structure, the axial gap with a distance corresponding to about one turn of α-helix (5 Å) is still present to allow the compression of the protofilament for flexible bending of the hook (Figure 3 upper middle panel). The mutual displacement of the chains involved in these 11-start tripartite interactions may be small enough to be achieved by conformational rearrangements of their side chains but an intersubunit sliding interaction similar to what occurs between the triangular loop of domain D1 and domain D2 of subunit −11 [25] may also occur upon bending of the hook.

3.4. *Role of the Longer β-Hairpin of Domain Dc in the Flagellar Rod Structure*

The flagellar rod is a drive shaft of the rotary motor, transmitting motor torque to the filament acting as a helical propeller through the hook as a universal joint. The rod is a helical tubular assembly of four proteins, FlgB, FlgC, FlgF, and FlgG, but its major part rotating inside the LP ring of the flagellar basal body is the distal part of the rod formed by FlgG [23]. The diameter of the FlgG rod is only 13 nm, markedly smaller than 18 nm of the hook, but the rod is much more rigid as a drive shaft than the hook working as a universal joint. Interestingly, however, *Salmonella* FlgG and FlgE share a highly homologous sequence with 32% identity by BLAST search [40], except that the FlgG sequence is shorter than FlgE by the absence of 141 residues (residues 144–284) in the central region of the FlgE sequence corresponding to domain D2, and so FlgG consists of domains D0-Dc and D1 and is missing D2. That is why the diameter of the rod is smaller than that of the hook.

On the other hand, FlgG has a region of 18 extra residues in the domain Dc region, suggesting that the β-hairpin of domain Dc is longer than that of FlgE (Figure 4a). FlgE of *C. jejuni* also has a similar sequence to *Salmonella* FlgG, with 17 extra residues in this region compared with *Salmonella* FlgE (Figure 4b), suggesting a longer β-hairpin in *Campylobacter* FlgE than *Salmonella* FlgE as well. In agreement with this prediction, the recent cryoEM structure of the hook isolated from *C. jejuni* showed an L-shaped, long β-hairpin named L-stretch in domain Dc of FlgE [28], which is much longer than that of *Salmonella* FlgE. So we used this structure to build a homology model of the β-hairpin of *Salmonella* FlgG and inserted it into the atomic model of FlgG in the polyrod solved by cryoEM analysis [23] in order to compare the differences in these three structures.

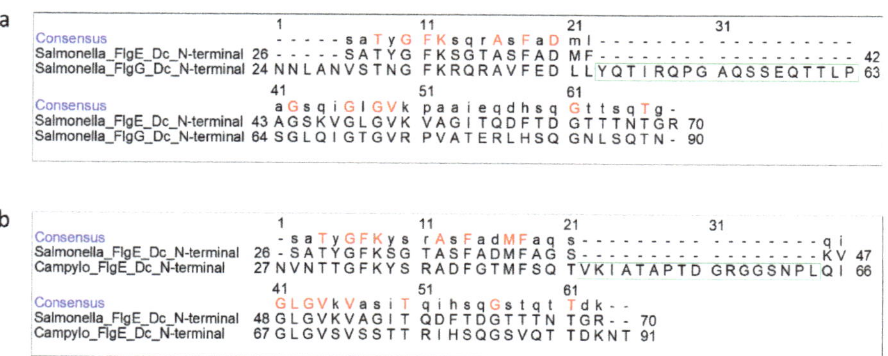

Figure 4. Pairwise sequence alignment of FlgE and FlgG in the domain Dc region. (**a**) Sequence alignment between *Salmonella* FlgE and FlgG, and (**b**) between *Salmonella* FlgE and *Campylobacter* FlgE. The 18 and 17 residue insertions in *Salmonella* FlgG and *Campylobacter* FlgE, respectively, are boxed in green.

Pairwise comparisons by superposing the D0 domains are shown in Figure 5 with *Salmonella* FlgE in blue, *Campylobacter* FlgE in red and *Salmonella* FlgG in green. The relative disposition of the β-hairpin to the N- and C-terminal helices of domain D0 is nearly the same between *Salmonella* FlgE and *Campylobacter* FlgE (Figure 5 left) and is slightly different between *Salmonella* FlgG and *Campylobacter* FlgE (Figure 5 middle) and between *Salmonella* FlgG and FlgE (Figure 5 right), but these differences are mainly in the distal portion of the β-hairpin. The model of the FlgG rod, which was missing domain Dc in the previous study [23] but is now completed with the long β-hairpin, is shown in Figure 6. The model is shown in different views and slices similarly to Figure 1 to visualize the structure and helical array of each of the three domains in different colors: D1 in yellow; Dc in magenta; and D0 in light blue, from left to right. The distinct structural feature of the rod compared to that of the hook shown in Figure 1 is that the tip of the β-hairpin of domain Dc is projecting out (Figure 6 middle). If we look at the β-hairpin structures and their locations and interactions with other domains in *Salmonella* hook, *Salmonella* rod and *Campylobacter* hook, as presented in Figure 7, it is clear that the long β-hairpins of *Salmonella* FlgG and *Campylobacter* FlgE are projecting out into the outer layer of their tubular structures composed of the D1 domains and are filling the gap formed in the helical array of the D1 domains to further stabilize and rigidify the tubular structure.

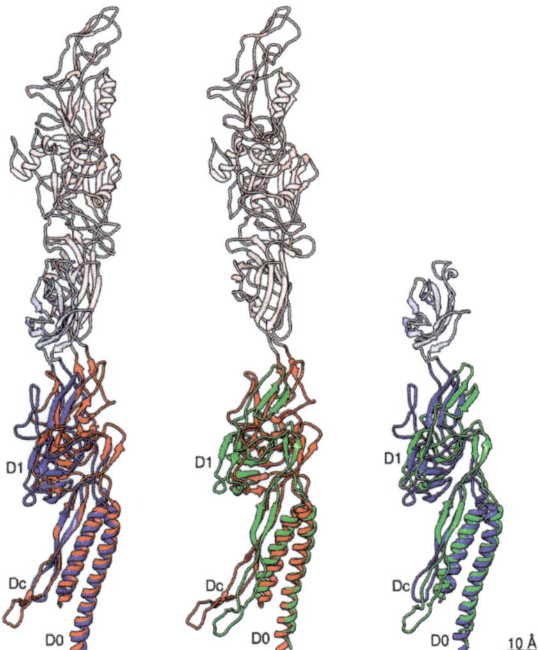

Figure 5. Pairwise structural comparison between *Salmonella* FlgE and FlgG and *Campylobacter* FlgE. *Salmonella* FlgE and *Campylobacter* FlgE are compared in the left panel, *Salmonella* FlgG and *Campylobacter* FlgE in the middle, and *Salmonella* FlgE and FlgG in the right. The Cα ribbon models are colored only for domains D0-Dc and D1, with *Salmonella* FlgE in blue, FlgG in green and *Campylobacter* FlgE in red. Domain D0 is used to superpose these molecules. The relative arrangements of domains D1 and D0-Dc between these three molecules are slightly different to one another as shown in this figure. The extra part of domain D1 present in FlgE and missing in FlgG is the triangular loop involved in the intersubunit sliding interactions between D1 and D2 along the protofilament for its compression and extension.

Although *Campylobacter* hook is curved in the native form and functions as a universal joint, it easily becomes straight and appears to be more rigid than *Salmonella* hook [28], and this is well explained by the tip of long β-hairpin filling the gap of the helical array of the D1 domains. *Campylobacter* hook needs to be more rigid than *Salmonella* hook, which is maybe because *Campylobacter* has two polar flagella and the hook needs to be rigid in order to keep orienting the flagellar filaments in the axis of the cell body to produce a thrust efficiently. The insertion of extra 18 residues of the *Salmonella* FlgG-specific sequence into *Salmonella* FlgE was expected to make *Salmonella* hook as rigid as *Salmonella* rod. The hook formed by this insertion mutant of FlgE actually became straight and rigid just like the rod, and the flagellar filaments on these mutant *Salmonella* cells were all spread apart and could not form a bundle to produce thrust for swimming even though individual motors were rotating rapidly [18].

Thus, the long β-hairpin of domain Dc plays an important role in the flagellar rod and hook structure, regulating and optimizing the structural stability and rigidity of the entire tubular assembly by changing its length for their required functions. The cryoEM structures of the flagellar rod and hook clearly revealed it, demonstrating the power of the cryoEM technique as a unique and essential tool for structural biology capable of visualizing native, functional structures of biological macromolecular assemblies.

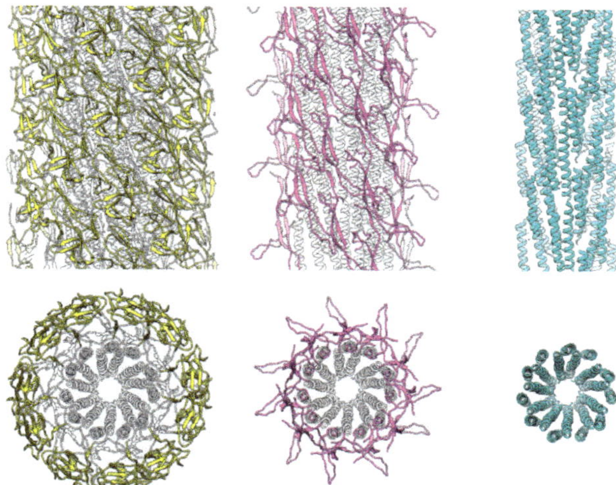

Figure 6. Structures and spatial arrangements of FlgG domains in the rod. The Cα ribbon models are presented in side and end-on views in the upper and lower panels, respectively, to visualize the structure and helical array of each of the three domains in different colors: D1, yellow; Dc, magenta; and D0, light blue, from left to right. The structure of FlgG is very similar to that of FlgE except for the length of the β-hairpin of domain Dc, with the extra length formed by the FlgG-specific 18 residue insertion projecting out nearly horizontally in the rod.

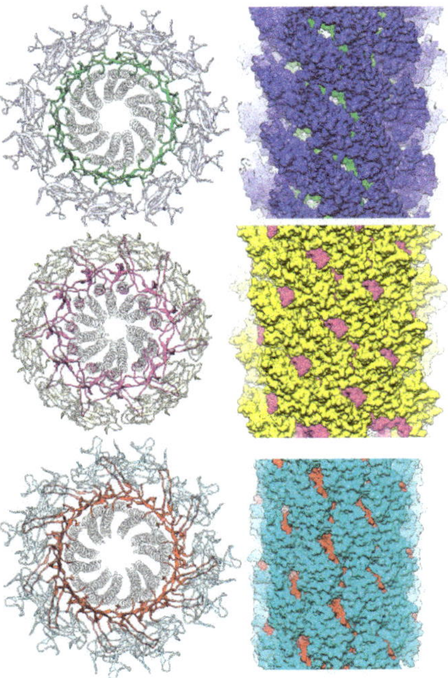

Figure 7. Structural comparison of *Salmonella* hook and rod and *Campylobacter* hook. The Cα ribbon models of *Salmonella* hook at the top, the rod in the middle, and *Campylobacter* hook at the bottom. The models are presented in the end-on and side views in the left and right panels, respectively, with only D0-Dc and D1 domains being shown for both hook structures. The Dc domains are highlighted in the left panels by different colors, and the D1 domains are also colored in the right panels. The gaps between the D1 domains are clear in the *Salmonella* hook structure at the top right, but the gaps are filled by the tip of the β-hairpin of domain Dc in the *Salmonella* rod and *Campylobacter* hook structures, further stabilizing and rigidifying their tubular structures.

4. Conclusions

We carried out the structural analysis of the flagellar hook isolated from *Salmonella* in the straight form by cryoEM helical image analysis and built an atomic model based on the 4.1 Å resolution map. This model revealed the entire structure of the hook component protein FlgE consisting of domains D0, D1 and D2 as well as the long β-hairpin connecting domains D0 and D1, which is involved in intricate intermolecular interactions to stabilize the tubular structure of the hook while maintaining its bending flexibility. This structural feature is distinct from that of the flagellar rod consisting of FlgG with a much longer β-hairpin, which further stabilizes and rigidifies the rod structure to work as the drive shaft of the flagellar motor.

Supplementary Materials: The following are available online at http://www.mdpi.com/2218-273X/9/9/462/s1, Figure S1: Fourier shell correlation of the 3D reconstruction.

Author Contributions: Conceptualization, T.K. and K.N.; Sample preparation, P.H. and T.M.; Data collection and analysis, P.H. and T.K.; Validation, P.H., T.K. and K.N.; Data curation, P.H., T.K. and K.N.; Writing—original draft preparation, P.H. and T.K.; Writing—review and editing, T.K., T.M. and K.N.; Visualization, P.H. and T.K.; Project administration, T.K. and K.N.; Funding acquisition, T.K. and K.N.

Funding: This work has been supported by JSPS KAKENHI Grant Number JP25000013 (to K.N.) and 18K06155 (to T.K.). This work has also been supported by Platform Project for Supporting Drug Discovery and Life Science Research (BINDS) from AMED under Grant Number JP19am0101117 (to K.N.) and by JEOL YOKOGUSHI Research Alliance Laboratories of Osaka University (to K.N.).

Acknowledgments: We thank Katsumi Imada for technical help in atomic model building and Tohru Minamino for discussion and critical reading of the manuscript.

Conflicts of Interest: The authors declare no conflict of interest.

References

1. Namba, K.; Vonderviszt, F. Molecular architecture of bacterial flagellum. *Q. Rev. Biophys.* **1997**, *30*, 1–65. [CrossRef] [PubMed]
2. Macnab, R.M. How bacteria assemble flagella. *Annu. Rev. Microbiol.* **2003**, *57*, 77–100. [CrossRef] [PubMed]
3. Berg, H.C. The rotary motor of bacterial flagella. *Annu. Rev. Biochem.* **2003**, *72*, 19–54. [CrossRef] [PubMed]
4. Minamino, T.; Imada, K.; Namba, K. Molecular motors of the bacterial flagella. *Curr. Opin. Struct. Biol.* **2008**, *18*, 693–701. [CrossRef] [PubMed]
5. Kojima, S.; Blair, D.F. The bacterial flagellar motor: Structure and function of a complex molecular machine. *Int. Rev. Cytol.* **2004**, *233*, 93–134. [CrossRef] [PubMed]
6. Lowe, G.; Meister, M.; Berg, H.C. Rapid rotation of flagellar bundles in swimming bacteria. *Nature* **1987**, *325*, 637–640. [CrossRef]
7. Kudo, S.; Magariyama, Y.; Aizawa, S.-I. Abrupt changes in flagellar rotation observed by laser dark-field microscopy. *Nature* **1990**, *346*, 677–680. [CrossRef] [PubMed]
8. Berg, H.C.; Anderson, R.A. Bacteria swim by rotating their flagellar filaments. *Nature* **1973**, *245*, 380–382. [CrossRef]
9. Silverman, M.; Simon, M. Flagellar rotation and the mechanism of bacterial motility. *Nature* **1974**, *249*, 73–74. [CrossRef]
10. Asakura, S. Polymerization of flagellin and polymorphism of flagella. *Adv. Biophys.* **1970**, *1*, 99–155.
11. Macnab, R.M.; Ornston, M.K. Normal-to-curly flagellar transitions and their role in bacterial tumbling. Stabilization of an alternative quaternary structure by mechanical force. *J. Mol. Biol.* **1977**, *112*, 1–30. [CrossRef]
12. Turner, L.; Ryu, W.S.; Berg, H.C. Real-time imaging of fluorescent flagellar filaments. *J. Bacteriol.* **2000**, *182*, 2793–2801. [CrossRef] [PubMed]
13. DePamphilis, M.L.; Adler, J. Purification of intact flagella from *Escherichia coli* and *Bacillus subtilis*. *J. Bacteriol.* **1971**, *105*, 376–383. [PubMed]
14. DePamphilis, M.L.; Adler, J. Fine structure and isolation of the hook-basal body complex of flagella from *Escherichia coli* and *Bacillus subtilis*. *J. Bacteriol.* **1971**, *105*, 384–395. [PubMed]
15. Kagawa, H.; Aizawa, S.I.; Asakura, S. Transformations in isolated polyhooks. *J. Mol. Biol.* **1979**, *129*, 333–336. [CrossRef]
16. Wagenknecht, T.; DeRosier, D.; Shapiro, L.; Weissborn, A. Three-dimensional reconstruction of the flagellar hook from *Caulobacter crescentus*. *J. Mol. Biol.* **1981**, *151*, 439–465. [CrossRef]
17. Hirano, T.; Yamaguchi, S.; Oosawa, K.; Aizawa, S. Roles of FliK and FlhB in determination of flagellar hook length in *Salmonella typhimurium*. *J. Bacteriol.* **1994**, *176*, 5439–5449. [CrossRef] [PubMed]
18. Hiraoka, K.D.; Morimoto, Y.V.; Inoue, Y.; Fujii, T.; Miyata, T.; Makino, F.; Minamino, T.; Namba, K. Straight and rigid flagellar hook made by insertion of the FlgG specific sequence into FlgE. *Sci. Rep.* **2017**, *7*, 46723. [CrossRef]
19. Fujii, T.; Matsunami, H.; Inoue, Y.; Namba, K. Evidence for the hook supercoiling mechanism of the bacterial flagellum. *Biophys. Physicobiol.* **2018**, *15*, 28–32. [CrossRef]
20. Sakai, T.; Inoue, Y.; Terahara, N.; Namba, K.; Minamino, T. A triangular loop of domain D1 of FlgE is essential for hook assembly but not for the mechanical function. *Biochem. Biophys. Res. Commun.* **2018**, *495*, 1789–1794. [CrossRef]
21. Homma, M.; Kutsukake, K.; Hasebe, M.; Iino, T.; Macnab, R.M. FlgB, FlgC, FlgF and FlgG. A family of structurally related proteins in the flagellar basal body of *Salmonella typhimurium*. *J. Mol. Biol.* **1990**, *211*, 465–477. [CrossRef]
22. Jones, C.J.; Macnab, R.M.; Okino, H.; Aizawa, S. Stoichiometric analysis of the flagellar hook-(basal-body) complex of *Salmonella typhimurium*. *J. Mol. Biol.* **1990**, *212*, 377–387. [CrossRef]

23. Fujii, T.; Kato, T.; Hiraoka, K.D.; Miyata, T.; Minamino, T.; Chevance, F.F.V.; Hughes, K.T.; Namba, K. Identical folds used for distinct mechanical functions of the bacterial flagellar rod and hook. *Nat. Commun.* **2017**, *8*, 14276. [CrossRef] [PubMed]
24. Minamino, T.; Kato, T.; Makino, F.; Horváth, P.; Miyata, T.; Namba, K. Electron microscopy of motor structure and possible mechanisms. In *Encyclopedia of Biophysics*, 2nd ed.; Roberts, G.C.K., Watts, A., Eds.; Springer: Basel, Switzerland, 2018. [CrossRef]
25. Samatey, F.A.; Matsunami, H.; Imada, K.; Nagashima, S.; Shaikh, T.R.; Thomas, D.R.; Chen, J.Z.; DeRosier, D.J.; Kitao, A.; Namba, K. Structure of the bacterial flagellar hook and implication for the molecular universal joint mechanism. *Nature* **2004**, *431*, 1062–1068. [CrossRef] [PubMed]
26. Shaikh, T.R.; Thomas, D.R.; Chen, J.Z.; Samatey, F.A.; Matsunami, H.; Imada, K.; Namba, K.; DeRosier, D.J. A partial atomic structure for the flagellar hook of *Salmonella typhimurium*. *Proc. Natl. Acad. Sci. USA* **2005**, *102*, 1023–1028. [CrossRef] [PubMed]
27. Fujii, T.; Kato, T.; Namba, K. Specific arrangement of alpha-helical coiled coils in the core domain of the bacterial flagellar hook for the universal joint function. *Structure* **2009**, *17*, 1485–1493. [CrossRef] [PubMed]
28. Matsunami, H.; Barker, C.S.; Yoon, Y.-H.; Wolf, M.; Samatey, F.A. Complete structure of the bacterial flagellar hook reveals extensive set of stabilizing interactions. *Nat. Commun.* **2016**, *7*, 13425. [CrossRef] [PubMed]
29. Aizawa, S.I.; Dean, G.E.; Jones, C.J.; Macnab, R.M.; Yamaguchi, S. Purification and characterization of the flagellar hook-basal body complex of *Salmonella typhimurium*. *J. Bacteriol.* **1985**, *161*, 836–849. [PubMed]
30. Zheng, S.Q.; Palovcak, E.; Armache, J.-P.; Verba, K.A.; Cheng, Y.; Agard, D.A. MotionCor2: Anisotropic correction of beam-induced motion for improved cryo-electron microscopy. *Nat. Methods* **2017**, *14*, 331–332. [CrossRef]
31. Zhang, K. Gctf: Real-time CTF determination and correction. *J. Struct. Biol.* **2016**, *193*, 1–12. [CrossRef] [PubMed]
32. He, S.; Scheres, S.H.W. Helical reconstruction in RELION. *J. Struct. Biol.* **2017**, *198*, 163–176. [CrossRef] [PubMed]
33. Webb, B.; Sali, A. Protein structure modeling with MODELLER. *Methods Mol. Biol.* **2014**, *1137*, 1–15. [CrossRef] [PubMed]
34. Wang, R.Y.-R.; Song, Y.; Barad, B.A.; Cheng, Y.; Fraser, J.S.; DiMaio, F. Automated structure refinement of macromolecular assemblies from cryo-EM maps using Rosetta. *eLife* **2016**, *5*, e17219. [CrossRef] [PubMed]
35. Emsley, P.; Lohkamp, B.; Scott, W.G.; Cowtan, K. Features and development of Coot. *Acta Crystallogr. D Biol. Crystallogr.* **2010**, *66*, 486–501. [CrossRef]
36. Pettersen, E.F.; Goddard, T.D.; Huang, C.C.; Couch, G.S.; Greenblatt, D.M.; Meng, E.C.; Ferrin, T.E. UCSF Chimera—A visualization system for exploratory research and analysis. *J. Comput. Chem.* **2004**, *25*, 1605–1612. [CrossRef]
37. Patterson-Delafield, J.; Martinez, R.J.; Stocker, B.A.; Yamaguchi, S. A new *fla* gene in *Salmonella typhimurium*–*flaR*—And its mutant phenotype-superhooks. *Arch Mikrobiol.* **1973**, *90*, 107–120. [CrossRef]
38. Furuta, T.; Samatey, F.A.; Matsunami, H.; Imada, K.; Namba, K.; Kitao, A. Gap compression/extension mechanism of bacterial flagellar hook as the molecular universal joint. *J. Struct. Biol.* **2007**, *157*, 481–490. [CrossRef]
39. Moriya, N.; Minamino, T.; Ferris, H.U.; Morimoto, Y.V.; Ashihara, M.; Kato, T.; Namba, K. Role of the Dc domain of the bacterial hook protein FlgE in hook assembly and function. *Biophysics* **2013**, *9*, 63–72. [CrossRef]
40. Altschul, S.F.; Gish, W.; Miller, W.; Myers, E.W.; Lipman, D.J. Basic local alignment search tool. *J. Mol. Biol.* **1990**, *215*, 403–410. [CrossRef]

Article

Structural and Functional Comparison of *Salmonella* Flagellar Filaments Composed of FljB and FliC

Tomoko Yamaguchi [1,2,†], Shoko Toma [1,†], Naoya Terahara [1,3], Tomoko Miyata [1], Masamichi Ashihara [1,‡], Tohru Minamino [1], Keiichi Namba [1,2,4,5,*] and Takayuki Kato [1,*,§]

1 Graduate School of Frontier Biosciences, Osaka University, 1-3 Yamadaoka, Suita, Osaka 565-0871, Japan; u523219i@fbs.osaka-u.ac.jp (T.Y.); tomax002@umn.edu (S.T.); terahara.19r@g.chuo-u.ac.jp (N.T.); miya@fbs.osaka-u.ac.jp (T.M.); masamichi.ashihara@thermofisher.com (M.A.); tohru@fbs.osaka-u.ac.jp (T.M.)

2 RIKEN Center for Biosystems Dynamics Research, 1-3 Yamadaoka, Suita, Osaka 565-0871, Japan

3 Faculty of Science and Engineering, Chuo University, Tokyo 112-8551, Japan

4 RIKEN SPring-8 Center, 1-3 Yamadaoka, Suita, Osaka 565-0871, Japan

5 JEOL YOKOGUSHI Research Alliance Laboratories, Osaka University, 1-3 Yamadaoka, Suita, Osaka 565-0871, Japan

* Correspondence: keiichi@fbs.osaka-u.ac.jp (K.N.); tkato@protein.osaka-u.ac.jp (T.K.); Tel.: +81-6-6879-4625 (K.N. & T.K.)

† These authors contributed equally to this work.

‡ Current address: Thermo Fisher Scientific K.K., 4-2-8 Shibaura, Minato-ku, Tokyo 108-0023, Japan.

§ Current address: Institute for Protein Research, Osaka University, 3-2 Yamadaoka, Suita, Osaka 565-0871, Japan.

Received: 27 December 2019; Accepted: 4 February 2020; Published: 6 February 2020

Abstract: The bacterial flagellum is a motility organelle consisting of a long helical filament as a propeller and a rotary motor that drives rapid filament rotation to produce thrust. *Salmonella enterica* serovar Typhimurium has two genes of flagellin, *fljB* and *fliC*, for flagellar filament formation and autonomously switches their expression at a frequency of 10^{-3}–10^{-4} per cell per generation. We report here differences in their structures and motility functions under high-viscosity conditions. A *Salmonella* strain expressing FljB showed a higher motility than one expressing FliC under high viscosity. To examine the reasons for this motility difference, we carried out structural analyses of the FljB filament by electron cryomicroscopy and found that the structure was nearly identical to that of the FliC filament except for the position and orientation of the outermost domain D3 of flagellin. The density of domain D3 was much lower in FljB than FliC, suggesting that domain D3 of FljB is more flexible and mobile than that of FliC. These differences suggest that domain D3 plays an important role not only in changing antigenicity of the filament but also in optimizing motility function of the filament as a propeller under different conditions.

Keywords: bacterial flagellar motility; flagellin; *Salmonella*; FljB; FliC; electron cryomicroscopy; viscosity; infection

1. Introduction

Salmonella infection is one of the four major causes of disease involving diarrheas in the world. *Salmonella enterica* serovar Typhimurium (hereafter *Salmonella*) has a wide range of hosts, and it infects not only mouse, which is its original host, but also humans. The infection occurs mainly via oral intake, and flagellar motility plays an important role in infection. The flagella enable bacteria to move through viscous environments such as mucosa, search for the host cell surface, and adhere to the cell membrane for infection [1,2].

The bacterial flagellum consists of three main parts: the basal body, which works as a rotary motor; the filament, which functions as a screw propeller; and the hook as a universal joint connecting the filament to the motor [3]. *Salmonella* has several peritrichous flagella, and the length of the filament is 10 to 15 μm long. When the cell swims straight, the motors rotate counterclockwise (CCW), and the normally left-handed supercoiled filaments form a bundle behind the cell to produce thrust. When the motors switch their rotation to the clockwise (CW) direction, the filaments switch to a right-handed supercoil in order for the bundle to fall apart so that the cell can change its orientation by tumbling to change the direction of swimming. *Salmonella* has two flagellin genes, *fljB* and *fliC*, and their expression is autonomously regulated to produce either FljB or FliC for filament formation at a frequency of 10^{-3}–10^{-4} per cell per generation. This is called phase variation [4], and it is thought to be a mechanism to enable escape from the host immune system by changing flagellar antigenicity.

It has been reported that FliC-expressing bacteria display a significant advantage for invasion of most epithelial cell lines of murine and human origin compared to FljB-expressing bacteria. Differences in the swimming behaviors near surfaces have also been observed, and FliC-expressing bacteria more frequently "stop" [5]. In order to understand the differences in their antigenicity and motility function, structural and functional characterization is necessary. However, structural information is available only for the FliC filament [6–8].

The three-dimensional (3D) structures of FliC and the FliC filament have been studied by X-ray crystallography and electron cryomicroscopy (cryoEM) helical image analysis, respectively [6–8]. *Salmonella* FliC from the strain SJW1103 consists of 494 amino acid residues, and the molecular mass is about 50 kDa. The molecule consists of four domains, D0, D1, D2, and D3, arranged from the inner core to the outer surface of the filament. Domains D0 and D1 form the inner core of the filament and are made of α-helical coiled coils. These domains play a critical role in forming the supercoiled structure of the filament as a helical propeller. In addition, a β-hairpin structure in domain D1 is considered to be important for switching the conformation of flagellin subunits between the two states to produce various types of supercoiled filaments in left- and right-handed forms for swimming and tumbling [6,8]. Domains D2 and D3 are found in the outer part of the filament structure. These two domains increase the stability of the filament structure as well as the drag force of the filament as a propeller by increasing the diameter [6,9]. The outermost domain, D3, is thought to contain epitopes for antibodies, determining the antigenicity of the flagella.

To understand the differences in the antigenicity between FljB and FliC, structural information for the FljB and FljB filaments is necessary. Functional characterization of cell motility is also necessary to investigate the potentially different physiological roles played by these two types of filament, if any.

In the present study, we therefore investigated differences in the FljB and FliC filament structures and their motility functions. A *Salmonella* strain expressing FljB showed a higher motility than the one expressing FliC under high-viscosity conditions. The structure of the FljB filament analyzed by cryoEM image analysis was nearly identical to that of the FliC filament, except for the position and orientation of the outermost domain D3. Domain D3 of FljB also showed a higher flexibility and mobility than that of FliC. These differences suggest that domain D3 plays an important role not only in changing antigenicity, but also in optimizing motility function of the filament as a propeller under different conditions. We have discussed the relationship between the structure and motility function by comparing FljB and FliC.

2. Materials and Methods

2.1. Salmonella Strains

Bacterial strains of *Salmonella enterica* serovar Typhimurium used in this study are listed in Table 1. SJW1103 lacks the *fljB* operon and so expresses only the FliC flagellin. To express FljB from the *fliC* promoter on the chromosome, the Δ*fliC*::*tetRA* allele was replaced by the *fljB* allele using the λ Red homologous recombination system [10], as described previously [11], to generate the SJW1103B strain. For cryoEM structural analyses, the strain expressing the flagellar filament of the R-type straight form (the right-handed helical symmetry) was used.

Table 1. Strains and plasmids used in this study.

Salmonella Strains	Relevant Characteristics	Source or Reference
SJW1103	FliC wild-type Δ*hin-fljB-fljA*	Yamaguchi et al. 1984 [12]
SJW1103ΔC	SJW1103 Δ*fliC::tetRA*	This study
SJW1103B	FljB wild-type Δ*fliC::fljB*	This study
SJW590	FljB_R-type straight filament *fljB(A461V)*, Δ*fliC*	This study

2.2. Swimming Motility Assay

Salmonella strains, SJW1103B (only expressing FljB) and SJW1103 (only expressing FliC), were pre-cultured in 5 mL of Luria–Bertani broth (LB, 1% (*w*/*v*) tryptone, 0.5% (*w*/*v*) yeast extract, 0.5% (*w*/*v*) NaCl) with overnight shaking at 37 °C. Bacterial growth was measured via optical density at 600 nm (OD_{600}).

A 5 μL measure of the culture medium was inoculated into 5 mL of fresh LB, and it was incubated for 6 h at 37 °C with shaking. The cells were diluted in the motility buffer (10 mM potassium phosphate pH 7.0, 0.1 mM EDTA, 10 mM sodium D-lactate). The viscosity of the motility buffer was adjusted by adding Ficoll PM400 (GE healthcare, Chicago, IL, USA) to a final concentration of 5%, 10%, or 15%. Swimming motility was observed using a bright-field microscope, CX-41 (Olympus, Tokyo, Japan) with a 40× objective (PlanC N 40× NA 0.65 Olympus) and a 1.25× variable magnification lens (U-CA 1.5 × Olympus) and recorded using a high-speed camera (GEV-B0620M-TC000 IMPREX, Boca Raton, FL, USA). The swimming speed was calculated using the software Move-tr2/2D (Library Co., Tokyo, Japan).

2.3. Fluorescence Labeling of Flagellin Antibodies

IgG antibodies against FliC and FljB were purified from rabbit serum using a Melon™ Gel IgG Spin Purification Kit (Thermo Fisher Scientific, Waltham, MA, USA). The buffer was changed to a phosphate-buffered salt solution (pH 7.4) simultaneously with concentration of IgG by using a YM-50 Centriprep centrifugal filter (Merck, Darmstadt, Germany). After adjusting the concentration of IgG to 1 mg/mL, the antibodies were labelled with an Alexa Fluor™ 594 Antibody Labeling Kit (Thermo Fisher Scientific) for the FliC antibody and with an Alexa Fluor™ 488 Antibody Labeling Kit (Thermo Fisher Scientific) for the FljB antibody.

2.4. Immunofluorescent Staining of the Flagellar Filament

SJW1103B and SJW1103 were cultured under the same conditions used for the swimming motility assay described above. The culture was diluted 50 to 100 times with the motility buffer. To label the flagellar filaments, the cells were first attached to the surface of a slide glass coated with polylysine by filling the culture medium under a cover slip for 5 min. The culture medium was then replaced with a 20 μL solution of fluorescently labeled antibodies at 25 μg/mL, and it was left for 5 to 10 min. The excess antibody dye was washed by gently flowing 40 μL of the motility buffer twice, and the filaments were then observed by fluorescence microscopy. These observations were performed under an IX-83 optical microscope with a UPlan-SApo 100× 1.4 NA objective lens (Olympus). The number of the flagellar filaments per cell was measured using ImageJ (National Institutes of Health, Bethesda, MD, USA).

2.5. Flagellar Filament Purification

SJW590 was pre-cultured in 30 mL LB with shaking overnight at 37 °C, and 15 mL of the culture was then inoculated into 1 L of fresh LB. The cells were grown until the OD_{600} value reached about 1.0. The culture was collected by centrifugation and resuspended in 10% sucrose buffer (10% (*w*/*v*) sucrose, 0.1 M Tris-HCl pH 8.0). EDTA (pH adjusted by HCl to 7.8) and lysozyme were added to final concentrations of 10 mM and 0.1 mg/mL, respectively. The suspension was stirred on ice for 1 h at 4 °C, and 0.1 M $MgSO_4$ and 10% (*w*/*v*) Triton X-100 were then added to final concentrations of 10 mM and 1% (*w*/*v*), respectively. After stirring for 1 h at 4 °C, 6 mL of 0.1 M EDTA (pH adjusted by NaOH to 11.0) was added. The solution was centrifuged at 15,000× *g*, and the supernatant was collected. The pH was adjusted to 10.9 with 5 N NaOH, and the sample solution was re-centrifuged at 15,000× *g* to remove undissolved membrane fractions. The supernatant was centrifuged at 67,000× *g* to collect the flagellar filament with the hook basal body attached (the filament) as a pellet. The sample was resuspended in 1 mL of Buffer C (10 mM Tris-HCl pH7.8, 5 mM EDTA, 1% (*w*/*v*) Triton X-100) and was centrifuged at 7300× *g*, and the supernatant was collected. The filament was purified by a sucrose density-gradient centrifugation method with a gradient of sucrose from 20% to 50%. The fraction containing the filament was collected and checked by SDS-PAGE. The fraction was 2-fold diluted by Buffer A (20 mM Tris-HCl pH7.8, 150 mM NaCl, 1 mM $MgCl_2$), and the sucrose was removed by centrifugation of the filament at 104,300× *g*. Finally, the filament was resuspended with 10–50 µL of Buffer A, and the sample solution was stored at 4 °C.

2.6. Negative Staining

A 5 µL aliquot of the sample solution was mixed with 2% PTA on Pala film and was placed on a thin carbon-coated, glow-discharged copper grid. The extra solution was removed from the grid by blotting, and the grid was dried for 1 h at room temperature. The sample was checked using a transmission electron microscope, JEM-1011 (JEOL, Akishima, Japan) with an accelerating voltage of 100 kV.

2.7. Electron Cryomicroscopy and Image Processing

A 3 µL aliquot of the sample solution was applied to a Quantifoil holey carbon grid R1.2/1.3 Mo 200 mesh (Quantifoil Micro Tools GmbH, Großlöbichau, Germany) with pretreatment of both sides of the grid by glow discharge. The grids were blotted and plunged into liquid ethane at the temperature of liquid nitrogen for rapid freezing [9,13] with Vitrobot Mark IV (Thermo Fisher Scientific). Grids were then transferred into electron cryomicroscopes with a cryostage cooled by liquid nitrogen. The frozen hydrated specimens of the FljB filament on the grid were observed using a Titan Krios electron cryomicroscope (Thermo Fisher Scientific) operated at an accelerating voltage of 300 kV. Dose-fractionated movie images were recorded using a direct electron detector camera, Falcon II (Thermo Fisher Scientific) and were automatically collected using EPU software (Thermo Fisher Scientific). Using a minimum dose system, all movies were taken with a total exposure of 2 s, an electron dose of 10.3 electron/Å^2 per frame, a defocus range of 0.2 to 1.9 µm, and a nominal magnification of 75,000×, corresponding to an image pixel size of 1.06 Å. All seven frames of the movie image were recorded with a frame rate of 0.1 s/frame, and the middle five frames from the 2nd to the 6th were used for image analysis. The defocus range was estimated by Gctf [14] after motion correction by RELION-3.0-β2 [15]. Using RELION-3.0-β2, the filament images were segmented and extracted in square boxes of 400 pixels with 90% overlap, and those segment images were aligned and analyzed. The overall resolution was estimated by Fourier Shell Correlation (FSC) = 0.143 (Figure S1). The summary of cryoEM data collection and statistics is shown in Supplementary Table S1.

The cryoEM 3D density map was deposited to the Electron Microscopy Data Bank under accession code EMD-9896, and the atomic model coordinates were deposited to the Protein Data Bank under accession code 6JY0.

3. Results

3.1. Motility Difference between Cells with the FljB and FliC Filaments

In order to investigate whether the two different types of flagellar filaments affected swimming motility, we first measured the swimming speed and the population of motile cells with the FljB and FliC filaments in motility buffer solutions of different viscosities adjusted by adding Ficoll. Optical microscopy was used for this observation. A wild-type *Salmonella* LT2 strain has two distinct flagellin genes, *fliC* and *fljB*, on the chromosome, and their expressions are switched autonomously. Therefore, we used two LT2 derivative strains, SJW1103 expressing only FliC and SJW1103B expressing only FljB, to examine structural and functional differences between these two flagellar filaments. These strains both have similar rates of growth (data not shown). The swimming speed of SJW1103 markedly decreased as the Ficoll concentration increased. When Ficoll was added to the final concentration of 10%, the average swimming speed decreased to 44% of that measured in the motility buffer without Ficoll. In contrast, the swimming speed of SJW1103B decreased only to 73% in the motility buffer with 10% Ficoll (Figure 1a). In addition, the decrease in the motile cell population of SJW1103B was much less than that of SJW1103. In the presence of 15% Ficoll in the motility buffer, 72% of SJW1103B cells were swimming while the motile fraction of SJW1103 cells was only 32% (Figure 1b). To examine the morphology, the length, and the number per cell of the flagellar filaments, the filaments were stained by immunofluorescence staining after 6 h of culture in LB and were observed by fluorescence microscopy. On average, most of the cells had three to five filaments, and their lengths were about 10 μm regardless of the strain (Figure 1c,d).

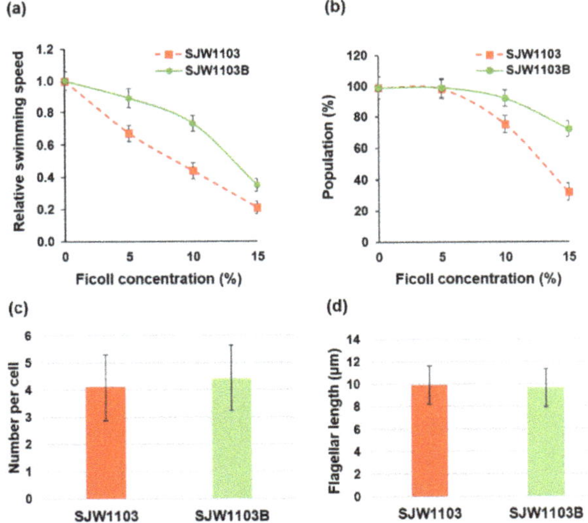

Figure 1. Motility and filament characteristics of *Salmonella* strains SJW1103 and SJW1103B. (**a**) Changes in the swimming speed and (**b**) changes in the swimming cell population under different viscosity conditions created by adding Ficoll to the motility buffer. (**c**) Number per cell and (**d**) length of the flagellar filaments. No significant difference was observed under either condition.

3.2. Structures of the FljB and FliC Filaments

We analyzed the structure of the FljB filament by cryoEM single-particle image analysis in one of the two mutant straight forms called the R-type. We used the R-type straight form to utilize the helical symmetry for image analysis as well as to compare the structure with the R-type straight filament formed by FliC. The 3D image of the FljB filament was reconstructed at 3.6 Å resolution from about

1.1 million segment images obtained from 2319 cryoEM movie images (Figure 2, Figure S1, Table S1). It showed a tubular structure formed by a helical assembly of subunit proteins with an outer diameter of 230 Å and an inner diameter of 20 Å and with a similar helical parameter to the R-type straight filament formed by FliC (Table 2). The 11 protofilaments could also be recognized as in the FliC filament structure [6–8,16]. As revealed previously [6,8,16], the subunit consisted of four domains, D0, D1, D2 and D3, arranged from the inner to the outer part of the tube. Domains D0 and D1 formed the inner tubes while domains D2 and D3 formed the outer part, with domain D3 exposed on the surface of the filament. The sequence regions forming these four domains are indicated in Supplementary Figure S2, in which the high sequence homology between FljB and FliC is also presented (FljB/FliC sequence alignment shows 76% identity and 1.4×10^{-66} *E*-value). The helical parameters of the FljB filament determined in this study were also identical to those of the R-type FliC filament [7], which is reasonable considering the high sequence homology between these two flagellins.

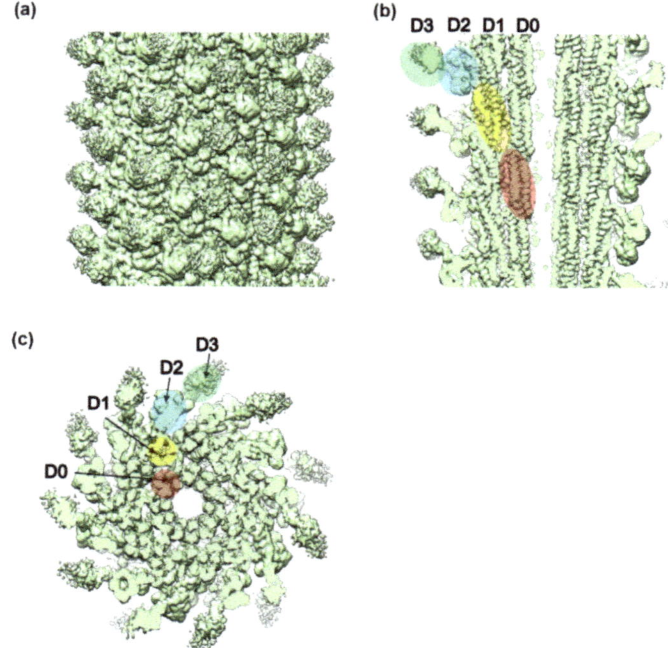

Figure 2. The 3D density map of the FljB filament resolved by cryoEM image analysis. (**a**) Side view of the filament surface, (**b**) longitudinal section along the filament axis, and (**c**) cross section viewed from the distal end. The four domains of FljB, D0, D1, D2, and D3, are labeled in different colors in (**b**,**c**).

Table 2. The helical parameters of the FljB and FliC R-type straight flagellar filaments.

	Rotation (Degree)	Axial Rise (Å)
FljB	65.81	4.69
FliC	65.84	4.71

To build the atomic model of the FljB filament, a homology model was first generated from the atomic model of the FliC filament (PDB ID: 1UCU) (Figure 3a) [8] using MODELLER [17], then fitted into the 3D density map and refined using UCSF Chimera [18], Coot (Crystallographic Object-Oriented Toolkit) [19], and Phenix [20] (Figure 3b). The model of FljB (PDB ID: 6JY0) in the filament structure contained residues 1–192 and 288–505, which covered domains D0, D1, and D2 (Figure S2). The missing residues 193–287, corresponding to domain D3, were not built because the resolution of the 3D map was too low to trace the chain (Figure 3b). In addition, the density level of domain D3 was much lower than other domains in FljB (Figure 3e), indicating that domain D3 was more mobile than other domains of FljB. This indicates that the two-chain hinge connection between domains D2 and D3 was relatively flexible, making domain D3 mobile. The structures of domains D0 and D1 were nearly identical between FljB and FliC, with the root mean square deviation of Cα atoms being 2.8 Å (Figure 3c). Domain D2 is composed of two subdomains, D2a and D2b, and while the D2b domains (residues 356–412 for FljB and 345–401 for FliC) ware also nearly identical between the two, the D2a domains (residues177–189, 294–355 for FljB and 177–189 and 284–344 for FliC) showed a small difference in their conformations due to a four residue insertion in FljB (Asp 310–Gly 314) (Figure 3c, Supplementary Figure S2).

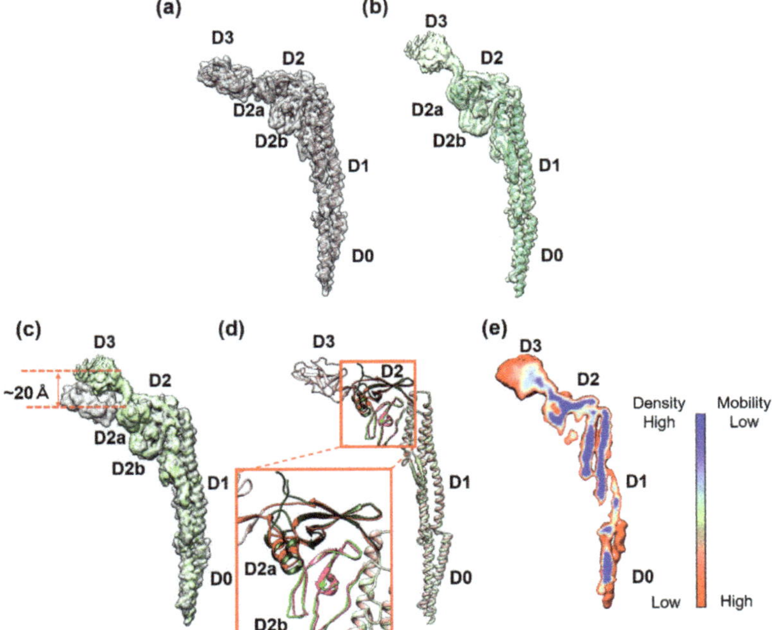

Figure 3. Comparison of subunit structures of the FliC and FljB filaments. The 3D map and Cα ribbon model of (**a**) FliC (PDB ID: 1UCU) and (**b**) FljB (PDB ID: 6JY0). The map of FljB was obtained by cryoEM image analysis, while that of FliC was calculated from the atomic model. The Cα ribbon model of domain D3 is not shown for FljB because the resolution of this domain was too low to trace the chain. (**c**) Superposition of the two maps with FljB in light green and FliC in gray. (**d**) The Cα ribbon model of FljB in light green is superimposed on that of FliC in light pink. The structures of FljB and FliC were nearly identical for domains D0, D1, and D2b, but the folding of part of domain D2a connecting to domain D3 was distinct between the two, which was responsible for the tilt of domain D3 of FljB, making its position higher than that of FliC by about 20 Å. (**e**) The color-coded density distribution of FljB in a longitudinal section of the molecule. The relatively low density of domain D3 indicated a high mobility of this domain.

The distinct difference in the two filament structures was the position and orientation of domain D3 relative to D2. When domains D0, D1, and D2 were superimposed between FljB and FliC, domain D3 of FljB was tilted from that of FliC by about 30°, making its position nearly 20 Å higher than that of FliC (Figure 3d). The two antiparallel chains connecting domains D2 and D3 were highly tilted in FljB to generate this difference (Figure 3c). Near this two-chain domain connection, FliC had close contacts between Ala 191, Val 283 and Asp 339, with a Cα distance between Ala 191 and Asp 339 of 4.9 Å, making the conformation of this two-chain domain connection stable (Figure 4a). Unlike FliC, FljB did not have these contacts, with the corresponding Cα distance between Ala 191 and Thr 350 being 18.4 Å (Figure 4b), possibly making the two-chain domain connection less stable and domain D3 oriented differently than FliC. As a result, domain D3 of FljB was more mobile than that of FliC, and the FljB and FliC filaments had different lateral intermolecular interactions between domains D2 and D3 of neighboring protofilaments.

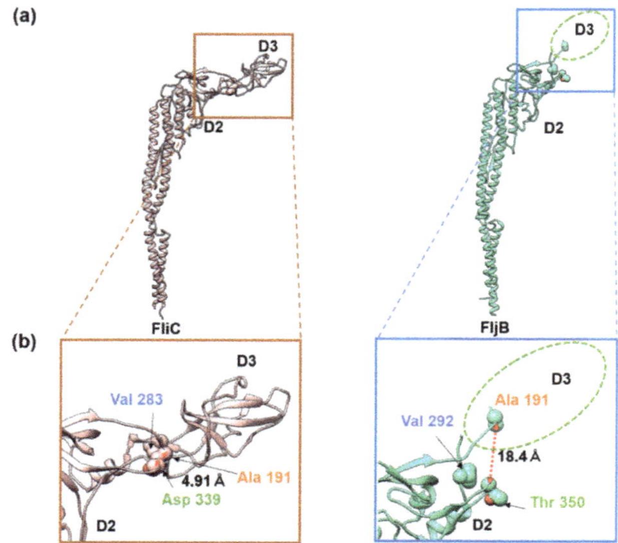

Figure 4. Interactions of residues stabilizing the position and orientation of domain D3 in FliC. (**a**) The Cα ribbon models of FliC (left) and FljB (right) with three residues important for stabilizing the position and orientation of domain D3 in space-filling representation. (**b**) Magnified view of the parts indicated by the boxes in (**a**). The ellipse in dashed line in the lower panel indicates the position of domain D3 of FljB. The corresponding residues of FliC and FljB are labeled in the same colors, orange (FliC Ala 191 and FljB Ala 191), blue (FliC Val 283 and FljB Val 292), and green (FliC Asp 339 and FljB Thr 350). In FliC, the three residues were closely packed each other by hydrophobic interactions, stabilizing the position and orientation of domain D3, while they were far apart in FljB, as indicated by the distances.

In the FljB filament structure, we newly identified an interesting axial intermolecular interaction between the D0 domains that contributes to the filament assembly and stability. Along each of the 11 protofilaments, the N-terminal five residues of the subunit 1 above (the distal side of the filament) were extended perpendicular to the filament axis before forming the N-terminal α-helix, and formed an antiparallel β-strand with the extended C-terminal chain connecting domains D0 and D1 of the subunit 2 below (the proximal side of the filament) (Figure 5). The residue Gln 2 of subunit 1 and Tyr 469 of subunit 2 formed a hydrogen bond between the main chains (Figure 5c). This explains why the N-terminus is more important than the C-terminus for the filament stability [21].

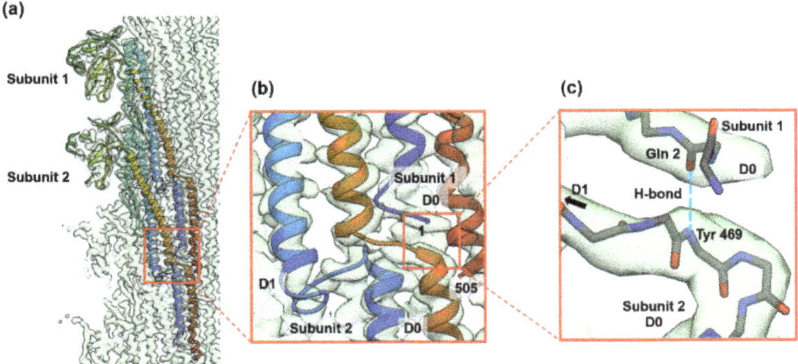

Figure 5. The folding and intersubunit interaction of the five N-terminal residues of FljB contributing to the assembly and stabilization of the filament structure. (**a**) The Cα ribbon models of two neighboring FljB subunits along the protofilament are shown in rainbow colors from the N-terminus in blue to the C-terminus in red. (**b**) The five N-terminal residues are extended, lying flat on the short extended chain connecting the C-terminal α-helices of domains D0 and D1 of a subunit below along the protofilament, forming an antiparallel β-strand to stabilize the axial intersubunit interactions. (**c**) The residue Gln 2 of subunit 1 and Tyr 469 of subunit 2 forms a hydrogen bond between the main chains (dashed line in right blue). In figure (**c**), only main chain backbone atoms are shown (nitrogen and oxygen atoms are colored blue and red, respectively).

4. Discussion

To investigate the potentially different physiological roles played by the two types of flagellar filaments formed by FljB and FliC, we carried out motility assays of two strains expressing either FljB or FliC and structural analysis of the FljB filament for comparison with the FliC filament.

There were no significant differences between strain SJW1103B, producing the FljB filament, and SJW1103, forming the FliC filament, either in the number of filaments per cell, their filament length, or their swimming speed in the motility buffer. However, their swimming speeds were clearly different under high-viscosity conditions, with a markedly smaller reduction in the swimming speed of SJW1103B than that of SJW1103 when the viscosity was increased by the addition of Ficoll in the motility buffer. These results indicated that the differences in the structure and dynamics of their flagellar filaments must be responsible for the difference in their motility.

We therefore examined the structure of the FljB filament by cryoEM single-particle image analysis in one of the two mutant straight forms called the R-type to utilize the helical symmetry. The overall structure was very similar to that of the FliC filament except for the position and orientation of the outermost domain, D3, exposed on the surface of the filament. Domain D3 of FljB also showed a higher flexibility and mobility than that of FliC. These differences suggest that domain D3 plays an important role not only in changing antigenicity but also in optimizing the motility function of the filament as a propeller under different conditions of solvent viscosity.

These differences in the relative position and dynamics of domain D3 were well correlated with the differences in amino acid sequence between FljB and FliC, which are found in regions Val 187–Gly 189 and Ala 284–Asn 285 and residues Asp 192 and Gln 282 of FliC, which form the two antiparallel chains connecting domains D2 and D3 (Supplementary Figures S2 and S3). In FliC, the position and orientation of domain D3 was stabilized by hydrophobic interactions between Ala 191, Val 283 and Asp 339 (Figure 4a). On the other hand, the distances between corresponding residues in FljB, Ala 191, Val 292, and Thr 350, were much longer than those of FliC, making the stabilizing hydrophobic interactions impossible (Figure 4b). As a result, the D3 domain of FljB was more mobile than that of FliC.

How the differences in the structure and dynamics of the D3 domains on the surface of the FljB and FliC filaments contribute to their different motility functions as a propeller is difficult to answer without actually examining the role of the more mobile domain D3 via fluid dynamic simulations with these two filament structures. There must be some advantages to a higher mobility of surface domains of the helical propeller in generating higher thrust under high-viscosity conditions, such as in mucosa, which pathogenic bacteria have to swim through to reach host cells. The sequence of FljB may have been optimized for this purpose by evolution.

5. Conclusions

In this study, we revealed that the flagellar filament formed by FljB has an advantage over the FliC filament by allowing cells to have higher motility under high-viscosity conditions. Comparing the two structures, the subunit structures and intersubunit packing interactions were well conserved except for the position and mobility of domain D3 on the surface of the filament. The advantage of the FljB filament in the swimming motility under highly viscous conditions may have an important role in infection when bacteria must go through viscous layers of mucosa on the surface of intestinal cells or keep a biofilm in good condition, with the cells with the FljB filaments acting as a nutrition deliverer [22,23]. However, in order to clarify the relationship between the filament structure and swimming motility of the cell, further computational analyses are needed.

Supplementary Materials: The supplementary materials are available online at http://www.mdpi.com/2218-273X/10/2/246/s1.

Author Contributions: Conceptualization, K.N. and T.K.; methodology, all authors; swimming analysis, S.T and N.T.; structural analysis, T.Y., S.T., T.M. (Tomoko Miyata) and T.K.; resources, S.T., N.T., T.M. (Tomoko Miyata) and T.M. (Tohru Minamino); writing—original draft preparation, T.Y., T.K. and K.N.; writing—review and editing, all authors; supervision, K.N. and T.K.; project administration, K.N.; funding acquisition, K.N., T.M. (Tomoko Miyata), T.M. (Tohru Minamino). All authors have read and agreed to the published version of the manuscript.

Funding: This work has been supported by JSPS KAKENHI Grant Number JP25000013 (to K.N.), JP18K06155 (to T.M. (Tomoko Miyata)), JP26293097 and JP19H03182 (to T.M. (Tohru Minamino)), and MEXT KAKENHI Grant Number JP15H01640 (to T.M. (Tohru Minamino)). This work has also been supported by Platform Project for Supporting Drug Discovery and Life Science Research (BINDS) from AMED under Grant Number JP19am0101117 to K.N., by the Cyclic Innovation for Clinical Empowerment (CiCLE) from AMED under Grant Number JP17pc0101020 to K.N. and by JEOL YOKOGUSHI Research Alliance Laboratories of Osaka University to K.N.

Acknowledgments: We thank Shigeru Yamaguchi for his kind gift of SJW590 and Yusuke Morimoto for technical help in identifying the mutation site of the *fljB* allele of SJW590.

Conflicts of Interest: The authors declare no conflicts of interest.

References

1. Haiko, J.; Westerlund-Wikström, B.; Haiko, J.; Westerlund-Wikström, B. The Role of the Bacterial Flagellum in Adhesion and Virulence. *Biology* **2013**, *2*, 1242–1267. [CrossRef] [PubMed]
2. Rossez, Y.; Wolfson, E.B.; Holmes, A.; Gally, D.L.; Holden, N.J. Bacterial Flagella: Twist and Stick, or Dodge across the Kingdoms. *PLoS Pathog.* **2015**, *11*, e1004483. [CrossRef] [PubMed]
3. Berg, H.C. The Rotary Motor of Bacterial Flagella. *Annu. Rev. Biochem.* **2003**, *72*, 19–54. [CrossRef] [PubMed]
4. Andrewes, F.W. Studies in group agglutination. II.—The absorption of agglutinin in the diphasic salmonellas. *J. Pathol. Bacteriol.* **1925**, *28*, 345–359. [CrossRef]
5. Horstmann, J.A.; Zschieschang, E.; Truschel, T.; de Diego, J.; Lunelli, M.; Rohde, M.; May, T.; Strowig, T.; Stradal, T.; Kolbe, M.; et al. Flagellin phase-dependent swimming on epithelial cell surfaces contributes to productive Salmonella gut colonisation. *Cell. Microbiol.* **2017**, *19*. [CrossRef] [PubMed]
6. Samatey, F.A.; Imada, K.; Nagashima, S.; Vonderviszt, F.; Kumasaka, T.; Yamamoto, M.; Namba, K. Structure of the bacterial flagellar protofilament and implications for a switch for supercoiling. *Nature* **2001**, *410*, 331–337. [CrossRef]
7. Yonekura, K.; Maki-Yonekura, S.; Namba, K. Complete atomic model of the bacterial flagellar filament by electron cryomicroscopy. *Nature* **2003**, *424*, 643–650. [CrossRef]

8. Maki-Yonekura, S.; Yonekura, K.; Namba, K. Conformational change of flagellin for polymorphic supercoiling of the flagellar filament. *Nat. Struct. Mol. Biol.* **2010**, *17*, 417–422. [CrossRef]
9. Mimori-Kiyosue, Y.; Yamashita, I.; Fujiyoshi, Y.; Yamaguchi, S.; Namba, K. Role of the outermost subdomain of Salmonella flagellin in the filament structure revealed by electron cryomicroscopy. *J. Mol. Biol.* **1998**, *284*, 521–530. [CrossRef]
10. Datsenko, K.A.; Wanner, B.L. One-step inactivation of chromosomal genes in Escherichia coli K-12 using PCR products. *Proc. Natl. Acad. Sci. USA* **2000**, *97*, 6640–6645. [CrossRef]
11. Hara, N.; Namba, K.; Minamino, T. Genetic Characterization of Conserved Charged Residues in the Bacterial Flagellar Type III Export Protein FlhA. *PLoS ONE* **2011**, *6*, e22417. [CrossRef] [PubMed]
12. Yamaguchi, S.; Fujita, H.; Sugata, K.; Taira, T.; Iino, T. Genetic Analysis of H2, the Structural Gene for Phase-2 Flagellin in Salmonella. *Microbiology* **1984**, *130*, 255–265. [CrossRef] [PubMed]
13. Schultz, P. Cryo-electron microscopy of vitrified specimens. *Q. Rev. Biophys.* **1988**, *21*, 129–228. [CrossRef]
14. Zhang, K. Gctf: Real-time CTF determination and correction. *J. Struct. Biol.* **2016**, *193*, 1–12. [CrossRef]
15. Zivanov, J.; Nakane, T.; Forsberg, B.O.; Kimanius, D.; Hagen, W.J.H.; Lindahl, E.; Scheres, S.H.W. New tools for automated high-resolution cryo-EM structure determination in RELION-3. *eLife* **2018**, *7*, 42166. [CrossRef]
16. Mimori, Y.; Yamashita, I.; Murata, K.; Fujiyoshi, Y.; Yonekura, K.; Toyoshima, C.; Namba, K. The structure of the R-type straight flagellar filament of Salmonella at 9 Å resolution by electron cryomicroscopy. *J. Mol. Biol.* **1995**, *249*, 69–87. [CrossRef]
17. Webb, B.; Sali, A. Comparative protein structure modeling using MODELLER. *Curr. Protoc. Bioinform.* **2006**, *15*, 1–6. [CrossRef]
18. Pettersen, E.F.; Goddard, T.D.; Huang, C.C.; Couch, G.S.; Greenblatt, D.M.; Meng, E.C.; Ferrin, T.E. UCSF Chimera—A visualization system for exploratory research and analysis. *J. Comput. Chem.* **2004**, *25*, 1605–1612. [CrossRef]
19. Emsley, P.; Lohkamp, B.; Scott, W.G.; Cowtan, K. Biological Crystallography Features and development of Coot. *Acta Crystallogr. D Biol. Crystallogr.* **2010**, *66*, 486–501. [CrossRef]
20. Liebschner, D.; Afonine, P.V.; Baker, M.L.; Bunkó, G.; Chen, V.B.; Croll, T.I.; Hintze, B.; Hung, L.-W.; Jain, S.; McCoy, A.J.; et al. Macromolecular structure determination using X-rays, neutrons and electrons: Recent developments in Phenix. *Acta Cryst.* **2019**, *75*, 861–877. [CrossRef]
21. Vonderviszt, F.; Aizawa, S.-I.; Namba, K. Role of the disordered terminal regions of flagellin in filament formation and stability. *J. Mol. Biol.* **1991**, *221*, 1461–1474. [CrossRef]
22. Flemming, H.-C.; Wingender, J.; Szewzyk, U.; Steinberg, P.; Rice, S.A.; Kjelleberg, S. Biofilms: An emergent form of bacterial life. *Nat. Rev. Microbiol.* **2016**, *14*, 563–575. [CrossRef] [PubMed]
23. Houry, A.; Gohar, M.; Deschamps, J.; Tischenko, E.; Aymerich, S.; Gruss, A.; Briandet, R. Bacterial swimmers that infiltrate and take over the biofilm matrix. *Proc. Natl. Acad. Sci. USA* **2012**, *109*, 13088–13093. [CrossRef] [PubMed]

Article

In Vitro Autonomous Construction of the Flagellar Axial Structure in Inverted Membrane Vesicles

Hiroyuki Terashima [1,2], Chinatsu Tatsumi [2], Akihiro Kawamoto [3,†], Keiichi Namba [3,4,5], Tohru Minamino [3] and Katsumi Imada [2,*]

1 Division of Biological Science, Graduate School of Science, Nagoya University, Furo-cho, Chikusa-Ku, Nagoya 464-8602, Japan; terashima.hiroyuki@h.mbox.nagoya-u.ac.jp
2 Department of Macromolecular Science, Graduate School of Science, Osaka University, 1-1 Machikaneyama, Toyonaka, Osaka 560-0043, Japan; tatsumic14@gmail.com
3 Graduate School of Frontier Biosciences, Osaka University, 1-3 Yamadaoka, Suita, Osaka 565-0871, Japan; kawamoto@protein.osaka-u.ac.jp (A.K.); keiichi@fbs.osaka-u.ac.jp (K.N.); tohru@fbs.osaka-u.ac.jp (T.M.)
4 RIKEN SPring-8 Center and Center for Biosystems Dynamic Research, 1-3 Yamadaoka, Suita, Osaka 565-0871, Japan
5 JEOL YOKOGUSHI Research Alliance Laboratories, Osaka University, 1-3 Yamadaoka, Suita, Osaka 565-0871, Japan
* Correspondence: kimada@chem.sci.osaka-u.ac.jp; Tel.: +81-6-6850-5455
† Current Affiliation: Institute for Protein Research, Osaka University, 3-2 Yamadaoka, Suita, Osaka 565-0871, Japan.

Received: 16 December 2019; Accepted: 9 January 2020; Published: 11 January 2020

Abstract: The bacterial flagellum is a filamentous organelle extending from the cell surface. The axial structure of the flagellum consists of the rod, hook, junction, filament, and cap. The axial structure is formed by axial component proteins exported via a specific protein export apparatus in a well-regulated manner. Although previous studies have revealed the outline of the flagellar construction process, the mechanism of axial structure formation, including axial protein export, is still obscure due to difficulties in direct observation of protein export and assembly in vivo. We recently developed an in vitro flagellar protein transport assay system using inverted membrane vesicles (IMVs) and succeeded in reproducing the early stage of flagellar assembly. However, the late stage of the flagellar formation process remained to be examined in the IMVs. In this study, we showed that the filament-type proteins are transported into the IMVs to produce the filament on the hook inside the IMVs. Furthermore, we provide direct evidence that coordinated flagellar protein export and assembly can occur at the post-translational level. These results indicate that the ordered construction of the entire flagellar structure can be regulated by only the interactions between the protein export apparatus, the export substrate proteins, and their cognate chaperones.

Keywords: bacterial flagellum; type III secretion system; flagellar specific export apparatus; inverted membrane vesicle; in vitro reconstitution; flagellar filament; *Salmonella typhimurium*

---

1. Introduction

The bacterial flagellum is a tubular organelle extending out from the cell surface, and is rotated by a nano-scale rotary motor embedded in the cytoplasmic membrane. Flagellar construction starts with the assembly of the basal structure containing the MS-ring, the C-ring, and the flagellar protein export apparatus in the cytoplasmic membrane, followed by assembly of the filamentous axial structure composed of the rod, the hook, the hook–filament junction, the filament, and the filament cap on the basal structure (Figure 1A) [1–3]. A certain copy number of component proteins assemble into each substructure in a specific order [1]. The axial proteins are transported via the flagellar export apparatus, which belongs to the type III secretion system family, into the central channel of the growing tubular axial structure and diffuse through it to the distal end, where they are incorporated into the structure [4,5]. The export apparatus consists of a transmembrane export gate composed of FlhA, FlhB, FliP, FliQ, and FliR, and a cytoplasmic ATPase complex composed of FliH, FliI, and FliJ [1–3]. The MS-ring and the C-ring function as a housing for the export gate and a sorting platform for the cytoplasmic ATPase complex, respectively [6,7].

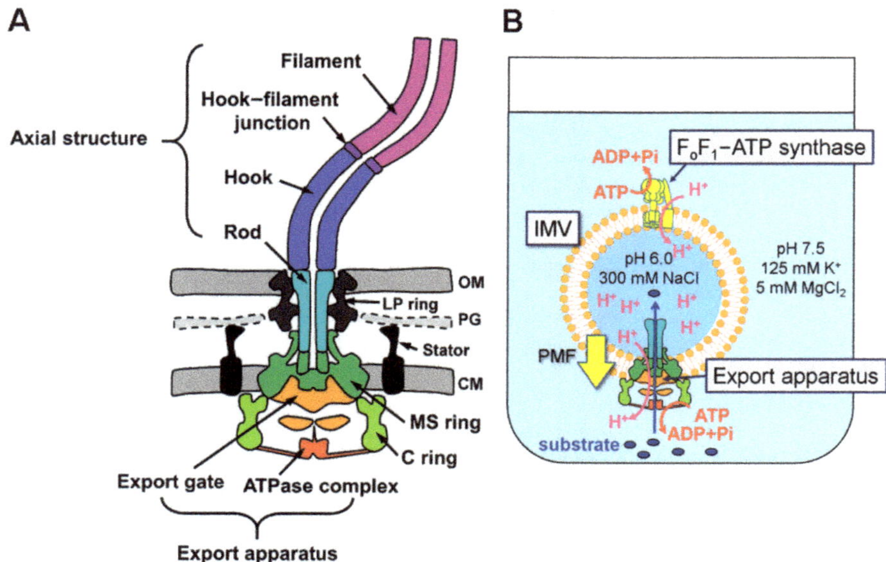

Figure 1. Schematic drawing of the *Salmonella* flagellum (**A**) and the in vitro transport assay system using the inverted membrane vesicles (IMVs) (**B**). (A) The sub-structures of the *Salmonella* flagellum are represented in the following colors: the transmembrane export gate, orange; the cytoplasmic ATPase complex, red; the MS-ring, green; the C-ring, light green; the rod, cyan; the hook, blue; the hook-filament junction, purple; the filament, magenta; the LP ring, gray; the stator, black. The core of the export apparatus consists of the export gate and the ATPase complex. The filamentous part composed of the rod, the hook, the hook–filament junction and the filament are called the flagellar axial structure. CM, the cytoplasmic membrane; PG, the peptidoglycan layer; OM, the outer membrane. (B) To apply the initial PMF to the IMVs, the IMVs were filled with 300 mM NaCl at pH 6.0 and suspended in solution with 125 mM K^+ and 5 mM $MgCl_2$ at pH 7.5. The export substrates, ATP-Mg^{2+}, the $FliH_2$/FliI complex, and FliJ were added to the assay mixture. To maintain PMF across the inverted membrane, endogenous F_oF_1-ATP synthase pumps proton into the IMVs by ATP hydrolysis energy. PMF and ATP hydrolysis energy generated by FliI ATPase drives the substrate protein transport into the IMVs.

Flagellar construction is a well-regulated process, in which the expression and export of flagellar axial proteins are coupled with the assembly state of the flagellum. The flagellar axial proteins are classified into two groups by the substrate-recognition mode of the flagellar protein export apparatus: one is the rod/hook-type substrate class, responsible for the assembly of the rod and hook structures, and the other is the filament-type substrate class, required for the construction of the hook–filament junction, the filament, and the filament cap. Before completion of the hook (the early stage of flagellar formation), only the rod/hook-type proteins are allowed to be exported, and the export of filament-type proteins is suppressed [8,9]. After the length of the hook has reached approximately 55 nm, the export of rod/hook-type proteins is stopped, and the filament-type proteins begin to be exported (the late stage of flagellar formation) [10–12]. Thus, the switching of the substrate specificity of the flagellar protein export apparatus from the rod/hook-type to the filament-type proteins is a crucial step in regulating the flagellar construction, as well as in controlling the hook length. The hook length is monitored by a secreted molecular ruler protein, FliK [13,14]. FliK is a rod/hook-type protein and is infrequently exported during hook assembly to measure the hook length using its N-terminal disordered region [15,16]. When the hook length is too short, FliK is secreted out into the extracellular media. When the hook length reaches approximately 55 nm, the C-terminal domain of FliK binds to FlhB, one of the export gate component proteins, to induce conformational changes of FlhB and FlhA to switch the substrate specificity of the export apparatus [17–22]. The filament-type proteins form a complex with their specific cognate chaperones, which prevent premature aggregation and/or proteolysis of their cognates and help them associate with the flagellar protein export apparatus in the cytoplasm [23–25]. The flagellar chaperones not only facilitate the docking of their cognate filament-type proteins to the protein export apparatus, but also regulate the export order through the interactions with FlhA, FliI, and FliJ [1,2]. Moreover, the flagellar chaperones are multifunctional proteins able to control the production of flagellar proteins as well as delivering their cognates to the export apparatus [26–28].

Genetic and biochemical studies have revealed that there are several morphological checkpoints involved in coordination of flagellar protein export and assembly, not only at the gene expression level but also at the post-translational level. However, the molecular mechanism of each process, including protein export, is still unclear because of difficulties in direct observation of protein export and assembly in vivo. To overcome this problem, we recently established an in vitro flagellar protein transport assay system using inverted membrane vesicles (IMVs) to quantitatively control and measure protein export and monitor the flagellar assembly process (Figure 1B) [29,30]. We demonstrated that the flagellar protein export apparatus in the IMVs maintains the export function for rod/hook-type proteins at a level similar to that in a living cell, and that ATP hydrolysis by FliI ATPase dramatically accelerates the export of rod/hook-type proteins. Proton-motive force (PMF) across the cytoplasmic membrane is a primary driving force for flagellar protein export [31–33], and the energy derived from ATP hydrolysis by FliI ATPase is thought to be required for the activation of the PMF-driven export gate complex [34]. However, our study with IMVs revealed that the energy of ATP hydrolysis is able to drive protein export even when PMF is absent [29]. Moreover, we successfully reproduced not only hook formation inside the IMVs, but also hook length control. However, we have not yet shown the export of the filament-type proteins via the switched export apparatus in the IMVs. Since the filament-type proteins require their specific cognate chaperones for efficient export [2], the protein export mechanism, including its regulation, would be expected to be different from that for the rod/hook-type proteins.

Here, we showed that the filament-type substrates (FlgK, FlgL, FliC, FliD) in complex with their cognate chaperones are transported into the interior of the IMVs after completion of the hook. These transported filament-type proteins assemble into the filament at the tip of the hook. Moreover, even when the hook-type proteins (FlgD, FlgE, FliK) and filament-type proteins in complex with substrate-specific flagellar chaperones were simultaneously added to the in vitro transport assay solutions, these proteins autonomously and sequentially assembled into the normal flagellar structure, indicating that the coupling of flagellar gene expression with assembly is not really required for well-ordered flagellar formation.

2. Materials and Methods

2.1. Bacteria Strains and Plasmids

Bacterial strains and plasmids used are listed in Table 1. The primers used in this study are shown in Supplementary Table S1. *Salmonella* and *Escherichia coli* cells were cultured in LB broth (1% (*w*/*v*) bactotryptone, 0.5% (*w*/*v*) yeast extract, 0.5% (*w*/*v*) NaCl). Chloramphenicol was added to a final concentration of 30 μg/mL. Ampicillin was added to a final concentration of 50 μg/mL.

Table 1. Bacterial strains and plasmids.

Strain or Plasmid	Genotype or Description	Reference
	E. coli strains	
DH5α	F^-, *mcrA*, Δ(*mrr-hsdRMS-mcrBC*), Φ80d*lacZ*ΔM15, Δ*lacX74*, *deoR*, *recA1*, *araD139*, Δ(*ara leu*)7697, *galU*, *galK*, λ^-, *rpsL*, *endA1*,*nupG*	
BL21(DE3)	F^- *ompT hsdS*$_B$ (r_B^- m_B^-) *gal dcm* (DE3)	Novagen
	Salmonella strains	
STH001	Δ*flhB* Δ*flgD* Δ*fliT*	[29]
	Plasmids	
pBAD33SD	Cm^r, pBAD33-based vector substituted NheI and EcoRI sites (GCTAGCGAATTC) into SD sequence (GCAGGAGGATTC)	[29]
pTrc99a	Amp^r, P_{trc} expression vector	
pTrc99aNde	pTrc99a-based vector substituted NcoI sites (cagACCATGgaa) into NdeI sequence (cagCATATGgaa)	This study
pET3c	Amp^r, T7 expression vector	Novagen
pET15b	Amp^r, T7 expression vector	Novagen
pITH103	pBAD33SD-*flhB* + *flhDC* (FlhB, FlhD/FlhC)	[29]
pITH105	pET15b-*flgD* (His-FlgD)	[29]
pITH106	pET15b-*flgE* (His-FlgE)	[29]
pMMIJ001	pET15b-*fliJ* (His-FliJ)	[35]
pMKM1702iH	pTrc99a-*his-fliI* + *fliH* (FliH/His-FliI), *his-fliI* was derived from the pET19b-based plasmid, pMM1901.	[36]
pITH107	pET15b-*fliK* (His-FliK)	[29]
pITH108	pET3c-*fliS* (FliS), the *fliS* gene amplified by PCR was cloned into the NdeI and BamHI sites of the pET3c vector	This study
pITH109	pET15b-*fliC* (His-FliC), the *fliC* gene amplified by PCR was inserted into the NdeI and BamHI sites of pET15b	This study
pITH110	pTrc99aNde-*fliS* + *his-fliC* (His-FliC/FliS), *fliS* derived from pITH108 and *his-fliC* derived from pITH109 were inserted into the NdeI and BamHI sites, and the XbaI and HindIII sites of pTrc99aNde, respectively	This study
pITH111	pET3c-*fliT* (FliT), the *fliT* gene amplified by PCR was cloned into the NdeI and BamHI sites of pET3c	This study
pITH112	pET15b-*fliD* (His-FliD), the *fliD* gene amplified by PCR was inserted into the NdeI and BamHI sites of pET15b	This study
pITH113	pTrc99aNde-*fliT* + *his-fliD* (His-FliD/FliT), *fliT* derived from pITH111 and *his-fliD* derived from pITH112 were subcloned into the NdeI and BamHI sites and the XbaI and HindIII sites of pTrc99aNde, respectively	This study
pMMGN110	pET22b-*flgN* (FlgN), The *flgN* gene amplified by PCR was cloned into the NdeI and BamHI sites of pET22b	[37]
pMMGN300	pET19b-*flgN* (His-FlgN), *flgN* derived from pMMGN110 was subcloned into the NdeI and BamHI sites of pET19b	This study
pMMGK130	pET15b-*flgK* (His-FlgK), the *flgK* gene amplified by PCR was inserted into the NdeI and BamHI sites of pET15b	[37]
pITH116	pET15b-*flgL* (His-FlgL), the *flgL* amplified by PCR was inserted into the NdeI and BamHI sites of pET15b	This study
pITH117	pTrc99aNde-*flgN* + *his-flgK* (His-FlgK/FlgN), *flgN* derived from pMMGN300 and *his-flgK* derived from pMMGK130 were subcloned into the NdeI and BamHI sites and the XbaI and HindIII sites of pTrc99aNde, respectively	This study
pITH118	pTrc99aNde-*flgN* + *his-flgL* (His-FlgL/FlgN), *flgN* derived from pMMGN300 and *his-flgL* derived from pITH116 were subcloned into the NdeI and BamHI sites and XbaI and HindIII sites of pTrc99aNde, respectively	This study

Amp^r, ampicillin-resistant; Cm^r, chloramphenicol-resistant.

2.2. Preparation of Inverted Membrane Vesicles

The inverted membrane vesicles (IMVs) were prepared according to the method described by Terashima et al. [29]. We prepared the IMVs from a *Salmonella* cell strain STH001 (ΔflhB, ΔflgD, and ΔfliT) harboring plasmid pITH103 (wild-type flhB and flhDC). The cells were cultured at 37 °C for 9 h, inoculated into 1 L of LB broth with 1/100 dilution and cultured at 30 °C for 1 h. L-arabinose was then added at the final concentration of 0.02% (*w/v*) for induction of FlhB and $FlhD_4/FlhC_2$, and the culture was continued at 18 °C for 12–16 h until OD_{600} reached around 1.5. The cells were collected by centrifugation and suspended in 75 mL of sucrose solution (10 mM Tris-HCl pH 8.0, 0.75 M sucrose), to which was added 22.5 mg of lysozyme powder. Next, 150 mL of 1.5 mM EDTA-NaOH pH 8.0 was poured into the cell suspension on ice to form spheroplasts. The cell suspension was stirred on ice for 1 h. The spheroplasts were collected at 5000× *g* for 10 min and suspended in 25 mL solution A (20 mM MES-NaOH pH 6.0, 300 mM NaCl) with a half tablet of protease inhibitor cocktail (cOmplete EDTA-free, Roche). In order to produce IMVs, the cell suspension was passed through high-pressure cell homogenizer (STANSTED) at 90 MPa. After centrifugation at 20,000× *g* for 10 min to remove cell debris, the membrane vesicles were precipitated by ultra-centrifugation at 100,000× *g* for 1 h. The crude IMVs were suspended in 1 mL of solution A and separated by sucrose density-gradient centrifugation (60% (*w/w*) 5 mL/50% (*w/w*) 9 mL/45% (*w/w*) 9 mL/40% (*w/w*) 6 mL stepwise gradient in the Beckman ultra-clear tube) at 60,000× *g* (SW32 Ti rotor: Beckman, Brea, CA, USA) for 16 h. A brown-colored layer fraction containing the IMVs was recovered, and the IMVs were precipitated by ultra-centrifugation at 100,000× *g* for 1 h. The precipitant was suspended in 900 µL of solution A. The suspension was divided into 300 µL aliquots, frozen with liquid nitrogen and stored at −80 °C. The frozen stock was thawed and filtered with a 0.8 µm polycarbonate filter (Nuclepore membrane PC MB 19 0.8U, GE healthcare, Chicago, IL, USA). The filtered solution was loaded onto a Sephadex G-50 fine column (GE healthcare) and eluted with solution B (125 mM KCl, 20 mM Tris-HCl pH 7.5). The concentration of IMVs was adjusted to give OD 600 nm = 0.1.

2.3. Protein Purification

FlgD, FlgE, FliK, the $FliH_2$/FliI complex, and FliJ were purified according to the method described by Terashima et al. [29]. The FliC/FliS, FliD/FliT, FlgK/FlgN, or FlgL/FlgN complex was expressed in *E. coli* BL21(DE3) cells from the plasmids pITH110, pITH113, pITH117, or pITH118, respectively. The cells harboring pITH110 or pITH117 were inoculated directly from the colonies onto the agar plate into LB broth and incubated overnight at 30 °C to express the proteins by leak-expression from pTrc99a-based plasmids. The cells harboring pITH113 or pITH118 were grown at 30 °C for overnight, inoculated into fresh LB broth with 1/100 dilution and cultured at 30 °C until the optical density at 600 nm reached 0.5–0.8. IPTG was then added to the final concentration of 0.1 mM, and the culture was continued at 18 °C for overnight. The cells expressing FliC/FliS, FliD/FliT, FlgK/FlgN, or FlgL/FlgN were suspended in cell-suspend solution (50 mM Tris-HCl pH 8.0, 500 mM NaCl) containing protease inhibitor cocktail (cOmplete EDTA-free, Roche, Rotkreuz, Switzerland), and disrupted by sonication. After removal of cell debris by low-speed centrifugation followed by filtration with a 0.45 µm cellulose acetate membrane filter device, the cell lysate was loaded to a HisTrap HP column (GE healthcare), and the hexahistidine-tagged proteins were then eluted by linear imidazole gradient. To chop off the hexahistidine tag, protein solutions were incubated with thrombin (GE healthcare) at room temperature for 3 h and then passed through HisTrap HP again to remove hexahistidine-tag-retained proteins. Finally, the protein solution was purified using a Superdex 200 column (GE healthcare) equilibrated with external solution (20 mM Tris-HCl pH 7.5, 125 mM KCl). The purity of the purified proteins was examined by SDS-PAGE and Coomassie Brilliant Blue staining (Supplementary Figure S1).

2.4. Transport Assay

Transport assay of the hook-type proteins was carried out as previously described [29]. Transport assay of the filament-type proteins were carried out using the IMVs after hook formation by the transport reaction. FlgD, FlgE, FliK, FliJ, and the $FliH_2$/FliI complex were added into the IMV solution

(20 mM Tris-HCl, pH 7.5, 125 mM KCl, 5 mM $MgCl_2$, and 1 mM dithiothreitol (DTT)) at the final concentration of 4 μM, 4 μM 4 μM, 0.25 μM, and 1.5 μM, respectively. After addition of ATP with a final concentration of 5 mM, the mixture was incubated at 37 °C for 1 h to form the hook in the IMVs and then was ultra-centrifuged at 100,000× *g* for 30 min. The precipitant was washed by external solution and the solution was removed. The precipitant was resuspended in 100 μL of external solution and used for the filament-type protein transport assay. The IMV solution was incubated with the filament-type proteins (2 μM), FliJ (0.25 μM), the $FliH_2$/FliI complex (1.5 μM), $MgCl_2$ (5 mM), and ATP (5 mM) at 37 °C for 2 h. The proteins transported in the IMVs were detected by immunoblotting.

2.5. Purification of the Hook–Basal Body from IMV

The hook–basal body complexes were purified according to the method described by Terashima et al. [29]. The IMV was solubilized by 0.1% (*v*/*v*) Triton X-100. The suspension was ultra-centrifuged at 150,000× *g* for 30 min. The filament–hook–basal body complex was precipitated and suspended in TET solution (10 mM Tris-HCl pH 8.0, 1 mM EDTA, 0.1% (*v*/*v*) Triton X100). In order to completely dissolve the membrane, the suspension was mixed with 1 mL of alkali solution (10% (*w*/*v*) sucrose, 0.1% (*v*/*v*) Triton X100, 0.1 M KCl, adjusted to pH 11.0 by KOH). The suspension was layered on 1 mL of 35% (*w*/*v*) sucrose solution prepared by dissolving sucrose in TET solution in an ultra-centrifuge tube. After incubation on ice for 30 min, the suspension was ultra-centrifuged at 38,000 rpm (Beckman TLA100.3 rotor) for 30 min to precipitate the filament–hook–basal body. The precipitates were suspended in TET solution and then observed by electron microscopy.

2.6. Dynamic Light Scattering

A total 12 uL of purified IMVs in solution B (OD 600 nm = 0.1) was injected into a low-volume quartz cuvette and measured using Zetasizer μV (Malvern Panalytical, Malvern, UK). The scattering data were analyzed using Zetasizer software (Malvern Panalytical).

2.7. Negative-Staining Electron Microscopy

Sample solutions were applied to carbon-coated copper grids and negatively stained with 2.0% (*w*/*v*) phosphotungstic acid or 2.0% (*w*/*v*) uranyl acetate. Images were observed with a JEM-1010 transmission electron microscope (JEOL, Tokyo, Japan) operating at 100 kV using a BIOSCAN model 792 CCD camera, a JEM-1011 transmission electron microscope (JEOL, Tokyo, Japan) operating at 100 kV using a TVIPS TemCam-F114 CCD camera or a TemCam-F415 CCD camera, or a JEM-2010 transmission electron microscope (JEOL, Tokyo, Japan) operating at 200 kV using a Orius SC200D model 833 CCD camera (Gatan, Pleasanton, CA, USA).

3. Results

3.1. In Vitro Protein Transport of the Filament-Type Proteins into the IMVs

We previously showed that FlgE, FlgD, and FliK, which belong to the rod/hook-type substrate class, are transported into the IMVs through the flagellar protein export apparatus to form the hook on the endogenous rod in the basal body inside the IMVs [29]. The transported FlgE molecules were assembled into the hook with the help of the FlgD cap, and the hook length was controlled to approximately 55 nm by the addition of FliK to the assay solutions. These results suggested that the FliK molecule alone is sufficient the hook length control, and terminates the export of the rod/hook-type proteins in the in vitro protein transport assay system [29]. However, we had not examined whether FliK actually induces the switching of substrate specificity of the flagellar protein export apparatus from the rod/hook-type to the filament-type proteins. We therefore studied the export of the filament-type proteins into the IMVs after the termination of rod/hook-type protein export triggered by FliK.

We first incubated IMVs with the substrate proteins required for hook formation (FlgD, FlgE, and FliK), the cytoplasmic ATPase complex proteins (the $FliH_2$/FliI complex and FliJ), ATP, and Mg^{2+} for

1 h at 37 °C to form the hook in the IMVs, as previously described [29]. The IMVs were precipitated by ultra-centrifugation and then used for in vitro transport assays for the filament-type proteins. The IMVs with the hook–basal bodies were suspended into transport assay mixtures containing the FlgN/FlgK, FlgN/FlgL, FliS/FliC, and FliT/FliD chaperone–substrate complexes and the components of the cytoplasmic ATPase complex. The mixtures were then incubated for 2 h at 37 °C after adding ATP and Mg^{2+}. FlgK, FlgL, FliC, and FliD were transported into the IMVs, but were not transported in the absence of either FliK or FlgE (Figure 2). These results indicate that both FliK and the presence of the completed hook structure are required for the export of the filament-type proteins into the IMVs. Therefore, we suggest that purified FliK alone triggers switching of the substrate specificity of the flagellar protein export apparatus upon completion of hook assembly.

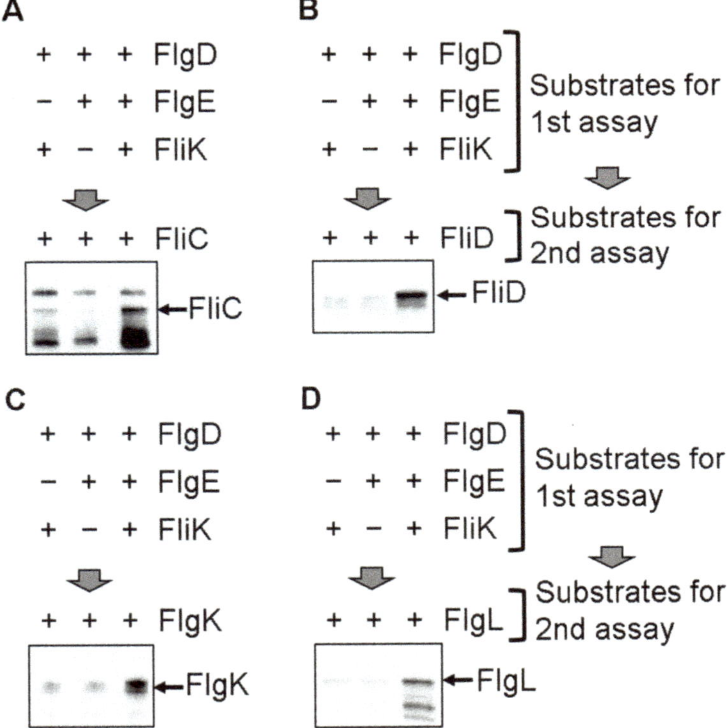

Figure 2. Hook formation is essential for the filament-type protein transport. The first reaction mixture for hook formation contained FlgD, the $FliH_2$/FliI complex, FliJ, and ATP at final concentrations of 4 μM, 1.5 μM 0.25 μM, and 5 mM, respectively. FlgE and FliK were added to the mixture at the final concentration of 4 μM, respectively. The first transport reaction was carried out at 37 °C for 1 h. After the reaction, the IMVs were precipitated by ultra-centrifugation (100,000× *g*, 30 min) and used for the second transport reaction to transport the filament-type substrates. Re-suspended IMVs were mixed in the second transport reaction mixture containing 1.5 μM of the $FliH_2$/FliI complex, 0.25 μM of FliJ, 5 mM of ATP–Mg, and 2 μM of the filament-type substrate–chaperone complex: FliC/FliS (**A**), FliD/FliT (**B**), FlgK/FlgN (**C**), or FlgL/FlgN (**D**).

3.2. Effect of the $FliH_2$/FliI Complex on Filament-Type Protein Export

The $FliH_2$/FliI complex greatly facilitates the export of rod/hook-type proteins such as FlgD and FlgE. The addition of the $FliH_2$/FliI complex to the final concentration of 1.5 μM to the assay solution increased the relative FlgD transport level 20-fold [29]. Therefore, we analyzed the impact of the

$FliH_2$/FliI complex on filament-type protein export. The transport levels of the filament-type substrates were significantly increased by adding 1.5 μM $FliH_2$/FliI complex, similarly to FlgD export (Figure 3). These results suggest that the $FliH_2$/FliI complex facilitates the export of filament-type proteins.

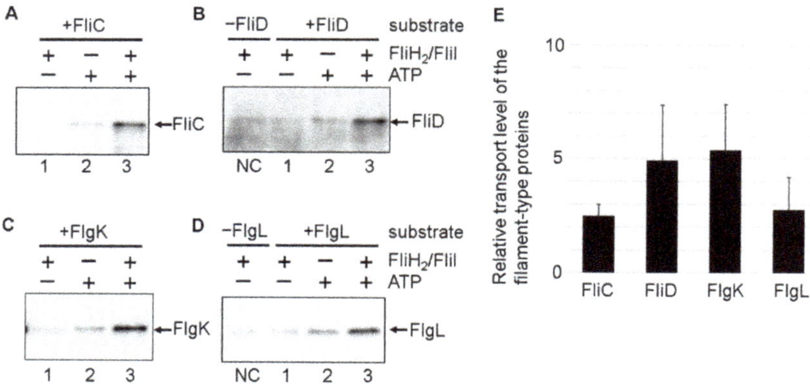

Figure 3. In vitro protein transport of the filament-type substrates into the IMVs. IMVs after completion of the hook were mixed with reaction mixture containing 0.25 μM of FliJ and 2 μM of the filament-type substrate–chaperone complex: FliC/FliS (**A**), FliD/FliT (**B**), FlgK/FlgN (**C**), and FlgL/FlgN (**D**). The transport assay was conducted with (+) or without (−) the $FliH_2$/FliI complex (1.5 μM) and ATP (5 mM). The bands in Lane 1 in (**B**,**C**) are cross-reacting bands because similar bands were detected in the assay not containing the substrates (Lane NC in (**B**,**C**)). (**E**) The transport levels of the filament-type protein relative to those without the $FliH_2$/FliI complex. The immunoblot band intensity was measured using Image J software. The relative transport level was calculated by dividing the band intensity of Lane 3 by that of Lane 2 after subtraction of that of Lane 1 (without ATP). Data from three independent experiments were averaged. Error bar represents standard deviation.

3.3. Filament Structure Formation on the Hook in the IMVs

We next examined whether the filament-type proteins exported into the IMVs formed the filament at the tip of the hook. We prepared the IMVs possessing the hook as shown above, and suspended them in an assay mixture containing the FlgN/FlgK, FlgN/FlgL, FliS/FliC, and FliT/FliD complexes (1 μM each), the $FliH_2$/FliI complex (1.5 μM), FliJ (0.25 μM), ATP (5 mM), and Mg^{2+} (5 mM). After incubation for 2 h at 37 °C, the IMVs were collected, washed and solubilized with detergent. Then the flagellar hook–basal bodies were precipitated by ultra-centrifugation, negatively stained with phosphotungstic acid, and observed by electron microscopy. We found the filament attached on the hook–basal body just like the filament–hook–basal body complex purified from *Salmonella* cells, indicating that the filament-type proteins and their cognate chaperones are sufficient to form the filament (Figure 4C) and no other soluble factor is needed for the protein export apparatus to coordinate the export of filament-type proteins with filament formation at the tip of the completed hook. The filament was not formed when either FliK or FlgE was absent in the reaction mixtures (Figure 4A,B), supporting the idea that the filament is formed only after completion of the hook. These results suggest that filament formation in the IMVs proceeded in the same way as in vivo.

We measured the length of the filament by electron microscopy. The filament length was widely distributed from 36 to 890 nm, with an average length of 287 nm (±166 nm), suggesting that the filament length was not controlled (Supplementary Figure S2). The filaments formed in the IMVs were much shorter than those extended from the *Salmonella* cell body [34], implying that the filament growth was restricted by the inner space of the IMVs. Dynamic light scattering (DLS) measurement showed that the Z-average hydrodynamic radius of IMVs was 275 nm ± 121, which was consistent with the average lengths of the filaments formed in the IMVs.

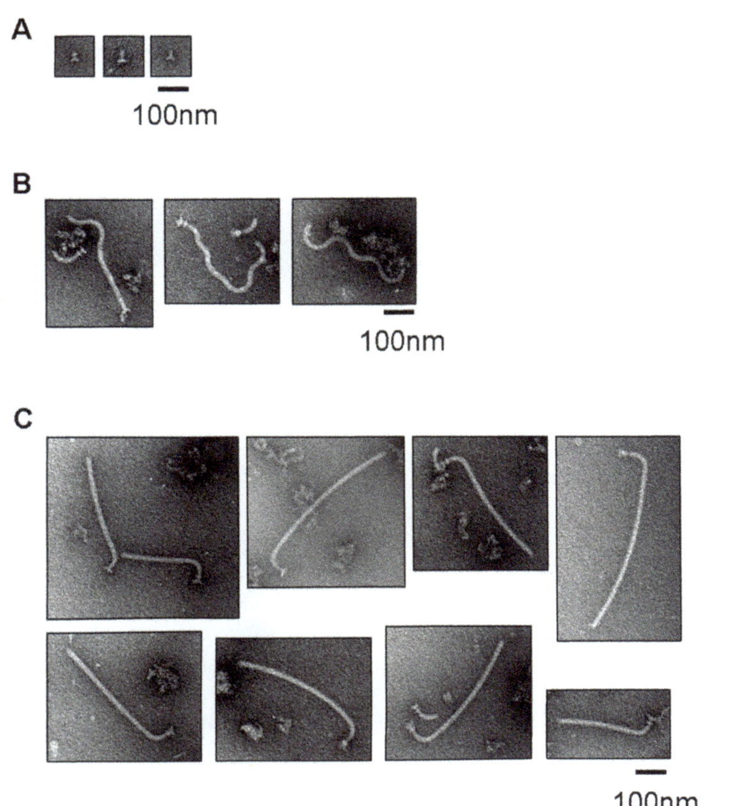

Figure 4. Negative-staining electron micrographs of the basal bodies after transport assay of the filament-type proteins purified from the IMVs. The basal bodies after transport assay using the IMVs prepared after first reaction with FlgD and FliK but without FlgE (**A**), with FlgD and FlgE but without FliK (**B**), and with FlgD, FlgE, and FliK (**C**). The concentrations of FlgD, FlgE, FliK, $FliH_2$/FliI complex, FliJ, and ATP–Mg in the first reaction mixture were 4 μM, 4 μM, 4 μM, 1.5 μM, 0.25 μM, and 5 mM, respectively. The concentrations of FliC/FliS, FliD/FliT, FlgK/FlgN, and FlgL/FlgN in the transport assay mixture for the filament-type proteins were 1 μM each and those of the $FliH_2$/FliI complex, FliJ, and ATP–Mg were the same as in the first reaction mixture.

3.4. Effect of Uncoupling of Flagellar Expression with Assembly on the Entire Assembly Process of the Hook–Filament Complex in the IMVs

We also examined whether the entire process of hook–filament construction proceeded inside the IMVs without adding any soluble components other than those that were used in the above experiment. We prepared a transport assay mixture containing FlgD, FlgE, FliK (1 μM each), the filament-type proteins in complex with their cognate chaperones (1 μM each), the ATPase complex proteins (0.25 μM FliJ and 1.5 μM $FliH_2$/FliI), ATP (5 mM), and Mg^{2+} (5 mM). The IMVs were incubated in the mixture for 4 h at 37 °C, collected by ultra-centrifugation, washed, and solubilized with detergent. The flagellar axial structure was then precipitated by ultra-centrifugation and observed by electron microscopy. The hook–filament structure was formed on the basal body, indicating that the entire process of hook–filament formation proceeded inside the IMVs with only the protein components preexistent in the IMVs and those added in the external solution (Figure 5 and Supplementary Figure S3). This suggests that the coupling of gene expression with the flagellar assembly process is not essential for ordered flagellar formation, although it may be important for efficient construction of the flagellum.

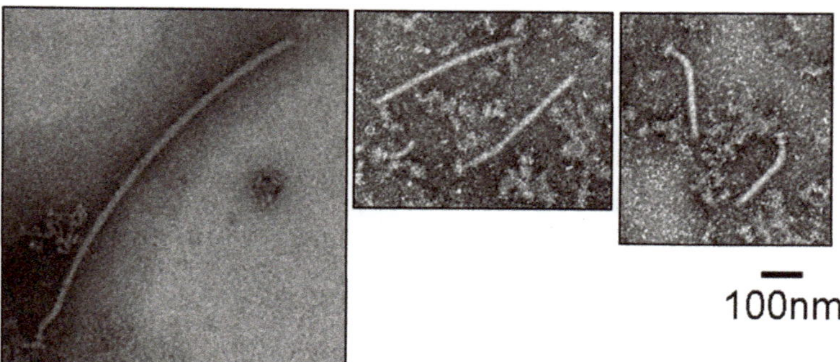

Figure 5. Autonomous hook–filament complex formation in the IMV. Negative-staining electron micrographs of the basal bodies after transport assay in the reaction mixture containing all the export proteins, their cognate chaperones, the export ATPase complex components, ATP, and Mg^{2+}.

4. Discussion

The flagellar hook–filament complex has been reconstructed in solution using purified FlgK, FlgL, FliC, and FliD monomers, and purified hook fragments as a template [38]. However, this is just spontaneous self-assembly and is completely different from in vivo flagellar assembly, in which the assembly process is coupled with flagellar axial protein export. To understand protein export and the following assembly process, we developed an in vitro protein transport assay system that can reproduce the export and assembly process of the bacterial flagellar axial proteins that occur in living cells [29]. Our system uses IMVs and purified proteins. In a series of studies, we have demonstrated that the flagellar protein export apparatus in the IMVs preserves the transport activity of flagellar axial proteins. The hook–filament complex was successfully constructed from FlgD (hook-capping protein), FlgE (hook protein), FlgK (first junction protein), FlgL (second junction protein), FliC (filament protein, flagellin), and FliD (filament-capping protein) transported into the interior of the IMVs. Moreover, the important functions and events, such as hook length control and substrate specificity switch, were nicely reproduced in the IMVs. Thus, the IMV-based system is a powerful tool for investigating the mechanisms of flagellar protein export and assembly.

Our in vitro experiments using the IMVs clearly revealed that the substrate specificity switching of the flagellar protein export apparatus is triggered by FliK alone, and that no other factors are needed for the switching event. When the hook was preassembled inside the IMVs, the filament was formed on the hook without adding any proteins other than the filament components, their cognate chaperones, and the ATPase complex proteins. Moreover, the hook–filament structures were constructed in the IMVs even when all the necessary proteins and factors were added to the external solution at once. These results indicate that coordinated flagellar protein export is not obligately linked to the regulated flagellar gene expression, although they are coupled to each other in the living cell, probably for a more efficient assembly process.

Our results also suggest that the entire flagellar construction process, including the ordered protein export and assembly, is regulated by only the interactions of the PMF-driven export gate with the cytoplasmic ATPase complex, the export substrates, and export chaperones, without the involvement of any other factors. Although the ordered assembly of flagellar axial proteins is achieved in principle by a template-structure driven mechanism based on specific protein–protein interactions, an ordered protein export that could be caused by different affinities of export substrates for the export apparatus may play a role in increasing the assembly efficiency. In fact, the binding affinity of the substrate–chaperone complexes for the export apparatus correlates with the export order of the filament-type proteins [39]. FlgN and FliT interact with FliJ, whereas FliS does not [40]. The cytoplasmic domain of FlhA, one of the export gate component proteins, shows the highest binding affinity for the FlgN/FlgK and FlgN/FlgL complexes, a medium affinity for the FliT/FliD complex, and the lowest affinity for the FliS/FliC complex [41–43]. The filament-type proteins are believed to be secreted in the following order: first FlgK, followed by FlgL, FliD, and FliC. Since the filament was constructed on the hook in the IMVs when only supplying the filament-type substrate–chaperone complexes, the affinity of the substrate–chaperone complexes for the export apparatus may determine the export order of the filament-type substrates. However, our results showed that the individual substrates are also secreted in the absence of their predecessors, suggesting that the in vitro export order is not strict for the filamentous type proteins. It might be possible that transcriptional and translational regulation is needed to strictly control the secretion order in vivo.

All the filament-type proteins showed an increase in the export level upon the addition of the $FliH_2$/FliI complex in the assay solution (Figure 3). In addition, we found some differences in FliH-I-dependent export level between the filament-type proteins. It has been reported that FlgN and FliT interact with the $FliH_2$/FliI complex whereas FliS does not, suggesting that the $FliH_2$/FliI complex may be required for more efficient and rapid export of FlgK, FlgL, and FliD than FliC in vivo [39,44–47]. FlgK, FlgL, and FliD need to be assembled on the hook prior to FliC for filament formation. In the absence of FliH and FliI, a large amount of FliC is leaked into culture media [48], probably due to inefficient assembly of FlgK and FlgL on the hook caused by a failure in the ordered protein export. Therefore, the slight difference in the $FliH_2$/FliI complex dependence of export may relate to the export order of the filament-type proteins.

The export signal of the substrate proteins is found in their disordered N-terminal region, although no significant common sequence has been identified in the region [49]. It has also been reported that the untranslated region of mRNA around the start codon of the substrates is also involved in substrate targeting to the export apparatus [50]. Our in vitro experiments, however, showed that the normal hook–filament structure was successfully constructed inside the IMVs in solution containing no mRNA. Thus, the substrate targeting signal of mRNA is not required for the export of the proteins necessary for hook–filament formation.

The flagellar export apparatus belongs to the type III secretion system family and shares high similarity with the bacterial pathogenic injectisome in its sequence, structure, and function. The injectisome serves to deliver virulence proteins called effectors into their host cells for their infection. The injectisome first exports the needle component proteins and switches the substrate specificity to deliver the effectors after completion of the needle part, just like the flagellum. Therefore, the IMV-based approach we presented here would be useful for studying virulence injectisomes as well.

5. Conclusions

We carried out in vitro flagellar protein transport assay to investigate the export process of the filament-type proteins and their assembly into the flagellar filament. The filament-type proteins were transported into the IMVs and formed the filament on the hook–basal body. The hook–filament structures were successfully formed inside the IMVs when all filament-type proteins, their cognate chaperone, FlgD, FlgE, FliK, the ATPase components, and ATP–Mg were simultaneously added into the assay solution. These results indicate that the coordinated flagellar construction is regulated only by the interactions between the flagellar protein export apparatus, the export substrate proteins, and their cognate chaperones.

Supplementary Materials: The following are available online at http://www.mdpi.com/2218-273X/10/1/126/s1, Table S1: Primer list for this study. Figure S1: SDS-PAGE analysis of purified proteins used in the in this study. Figure S2: Length distribution of the filament formed on the hook in the IMVs. Figure S3: Magnified electron microscopic image of the hook-filament junction region of the hook-filament complex constructed in the IMV.

Author Contributions: Conceptualization, H.T. and K.I.; methodology, H.T. and K.I.; validation, H.T., C.T. and K.I.; formal analysis, H.T., K.N., T.M. and K.I.; investigation, H.T., C.T. and A.K.; resources, H.T., K.N., T.M. and K.I.; data curation, H.T. and K.I.; writing—original draft preparation, H.T. and K.I.; writing—review and editing, H.T., K.N., T.M. and K.I.; visualization, H.T.; supervision, K.I.; project administration, K.I.; funding acquisition, H.T., K.N., T.M. and K.I. All authors have read and agree to the published version of the manuscript.

Funding: This work was supported in part by JSPS KAKENHI Grant Numbers JP18K07108 (to H.T.), JP15H02386 (to K.I.), JP21227006 and JP25000013 (to K.N.), and JP26293097 and JP19H03182 (to T.M.), and MEXT KAKENHI Grant Numbers JP23115008 (to K.I.), and JP15H01640 (to T.M.), and Platform Project for Supporting Drug Discovery and Life Science Research (BINDS) from AMED under Grant Number JP19am0101117 (to K.N.), and the Public Foundation of Chubu Science and Technology Center Grant (to H.T.). This work also supported by JEOL YOKOGUSHI Research Alliance Laboratories of Osaka University to K.N.

Acknowledgments: We thank S. Kobashi and M. Kinoshita for technical support in purification of proteins, R. Yonehara, A. Nakagawa and K. Togon for technical support in DLS, and K. Maki for technical support for EM observation.

Conflicts of Interest: The authors declare no conflict of interest.

References

1. Macnab, R.M. How bacteria assemble flagella. *Annu. Rev. Microbiol.* **2003**, *57*, 77–100. [CrossRef] [PubMed]
2. Minamino, T. Protein export through the bacterial flagellar type III export pathway. *Biochim. Biophys. Acta* **2014**, *1843*, 1642–1648. [CrossRef]
3. Terashima, H.; Kojima, S.; Homma, M. Flagellar motility in bacteria: Structure and function of flagellar motor. *Int. Rev. Cell. Mol. Biol.* **2008**, *270*, 39–85. [PubMed]
4. Iino, T. Assembly of *Salmonella flagellin* in vitro and in vivo. *J. Supramol. Struct.* **1974**, *2*, 372–384. [PubMed]
5. Zhao, Z.; Zhao, Y.; Zhuang, X.Y.; Lo, W.C.; Baker, M.A.B.; Lo, C.J.; Bai, F. Frequent pauses in *Escherichia coli* flagella elongation revealed by single cell real-time fluorescence imaging. *Nat. Commun.* **2018**, *9*, 1885. [CrossRef] [PubMed]
6. Chen, S.; Beeby, M.; Murphy, G.E.; Leadbetter, J.R.; Hendrixson, D.R.; Briegel, A.; Li, Z.; Shi, J.; Tocheva, E.I.; Müller, A.; et al. Structural diversity of bacterial flagellar motors. *EMBO J.* **2011**, *30*, 2972–2981. [CrossRef]
7. Kawamoto, A.; Morimoto, Y.V.; Miyata, T.; Minamino, T.; Hughes, K.T.; Kato, T.; Namba, K. Common and distinct structural features of *Salmonella* injectisome and flagellar basal body. *Sci. Rep.* **2013**, *3*, 3369. [CrossRef]
8. Minamino, T.; Doi, H.; Kutsukake, K. Substrate specificity switching of the flagellum-specific export apparatus during flagellar morphogenesis in *Salmonella typhimurium*. *Biosci. Biotechnol. Biochem.* **1999**, *63*, 1301–1303. [CrossRef]
9. Hirano, T.; Minamino, T.; Namba, K.; Macnab, R.M. Substrate specificity classes and the recognition signal for *Salmonella* type III flagellar export. *J. Bacteriol.* **2003**, *185*, 2485–2492. [CrossRef]
10. Hirano, T.; Yamaguchi, S.; Oosawa, K.; Aizawa, S. Roles of FliK and FlhB in determination of flagellar hook length in *Salmonella typhimurium*. *J. Bacteriol.* **1994**, *176*, 5439–5449. [CrossRef]
11. Moriya, N.; Minamino, T.; Hughes, K.T.; Macnab, R.M.; Namba, K. The type III flagellar export specificity switch is dependent on FliK ruler and a molecular clock. *J. Mol. Biol.* **2006**, *359*, 466–477. [CrossRef] [PubMed]
12. Shibata, S.; Takahashi, N.; Chevance, F.F.; Karlinsey, J.E.; Hughes, K.T.; Aizawa, S. FliK regulates flagellar hook length as an internal ruler. *Mol. Microbiol.* **2007**, *64*, 1404–1415. [CrossRef] [PubMed]
13. Kutsukake, K.; Minamino, T.; Yokoseki, T. Isolation and characterization of FliK-independent flagellation mutants from *Salmonella typhimurium*. *J. Bacteriol.* **1994**, *176*, 7625–7629. [CrossRef] [PubMed]
14. Williams, A.W.; Yamaguchi, S.; Togashi, F.; Aizawa, S.I.; Kawagishi, I.; Macnab, R.M. Mutations in *fliK* and *flhB* affecting flagellar hook and filament assembly in *Salmonella typhimurium*. *J. Bacteriol.* **1996**, *178*, 2960–2970. [CrossRef] [PubMed]
15. Minamino, T.; González-Pedrajo, B.; Yamaguchi, K.; Aizawa, S.I.; Macnab, R.M. FliK, the protein responsible for flagellar hook length control in *Salmonella*, is exported during hook assembly. *Mol. Microbiol.* **1999**, *34*, 295–304. [CrossRef]
16. Erhardt, M.; Singer, H.M.; Wee, D.H.; Keener, J.P.; Hughes, K.T. An infrequent molecular ruler controls flagellar hook length in *Salmonella enterica*. *EMBO J.* **2011**, *30*, 2948–2961. [CrossRef]
17. Hirano, T.; Shibata, S.; Ohnishi, K.; Tani, T.; Aizawa, S. N-terminal signal region of FliK is dispensable for length control of the flagellar hook. *Mol. Microbiol.* **2005**, *56*, 346–360. [CrossRef]
18. Minamino, T.; Ferris, H.U.; Moriya, N.; Kihara, M.; Namba, K. Two parts of the T3S4 domain of the hook-length control protein FliK are essential for the substrate specificity switching of the flagellar type III export apparatus. *J. Mol. Biol.* **2006**, *362*, 1148–1158. [CrossRef]

19. Kinoshita, M.; Aizawa, S.-I.; Inoue, Y.; Namba, K.; Minamino, T. The role of intrinsically disordered C-terminal region of FliK in substrate specificity switching of the bacterial flagellar type III export apparatus. *Mol. Microbiol.* **2017**, *105*, 572–588. [CrossRef]
20. Terahara, N.; Inoue, Y.; Kodera, N.; Morimoto, Y.V.; Uchihashi, T.; Imada, K.; Ando, T.; Namba, K.; Minamino, T. Insight into structural remodeling of the FlhA ring responsible for bacterial flagellar type III protein export. *Sci. Adv.* **2018**, *4*, eaao7054. [CrossRef]
21. Inoue, Y.; Ogawa, Y.; Kinoshita, M.; Terahara, N.; Shimada, M.; Kodera, N.; Ando, T.; Namba, K.; Kitao, A.; Imada, K.; et al. Structural insight into the substrate specificity switch mechanism of the type III protein export apparatus. *Structure* **2018**, *27*, 965–976. [CrossRef] [PubMed]
22. Minamino, T.; Inoue, Y.; Kinoshita, M.; Namba, K. FliK-driven conformational rearrangements of FlhA and FlhB are rquired for export switching of the flagellar potein export apparatus. *J. Bacteriol.* **2020**, 202. [CrossRef]
23. Fraser, G.M.; Bennett, J.C.; Hughes, C. Substrate-specific binding of hook-associated proteins by FlgN and FliT, putative chaperones for flagellum assembly. *Mol. Microbiol.* **1999**, *32*, 569–580. [CrossRef] [PubMed]
24. Auvray, F.; Thomas, J.; Fraser, G.M.; Hughes, C. Flagellin polymerisation control by a cytosolic export chaperone. *J. Mol. Biol.* **2001**, *308*, 221–229. [CrossRef] [PubMed]
25. Aldridge, P.; Karlinsey, J.; Hughes, K.T. The type III secretion chaperone FlgN regulates flagellar assembly via a negative feedback loop containing its chaperone substrates FlgK and FlgL. *Mol. Microbiol.* **2003**, *49*, 1333–1345. [CrossRef] [PubMed]
26. Yokoseki, T.; Iino, T.; Kutsukake, K. Negative regulation by *fliD*, *fliS*, and *fliT* of the export of the flagellum-specific anti-sigma factor, FlgM, in *Salmonella typhimurium*. *J. Bacteriol.* **1996**, *178*, 899–901. [CrossRef] [PubMed]
27. Karlinsey, J.E.; Lonner, J.; Brown, K.L.; Hughes, K.T. Translation/secretion coupling by type III secretion systems. *Cell* **2000**, *102*, 487–497. [CrossRef]
28. Yamamoto, S.; Kutsukake, K. FliT acts as an anti-$FlhD_2C_2$ factor in the transcriptional control of the flagellar regulon in *Salmonella enterica* serovar *typhimurium*. *J. Bacteriol.* **2006**, *188*, 6703–6708. [CrossRef]
29. Terashima, H.; Kawamoto, A.; Tatsumi, C.; Namba, K.; Minamino, T.; Imada, K. In vitro reconstitution of functional type III protein export and insights into flagellar assembly. *MBio* **2018**, 9. [CrossRef]
30. Terashima, H.; Imada, K. Novel insight into an energy transduction mechanism of the bacterial flagellar type III protein export. *Biophys. Phys.* **2018**, *15*, 173–178. [CrossRef]
31. Minamino, T.; Namba, K. Distinct roles of the FliI ATPase and proton motive force in bacterial flagellar protein export. *Nature* **2008**, *451*, 485–488. [CrossRef] [PubMed]
32. Paul, K.; Erhardt, M.; Hirano, T.; Blair, D.F.; Hughes, K.T. Energy source of flagellar type III secretion. *Nature* **2008**, *451*, 489–492. [CrossRef] [PubMed]
33. Minamino, T.; Morimoto, Y.V.; Hara, N.; Namba, K. An energy transduction mechanism used in bacterial flagellar type III protein export. *Nat. Commun.* **2011**, *2*, 475. [CrossRef] [PubMed]
34. Minamino, T.; Morimoto, Y.V.; Kinoshita, M.; Aldridge, P.D.; Namba, K. The bacterial flagellar protein export apparatus processively transports flagellar proteins even with extremely infrequent ATP hydrolysis. *Sci. Rep.* **2014**, *4*, 7579. [CrossRef]
35. Ibuki, T.; Imada, K.; Minamino, T.; Kato, T.; Miyata, T.; Namba, K. Common architecture of the flagellar type III protein export apparatus and F- and V-type ATPases. *Nat. Struct. Mol. Biol.* **2011**, *18*, 277–282. [CrossRef]
36. Imada, K.; Minamino, T.; Uchida, Y.; Kinoshita, M.; Namba, K. Insight into the flagella type III export revealed by the complex structure of the type III ATPase and its regulator. *Proc. Natl. Acad. Sci. USA* **2016**, *113*, 3633–3638. [CrossRef]
37. Kinoshita, M.; Nakanishi, Y.; Furukawa, Y.; Namba, K.; Imada, K.; Minamino, T. Rearrangements of α-helical structures of FlgN chaperone control the binding affinity for its cognate substrates during flagellar type III export. *Mol. Microbiol.* **2016**, *101*, 656–670. [CrossRef]
38. Ikeda, T.; Asakura, S.; Kamiya, R. Total reconstitution of *Salmonella* flagellar filaments from hook and purified flagellin and hook-associated proteins in vitro. *J. Mol. Biol.* **1989**, *209*, 109–114. [CrossRef]
39. Minamino, T. Hierarchical protein export mechanism of the bacterial flagellar type III protein export apparatus. *FEMS Microbiol. Lett.* **2018**, *365*, 117. [CrossRef]
40. Evans, L.D.; Stafford, G.P.; Ahmed, S.; Fraser, G.M.; Hughes, C. An escort mechanism for cycling of export chaperones during flagellum assembly. *Proc. Natl. Acad. Sci. USA* **2006**, *103*, 17474–17479. [CrossRef]

41. Minamino, T.; Kinoshita, M.; Hara, N.; Takeuchi, S.; Hida, A.; Koya, S.; Glenwright, H.; Imada, K.; Aldridge, P.D.; Namba, K. Interaction of a bacterial flagellar chaperone FlgN with FlhA is required for efficient export of its cognate substrates. *Mol. Microbiol.* **2012**, *83*, 775–788. [CrossRef] [PubMed]
42. Kinoshita, M.; Hara, N.; Imada, K.; Namba, K.; Minamino, T. Interactions of bacterial flagellar chaperone-substrate complexes with FlhA contribute to co-ordinating assembly of the flagellar filament. *Mol. Microbiol.* **2013**, *90*, 1249–1261. [CrossRef] [PubMed]
43. Bange, G.; Kümmerer, N.; Engel, C.; Bozkurt, G.; Wild, K.; Sinning, I. FlhA provides the adaptor for coordinated delivery of late flagella building blocks to the type III secretion system. *Proc. Natl. Acad. Sci. USA* **2010**, *107*, 11295–11300. [CrossRef] [PubMed]
44. Thomas, J.; Stafford, G.P.; Hughes, C. Docking of cytosolic chaperone-substrate complexes at the membrane ATPase during flagellar type III protein export. *Proc. Natl. Acad. Sci. USA* **2004**, *101*, 3945–3950. [CrossRef]
45. Imada, K.; Minamino, T.; Kinoshita, M.; Furukawa, Y.; Namba, K. Structural insight into the regulatory mechanisms of interactions of the flagellar type III chaperone FliT with its binding partners. *Proc. Natl. Acad. Sci. USA* **2010**, *107*, 8812–8817. [CrossRef]
46. Minamino, T.; Kinoshita, M.; Imada, K.; Namba, K. Interaction between FliI ATPase and a flagellar chaperone FliT during bacterial flagellar protein export. *Mol. Microbiol.* **2012**, *83*, 168–178. [CrossRef]
47. Sajó, R.; Liliom, K.; Muskotál, A.; Klein, A.; Závodszky, P.; Vonderviszt, F.; Dobó, J. Soluble components of the flagellar export apparatus, FliI, FliJ, and FliH, do not deliver flagellin, the major filament protein, from the cytosol to the export gate. *Biochim. Biophys. Acta* **2014**, *1843*, 2414–2423. [CrossRef]
48. Inoue, Y.; Morimoto, Y.V.; Namba, K.; Minamino, T. Novel insights into the mechanism of well-ordered assembly of bacterial flagellar proteins in *Salmonella*. *Sci. Rep.* **2018**, *8*, 1787. [CrossRef]
49. Végh, B.M.; Gál, P.; Dobó, J.; Závodszky, P.; Vonderviszt, F. Localization of the flagellum-specific secretion signal in *Salmonella* flagellin. *Biochem. Biophys. Res. Commun.* **2006**, *345*, 93–98. [CrossRef]
50. Singer, H.M.; Erhardt, M.; Hughes, K.T. Comparative analysis of the secretion capability of early and late flagellar type III secretion substrates. *Mol. Microbiol.* **2014**, *93*, 505–520. [CrossRef]

Article

The Homologous Components of Flagellar Type III Protein Apparatus Have Acquired a Novel Function to Control Twitching Motility in a Non-Flagellated Biocontrol Bacterium

Alex M. Fulano [1], Danyu Shen [1], Miki Kinoshita [2], Shan-Ho Chou [3] and Guoliang Qian [1,*]

1 College of Plant Protection (Laboratory of Plant Immunity; Key Laboratory of Integrated Management of Crop Diseases and Pests), Nanjing Agricultural University, Nanjing 210095, China; fluxali.alex@gmail.com (A.M.F.); shendanyu@njau.edu.cn (D.S.)

2 Graduate School of Frontier Biosciences, Osaka University, 1-3 Yamadaoka, Suita, Osaka 565-0871, Japan; miki@fbs.osaka-u.ac.jp

3 Institute of Biochemistry, and NCHU Agricultural Biotechnology Center, National Chung Hsing University, Taichung 402, Taiwan; shchou@nchu.edu.tw

* Correspondence: glqian@njau.edu.cn; Tel.: +86-25-84396109

Received: 27 March 2020; Accepted: 5 May 2020; Published: 7 May 2020

Abstract: The bacterial flagellum is one of the best-studied surface-attached appendages in bacteria. Flagellar assembly in vivo is promoted by its own protein export apparatus, a type III secretion system (T3SS) in pathogenic bacteria. *Lysobacter enzymogenes* OH11 is a non-flagellated soil bacterium that utilizes type IV pilus (T4P)-driven twitching motility to prey upon nearby fungi for food. Interestingly, the strain OH11 encodes components homologous to the flagellar type III protein apparatus (FT3SS) on its genome, but it remains unknown whether this FT3SS-like system is functional. Here, we report that, despite the absence of flagella, the FT3SS homologous genes are responsible not only for the export of the heterologous flagellin in strain OH11 but also for twitching motility. Blocking the FT3SS-like system by in-frame deletion mutations in either *flhB* or *fliI* abolished the secretion of heterologous flagellin molecules into the culture medium, indicating that the FT3SS is functional in strain OH11. A deletion of *flhA*, *flhB*, *fliI*, or *fliR* inhibited T4P-driven twitching motility, whereas neither that of *fliP* nor *fliQ* did, suggesting that FlhA, FlhB, FliI, and FliR may obtain a novel function to modulate the twitching motility. The flagellar FliI ATPase was required for the secretion of the major pilus subunit, PilA, suggesting that FliI would have evolved to act as a PilB-like pilus ATPase. These observations lead to a plausible hypothesis that the non-flagellated *L. enzymogenes* OH11 could preserve FT3SS-like genes for acquiring a distinct function to regulate twitching motility associated with its predatory behavior.

Keywords: flagellar type III apparatus; type IV pilus; non-flagellated bacteria; *Lysobacter*; twitching motility

---

1. Introduction

Lysobacter enzymogenes is a Gram-negative, environmentally ubiquitous bacterium [1]. It was shown that this bacterium produces numerous anti-infectious metabolites and extracellular lytic enzymes [1–4]. A distinct feature of *L. enzymogenes* is the evolutionary loss of a surface-attached flagellum, due to the lack of multiple flagellar biogenesis genes such as the *fliC* gene encoding the flagellin subunit [5]. This non-flagellated bacterium exhibits a twitching behavior in natural niches that is powered by type IV pilus (T4P) [6]. As a powerful agent against crop fungal pathogens, *L. enzymogenes* deploys the T4P-driven twitching motility to move towards ecologically relevant, filamentous fungi to prey on them as foods [2,7]. In the model strain OH11, we previously discovered that numerous pilus structural component proteins, including the major pilus subunit, PilA and the motor proteins PilB, and the outer membrane secretin PilQ, are required for the biogenesis of T4P and the function of twitching motility [7].

Flagellated bacteria usually use the flagellum consisting of a filament (helical propeller), a hook (universal joint), and a basal body (rotary motor), to migrate towards more suitable conditions and to escape from undesirable environments for ecological adaption and survival [8–12]. Flagellar assembly is a complicated process involving many flagellar building blocks exported beyond the cellular membranes. This assembly process is dependent upon the flagellar type III protein export apparatus (FT3SS) [13]. The FliI ATPase energizes the unfolding of substrates and disassembly of substrate/chaperone complexes to load them onto the export gate [13]. The export itself mainly works in a proton-motive force (PMF)-driven manner [14]. However, protein export is also possible in the absence of FliI, although less efficient [15]. Five highly conserved inner membrane proteins (FlhA, FlhB, FliP, FliQ, and FliR) form the transmembrane export gate complex that collaborates with the cytoplasmic ATPase ring complex (formed by FliH, FliI, and FliJ) for energy transduction to transport flagellar proteins from the cytoplasm to the distal end of the growing flagellar structure [10,14–19]. Two of the transmembrane export gate complex components, FlhA and FlhB, are involved in substrate (hook-type and filament-type proteins) specificity switching [20–23]. In general, FT3SS components are commonly distributed among flagellated bacteria and are mainly responsible for flagellar protein export, thereby playing an indispensable role inflagellar-driven motility [22]. Recently, a divergent function of FT3SS has been observed in *Bacillus subtilis*, a flagellated, Gram-positive bacterium. In this bacterium, the FT3SS gene products participate in forming flagella-independent nanotubes that are involved in cell-cell exchange of proteins or plasmids [24].

The non-flagellated strain OH11 encodes flagellar FT3SS-like genes on its genome. To clarify whether FT3SS is functional in the non-flagellated strain OH11, we disrupted each homologous gene and analyzed whether deletion mutant strains can transport heterologous FliC molecules (also known as flagellin) derived from *Xanthomonas oryzae*, a flagellated, closely related species of *L. enzymogenes*. We show that mutations in two selected FT3SS-like genes in the wild-type OH11 blocked the secretion of the heterologous flagellin molecules. These findings suggest that despite the lack of a flagellum, strain OH11 still seems to retain anFT3SS-like system that could mediate the secretion of the heterologous flagellar proteins with functions similar to those of canonical flagellar FT3SS. We further show that four FT3SS-like proteins (FlhA, FlhB, FliI, and FilR) in strain OH11 are required for T4P-driven twitching motility. In particular, the flagellar FliI homolog in strain OH11 acts as a PilB-like ATPase to facilitate the secretion of PilA. Our results demonstrate that several FT3SS-like genes that are potentially vestigial in the non-flagellated *L. enzymogenes* appear to be required for T4P-driven twitching motility, highlighting the functional divergence of the FT3SS genes in flagellated and non-flagellated bacteria. Our findings also build on recent work from others that FT3SS play roles in bacterial physiology beyond flagella production.

2. Materials and Methods

2.1. Bacterial Strains, Plasmids and Culture Conditions

The bacterial strains and plasmids used in this study are listed in Table S1. *Escherichia coli* strain DH5α was used for vector construction and was grown in Luria Bertani (LB) broth at 37 °C. Unless otherwise stated, the wild-type *L. enzymogenes* OH11 and its derivatives were grown in LB medium at 28 °C. When required, the medium was amended with gentamicin (Gm) and kanamycin (Km) atfinal concentrations of 150 µg/mL and 100 µg/mL, respectively.

2.2. Genetic Manipulation

The in-frame deletion mutants of the FT3SS genes of strain OH11 have been generated and stored in the laboratory. Recombinant plasmids for complementation were constructed according to our earlier reports [25,26]. In summary, the *flhA*, *flhB*, *fliI*, and *fliR* DNA fragments, each containing full-length gene and its predicated promoter region, were amplified by PCR with different conjugated primer pairs (Table S2). Promoter prediction analysis was conducted with prediction programs [27]. Each amplified DNA fragment was cloned into the broad-host vector pBBR1-MCS5. The resulting recombinant plasmids, pBBR-*flhA*, pBBR-*flhB*, pBBR-*fliI*, and pBBR-*fliR* were individually transformed into competent cells of Δ*flhA*, Δ*flhB*, Δ*fliI*, and Δ*fliR* by electroporation, respectively. The resulting clones were screened by colony PCR and further validated by sequencing. The same approach was used in constructing pBBR-*pilA*, which was introduced to various strains including wild-type *L. enzymogenes* OH11 as listed in Table S1.

For cross complementation, each predicted FT3SS-like gene from strain OH11 was amplified by PCR with the primers listed in Table S2 and cloned into the vector of pTrc99A (Table S1). The resultant recombinant plasmids were individually introduced into the *Salmonella* mutants lacking the respective flagellar counterpart genes (Table S1). The flagellum-driven swimming motility assay was performed as described in [14]. For the expression of heterologous flagellar protein in strain OH11, the broad-host vector, pBBR1-MCS5 was cloned with the $fliC_{Xoo}$ gene from *Xanthomonas oryzae* pv. *oryzae* PX099A (GenBank accession no. NC_010717.2) fused with a Flag tag (Table S1). The resulting plasmid was transformed into respective competent cells of in-frame deletion mutants of *L. enzymogenes* OH11 by electroporation. The resulting clones were screened by colony PCR and further validated by sequencing.

2.3. Twitching Motility Assay

Twitching motility assays were carried out as described previously [6,7]. In brief, *L. enzymogenes* OH11 and its derivatives were grown in 1/20 tryptic soy broth (TSB) agar to a cell density of OD_{600} 1. We aseptically moistened a piece of blotting paper by flushing 750 µL of sterile de-ionized water on the left and right side of a glass slide. Then 20 µL of 5% TSA containing 1.8% agar was evenly spread onto a sterilized microscope slide placed on a humid blotter. The edge of a sterilized cover-slip was dropped into 1000 µL bacterial cell suspension in another glass Petri dish and then gently pressed onto the surface of the medium to create a thin inoculation line. After 24 h of incubation at 28 °C, the margin of the bacterial culture on the glass slide was observed under a microscope at ×640 magnification with images captured using an Olympus DP72 Camera (Olympus, Center Valley, PA, USA). The presence of individual mobile cells or small clusters of cells growing outwardly from the main colony wasan indication of twitching motility. Three independent experiments with each involving three replicates were carried out.

2.4. Immunoblotting

The wild-type OH11 and its derivative strains were grown to an OD_{600} 1.5. For detecting secretion of PilA-Flag, a protocol from a recent study [28] was adopted with slight modifications. The culture samples, each measuring 40 mL, were centrifuged for 30 min at a speed of 6000 rpm at 4 °C to separate cells and culture supernatants. Supernatant fractions were then passed through 0.22-mm filters, with 10% *v/v* trichloroacetic acid added to the filtered supernatants, which were kept at 4 °C overnight to precipitate secreted proteins. The resultant supernatant precipitates were pelleted by centrifugation at 12,000 rpm for 20 min at 4 °C and subsequently washed in ice-cold acetone three times (centrifugation at 12,000 rpm for 15 min at 4 °C). The harvested supernatant precipitates were re-suspended in 1 mL sterile distilled water before subjecting them to centrifugation at 12,000 rpm for 15 min at 4 °C. For all cellular proteins in this study, we pelleted bacterial cells by centrifugation. The pellets were suspended in 2× SDS buffer to a final volume of 60 µL before addition of SDS-PAGE loading buffer. Both supernatant precipitates and cellular fractions were subjected to sodium dodecyl sulfate-polyacrylamide gel electrophoresis (SDS-PAGE) preceding Western blotting assay.

The proteins were transferred onto polyvinylidene difluoride (PVDF) membrane using a semi-dry blot machine (Bio-RAD, Hercules, CA, USA). The membrane was blocked in 5% (*w/v*) skim milk in 1× TBS for 1 h at room temperature or overnight at 4 °C with gentle constant rocking. This was then incubated with monoclonal antibody specific for the Flag tag (Anti-DYKDDDDK-Tag Mouse mAb,

Abmart) at 1:10,000 dilution at room temperature for 1 h. Secondary HRP-conjugated goat anti-mouse antibody (Abmart) at the same dilution factor as the above primary antibody was applied after washing the membranes in TBST buffer (50 mM Tris, 150 mM NaCl, 0.05% (*v*/*v*) Tween 20, pH 7.4) three times at an interval of 10 min. For detecting expression of FT3SS components, we used monoclonal antibodies at 1:20,000 dilution. The secondary antibody used for the above primary monoclonal antibodies was PRP-Goat Anti-Rabbit IgG (H + C) from Jackson Immuno Research (West Grove, PA, USA). Three independent experiments with each involving three replicates were carried out.

3. Results

3.1. The Non-Flagellated L. enzymogenes OH11 Encodes FT3SS-Like Genes

The plant pathogenic *Xanthomonas oryzae* is a flagellated, taxonomically close bacterium with *L. enzymogenes* [29]. Using its reported FT3SS counterparts as queries, we performed a local BLASTP search in the genome of strain OH11 and identified the respective counterparts (Figure 1A). Interestingly, the predicted FT3SS-like genes did not form an operon in the genome of strain OH11 (Figure 1A). To characterize whether these FT3SS-like genes have similar or distinct functions with their flagellar counterparts in regulating flagellar motility, we carried out a cross-complementation assay, in which each FT3SS-like gene from strain OH11 was cloned and introduced into the *Salmonella* mutants lacking the respective flagellar counterpart genes. We found that the FliR homolog restored the flagellum-mediated swimming motility of the *Salmonella fliR* null mutant to some extent after prolonged incubation at 30 °C, whereas the other homolog did not (Figure 1B; Figure S1), indicating that the FliR homolog is partially functional in *Salmonella* cells. This suggests that the FliR homolog forms a PMF-driven transmembrane export gate complex along with the *Salmonella* FlhA, FlhB, FliP and FliQ proteins.

To clarify whether FT3SS is functional in the non-flagellated strain OH11, we introduced the heterologous *fliC* gene, which encodes the flagellar filament protein named flagellin, into the non-flagellated strain OH11 to see whether the FT3SS-like genes could play a role in mediating flagellin export because this is the well-known function played by the canonical flagellar FT3SS genes [23]. For this purpose, we selected the FliC protein (PXO_06154, referred to here as $FliC_{Xoo}$) from *X. oryzae* to generate a recombinant $fliC_{Xoo}$ product fused with a C-terminal Flag tag, which was cloned into a vector to drive its expression under a constitutive promoter. Using Western blotting with a specific anti-Flag antibody, we indeed found that the $FliC_{Xoo}$-Flag fusion protein could be secreted into the culture supernatant of the wild-type OH11, while its secretion was completely abolished in either *flhB* or *fliI* knock-out strain. Cellular presence of $FliC_{Xoo}$-Flag was detected at similar levels in all tested strains withthe RNA polymerase ß-subunit serving as a control (Figure 1C). These results suggest that the non-flagellated *L. enzymogenes* OH11 produces a functional FT3SS that facilitates the export of the test heterologous $FliC_{Xoo}$ protein. It is noteworthy that besides the FliR homolog, other predicted FT3SS-like proteins, FlhA, FlhB, FliP, FliQ and FliI, which failed to exert the flagellar protein export function in *Salmonella* (Figure S1), might have undergone functional divergence to match the native trait of strain OH11 in missing flagellum.

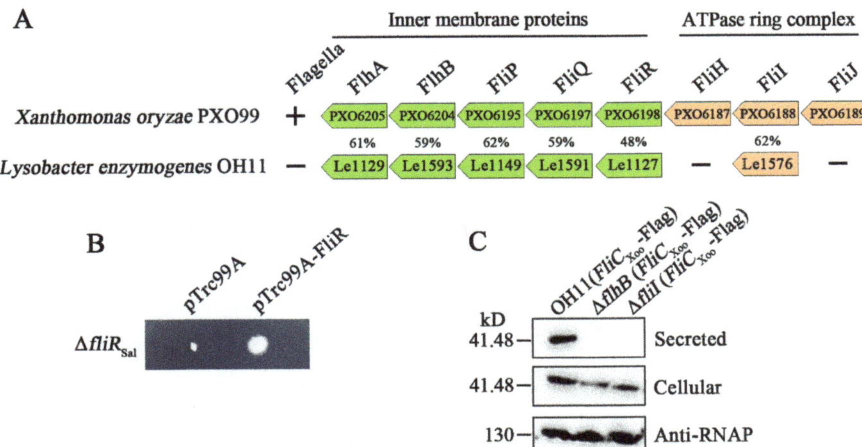

Figure 1. The non-flagellated *Lysobacter enzymognes* OH11 likely carried a functional flagellar type III apparatus (FT3SS) that was able to mediate the export of heterologous flagellar protein. (**A**) Presence of flagellar type III protein apparatus (FT3SS) homologs in the non-flagellated *Lysobacter enzymogenes* OH11. The FT3SS genes characterized in the flagellated *Xanthomonas oryze* PXO99A, a closely related species of *L. enzymogenes* that was used as reference. The sequence similarity between each pair of proteins was provided accordingly. + and −, flagella present and absent, respectively (**B**) The FliR homolog of strain OH11 partially restored the swimming motility of the *Salmonella* $fliR_{Sal}$ null mutant. The photograph was taken after incubation of the transformed strain at 30 °C for 24 h. (**C**) Secretion of the heterologous $FliC_{Xoo}$was dependent on the presence of FT3SS genes in strain OH11. The coding gene of $FliC_{Xoo}$ from *X. oryzae* PXO99A was fused with a Flag-tag and introduced into the wild-type OH11 and two selected FT3SS-defective mutants (Δ*fllhB* and Δ*fliI*). The signal of the $FliC_{Xoo}$-Flag fusion protein was detected in the culture supernatant (defined as "Secreted") of wild-type OH11 but not the test mutant strains, while the intracellular presence (defined as "Cellular") of $FliC_{Xoo}$-Flag was observed in both wild type and two FT3SS-defective mutants.

3.2. Several FT3SS-Like Genes in the Non-Flagellated L. enzymogenes OH11 Play Novel Functions to Affect Twitching Motility

The T4P pili facilitate twitching motility of *L. enzymogenes* OH11 on solid surfaces. The wild-type OH11 does not produce functional flagella because of the lack of flagellar genes (i.e., the flagellin gene). Thus, the motility of this strain (OH11) is not driven by the flagella. However, we found that the FT3SS is functional in the OH11 strain, raising the possibility that the FT3SS might contribute to the other motility systems (twitching motility) of strain OH11. Therefore, we next tested whether, upon the loss of flagellar motility, those remaining FT3SS-like genes in the non-flagellated strain OH11 would have an impact on twitching behavior. To test this, we individually generated in-frame deletion mutants of the FT3SS genes in the wild-type OH11 background and tested whether each gene disruption affectedtwitching motility on solid surfaces. As shown in Figure 2, the individual deletion mutation of *flhA*, *flhB*, *fliI*, and *fliR* in wild-type OH11 inhibited twitching motility, since no mobile cells at the colony margin of each mutant could be observed. Introduction of plasmid-borne *flhA*, *flhB*, *fliI*, or *fliR* back to each respective mutant rescued twitching motility. In contrast, the *fliP* and *fliQ* deletion mutants still exhibited wild-type twitching motility. These results suggest that the FlhA, FlhB, FliI and FliR homologs in the non-flagellated strain OH11 may have obtained a novel function to control twitching motility during the evolutionary process, whereas FliP and FliQ did not.

Figure 2. Involvement of the *FT3SS* genes in twitching motility in *L. enzymognes* OH11. In 1/10 tryptic soy broth (TSB) agar medium, wild-type OH11 exhibit twitching motility, as evidenced by the appearance of mobile cells at the colony margin, while the mutant strains, including Δ*flhA*, Δ*fllhB*, Δ*fliI*, and Δ*fliR* reveal no such capability. Complementing plasmid-borne *flhA*, *flhB*, *fliI*, or *fliR* gene back to each respective mutant rescued twitching motility. However, the Δ*fliP* and Δ*fliQ* mutants still show wild-type twitching motility.

3.3. The FliI Homolog Affects PilA Secretion in L. enzymogenes OH11

What is the mechanism by which the FT3SS genes modulate twitching motility in the non-flagellated strain OH11? In our earlier study [7], we showed that the cellular expression and the secretion of PilA, the major pilin, isessential for the formation of twitching motility in strain OH11, and we thus investigated whether the FT3SS components were utilized in such a manner to affect twitching motility. For this purpose, a plasmid containing the PilA-Flag fusion was introduced into wild-type OH11 and its deletion mutant derivatives. The transformed strains were cultivated in 1/10 TSB with cells collected at OD_{600} 1.5, followed by Western blotting via anti-Flag antibody. The secretion of the PilA-Flag fusion into the culture medium was not detected in the *fliI* mutant, although its intracellular amount was similar tothat of wild type. Under the same test conditions, the PilA-Flag fusion was detected in the culture supernatants of the *flhA*, *flhB* and *fliR* mutants in a way similar to thewild-type strain OH11 (Figure 3A). As controls, the presence of the PilA-Flag fusion was also detected in the culture supernatant and cellular fractions of the *fliQ* and *fliP* mutants (Figure 3A), both of which produced the wild-type twitching motility (Figure 2). These results imply that while FlhA, FlhB, FliI, and FliR all participate in the formation of twitching motility, only FliI has a remarkable impact on the secretion of PilA-Flag.

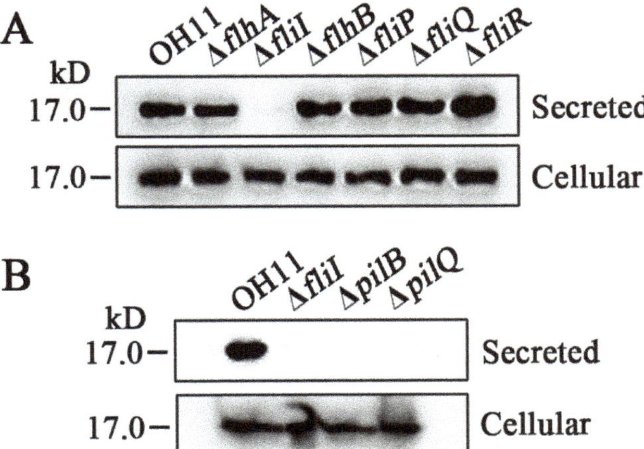

Figure 3. Differential contribution of FT3SS components to the secretion of PilA, the major pilin of type IV pilus in the non-flagellated *L. enzymognes* OH11. (**A**) Western blotting showing inactivation of FliIfully blocked PilA secretion. The coding gene of PilA was fused with a Flag-tag and introduced into wild-type OH11 and the FT3SS-defective mutants (Δ*flhA*, Δ*fllhB*, Δ*fliI*, Δ*fliP*, Δ*fliQ*, or Δ*fliR*). Strains were cultivated in 1/10 TSB, and cells were collected at OD_{600}, 1.5. Except for the *fliI* mutant (Δ*fliI*), the signal of PilA-Flag determined by anti-Flag antibody was detected both in the culture supernatant of wild-type OH11 and the remaining five mutants, while cellular presence of PilA-Flag was observed in all test strains. (**B**) Western blot verification of PilA secretion was dependent both on the FliI and on two other pilus apparatus proteins, PilB (a pilus ATPase that provides energy via ATP hydrolysis to promote pilus extension) and PilQ (which forms an outer membrane secretin pore comprising multiple monomers to facilitate pilus extraction).

To validate the above findings, we further chose two *L. enzymogenes* T4P-defective mutants, Δ*pilB* and Δ*pilQ*, as additional strains, since both fail to produce surface-attached T4P according to our earlier study [7]. As expected, we could only detect the intracellular presence of PilA-Flag in both mutants, but no signal was observed in their culture supernatants (Figure 3B), which agreed with those of the *fliI*mutant (Figure 3B). These results lead to a plausible hypothesis that the flagellar FliI ATPase in the non-flagellated strain OH11 has likely evolved to act as PilB-like ATPase to promote PilA secretion, there by contributing totwitching motility in *L. enzymogenes*.

4. Discussion

In bacterial physiology, the bacterial flagellum is one of the best-studied surface-attached appendages, whose assembly relies on a wide array of flagellar proteins [10,22]. The translocation of flagellar proteins from cytoplasm across the cytoplasmic membrane is governed by FT3SS [14], which is generally recognized as the main factor in controlling flagellar assembly and hence affects flagellum-driven motility in flagellated bacteria [14–22]. Here, we demonstrated an intriguing case showing that the FT3SS-like geneswere encodedon the genome of a non-flagellated environmentally ubiquitous bacterium and acquired novel functions to enhance twitching motility. Such findings expand our current knowledge on the functional evolution or divergence of the conserved FT3SS genes from flagellated to non-flagellated bacteria, which may provoke the interests of scientists studying various bacterial species with and without flagella.

However, one may question why the FT3SS-like genesare evolutionarily retained in the non-flagellated *L. enzymogenes*. In this bacterium, our experimental data indicate that the retained FT3SS-like gene products possibly assemble into a functional FT3SS system that is able to mediate the export of the heterologous flagellin, as with the case played by the canonical *FT3SS* genes in flagellated bacteria [12]. This finding raises a possibility that the FT3SS-like genes in the non-flagellated

L. enzymogenes might have a coordinated role in mediating export of native, yet unidentified flagellar or non-flagellar proteins. We are currently testing this hypothesis. The findings that several FT3SS-like genes in strain OH11 playing a role in boosting twitching motility is interesting and ecologically relevant, because such existing twitching behavior may help the non-flagellated *L. enzymogenes* move and prey upon fungi for nutrients.

One may further ask how the FT3SS-lilke genes diverge to affect twitching motility in the non-flagellated *L. enzymogenes*. As documented previously [7], twitching motility is controlled by the T4P, which is assembled by the major pilin, PilA, and other minor pilins. PilB is an ATPase to hydrolyze ATP to drive pilus assembly, forming a pilus that extends to the cell surface via an outer membrane secretin pore comprised of multiple PilQ monomers [29–31]. The non-flagellated *L. enzymogenes* also utilizes this similar strategy to perform twitching motility, because it has been shown that the inactivation of either PilB or PilQ abolishes the secretion of PilA, thereby inhibiting twitching motilityof strain OH11 [7]. Here, we showed that among the four FT3SS components (FlhA, FlhB, FliI, and FliR), FliI is the only component that is directly involved in PilA secretion. In flagellated bacteria, FliI is the Walker-type ATPase of FT3SS which hydrolyzes ATP to allow the transmembrane export gate complex made up of FlhA, FlhB, FliP, FliQ and FliR to drive the export of flagellar proteins from the cytoplasm to the distal end of growing flagellar structure in a PMF-dependent manner [14,16,19]. The crystal structure of the *Salmonella* FliI shows extensive structural similarities to α and β subunits of F1-ATPase [32,33]. The FliI ATPase forms a homo-hexamer to hydrolyze ATP at an interface between FliI subunits in a way similar to F1-ATPase [34]. PilB and PilQ also form a hexameric ring-like structure as seen in F1-ATPase and FliI [35]. Moreover, bioinformatic analyses via the online software [36] led to the observation that the *L. enzymogenes* FliI homolog belonged to the F1-ATPase family; it possessed a nucleotide-binding domain and a beta-barrel domain, ranging from the sequence positions of 162 to 372 and 40 to 106, respectively. This information, along with our present findings, suggests that other than the canonical PilB, FliI in the non-flagellated *L. enzymogenes* has evolved to act as the second ATPase hexamer to contribute to twitching motility of *L. enzymogenes* via affecting PilA secretion. In this regard, the FliI–PilA pair is postulated to act as a bridge to indirectly link FT3SS with pilus assembly in the non-flagellated *L. enzymogenes*. This finding is also in agreement with a recent study showing that different flagellar gene mutations affect levels of pilus production in *Caulobacter crescentus*, another flagellated bacterium [37]. However, under the current test conditions, this mode of action was possibly not applied by the remaining three FT3SS components (FlhA, FlhB, and FliR) in *L. enzymogenes*, because although they were required for twitching motility of the strain OH11, they did not in fact affect PilA secretion. This datum reveals the possibility that an additionally uncharacterized mechanism might exist for FT3SS components to control the twitching motility in *L. enzymogenes*.

Overall, we provided some intriguing evidence showing how some FT3SS-like genes have adopted altered function during the evolutional process to confer adaptive advantage for a non-flagellated bacterium. This study thus increases our understanding ofthe functional divergence of flagellar genes from flagellated to non-flagellated bacteria.

5. Conclusions

In the present study, we report that several FT3SS-like genes—which are potentially vestigial in the non-flagellated biocontrol bacterium, *L. enzymogenes* OH11—appear to be required for type-IV-pilus-driven twitching motility. We furthermore suggest that, upon the loss of flagella-driven motility, several of the remaining FT3SS-like genes have acquired a novel function to control twitching motility. The flagellar FliI ATPase in particular would have evolved to act as a PilB-like ATPase involved in PilA secretion. These findings reveal that the FT3SS genes have undergone functional divergence between flagellated and non-flagellated bacteria, supporting the recent notion that FT3SS plays roles in bacterial physiology beyond flagella production.

Supplementary Materials: The following are available online at http://www.mdpi.com/2218-273X/10/5/733/s1, Figure S1: Cross-complementation of the *L. enzymogenes* FT3SS-like genes in the *Salmonella* mutants lacking the respective flagellar counterparts, Table S1: Strains and plasmids used in this study title, Table S2: Primers used in this study.

Author Contributions: G.Q. conceived the project with input from all authors. G.Q. and A.M.F. designed experiments. A.M.F. and M.K. performed experiments. D.S. performed the bioinformatics analysis. G.Q., D.S., M.K., and A.M.F. analyzed data. A.M.F., S.-H.C., and G.Q. wrote the manuscript. G.Q supervised the project. All authors have read and agreed to the published version of the manuscript.

Funding: This research was funded by the Natural Science Foundation of Jiangsu Province (BK20190026; BK20181325), the National Natural Science Foundation of China (31872016), the Fundamental Research Funds for the Central Universities (KJJQ202001; KYYJ201904;KYT201805 and KYTZ201403), Jiangsu Agricultural Sciences and Technology Innovation Fund [CX(18)1003] and Innovation Team Program for Jiangsu Universities (2017) and by JSPS KAKENHI Grant Numbers JP18K14638 (to M.K.).

Conflicts of Interest: The authors declare no conflict of interest.

References

1. Christensen, P.; Cook, F.D. Lysobacter, a New Genus of Nonfruiting, Gliding Bacteria with a High Base Ratio. *Int. J. Syst. Bacteriol.* **1978**, *28*, 367–393. [CrossRef]
2. Kobayashi, D.Y.; Reedy, R.M.; Palumbo, J.D.; Zhou, J.-M.; Yuen, G.Y. A clp Gene Homologue Belonging to the Crp Gene Family Globally Regulates Lytic Enzyme Production, Antimicrobial Activity, and Biological Control Activity Expressed by Lysobacterenzymogenes Strain C3. *Appl. Environ. Microbiol.* **2005**, *71*, 261–269. [CrossRef]
3. Xie, Y.; Wright, S.; Shen, Y.-M.; Du, L. Bioactive natural products from Lysobacter. *Nat. Prod. Rep.* **2012**, *29*, 1277–1287. [CrossRef] [PubMed]
4. Panthee, S.; Hamamoto, H.; Paudel, A.; Sekimizu, K. Lysobacter species: a potential source of novel antibiotics. *Arch. Microbiol.* **2016**, *198*, 839–845. [CrossRef] [PubMed]
5. De Bruijn, I.; Cheng, X.; De Jager, V.; Expósito, R.G.; Watrous, J.D.; Patel, N.; Postma, J.; Dorrestein, P.C.; Kobayashi, D.Y.; Raaijmakers, J.M. Comparative genomics and metabolic profiling of the genus Lysobacter. *BMC Genom.* **2015**, *16*, 991. [CrossRef] [PubMed]
6. Zhou, X.; Qian, G.L.; Chen, Y.; Du, L.C.; Liu, F.Q.; Yuen, G.Y. PilG is involved in the regulation of twitching motility and antifungal antibiotic biosynthesis in the biological control agent *Lysobacterenzymogenes*. *Phytopathology* **2015**, *105*, 1318–1324. [CrossRef] [PubMed]
7. Xia, J.; Chen, J.; Chen, Y.; Qian, G.; Liu, F. Type IV pilus biogenesis genes and their roles in biofilm formation in the biological control agent Lysobacterenzymogenes OH11. *Appl. Microbiol. Biotechnol.* **2017**, *102*, 833–846. [CrossRef] [PubMed]
8. Patnaik, P.R. Robustness analysis of the E.coli chemosensory system to perturbations in chemoattractant concentrations. *Bioinformatics* **2007**, *23*, 875–881. [CrossRef]
9. Porter, S.L.; Wadhams, G.H.; Armitage, J.P. Signal processing in complex chemotaxis pathways. *Nat. Rev. Genet.* **2011**, *9*, 153–165. [CrossRef]
10. Fukumura, T.; Makino, F.; Dietsche, T.; Kinoshita, M.; Kato, T.; Wagner, S.; Namba, K.; Imada, K.; Minamino, T. Assembly and stoichiometry of the core structure of the bacterial flagellar type III export gate complex. *PLoS Boil.* **2017**, *15*, e2002281. [CrossRef]
11. Gu, H. Role of Flagella in the Pathogenesis of Helicobacter pylori. *Curr. Microbiol.* **2017**, *74*, 863–869. [CrossRef]
12. Nakamura, S.; Minamino, T. Flagella-Driven Motility of Bacteria. *Biomoecules* **2019**, *9*, 279. [CrossRef] [PubMed]
13. Minamino, T. Protein export through the bacterial flagellar type III export pathway. *Biochim. Biophys. Acta (BBA) Bioenerg.* **2014**, *1843*, 1642–1648. [CrossRef] [PubMed]
14. Minamino, T.; Namba, K. Distinct roles of the FliI ATPase and proton motive force in bacterial flagellar protein export. *Nature* **2008**, *451*, 485–488. [CrossRef] [PubMed]
15. Minamino, T.; Morimoto, Y.V.; Kinoshita, M.; Aldridge, P.D.; Namba, K. The bacterial flagellar protein export apparatus processively transports flagellar proteins even with extremely infrequent ATP hydrolysis. *Sci. Rep.* **2014**, *4*, 7579. [CrossRef]
16. Minamino, T.; Morimoto, Y.V.; Hara, N.; Namba, K. An energy transduction mechanism used in bacterial flagellar type III protein export. *Nat. Commun.* **2011**, *2*, 475. [CrossRef]

17. Diepold, A.; Armitage, J.P. Type III secretion systems: the bacterial flagellum and the injectisome. *Philos. Trans. R. Soc. B Boil. Sci.* **2015**, *370*, 20150020. [CrossRef]
18. Dietsche, T.; Mebrhatu, M.T.; Brunner, M.J.; Abrusci, P.; Yan, J.; Franz-Wachtel, M.; Schärfe, C.; Zilkenat, S.; Grin, I.; Galán, J.E.; et al. Structural and Functional Characterization of the Bacterial Type III Secretion Export Apparatus. *PLoS Pathog.* **2016**, *12*, e1006071. [CrossRef]
19. Morimoto, Y.V.; Kami-Ike, N.; Miyata, T.; Kawamoto, A.; Kato, T.; Namba, K.; Minamino, T. High-Resolution pH Imaging of Living Bacterial Cells To Detect Local pH Differences. *mBio* **2016**, *7*, e01911–e01916. [CrossRef]
20. Kuhlen, L.; Abrusci, P.; Johnson, S.; Gault, J.; Deme, J.; Caesar, J.; Dietsche, T.; Mebrhatu, M.T.; Ganief, T.; Macek, B.; et al. Structure of the Core of the Type Three Secretion System Export Apparatus. *Nat. Struct. Mol. Boil.* **2018**, *25*, 583–590. [CrossRef]
21. Xing, Q.; Shi, K.; Portaliou, A.; Rossi, P.; Economou, A.; Kalodimos, C.G. Structures of chaperone-substrate complexes docked onto the export gate in a type III secretion system. *Nat. Commun.* **2018**, *9*, 1773. [CrossRef]
22. Fabiani, F.D.; Renault, T.T.; Peters, B.; Dietsche, T.; Galvez, E.; Guse, A.; Freier, K.; Charpentier, E.; Strowig, T.; Franz-Wachtel, M.; et al. A flagellum-specific chaperone facilitates assembly of the core type III export apparatus of the bacterial flagellum. *PLoS Boil.* **2017**, *15*, e2002267. [CrossRef] [PubMed]
23. Minamino, T. Hierarchical protein export mechanism of the bacterial flagellar type III protein export apparatus. *FEMS Microbiol. Lett.* **2018**, *365*, 117. [CrossRef] [PubMed]
24. Bhattacharya, S.; Baidya, A.K.; Pal, R.R.; Mamou, G.; Gatt, Y.E.; Margalit, H.; Rosenshine, I.; Ben-Yehuda, S. A Ubiquitous Platform for Bacterial Nanotube Biogenesis. *Cell Rep.* **2019**, *27*, 334–342. [CrossRef] [PubMed]
25. Qian, G.; Wang, Y.; Liu, Y.; Xu, F.; He, Y.-W.; Du, L.; Venturi, V.; Fan, J.; Hu, B.; Liu, F. Lysobacterenzymogenes Uses Two Distinct Cell-Cell Signaling Systems for Differential Regulation of Secondary-Metabolite Biosynthesis and Colony Morphology. *Appl. Environ. Microbiol.* **2013**, *79*, 6604–6616. [CrossRef] [PubMed]
26. Qian, G.; Xu, F.; Venturi, V.; Du, L.; Liu, F. Roles of a solo LuxR in the biological control agent Lysobacterenzymogenes strain OH11. *Phytopathology* **2014**, *104*, 224–231. [CrossRef]
27. Reese, M.G. Application of a time-delay neural network to promoter annotation in the Drosophila melanogaster genome. *Comput Chem.* **2001**, *26*, 51–56. [CrossRef]
28. Green, E.R.; Mecsas, J. Bacterial Secretion Systems: An Overview. *Microbiol. Spectr.* **2016**, *4*, 215–239. [CrossRef]
29. Lim, S.-H.; So, B.-H.; Wang, J.-C.; Song, E.-S.; Park, Y.-J.; Lee, B.-M.; Kang, H.-W. Functional analysis of pilQ gene in Xanthomanasoryzaepv. oryzae, bacterial blight pathogen of rice. *J. Microbiol.* **2008**, *46*, 214–220. [CrossRef]
30. Wilson, M.M.; Bernstein, H.D.; Information, P.E.K.F.C. Surface-Exposed Lipoproteins: An Emerging Secretion Phenomenon in Gram-Negative Bacteria. *Trends Microbiol.* **2016**, *24*, 198–208. [CrossRef]
31. Desvaux, M.; Candela, T.; Serror, P. Surfaceome and Proteosurfaceome in Parietal Monoderm Bacteria: Focus on Protein Cell-Surface Display. *Front. Microbiol.* **2018**, *9*, 100. [CrossRef]
32. Imada, K.; Minamino, T.; Tahara, A.; Namba, K. Structural similarity between the flagellar type III ATPase FliI and F1-ATPase subunits. *Proc. Natl. Acad. Sci. USA* **2007**, *104*, 485–490. [CrossRef]
33. Imada, K.; Minamino, T.; Uchida, Y.; Kinoshita, M.; Namba, K. Insight into the flagella type III export revealed by the complex structure of the type III ATPase and its regulator. *Proc. Natl. Acad. Sci. USA* **2016**, *113*, 3633–3638. [CrossRef]
34. Kazetani, K.-I.; Minamino, T.; Miyata, T.; Kato, T.; Namba, K. ATP-induced FliIhexamerization facilitates bacterial flagellar protein export. *Biochem. Biophys. Res. Commun.* **2009**, *388*, 323–327. [CrossRef]
35. Chang, Y.W.; Rettberǵ, L.A.; Treuner-Lanǵe, A.; Iwasa, J.; Søǵaard-Andersen, L.; Jensen, G.J. Architecture of the type Iva pilus machine. *Science* **2016**, *351*, aad2001. [CrossRef]
36. Razia, M.; Raja, K.R.; Padmanaban, K.; Sivaramakrishnan, S.; Chellapandi, P. A phylogenetic approach for assigning function of hypothetical proteins in *Photorhabdus luminescens* Subsp. *laumondii* TT01 genome. *J. Comput. Sci. Syst. Biol.* **2010**, *3*, 021–029.
37. Ellison, C.K.; Rusch, D.B.; Brun, Y.V. Flagellar Mutants Have Reduced Pilus Synthesis in Caulobactercrescentus. *J. Bacteriol.* **2019**, *201*, e00031-19. [CrossRef]

Article

GFP Fusion to the N-Terminus of MotB Affects the Proton Channel Activity of the Bacterial Flagellar Motor in *Salmonella*

Yusuke V. Morimoto [1,2,*], Keiichi Namba [2,3,4] and Tohru Minamino [2]

1 Department of Physics and Information Technology, Faculty of Computer Science and Systems Engineering, Kyushu Institute of Technology, 680-4 Kawazu, Iizuka, Fukuoka 820-8502, Japan
2 Graduate School of Frontier Biosciences, Osaka University, 1-3 Yamadaoka, Suita, Osaka 565-0871, Japan; keiichi@fbs.osaka-u.ac.jp (K.N.); tohru@fbs.osaka-u.ac.jp (T.M.)
3 RIKEN Spring-8 Center & Center for Biosystems Dynamics Research (BDR), 1-3 Yamadaoka, Suita, Osaka 565-0871, Japan
4 JEOL YOKOGUSHI Research Alliance Laboratories, Osaka University, 1-3 Yamadaoka, Suita, Osaka 565-0871, Japan
* Correspondence: yvm001@bio.kyutech.ac.jp; Tel.: +81-948-29-7833

Received: 30 July 2020; Accepted: 27 August 2020; Published: 29 August 2020

Abstract: The bacterial flagellar motor converts the energy of proton flow through the MotA/MotB complex into mechanical works required for motor rotation. The rotational force is generated by electrostatic interactions between the stator protein MotA and the rotor protein FliG. The Arg-90 and Glu-98 from MotA interact with Asp-289 and Arg-281 of FliG, respectively. An increase in the expression level of the wild-type MotA/MotB complex inhibits motility of the *gfp-motB fliG(R281V)* mutant but not the *fliG(R281V)* mutant, suggesting that the MotA/GFP-MotB complex cannot work together with wild-type MotA/MotB in the presence of the *fliG(R281V)* mutation. However, it remains unknown why. Here, we investigated the effect of the GFP fusion to MotB at its N-terminus on the MotA/MotB function. Over-expression of wild-type MotA/MotB significantly reduced the growth rate of the *gfp-motB fliG(R281V)* mutant. The over-expression of the MotA/GFP-MotB complex caused an excessive proton leakage through its proton channel, thereby inhibiting cell growth. These results suggest that the GFP tag on the MotB N-terminus affects well-regulated proton translocation through the MotA/MotB proton channel. Therefore, we propose that the N-terminal cytoplasmic tail of MotB couples the gating of the proton channel with the MotA–FliG interaction responsible for torque generation.

Keywords: bacterial flagellar motor; proton motive force; ion channel; torque generation; fluorescent protein

---

1. Introduction

Many bacteria are propelled by rotating flagella to swim in liquid environments. The basal body is located at the base of the flagellar filament acting as a helical propeller and works as a rotary motor powered by the electrochemical potential difference of cations, such as proton and sodium ion, across the membrane that translocate those cations through the transmembrane channel of the stator complex associated around the rotor [1,2].

The proton-driven flagellar motor of *Salmonella enterica* generates the rotational force through processive interactions between the rotor and multiple stator units [3–7]. A ring-like structure, the basal body MS–C ring complex, functions as a bi-directional rotor, and the switch proteins FliG, FliM, and FliN form the C ring just below the MS ring formed by a transmembrane protein, FliF [8]. The C ring is a switching device that allows the *Salmonella* motor to spin counterclockwise (CCW) and clockwise (CW) [9]. The stator complex is composed of two transmembrane proteins, MotA and MotB, and acts as a transmembrane proton channel that couples the proton flow though the channel with torque generation [10–12]. At least 11 stator units can associate around the rotor, but they show rapid exchanges between rotor-associated and freely diffusing forms during motor rotation [13,14]. The flagellar motor autonomously controls the number of functional stator units around the rotor in response to changes in the environment [15].

MotA has four transmembrane (TM) helices (TM1–TM4), two short periplasmic loops, and a relatively large cytoplasmic loop ($MotA_C$) between TM2 and TM3 and the C-terminal cytoplasmic tail. $MotA_C$ contains highly conserved charged residues, Arg-90 and Glu-98 [16,17]. The motility defect of the *motA(R90E)* and *motA(E98K)* mutants are partially restored by the *fliG(D289K)* and *fliG(R281V)* mutations, respectively, suggesting that electrostatic interactions between Arg-90 of MotA and Asp-289 of FliG and between Glu-98 of MotA and Arg-281 of FliG are responsible for flagellar motor rotation [17].

The TM helix of MotB (MotB-TM) form a proton channel along with the TM3 and TM4 helices of MotA. The highly conserved Asp-33 residue of *Salmonella* MotB, which is located in MotB-TM, is the most important residue involved in the proton influx through the MotA/MotB proton channel complex, and the *motB(D33N)* mutation destroys the proton channel activity of the MotA/MotB complex, thereby conferring a loss-of-motility phenotype [18–21]. The N-terminal tail of MotB ($MotB_{NCT}$) exists in the cytoplasm, and the large C-terminal domain containing a peptidoglycan binding (PGB) motif ($MotB_{PGB}$) is located in the periplasm (Figure 1) [22–24]. A flexible linker connecting MotB-TM and $MotB_{PGB}$ contains a plug segment that binds to the proton channel to suppress premature proton translocation through the MotA/MotB proton channel until the MotA/MotB complex becomes an active stator unit around the rotor [25,26]. This flexible linker also suppresses the peptidoglycan binding activity of the MotA/MotB complex until the MotA/MotB complex encounters the rotor, and the interactions between MotA and FliG are postulated to trigger a structural transition of the N-terminal portion of $MotB_{PGB}$ from a compact to an extended conformation, allowing $MotB_{PGB}$ to reach the peptidoglycan (PG) layer for binding [24,27,28]. $MotB_{NCT}$ is critical for the MotB function [29,30], although its role in flagellar motor rotation remains unknown.

Live cell imaging techniques using a fluorescent protein are widely used to elucidate the rotation mechanism of the flagellar motor [13,31–36]. A fusion of a green fluorescent protein (GFP) to the N-terminus of MotB does not affect the MotB function much, although the cell motility is not at the wild-type level. Using this functional GFP-MotB fusion, the assembly mechanism of the MotA/MotB stator complex was extensively analyzed in various genetic backgrounds (Figure 1), and we found that the interaction between Arg-90 of MotA and Asp-289 of FliG is more important for proper positioning of the MotA/MotB complex relative to the rotor whereas the interaction between Glu-98 of MotA and Arg-281 of FliG is more critical for torque generation [13,34,36]. However, the fusion of a fluorescent protein to motor component proteins sometimes affects the motor function significantly depending on the fusion sites of the target proteins. For example, simply changing the type of fluorescent protein fused to the N terminus of MotB changes the frequency of directional switching of the flagellar motor [37].

The *Salmonella gfp-motB fliG(D289K)* mutant is non-motile [36]. In contrast, the *gfp-motB fliG(R281V)* mutant is motile with its average swimming speed being about two-thirds of the *gfp-motB* cells [36]. The expression of wild-type MotA/MotB complex restores the motility of the *gfp-motB fliG(D289K)* mutant to about 70% of that of the *gfp-motB* cells, but inhibits the motility of the *gfp-motB fliG(R281V)* mutant while not affecting the motility of the *fliG(R281V)* mutant in the absence of GFP-tagged MotB [36]. These observations suggest that the MotA/GFP-MotB complex cannot work with the wild-type MotA/MotB complex when the *fliG(R281V)* mutation is present. However, it remains unknown how the fusion of GFP to the N-terminus of MotB affects the motor function.

To clarify this question, we analyzed the multicopy effect of the MotA/GFP-MotB complex on intracellular pH to determine whether the GFP tag affects the proton channel activity of the MotA/MotB complex. We found that the over-expression of the MotA/GFP-MotB complex reduces the intracellular pH, thereby causing a growth defect. We also found that $MotB_{NCT}$ is close to both $MotA_C$ and FliG and that a fusion of GFP to the N-terminus of MotB facilitates the MotA/MotB proton channel activity regardless of whether or not the MotA/MotB complex being a functionally active stator unit in the motor.

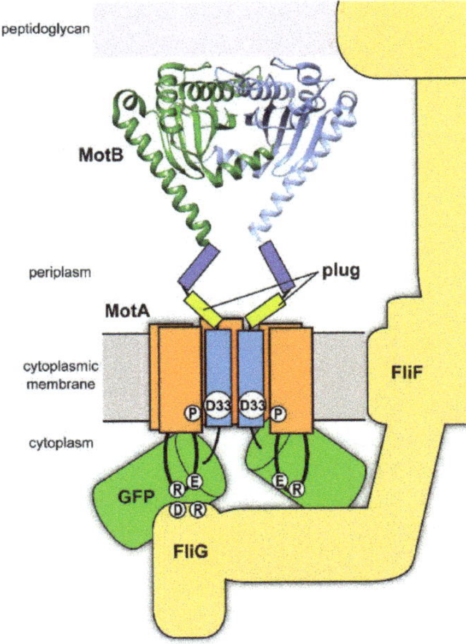

Figure 1. Schematic illustration of the MotA/GFP-MotB complex assembled into the flagellar motor. The N-terminus of MotB faces the cytoplasmic side. MotB-Asp33 (D33) and MotA-P173 (P) residues are located in the proton pathway and play important roles in the energy coupling mechanism of the flagellar motor. Highly charged Arg-90 (R) and Glu-98 (E) residues in the cytoplasmic loop of MotA interact with Asp-289 (D) and Arg-281 (R) of FliG. A plug segment of MotB (plug) suppresses proton leakage through the MotA/MotB complex. The ribbon diagram shows the crystal structure of the PGB domain of MotB (PDB ID: 2ZVY) forming a dimer; its dimerization is critical for the MotB function.

2. Materials and Methods

2.1. Bacterial Strains, Plasmids and Media

Bacterial strains and plasmids used in this study are listed in Table 1. To construct a plasmid encoding MotA and GFP-MotB, the *motA* and *gfp-motB* genes were amplified by PCR from the chromosomal DNA of the *Salmonella* YVM003 strain, followed by DNA digestion by restriction enzymes, PstI and HindIII, and finally the insertion of the PCR product into the PstI and HindIII sites of the pBAD24 vector. Procedures for DNA manipulations and DNA sequencing were carried out as described previously [38]. L-broth (LB) and motility medium were prepared as described previously [39,40].

Table 1. Bacterial strains and plasmids.

Strain or Plasmid	Relevant Characteristics	Reference
Salmonella		
SJW1103	Wild-type for motility and chemotaxis	[41]
SJW1368	Δ(*cheW-flhD*); master operon mutant	[42]
YVM003	*gfp-motB*	[34]
YVM034	*gfp-motB fliG(R281V)*	[36]
YVM036	*motA(E98K) gfp-motB fliG(R281V)*	[36]
YVM046	*fliG(R281V)*	[36]
YVM047	*motA(E98K)*	[36]
YVM048	*motA(E98K) fliG(R281V)*	[36]
Plasmid		
pBAD24	Expression vector	[43]
pYC20	pBAD24/MotA+MotB	[26]
pYC20(E98K)	pBAD24/MotA(E98K)+MotB	[34]
pYC109	pBAD24/MotA+MotB(Δ52–71)	[26]
pYVM042	pBAD24/MotA+GFP-MotB	This study
pYVM042(E98K)	pBAD24/MotA(E98K)+GFP-MotB	This study
pBAD-mNectarine	pBAD/His-mNectarine (addgene #21717)	[44]

2.2. Cell Growth

Overnight cultures of *Salmonella* cells grown at 30 °C in LB containing 100 µg/mL ampicillin were diluted 100-fold into fresh LB containing 100 µg/mL ampicillin and 0.2% (*w*/*v*) arabinose, and the cells were grown at 30 °C for 6 h with shaking. The cell growth was monitored at an optical density of 600 nm (OD_{600}) every hour. The growth profiles were measured at least three times.

2.3. Measurements of Free-Swimming Speeds of Bacterial Cells

For analyses of swimming speeds and swimming fractions, *Salmonella* cells were observed under a phase contrast microscope (IX73, Olympus, Tokyo, Japan) at room temperature. The swimming speed of individual motile cells was analyzed as described previously [45]. Statistical analyses were performed by two-tailed Student's *t*-test using Prism 7.0c software (GraphPad, CA, USA).

2.4. Intracellular pH Measurement

Intracellular pH of *Salmonella* cells was detected using a pH-sensitive red fluorescent protein, mNectarine [44]. SJW1103 and YVM003 were transformed with the pBAD-mNectarine plasmid and grown overnight in LB containing 0.2% (*w*/*v*) arabinose and 100 µg/mL ampicillin at 30 °C. These overnight cultures were measured using a fluorescence spectrophotometer (RF-5300PC, Shimadzu, Kyoto, Japan) with an excitation wavelength at 540 nm and emission at 575 nm as described previously [46].

3. Results

3.1. Effect of MotA/MotB Over-Expression on Cell Growth in *gfp-motB fliG(R281V)* Strain

The proton channel activity of the MotA/MotB complex is suppressed by the plug segment in the MotB linker region when the MotA/MotB complex freely diffuses in the cytoplasmic membrane. Therefore, the expression of the MotA/MotB complex, lacking the plug segment, severely inhibits not only the cell growth, but also motility by reducing the intracellular pH [25,26]. When the intracellular pH decreases, the dissociation rate of protons from the cytoplasmic entrance of the MotA/MotB proton channel into the cytoplasm is reduced significantly, resulting in a slower torque generation cycle of the motor to cause severely impaired motility [47]. These observations lead to a plausible hypothesis that the motility inhibition of the *gfp-motB fliG(R281V)* strain caused by over-expression of the MotA/MotB complex [36] may be a consequence of reduction in intracellular pH caused by undesirable proton flow through the active MotA/MotB proton channel complex in the GFP-MotB/FliG(R281V) motor. To test this hypothesis, we investigated the effect of over-expression of the MotA/MotB complex on the cell growth in the *fliG(R281V)* mutant background. The *Salmonella* wild-type, *fliG(R281V)*, *gfp-motB*, and *gfp-motB fliG(R281V)* strains were transformed with pYC20 encoding MotA and MotB on the pBAD24 vector. These four transformants were grown in LB containing 0.2% arabinose to monitor the cell growth. The over-expression of the MotA/MotB complex did not affect the growth rate of the wild-type and *fliG(R281V)* mutant cells (Figure 2a). Conversely, in the strain expressing GFP-MotB, the *fliG(R281V)* mutation caused a significant delay in the cell growth upon over-expression of the MotA/MotB complex (Figure 2b), indicating that the combination of the GFP tagging to MotB and the *fliG(R281V)* mutation strongly affect the cell growth. This raises the possibility that the interaction of $MotB_{NCT}$ with FliG may also control the gating of the MotA/MotB proton channel during flagellar motor rotation.

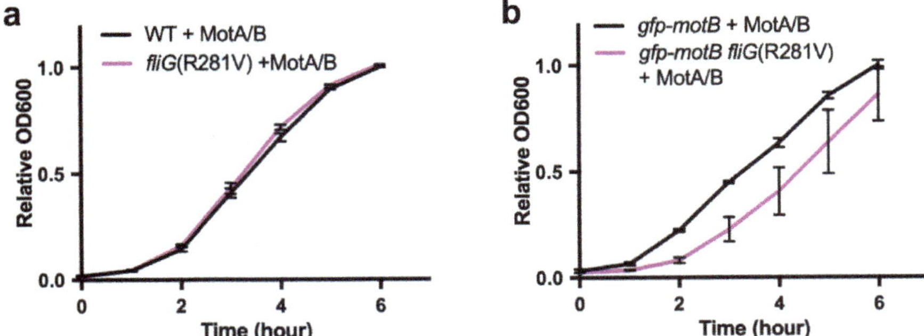

Figure 2. Multicopy effect of the MotA/MotB complex on cell growth. (**a**) Growth curves of the wild-type strain SJW1103 carrying pYC20 (WT + MotA/B, black line) and YVM046 carrying pYC20 (*fliG(R281V)* + MotA/B, magenta line). The values of the optical density at 600 nm were normalized and obtained at 6 h in SJW1103/pYC20. (**b**) Growth curves of the YVM003 carrying pYC20 (*gfp-motB* + MotA/B, black line) and YVM034 carrying pYC20 (*gfp-motB fliG(R281V)* + MotA/B, magenta line). The values of the optical density were normalized and obtained at 6 h in YVM003/pYC20. The cells were grown in LB medium containing 0.2% arabinose and 100 μg/mL ampicillin at 30 °C with shaking. Error bars represent standard deviations.

3.2. Effect of the GFP Tagging on the Proton Channel Activity of the MotA/MotB Complex

Next, we investigated whether the GFP tag affects the proton channel activity of the MotA/MotB channel complex incorporated into the flagellar motor in the presence of the *fliG(R281V)* mutation. To test this question, we first analyzed the multicopy effect of the MotA/GFP-MotB on the cell growth in the absence of the flagellar motor to clarify whether the interaction of MotA with FliG(R281V) is responsible for facilitating the proton channel activity of the MotA/GFP-MotB complex in the motor. We over-expressed the MotA/GFP-MotB complex in a flagellar master operon deletion mutant strain, SJW1368, in which no flagellar, motility, and chemotaxis genes are expressed, to measure the proton channel activity of the MotA/GFP-MotB complex by itself. Complexes of wild-type MotA/MotB and MotA/MotB (Δ52–71) lacking the plug segment were used as the negative and positive controls, respectively. Because residues 52–71 of MotB act as the plug that suppresses the proton channel activity of the MotA/MotB complex until the MotA/MotB complex encounters a rotor to become an active stator unit in the flagellar motor (Figure 1), deletion of the plug segment was predicted to result in a marked decrease in the intracellular pH due to massive proton leakage into the cytoplasm, thereby arresting the cell growth [25,26]. Consistently, the expression of MotA/MotB (Δ52–71) drastically interfered with the cell growth (Figure 3a). Although not as much as the plug deletion mutant, a significant growth inhibition was observed in the cells expressing the MotA/GFP-MotB complex compared to the cells expressing the wild-type MotA/MotB complex (Figure 3a), suggesting the possibility that the GFP tag to MotB facilitates excessive proton flow through the MotA/MotB proton channel even in the presence of the plug segment of MotB.

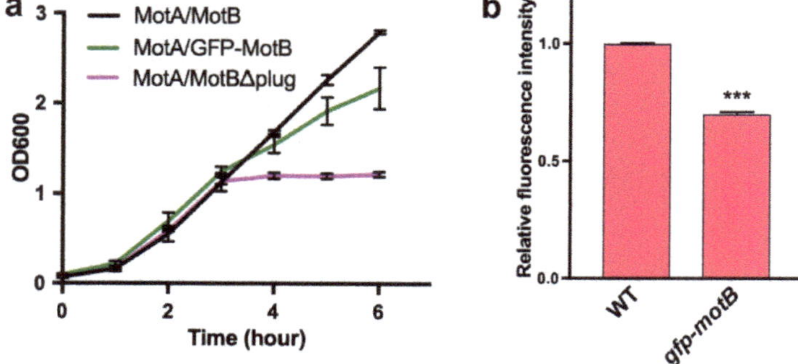

Figure 3. Effect of the GFP tag attached to the N-terminus of MotB on the proton channel activity of the MotA/MotB complex. (**a**) Growth curves of the SJW1368 strain carrying pYC20 (MotA/MotB, black line), pYVM042 (MotA/GFP-MotB, green line) or pYC109 [MotA/MotB (Δplug), magenta line]. The cells were grown in LB at 30 °C with shaking for 3 h, and then arabinose was added to a final concentration of 0.2%. Error bars represent standard deviations. (**b**) Fluorescence intensities of mNectarine in SJW1103 (WT) harboring pBAD-mNectarine and YVM003 (*gfp-motB*) carrying pBAD-mNectarine. The cells were grown overnight in LB at 30 °C with shaking, and then the fluorescence intensities of mNectarine were measured using a fluorescence spectrophotometer. The fluorescence intensities measured in the YVM003 strain were normalized to those of the SJW1103 strain. Statistical analysis was carried out using a two-tailed *t*-test (*** $p < 0.001$). Error bars represent standard deviations.

To further clarify the cause of the growth defect, we decided to measure intracellular pH changes by the expression of the MotA/GFP-MotB complex using a pH-sensitive red fluorescent protein, mNectarine. The fluorescence intensity of mNectarine increases significantly with an increase in the surrounding pH from 5.5 to 8.0 [45]. To avoid difficulties in interpreting results due to the over-expression of the membrane protein complex, we expressed mNectarine in the wild-type and *gfp-motB* strains, in which the wild-type MotA/MotB and MotA/GFP-MotB complexes are expressed from their promoter on the chromosomal DNA, and then we measured the fluorescence intensity of mNectarine using a fluorescence spectrophotometer. The fluorescence intensity of mNectarine was significantly lower in the *gfp-motB* cells than the wild-type (Figure 3b), indicating that the GFP tag facilitates the proton channel activity of the MotA/MotB complex even in the presence of the plug segment of MotB, thereby decreasing intracellular pH at the chromosomal expression level. Therefore, we propose that $MotB_{NCT}$ plays an important role in well-coordinated gating of the MotA/MotB proton channel.

3.3. *Multicopy Effect of Mutant MotA/MotB Complexes on Swimming Motility*

As described above, a fusion of the GFP tag to the N-terminus of MotB causes excessive proton flow through the MotA/MotB proton channel even in the presence of the plug segment of MotB (Figure 3). Consistently, over-expression of the MotA/GFP-MotB complex reduced the swimming speed of wild-type cells by about 20% ($p < 0.01$) but did not reduce the percentage of motile cells (Figure 4a), possibly due to a decrease in the intracellular pH that reduces the rotational speed of the flagellar motor [47]. It has been shown that the *motA(R90E)* and *motA(E98K)* mutations significantly reduce the efficiency of stator assembly into a motor [34,36] and that an increase in the expression level of the MotA(R90E)/MotB complex restores motility of the *motA(R90E)* mutant to about 60% of the wild-type level [34]. Therefore, we also examined the multicopy effect of the MotA(E98K)/MotB complex on the *motA(E98K)* mutant strain. Only 20% of the cells of the *motA(E98K)* mutant strain showed motility at a markedly reduced level even when the MotA(E98K)/MotB complex was highly expressed by adding 0.2% arabinose (Figure 4b). The over-expression of MotA(E98K)/GFP-MotB complex had no restoration effect on the swimming motility of the *motA(E98K)* mutant (Figure 4c). Because the MotA(E98K)/GFP-MotB complex retains the ability to assemble into the motor to a significant degree [36], this raises the possibility that the MotA/MotB complex with the *motA(E98K)* mutation cannot fully activate its proton channel activity and that the GFP tag to MotB affects such an activation mechanism.

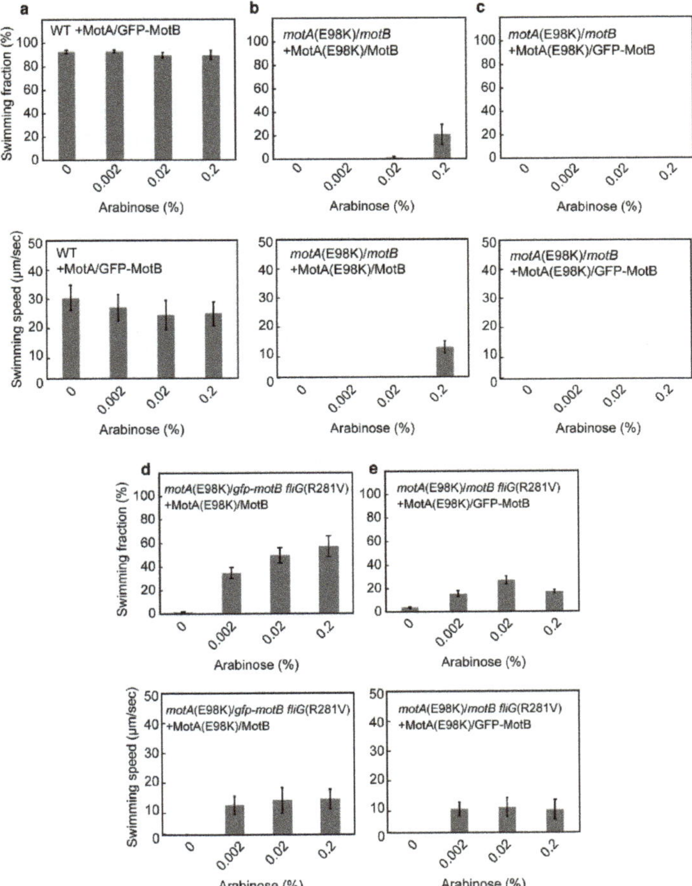

Figure 4. Multicopy effect of MotA/MotB mutants on swimming motility in liquids. The fraction and speed of free-swimming cells were measured for SJW1103 carrying pYVM042 (**a**), YVM47 carrying pYC20(E98K) (**b**), or pYVM042(E98K) (**c**), YVM036 carrying pYC20(E98K) (**d**), and YVM048 carrying pYVM042(E98K) (**e**). Swimming fraction is the fraction of swimming cells. Swimming speed is the average speed of more than 30 cells, and error bars are the standard deviations. If the fraction of motile cells was less than 5% of the total cells, the swimming speed is zero. The cells were incubated at 30 °C for 5 h in LB with 0.2, 0.02, or 0.002% arabinose. Measurements were recorded at around 23 °C.

We found that the co-expression of MotB and GFP-MotB has a considerable impact on cell motility and growth in the presence of the *fliG(R281V)* mutation (Figure 2b) [36]. Since it has been reported that the *fliG(R281V)* mutation restores the flagellar motor function of the *motA(E98K)* mutant to a considerable degree [17,36], we next investigated whether this *fliG* suppressor mutation affects the multicopy effect of the MotA(E98K)/MotB complex on the motility of the *motA(E98K) gfp-motB* cells. The fraction of motile cells of the *motA(E98K) fliG(R281V)* strain was less than 5% in the presence or absence of the GFP tag to MotB at the chromosomal expression level of the MotA/MotB stator complex (the left-most bars in the upper panels of Figure 4d,e). However, when the expression level of MotA(E98K)/MotB was increased by adding arabinose, the motile fraction of *motA(E98K) gfp-motB fliG(R281V)* cells increased to about 60% (Figure 4d, upper panel), indicating that the MotA(E98K)/MotB complex is installed into the motor, allowing the *motA(E98K) gfp-motB fliG(R281V)* cells to become motile. In contrast, the average swimming speed of motile cells was unchanged by an increase in the expression level of MotA(E98K)/MotB (Figure 4d, lower panel), suggesting that the interaction between Glu-98 of MotA and Arg-281 of FliG is critical for torque generation as previously proposed [36]. Conversely, MotA(E98K)/GFP-MotB over-expression in the *motA(E98K) fliG(R281V)* mutant cells with 0.02% arabinose increased the motile fraction only to about 30% ($p < 0.01$), and an even higher level of MotA(E98K)/GFP-MotB expression with 0.2% arabinose reduced it again to about 20% ($p < 0.01$; Figure 4e), whereas the average swimming speed did not change over a wide range of the expression level of MotA(E98K)/GFP-MotB. These results suggested that the GFP tag may weaken the interaction between MotA(E98K) and FliG(R281V). Because the GFP tag to MotB increases the proton channel activity of the MotA/MotB complex (Figure 3b), we propose that the interaction between Glu-98 of MotA and Arg-281 of FliG play an important role in the activation mechanism of the MotA/MotB proton channel and that physical communications between $MotA_C$ and $MotB_{NCT}$ promote the opening of the cytoplasmic side of the proton channel of the MotA/MotB complex.

4. Discussion

Electrostatic interactions between $MotA_C$ and FliG are critical not only for efficient stator assembly around the rotor, but also for triggering the detachment of the plug segment from the proton channel to allow $MotB_{PGB}$ to reach and bind to the PG layer, thereby activating the MotA/MotB proton channel for the MotA/MotB complex to become an active stator unit in the motor [24–26,28]. The interaction between MotAc and FliG directly transmits a mechanical signal to the MotA/MotB proton channel to regulate its proton channel activity and the affinity of the stator to the rotor to control the number of active stator units around the rotor in response to changes in external loads [48–50]. However, its mechanism still remains unknown. Here, we showed that a fusion of GFP to the N-terminus of MotB facilitates the proton channel activity of the MotA/MotB complex (Figure 3). This GFP tag partially inhibits the motility of the *motA(E98K) fliG(R281V)* mutant cells when the MotA(E98K)/GFP-MotB complex is over-expressed (Figure 4e), suggesting that $MotB_{NCT}$ is close to $MotA_C$ and FliG. Therefore, we propose that the interaction between $MotA_C$ and FliG may induce a conformational change of $MotB_{NCT}$ to open the cytoplasmic side of the proton channel of the MotA/MotB complex for flagellar motor rotation. The *gfp-motB fliG(R281V)* strain is functional, but the over-expression of the MotA/MotB complex causes a non-motile phenotype on this strain, suggesting that the MotA/GFP-MotB complex cannot cooperatively work along with the wild-type in the presence of the *fliG(R281V)* mutation [36]. The *fliG(R281V)* mutation restores the motility of the *motA(E98K)* mutant [17,36]. The *motA(E98K)* mutation does not interfere with the proton channel activity of unplugged MotA/MotB complex while the plugged proton channel cannot be unplugged and activated due to the loss of interaction between MotA and FliG by this MotA mutation [36]. These observations suggest that the MotA/MotB complex with MotA(E98K) mutation cannot activate its proton channel when it encounters FliG in the rotor, whereas the FliG(R281V) mutation allows the proper interaction between $MotA_C$ and FliG for the MotA(E98K)/MotB complex to open the channel to become an active stator in the motor. Therefore, we propose that the interaction between FliG and MotA transmits the mechanical signal via $MotB_{NCT}$ to the proton channel, thereby inducing the dissociation of the plug segment from the proton channel to activate the MotA/MotB complex as a stator unit.

5. Conclusions

Our results suggest that $MotB_{NCT}$ is close to $MotA_C$ and FliG and that a fusion of GFP to the N-terminus of MotB facilitates the MotA/MotB proton channel activity regardless of the MotA/MotB complex becoming a functionally active stator unit in the motor. These observations suggest that the interaction between $MotA_C$ and FliG induces a conformational rearrangement of $MotB_{NCT}$, thereby activating the proton channel of the MotA/MotB complex when placed around the rotor.

Author Contributions: Conceptualization, Y.V.M., K.N., and T.M.; methodology, Y.V.M. and T.M.; validation, Y.V.M. and T.M.; formal analysis, Y.V.M.; investigation, Y.V.M.; writing—original draft preparation, Y.V.M.; writing—review and editing, Y.V.M., K.N., and T.M.; funding acquisition, Y.V.M., K.N., and T.M. All authors have read and agreed to the published version of the manuscript.

Funding: This work was supported in part by JSPS KAKENHI Grant Numbers JP18K06159 (to Y.V.M.), JP25000013 (to K.N.), and JP19H03182 (to T.M.).

Acknowledgments: We acknowledge Masahiro Ueda and Takuo Yasunaga for the continuous support and encouragement.

Conflicts of Interest: The authors declare no conflict of interest.

References

1. Minamino, T.; Namba, K. Self-assembly and type III protein export of the bacterial flagellum. *J. Mol. Microbiol. Biotechnol.* **2004**, *7*, 5–17. [CrossRef]
2. Minamino, T.; Imada, K. The bacterial flagellar motor and its structural diversity. *Trends Microbiol.* **2015**, *23*, 267–274. [CrossRef]
3. Berg, H.C. The rotary motor of bacterial flagella. *Annu. Rev. Biochem.* **2003**, *72*, 19–54. [CrossRef]
4. Kojima, S.; Blair, D.F. The bacterial flagellar motor: Structure and function of a complex molecular machine. *Int. Rev. Cytol.* **2004**, *233*, 93–134. [CrossRef] [PubMed]
5. Sowa, Y.; Berry, R.M. Bacterial flagellar motor. *Q. Rev. Biophys.* **2008**, *41*, 103–132. [CrossRef] [PubMed]
6. Morimoto, Y.V.; Minamino, T. Structure and function of the bi-directional bacterial flagellar motor. *Biomolecules* **2014**, *4*, 217–234. [CrossRef] [PubMed]
7. Nakamura, S.; Minamino, T. Flagella-driven motility of bacteria. *Biomolecules* **2019**, *9*, 279. [CrossRef]
8. Francis, N.R.; Sosinsky, G.E.; Thomas, D.; DeRosier, D.J. Isolation, characterization and structure of bacterial flagellar motors containing the switch complex. *J. Mol. Biol.* **1994**, *235*, 1261–1270. [CrossRef]
9. Minamino, T.; Kinoshita, M.; Namba, K. Directional switching mechanism of the bacterial flagellar motor. *Comput. Struct. Biotechnol. J.* **2019**, *17*, 1075–1081. [CrossRef]
10. Braun, T.F.; Blair, D.F. Targeted disulfide cross-linking of the MotB protein of *Escherichia coli*: Evidence for two H^+ channels in the stator Complex. *Biochemistry* **2001**, *40*, 13051–13059. [CrossRef]
11. Braun, T.F.; Al-Mawsawi, L.Q.; Kojima, S.; Blair, D.F. Arrangement of core membrane segments in the MotA/MotB proton-channel complex of *Escherichia coli*. *Biochemistry* **2004**, *43*, 35–45. [CrossRef] [PubMed]
12. Kojima, S.; Blair, D.F. Solubilization and purification of the MotA/MotB complex of *Escherichia coli*. *Biochemistry* **2004**, *43*, 26–34. [CrossRef] [PubMed]
13. Leake, M.C.; Chandler, J.H.; Wadhams, G.H.; Bai, F.; Berry, R.M.; Armitage, J.P. Stoichiometry and turnover in single, functioning membrane protein complexes. *Nature* **2006**, *443*, 355–358. [CrossRef] [PubMed]
14. Reid, S.W.; Leake, M.C.; Chandler, J.H.; Lo, C.J.; Armitage, J.P.; Berry, R.M. The maximum number of torque-generating units in the flagellar motor of *Escherichia coli* is at least 11. *Proc. Natl. Acad. Sci. USA* **2006**, *103*, 8066–8071. [CrossRef] [PubMed]
15. Minamino, T.; Terahara, N.; Kojima, S.; Namba, K. Autonomous control mechanism of stator assembly in the bacterial flagellar motor in response to changes in the environment. *Mol. Microbiol.* **2018**, *109*, 723–734. [CrossRef]
16. Zhou, J.; Blair, D.F. Residues of the cytoplasmic domain of MotA essential for torque generation in the bacterial flagellar motor. *J. Mol. Biol.* **1997**, *273*, 428–439. [CrossRef] [PubMed]
17. Zhou, J.; Lloyd, S.A.; Blair, D.F. Electrostatic interactions between rotor and stator in the bacterial flagellar motor. *Proc. Natl. Acad. Sci. USA* **1998**, *95*, 6436–6441. [CrossRef] [PubMed]

18. Sharp, L.L.; Zhou, J.; Blair, D.F. Tryptophan-scanning mutagenesis of MotB, an integral membrane protein essential for flagellar rotation in *Escherichia coli*. *Biochemistry* **1995**, *34*, 9166–9171. [CrossRef]
19. Zhou, J.; Sharp, L.L.; Tang, H.L.; Lloyd, S.A.; Billings, S.; Braun, T.F.; Blair, D.F. Function of protonatable residues in the flagellar motor of *Escherichia coli*: A critical role for Asp 32 of MotB. *J. Bacteriol.* **1998**, *180*, 2729–2735. [CrossRef]
20. Kojima, S.; Blair, D.F. Conformational change in the stator of the bacterial flagellar motor. *Biochemistry* **2001**, *40*, 13041–13050. [CrossRef]
21. Che, Y.S.; Nakamura, S.; Kojima, S.; Kami-ike, N.; Namba, K.; Minamino, T. Suppressor analysis of the MotB(D33E) mutation to probe bacterial flagellar motor dynamics coupled with proton translocation. *J. Bacteriol.* **2008**, *190*, 6660–6667. [CrossRef] [PubMed]
22. De Mot, R.; Vanderleyden, J. The C-terminal sequence conservation between OmpA-related outer membrane proteins and MotB suggests a common function in both gram-positive and gram-negative bacteria, possibly in the interaction of these domains with peptidoglycan. *Mol. Microbiol.* **1994**, *12*, 333–334. [CrossRef] [PubMed]
23. Kojima, S.; Furukawa, Y.; Matsunami, H.; Minamino, T.; Namba, K. Characterization of the periplasmic domain of MotB and implications for its role in the stator assembly of the bacterial flagellar motor. *J. Bacteriol.* **2008**, *190*, 3314–3322. [CrossRef]
24. Kojima, S.; Imada, K.; Sakuma, M.; Sudo, Y.; Kojima, C.; Minamino, T.; Homma, M.; Namba, K. Stator assembly and activation mechanism of the flagellar motor by the periplasmic region of MotB. *Mol. Microbiol.* **2009**, *73*, 710–718. [CrossRef] [PubMed]
25. Hosking, E.R.; Vogt, C.; Bakker, E.P.; Manson, M.D. The *Escherichia coli* MotAB proton channel unplugged. *J. Mol. Biol.* **2006**, *364*, 921–937. [CrossRef]
26. Morimoto, Y.V.; Che, Y.S.; Minamino, T.; Namba, K. Proton-conductivity assay of plugged and unplugged MotA/B proton channel by cytoplasmic pHluorin expressed in *Salmonella*. *FEBS Lett.* **2010**, *584*, 1268–1272. [CrossRef] [PubMed]
27. Terahara, N.; Kodera, N.; Uchihashi, T.; Ando, T.; Namba, K.; Minamino, T. Na^{+}-induced structural transition of MotPS for stator assembly of the *Bacillus* flagellar motor. *Sci. Adv.* **2017**, *3*, eaao4119. [CrossRef]
28. Kojima, S.; Takao, M.; Almira, G.; Kawahara, I.; Sakuma, M.; Homma, M.; Kojima, C.; Imada, K. The helix rearrangement in the periplasmic domain of the flagellar stator B subunit activates peptidoglycan binding and ion influx. *Structure* **2018**, *26*, 590–598. [CrossRef]
29. Togashi, F.; Yamaguchi, S.; Kihara, M.; Aizawa, S.I.; Macnab, R.M. An extreme clockwise switch bias mutation in *fliG* of *Salmonella typhimurium* and its suppression by slow-motile mutations in *motA* and *motB*. *J. Bacteriol.* **1997**, *179*, 2994–3003. [CrossRef]
30. Muramoto, K.; Macnab, R.M. Deletion analysis of MotA and MotB, components of the force-generating unit in the flagellar motor of *Salmonella*. *Mol. Microbiol.* **1998**, *29*, 1191–1202. [CrossRef]
31. Fukuoka, H.; Wada, T.; Kojima, S.; Ishijima, A.; Homma, M. Sodium-dependent dynamic assembly of membrane complexes in sodium-driven flagellar motors. *Mol. Microbiol.* **2009**, *71*, 825–835. [CrossRef] [PubMed]
32. Delalez, N.J.; Wadhams, G.H.; Rosser, G.; Xue, Q.; Brown, M.T.; Dobbie, I.M.; Berry, R.M.; Leake, M.C.; Armitage, J.P. Signal-dependent turnover of the bacterial flagellar switch protein FliM. *Proc. Natl. Acad. Sci. USA* **2010**, *107*, 11347–11351. [CrossRef] [PubMed]
33. Fukuoka, H.; Inoue, Y.; Terasawa, S.; Takahashi, H.; Ishijima, A. Exchange of rotor components in functioning bacterial flagellar motor. *Biochem. Biophys. Res. Commun.* **2010**, *394*, 130–135. [CrossRef] [PubMed]
34. Morimoto, Y.V.; Nakamura, S.; Kami-ike, N.; Namba, K.; Minamino, T. Charged residues in the cytoplasmic loop of MotA are required for stator assembly into the bacterial flagellar motor. *Mol. Microbiol.* **2010**, *78*, 1117–1129. [CrossRef] [PubMed]
35. Lele, P.P.; Hosu, B.G.; Berg, H.C. Dynamics of mechanosensing in the bacterial flagellar motor. *Proc. Natl. Acad. Sci. USA* **2013**, *110*, 11839–11844. [CrossRef] [PubMed]
36. Morimoto, Y.V.; Nakamura, S.; Hiraoka, K.D.; Namba, K.; Minamino, T. Distinct roles of highly conserved charged residues at the MotA-FliG interface in bacterial flagellar motor rotation. *J. Bacteriol.* **2013**, *195*, 474–481. [CrossRef]
37. Heo, M.; Nord, A.L.; Chamousset, D.; van Rijn, E.; Beaumont, H.J.E.; Pedaci, F. Impact of fluorescent protein fusions on the bacterial flagellar motor. *Sci. Rep.* **2017**, *7*, 12583. [CrossRef]

38. Hara, N.; Namba, K.; Minamino, T. Genetic characterization of conserved charged residues in the bacterial flagellar type III export protein FlhA. *PLoS ONE* **2011**, *6*, e22417. [CrossRef]
39. Minamino, T.; Macnab, R.M. Components of the *Salmonella* flagellar export apparatus and classification of export substrates. *J. Bacteriol.* **1999**, *181*, 1388–1394. [CrossRef]
40. Minamino, T.; Imae, Y.; Oosawa, F.; Kobayashi, Y.; Oosawa, K. Effect of intracellular pH on rotational speed of bacterial flagellar motors. *J. Bacteriol.* **2003**, *185*, 1190–1194. [CrossRef]
41. Yamaguchi, S.; Fujita, H.; Sugata, K.; Taira, T.; Iino, T. Genetic analysis of H2, the structural gene for phase-2 flagellin in *Salmonella*. *J. Gen. Microbiol.* **1984**, *130*, 255–265. [CrossRef] [PubMed]
42. Ohnishi, K.; Ohto, Y.; Aizawa, S.; Macnab, R.M.; Iino, T. FlgD is a scaffolding protein needed for flagellar hook assembly in *Salmonella typhimurium*. *J. Bacteriol.* **1994**, *176*, 2272–2281. [CrossRef]
43. Guzman, L.M.; Belin, D.; Carson, M.J.; Beckwith, J. Tight regulation, modulation, and high-level expression by vectors containing the arabinose PBAD promoter. *J. Bacteriol.* **1995**, *177*, 4121–4130. [CrossRef] [PubMed]
44. Johnson, D.E.; Ai, H.-W.; Wong, P.; Young, J.D.; Campbell, R.E.; Casey, J.R. Red fuorescent protein pH biosensor to detect concentrative nucleoside transport. *J. Biol. Chem.* **2009**, *284*, 20499–20511. [CrossRef] [PubMed]
45. Morimoto, Y.V.; Namba, K.; Minamino, T. Measurements of fee-swimming speed of motile *Salmonella* cells in liquid media. *Bio-Protoc.* **2017**, *7*, e2093. [CrossRef]
46. Nakamura, S.; Morimoto, Y.V.; Kudo, S. A lactose fermentation product produced by *Lactococcus lactis* subsp. *lactis* acetate inhibits the motility of flagellated pathogenic bacteria. *Microbiology* **2015**, *161*, 701–707. [CrossRef] [PubMed]
47. Nakamura, S.; Kami-ike, N.; Yokota, J.P.; Kudo, S.; Minamino, T.; Namba, K. Effect of intracellular pH on the torque-speed relationship of bacterial proton-driven flagellar motor. *J. Mol. Biol.* **2009**, *386*, 332–338. [CrossRef]
48. Castillo, D.J.; Nakamura, S.; Morimoto, Y.V.; Che, Y.S.; Kami-ike, N.; Kudo, S.; Minamino, T.; Namba, K. The C-terminal periplasmic domain of MotB is responsible for load-dependent control of the number of stators of the bacterial flagellar motor. *Biophysics* **2013**, *9*, 173–181. [CrossRef]
49. Che, Y.S.; Nakamura, S.; Morimoto, Y.V.; Kami-Ike, N.; Namba, K.; Minamino, T. Load-sensitive coupling of proton translocation and torque generation in the bacterial flagellar motor. *Mol. Microbiol.* **2014**, *91*, 175–184. [CrossRef]
50. Pourjaberi, S.N.S.; Terahara, N.; Namba, K.; Minamino, T. The role of a cytoplasmic loop of MotA in load-dependent assembly and disassembly dynamics of the MotA/B stator complex in the bacterial flagellar motor. *Mol. Microbiol.* **2017**, *106*, 646–658. [CrossRef]

Article

MotP Subunit is Critical for Ion Selectivity and Evolution of a K^+-Coupled Flagellar Motor

Shun Naganawa [1] **and Masahiro Ito** [1,2,*]

[1] Graduate School of Life Sciences, Toyo University, Oura-gun, Gunma 374-0193, Japan; success630408@gmail.com

[2] Bio-Nano Electronics Research Centre, Toyo University, 2100 Kujirai, Kawagoe Saitama 350-8585, Japan

* Correspondence: masahiro.ito@toyo.jp; Tel.: +81-276-82-9202

Received: 22 March 2020; Accepted: 23 April 2020; Published: 29 April 2020

Abstract: The bacterial flagellar motor is a sophisticated nanomachine embedded in the cell envelope. The flagellar motor is driven by an electrochemical gradient of cations such as H^+, Na^+, and K^+ through ion channels in stator complexes embedded in the cell membrane. The flagellum is believed to rotate as a result of electrostatic interaction forces between the stator and the rotor. In bacteria of the genus *Bacillus* and related species, the single transmembrane segment of MotB-type subunit protein (MotB and MotS) is critical for the selection of the H^+ and Na^+ coupling ions. Here, we constructed and characterized several hybrid stators combined with single Na^+-coupled and dual Na^+- and K^+-coupled stator subunits, and we report that the MotP subunit is critical for the selection of K^+. This result suggested that the K^+ selectivity of the MotP/MotS complexes evolved from the single Na^+-coupled stator MotP/MotS complexes. This finding will promote the understanding of the evolution of flagellar motors and the molecular mechanisms of coupling ion selectivity.

Keywords: *alkaliphiles*; Mot complex; potassium ion; flagellar motor; evolution; *Bacillus*

---

1. Introduction

Many motile bacteria have a spiral flagellum as a locomotor and move in the environment by rotating one or more flagellar bundles. The bacterial flagellar motor is a high-performance nanomachine rotating at high speed [1–3]. The flagellum is rotated in the counterclockwise direction for smooth swimming and is rotated in the clockwise direction to change the direction [4,5]. Bacterial flagella consist of three parts: a base body corresponding to the motor part embedded in the membrane, flagellar filament equivalent to a propeller extended long outside of the cell body, and a hook connecting the base body and flagellar filament [6]. Bacterial flagella are composed of approximately 30 kinds of proteins, and they form a supercomplex [7]. The basal body of the flagellar motor consists of a rotor and a stator. The electrostatic interaction between the rotor and the stator is the driving force of flagellar rotation [4,8–10]. The stator complex is composed of two subunits (MotA-type and MotB-type) formed at a ratio of 4:2, functions as an ion channel and anchors to the cell wall through a putative peptidoglycan-binding (PGB) motif in the periplasmic domain of a MotB-type protein. Mot complexes are arranged in a ring of membrane-embedded complexes surrounding each flagellum [1,2]. A typical number of such complexes surrounding the basal motor appears to be at least 11 [11]. In general, *Escherichia coli* MotA/MotB complex is an H^+ driven flagellar motor. In contrast, marine *Vibrio* species PomA/PomB complex and alkaliphilic *Bacillus* MotP/MotS complex are Na^+ driven flagellar motors, respectively [12–15]. Each stator complex has homology to each other. *E. coli* and alkaliphilic *Bacillus pseudofirmus* have only one type of Mot complex in the motor [16–18]. On the other hand, *Bacillus subtilis* and *Shewanella oneidensis* can have two distinct Mot complexes in the motor [19,20]. However, in 2008, the alkaliphilic *Bacillus clausii* KSM-K16 was identified as the first bacterium to have a single stator rotor that uses both H^+ and Na^+ for ion coupling depending on the Ph [21].

Mutations that convert the bifunctional stator to each single stator type have been demonstrated, and the same approach was utilized to generate dual-ion use stators from the two single ion-use stators of *B. subtilis* [21]. Subsequent findings have shown that alkaliphilic *Bacillus alcalophilus* AV1934 uses Na^+, K^+, and Rb^+ [22] and that *Paenibacillus* sp. TCA20 uses Ca^{2+} and Mg^{2+} as coupling ions for flagellar rotation [3,23].

Previously, the transmembrane segment of the MotB-type subunit was proposed to be critical for the coupling ion selectivity of the stator when H^+ or Na^+ were used as coupling ions [24,25]. At the N-terminal side of the single transmembrane region of the MotB-type subunit, there is an aspartic acid residue that functions as a universally conserved putative coupling ion binding site [26]. The amino acid residue located ten amino acids downstream from the aspartic acid residue is presumed to be critical for coupling ion selectivity [21]. The amino acid residue at position is highly conserved as a valine residue in the H^+-coupled MotB subunit and a leucine residue in the Na^+-coupled MotS and PomB subunits [21]. However, it has been reported that the *B. alcalophilus* Na^+- and K^+-coupled MotS subunit is conserved as a methionine residue (MotS_Met33) at the same position, and the MotS_Met33Leu substituted stator of *B. alcalophilus* lost its original potassium-coupling capacity. Therefore, the methionine residue is critical for K^+ selectivity [22].

There have been no reports of bacteria that can utilize both Na^+ and K^+ as the coupling ions for a flagellar motor except *B. alcalophilus*. It is premature to draw conclusions about all the K^+ selection mechanisms in one reported example. Details on the coupling of the K^+ selective mechanism are still poor. Finding another example that can utilize K^+ for flagellar rotation is important for elucidating the coupling ion selectivity mechanism. Therefore, we investigated novel bacteria that can use K^+ as a coupling ion for a flagellar motor and identified the stator of the alkaliphilic bacterium *Bacillus trypoxylicola*, which was isolated from the intestines of Japanese beetle larvae [27]. This bacterium was the second example that could utilize both Na^+ and K^+ as a coupling ion for the flagellar motor.

In this study, we analyzed the differences in the ion selectivity mechanisms of the flagellar motor stator between the single Na^+-coupled and dual Na^{+-} and K^+-coupled stators.

2. Materials and Methods

2.1. Bacterial Strains and Plasmids

The bacterial strains and plasmids used in this study are listed in Tables 1 and 2, respectively. The primers used in this study are listed in Table S1.

Table 1. The bacterial strains used in this study.

Strain	Description	Source or Reference
Bacillus trypoxylicola	wild type (NBRC 102646)	[27]
Bacillus alcalophilus	wild type (JCM5652)	[28]
Bacillus pseudofirmus OF4	wild type	[29]
Escherichia coli strains		
W3110	F⁻ *IN (rrnD-rrnE)1*	R. Aono
DH5αMCR	F⁻ *mcrA*Δ*1 (mrr-hsd RMS-mcrBC) Φ80dlacZ* Δ*(lacZYAargF) U169 deoR recA1 endA1 supE44* λ*thi-1 gyr-496 relA1*	Stratagene
RP6665	Δ*motAB*	J. S. Parkinson
Bt-PS	RP6665, pBAD24 + *motPS* from *B. trypoxylicola*	This study
Ba-PS	RP6665, pBAD24 + *motPS* from *B. alcalophilus*	This study
TK2420	F⁻ *thi rha lacZ nagA* Δ*(kdpFAB)* Δ*(trk-mscL) trkD1*	[30]

Table 1. *Cont.*

Strain	Description	Source or Reference
TK-BtPS	TK2420, pBAD24 + *motPS* from *B. trypoxylicola*	This study
TK-BaPS	TK2420, pBAD24 + *motPS* from *B. alcalophilus*	[22]
TK-pBAD	TK2420, pBAD24	[22]
HB101	*supE44, Δ(mcrC-mrr), recA13, ara-14, proA2, lacY1, galK2, rpsL20, xyl-5, mtl-1, leuB6, thi-1*	Takara Bio
HB-pBAD	HB101, pBAD24	[22]
Bacillus subtilis strains		
BR151MA	*lys3 trpC2* (wild type)	[25]
ΔABΔPS	BR151MA Δ*motAB* Δ*motPS*	[31]
TTPS	ΔABΔPS *lacA*::P_{xylA}-*motPS* from *B. trypoxylicola*	This study
OF4PS	ΔABΔPS *lacA*::P_{xylA}-*motPS* from *B. pseudofirmus*	[23]
AAPS	ΔABΔPS *lacA*::P_{xylA}-*motPS* from *B. alcalophilus*	This study
TPPS	ΔABΔPS *lacA*::P_{xylA}-TP-*motPS* (*motP* from *B. trypoxylicola, motS* from *B. pseudofirmus*)	This study
PTPS	ΔABΔPS *lacA*::P_{xylA}-PT-*motPS* (*motP* from *B. pseudofirmus, motS* from *B. trypoxylicola*)	This study
PAPS	ΔABΔPS *lacA*::P_{xylA}-PA-*motPS* (*motP* from *B. pseudofirmus, motS* from *B. alcalophilus*)	This study
APPS	ΔABΔPS *lacA*::P_{xylA}-AP-*motPS* (*motP* from *B. alcalophilus, motS* from *B. pseudofirmus*)	This study
TAPS	ΔABΔPS *lacA*::P_{xylA}-TA-*motPS* (*motP* from *B. trypoxylicola, motS* from *B. alcalophilus*)	This study
ATPS	ΔABΔPS *lacA*::P_{xylA}-AT-*motPS* (*motP* from *B. alcalophilus, motS* from *B. trypoxylicola*)	This study

Table 2. The plasmids used in this study.

Plasmid	Description	Source or Reference
pGEM7zf (+)	Cloning vector; Ap^R	Promega
pGEM-T Easy	TA-cloning vector; Ap^R	Promega
pBAD24	Expression vector; Ap^R; P_{BAD} promoter	[32]
pAX01	*lacA* integration vector with Em^R gene and P_{xylA} promoter upstream of multiple cloning site	[33]
pGEM-T-BtPS	pGEM-T Easy + *motPS* from *B. trypoxylicola*	This study
pGEM-BtPS	pGEM7zf (+) + *motPS* from *B. trypoxylicola*	This study
pGEM-BpPS	pGEM7zf (+) + *motPS* from *B. pseudofirmus*	This study
pGEM-BaPS	pGEM7zf (+) + *motPS* from *B. alcalophilus*	This study
pGEM-tpPS	pGEM7zf (+) +tp-*motPS* (*motP* from *B. trypoxylicola, motS* from *B. pseudofirmus*)	This study
pGEM-taPS	pGEM7zf (+) +ta-*motPS* (*motP* from *B. trypoxylicola, motS* from *B. alcalophilus*)	This study
pGEM-ptPS	pGEM7zf (+) +pt-*motPS* (*motP* from *B. pseudofirmus, motS* from *B. trypoxylicola*)	This study

Table 2. *Cont.*

Plasmid	Description	Source or Reference
pGEM-paPS	pGEM7zf (+) +pt-*motPS* (*motP* from *B. pseudofirmus*, *motS* from *B. alcalophilus*)	This study
pGEM-atPS	pGEM7zf (+) +at-*motPS* (*motP* from *B. alcalophilus*, *motS* from *B. trypoxylicola*)	This study
pGEM-apPS	pGEM7zf (+) +at-*motPS* (*motP* from *B. alcalophilus*, *motS* from *B. pseudofirmus*)	This study
pBAD-BtPS	pBAD24 + *motPS* from *B. trypoxylicola*	This study
pBAD-BaPS	pBAD24 + *motPS* from *B. alcalophilus*	[22]
pAX-BtPS	pAX01 + *motPS* from *B. trypoxylicola*	This study
pAX-BaPS	pAX01 + *motPS* from *B. alcalophilus*	This study
pAX-tpPS	pAX01+tp-*motPS* (*motP* from *B. trypoxylicola*, *motS* from *B. pseudofirmus*)	This study
pAX-taPS	pAX01+ta-*motPS* (*motP* from *B. trypoxylicola*, *motS* from *B. alcalophilus*)	This study
pAX-ptPS	pAX01+pt-*motPS* (*motP* from *B. pseudofirmus*, *motS* from *B. trypoxylicola*)	This study
pAX-paPS	pAX01+pa-*motPS* (*motP* from *B. pseudofirmus*, *motS* from *B. alcalophilus*)	This study
pAX-atPS	pAX01+at-*motPS* (*motP* from *B. alcalophilus*, *motS* from *B. trypoxylicola*)	This study
pAX-apPS	pAX01+ta-*motPS* (*motP* from *B. alcalophilus*, *motS* from *B. pseudofirmus*)	This study

2.2. *Cloning of the motP/motS Genes with the Pre- and Post-Regions of B. trypoxylicola*

Since *B. trypoxylicola* NBRC102646 has no whole-genome sequence information at the beginning of the study in 2014, the primers used for PCR were designed based on the gene sequence in the region before and after the closely related *B. alcalophilus* stator gene. The *B. trypoxylicola motPS* genes were amplified by PCR using *B. trypoxylicola* chromosomal DNA as the template with Ba-ccpA-1-F and Ba-acuC-2-R primers (Figure 1A and Figure S1). GoTaq Green Master Mix (Promega, Madison, Wisconsin, USA) was used for PCR, and the reaction was performed according to the manufacturer's protocol. The amplified 2215 bp PCR product was purified using a QIAquick Gel Extraction Kit (QIAGEN, Hinden, Germany) and ligated to a pGEM-T Easy Vector (Promega) using T4 DNA Ligase (New England Biolabs, Ipswich, MA, USA). The ligation reaction and composition were performed according to the instructions. The reaction solution was added to competent *E. coli* DH5α MCR cells prepared by the rubidium chloride method [34], and transformation was performed by the heat-shock method. The transformed cells were spread on LB plates supplemented with ampicillin to a final concentration of 100 μg/mL and cultured at 37 °C overnight. The desired plasmid was extracted from the colonies on the plates by a QIAprep Spin Miniprep Kit (QIAGEN, Germany) according to the instructions, and the desired plasmid was obtained and named pGEM-T-BtPS. In the following molecular biology experiments, the same method as described above was used unless otherwise specified.

Next, the *B. trypoxylicola motPS* genes were amplified by PCR using pGEM-T-BtPS as a template with the Bt-motP-EcoRI-F and Bt-motS-XbaI-R primers (Figure 1A and Figure S1). Phusion High-Fidelity DNA Polymerase (New England Biolabs, Ipswich, MA, USA) was used for PCR. The PCR composition and reaction conditions were performed in accordance with the Phusion High-Fidelity DNA Polymerase protocol. The amplified 2123 bp PCR product was purified, and the PCR product was ligated to *Sma*I-digested pGEM7zf (+). The ligation mixture was transformed into *E. coli* DH5α MCR, and the transformed cells were spread on LB plates supplemented with ampicillin to a final concentration of 100 µg/mL and cultured at 37 °C overnight. The desired plasmid was obtained and named pGEM-BtPS. The DNA sequence was deposited into DNA Data Bank of Japan (DDBJ), and the accession number is LC532380.1.

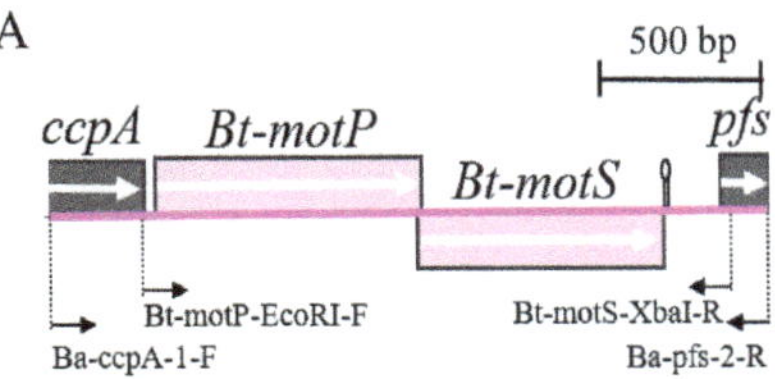

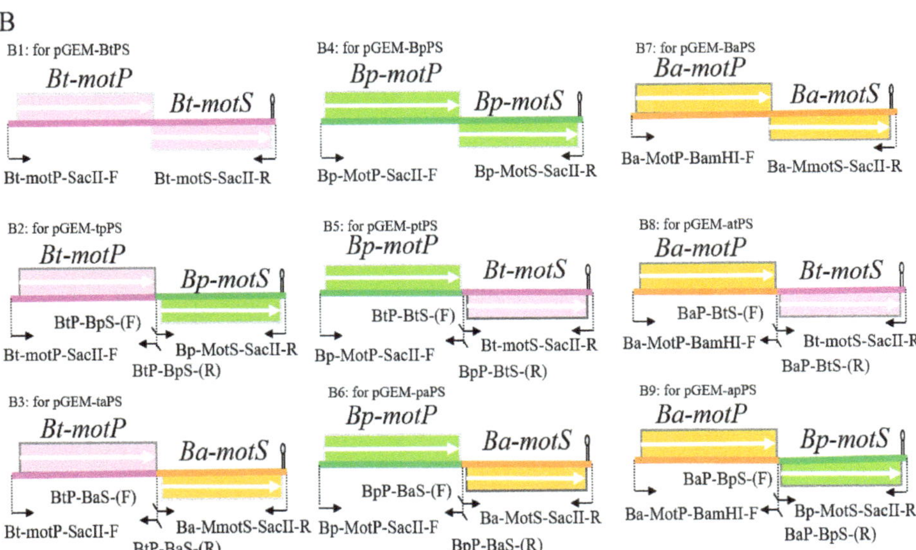

Figure 1. Schematic diagram of the primers used for hybrid stator construction and PCR. (**A**) Schematic diagram of the *motP/motS* locus of *B. trypoxylicola* chromosome and the primers used for PCR. (**B**) Schematic diagram of amplification of *Bt-motP/motS*, BP-*motP/motS*, *Ba-motP/motS*, and a series of hybrid *motPS* that indicate the primers used for cloning downstream of P_{xylA} promoter of integration plasmid pAX01. The detailed protocol for construction of each PCR product is described in Materials and Methods. The mark after each *motS* gene indicates the position of the terminator.

2.3. Cloning of the motP/motS Genes of B. trypoxylicola into pBAD24

To clone *B. trypoxylicola motPS* genes into the downstream region of the arabinose-inducible promoter, pGEM-BtPS was digested with EcoRI and XbaI to cut out the *motPS* genes and ligated to *Eco*RI- and *Xba*I-digested pBAD24. The ligation mixture was transformed into *E. coli* DH5α MCR, and the transformed cells were applied to LB plates supplemented with ampicillin to a final concentration of 100 µg/mL and cultured at 37 °C overnight. The desired plasmid was obtained and named pBAD-BtPS.

2.4. Cloning of the Hybrid Stator Gene into the pGEM7zf (+) Vector and Integration Vector pAX01 for Bacillus Subtilis

B. trypoxylicola-derived Bt-MotP and Bt-MotS, *B. alcalophilus*-derived Ba-MotP and Ba-MotS and *B. pseudofirmus*-derived Bp-MotP and Bp-MotS were each replaced at the subunit level, and hybrid stators were constructed (Figure 1B).

First, each Bt-*motP*, Bt-*motS*, Ba-*motP*, Ba-*motS*, Bp-*motP*, and Bp-*motS* was separately amplified by PCR using each chromosomal DNA as a template. Each amplified gene product was ligated to each other using a 2nd PCR for the Gene SOEing method [35] to form a different combination of *motP* and *motS*. This popular method is based on PCR, recombines DNA sequences independently of restriction sites, and directly produces mutant DNA fragments in vitro. The combination of the primers used for PCR is shown in Figure 1B, and the information on the primers is shown in Table S1. Phusion High-Fidelity DNA Polymerase (New England Biolabs, Ipswich, MA, USA) was used for the PCR. The PCR composition and reaction conditions were performed in accordance with the Phusion High-Fidelity DNA Polymerase protocol. The PCR product was purified in the same manner. The cloning vector pGEM7zf (+) was digested with *Sma*I. The *Sma*I-digested pGEM7zf (+) and each PCR product were ligated in the same manner. The ligation reaction solution was transformed into *E. coli* DH5α MCR. The transformed cells were applied to S-Gal/LB agar (Sigma-Aldrich, St. Louis, MO, USA) supplemented with ampicillin to a final concentration of 100 μg/mL and cultured at 37 °C overnight. The desired plasmids were obtained, and each constructed plasmid was named pGEM-BtPS (Figure 1(B1)), pGEM-tpPS (Figure 1(B2)), pGEM-taPS (Figure 1(B3)), pGEM-ptPS (Figure 1(B5)), pGEM-paPS (Figure 1(B6)), pGEM-BaPS (Figure 1(B7)), pGEM-atPS (Figure 1(B8)), and pGEM-apPS (Figure 1(B9)). Sequence analysis confirmed that each obtained plasmid was free from mutations.

Under the control of the P_{xylA} promoter, each *mot* gene constructed above was cloned into pAX01, a plasmid vector for integrating a foreign gene into the *lacA* region of the *B. subtilis* chromosome. pGEM-BtPS, pGEM-tpPS, pGEM-taPS, pGEM-ptPS, pGEM-paPS, and pAX01 were digested with *Sac*II. pGEM-BaPS, pGEM-atPS, pGEM-apPS, and pAX01 were digested with *Bam*HI and *Sac*II. The ligation solution was added to competent *E. coli* DH5α MCR cells in the same manner. The transformed cells were spread on an LB plate supplemented with ampicillin to a final concentration of 100 μg/mL and cultured at 37 °C overnight. The desired plasmids were obtained, and each constructed plasmid was named pAX-BtPS, pAX-tpPS, pAX-taPS, pAX-ptPS, pAX-paPS, pAX-BaPS, pAX-atPS, and pAX-atPS. pGEM-BpPS, and pAX-BpPS were not constructed because strain OF4PS had been constructed previously [23].

2.5. Construction of B. subtilis Integration Mutants Expressing the Hetero Hybrid Stator

Competent cells of the *B. subtilis* ΔABΔPS strain were prepared using the conventional method with Spizizen medium [36]. pAX-tpPS, pAX-ptPS, pAX-paPS, pAX-apPS, pAX-taPS, pAX-atPS pAX-BtPS, and pAX-BaPS were added to competent cells, and transformation was performed by the method of Spizizen et al. [36]. The transformed cells were spread onto S-Gal/LB agar plates supplemented with erythromycin to a final concentration of 1 μg/mL and cultured at 37 °C overnight. Transformants were selected from the colonies by color selection. Each positive colony was designated as strains TPPS, PTPS, PAPS, APPS, TAPS, ATPS, TTPS, and AAPS. Each transformant was cultured for 16 h at 37 °C and 200 rpm in LB supplemented with erythromycin to a final concentration of 1 μg/mL. Each genomic DNA sample was prepared from the culture using the UltraClean Microbial DNA Kit (QIAGEN, Hinden, Germany). The experiment was performed according to the manufacturer's protocol. The genomic DNA of the TPPS, PTPS, PAPS, APPS, TAPS, ATPS, TTPS, and AAPS strains were used as templates, the stator gene region of each constructed strain was amplified by PCR with suitable primers. Phusion High-Fidelity DNA Polymerase was used for PCR. The PCR product was purified, and DNA sequence analysis was performed with the suitable primer set; it was confirmed that there was no mutation in the stator gene of the obtained integration mutants.

2.6. Growth Media and Growth Conditions for Growth and Swimming Assays

B. trypoxylicola, *B. alcalophilus*, and *B. pseudofirmus* were cultured in alkaline complex medium [21] (89 mM K_2HPO_4, 33 mM KH_2PO_4, 1 mM citric acid monohydrate, 0.4 mM $MgSO_4$, 5% peptone, 2% yeast extract, 100 mM Na_2CO_3, 28 mM glucose) at 30 °C for 16 h with shaking. This was used as a preculture. Two milliliters of Tris medium [23] (30 mM Tris base, 7 mM citric acid monohydrate, 0.05% (*w*/*v*) yeast extract, 50 mM glucose and 1% (*v*/*v*) trace elements [37] pH 9.0) was inoculated so that the OD_{600} became 0.01, and the culture was grown at 30 °C for 16 h with shaking. The final OD_{600} of each culture was measured. Three independent experiments were conducted.

For *B. trypoxylicola*, *B. alcalophilus*, and *B. pseudofirmus* swimming speed assays in liquid, each strain was precultured in alkaline complex medium [21] (89 mM K_2HPO_4, 33 mM KH_2PO_4, 1 mM citric acid monohydrate, 0.4 mM $MgSO_4$, 5% peptone, 2% yeast extract, 100 mM Na_2CO_3, 28 mM glucose) at 30 °C for 16 h with shaking. The culture was inoculated at an OD_{600} of 0.01 with fresh medium and cultured with shaking at 30 °C for approximately 7 to 8 h. The cultures were measured in 30 mM Tris-HCl containing 5 mM glucose and several NaCl or KCl concentrations at pH 9.0. The results represent the average swimming speed of 30 independent cells from three independent experiments. The error bars indicate the standard deviations.

To observe the effect of sodium and potassium ions on the swimming speed of *E. coli* mutant strains, the Bt-PS, Ba-PS, and ΔAB mutant strains were grown for 7 h at 30 °C in LB medium containing 0.1% arabinose with shaking. Cells were suspended in 1 mL of swimming buffer (pH 7.0) plus several different concentrations of NaCl or KCl and then incubated at 30 °C for 10 min. The swimming buffer contained 30 mM N-Tris(hydroxymethyl)methyl-2-aminoethanesulfonic acid (TES), 5 mM glucose, 0.1% arabinose, adjusted to pH 7.0 with *N*-methyl-D-glucamine. The results represent the average swimming speed of 30 independent cells from three independent experiments.

E. coli W3110 (wild-type) and Bt-PS were grown in liquid LB medium at 30 °C for 16 h with shaking. Each strain was inoculated into the same medium inoculated at an OD_{600} adjusted to 0.01 with fresh medium and cultured with shaking at 30 °C for approximately 6 to 7 h. If necessary, ampicillin was added to a final concentration of 0.1 mg/mL. For swimming speed assays of *B. pseudofirmus* OF4, strain OF4 was grown in liquid MYE medium (pH 10.5) with shaking at 30 °C for 7 h. Cells were suspended in 1 mL of swimming buffer (pH 7.0) plus several carbonyl cyanide m-chlorophenyl hydrazine (CCCP) concentrations, 5-(*N*-ethyl-*N*-isopropyl)-amiloride (EIPA) concentrations, or valinomycin concentrations and incubated at 30 °C for 10 min. The swimming buffer contained 30 mM TES, 5 mM glucose, 0.1% arabinose, 100 mM NaCl or 100 mM KCl, adjusted to pH 7.0 with *N*-methyl-D-glucamine. The results represent the average swimming speed of 30 independent cells from three independent experiments.

2.7. Swimming Assay of Mutant Strains Expressing Hybrid Stators

For the measurement of swimming speed, *B. subtilis* (TPPS, PTPS, PAPS, APPS, TAPS, ATPS, TTPS, AAPS, and OF4PS) cells were aerobically grown on Spizizen I medium (30) (Spizizen salts, 0.5% glucose) supplemented with 1 μg/mL erythromycin, 10 μg/mL tryptophan and lysine at 37 °C for 16 h. The culture was inoculated into 20 mL of fresh medium supplemented with 1% (*w*/*v*) xylose at an OD_{600} of 0.01 and aerobically grown at 37 °C for approximately 6 to 7 h. Cells were suspended in 1 mL of swimming medium (pH 8.0) plus several NaCl concentrations plus KCl concentrations and were incubated at 37 °C for 10 min. The swimming medium contained 0.04% tryptone, 0.02% yeast extract, and 5 mM glucose, adjusted to pH 8.0 with *N*-methyl-D-glucamine. The results represent the average swimming speed of 30 independent cells from three independent experiments. Details of the swim analysis procedure are described in a separate section.

2.8. Cell Harvest Method for Swimming Assay and Swimming Video Recording Analysis

One hundred microliters of the culture broth were harvested by filtration on OMNIPORE membrane filters (0.45 μm), which were sandwiched between filter holders (SANSYO, Tokyo, Japan) connected to an aspirator MCV-20PS (ULVAC, Chigasaki, Japan), and suction was performed at 0.06 MPa. The filter was washed three times with 2 mL of the indicated buffer. The filter was placed in a 14 mL culture tube, suspended in 2 mL of the same buffer and incubated at 30 °C and 200 rpm for 10 min. Cell motility was observed under a dark-field microscope (Leica microscope DMLB100) and recorded in high definition with a digital color camera (model DFC310FX; Leica Microsystems, Tokyo, Japan). The swimming speed was determined with two-dimensional (2D) movement measurement capture 2 two-dimensional particle tracking velocimetry (2D-PTV) software (DigiMo, Tokyo, Japan). In addition, the swimming analysis was independently performed at least three times, the swimming speed of a total of 90 or more cells was measured, and the results were evaluated from the average value.

2.9. Measurement of Intracellular Potassium Concentration of *E. coli* HB-pBAD, TK-pBAD, TK-BaPS, and TK-BtPS

E. coli HB-pBAD, TK-pBAD, TK-BaPS and TK-BtPS cells were grown at 30 °C and 200 rpm for 16 h in LBK medium [30] (1% tryptone, 0.5% yeast extract, 83 mM KCl, pH 7.5) with 0.2% arabinose plus 100 μg/mL ampicillin. Then, each preculture was inoculated into modified TK2420 medium [38] (33.6 mM MOPS, 1 mM K_2HPO_4, 1.1 mM citric acid, 7.6 mM $(NH_4)_2SO_4$, 6 mM $FeSO_4$, 830 mM $MgSO_4$, 10 mM glucose, 1 mg/mL thiamine and 0.2% arabinose, pH 7.0) with 0.2% arabinose, 100 μg/mL ampicillin plus 7 mM or 12 mM KCl, so that $OD_{600\ nm}$ was adjusted to 0.01. Growth was monitored hourly at $OD_{600\ nm}$. The cells were harvested by centrifugation (10,000 rpm (9100× *g*), 3 min, 25 °C) and washed with suspension in the same medium. Then, the cells were resuspended in 5 mL of 500 mM sucrose solution, centrifuged (10,000 rpm (9100× *g*), 3 min, 25 °C), and the supernatant was removed. This operation was repeated twice. Each cell protein was measured by the Lowry method using 100 μL of the cell suspension. The rest of the suspension was harvested and resuspended in 5 mL of 5% trichloroacetic acid (TCA) solution and shaken at 200 rpm at room temperature for 20 min. After shaking, centrifugation was performed at 10,000 rpm (9100× *g*) at 4 °C for 10 min, and 1 mL of the supernatant was collected. The supernatant was diluted 10 times and 100 times, and the K^+ concentration was measured using a digital flame photometer ANA-135 (Tokyo Koden, Japan); a 50 ppm KCl solution was used as a standard. The intracellular K^+ concentration (mM) was calculated using 1 mg of protein as a cell volume of 3 mL [30,39].

2.10. Phylogenetic Analysis and Multiple Alignment of the Transmembrane Domain Region of the *B. trypoxylicola* MotS Subunit

A phylogenetic analysis of the flagellar motor stator of the genus *Bacillus* was performed using DDBJ's ClustalW (http://clustalw.ddbj.nig.ac.jp/index.php?lang=en) for alignment and the neighbor-joining method (NJ method) TreeView (http://code.google.com/p/treeviewx). The accession numbers of the strains used for analysis are as follows. *E. coli* MotB [EC_MotB (POAF06.1)], *B. clausii* KSM-K16 MotB [BC-MotB (BAD64519.9)], *B. subtilis* MotB [BS-MotB (CAB13241.1)], *B. subtilis* MotS [BS-MotS (CAB14950.1)], *B. alcalophilus* AV1934 MotS [BA-MotS (KGA96617.1)], *B. trypoxylicola* MotS [BT-MotS (CAB14950.1)], *B. pseudofirmus* OF4 MotS [BP-MotS (ADC48829.1)], *Bacillus halodurans* C-125 MotS [BH-MotS (BAB06958.1)], *Bacillus oceanisediminis* MotS [BO-MotS (WP_019383049)], *Bacillus cereus* MotB [BCS-MotB (KMN69571)], *Bacillus megaterium* MotB [BM-MotB (WP_013056745)], *Bacillus flexus* MotB [BF-MotB (WP_025909154)], and *Bacillus thermoamylovorans* MotB [BTH-MotB (WP_034770454)].

3. Results

3.1. Na^+- or K^+-Dependent Growth Capacities of Alkaliphiles, *B. trypoxylicola*, *B. alcalophilus* and *B. pseudofirmus* at pH 9.0

The growth of *B. trypoxylicola*, *B. alcalophilus* and *B. pseudofirmus* in Tris medium with several added Na^+ and K^+ concentrations was compared (Figure 2). As described previously, *B. alcalophilus* and *B. pseudofirmus* showed Na^+- and K^+-dependent and only Na^+-dependent growth respectively in Tris medium at pH 9.0 [22,40]. *B. trypoxylicola* showed only K^+-dependent growth. This result indicated that *B. trypoxylicola* prefers to use K^+ rather than Na^+ for growth in alkaline environments.

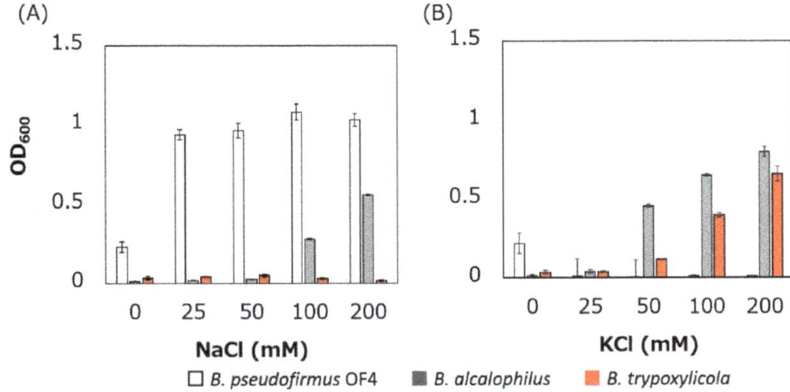

Figure 2. Effect of sodium and potassium ions on the growth of *B. trypoxylicola*, *B. alcalophilus*, and *B. pseudofirmus* OF4. Growth of bacterial cultures at 30 °C in Tris medium (pH 9) containing various concentrations of NaCl (**A**) or KCl (**B**) was monitored at OD_{600}. The results are the averages of three independent experiments, and the error bars represent the standard deviations.

3.2. Swimming Assay of *B. trypoxylicola*, *B. alcalophilus*, and *B. pseudofirmus* under Various K^+ and Na^+ Concentrations

The swimming speeds of *B. trypoxylicola*, *B. alcalophilus*, and *B. pseudofirmus* in swimming assay buffer with different K^+ and Na^+ concentrations were compared (Figure 3). Similar to previously reported results, *B. alcalophilus* swimming showed NaCl and KCl concentration dependence, and emphB. pseudofirmus swimming showed NaCl concentration dependence [22,40]. *B. trypoxylicola* swimming was observed under the same concentrations of NaCl and KCl as those used for *B. alcalophilus*. To summarize these results, it was suggested that *B. trypoxylicola* has a flagellar motor coupled to both K^+ and Na^+.

3.3. Identification of MotP/MotS Operon of *B. trypoxylicola*

The draft genome of *B. trypoxylicola* NBRC102646 was retrieved from the DDBJ/EMBL/GenBank databases (accession number BCWA00000000.1). The annotation of the draft genome sequence shows that *B. trypoxylicola* has a *motP/motS* operon: Bt-*motP*/Bt-*motS*. The GenBank accession numbers and numbers of amino acids in Bt-MotP and Bt-MotS are WP_045481010.1 and 267 aa and WP_061950140.1 and 247 aa, respectively. The amino acid sequence was compared with other stator sequences of *Bacillus* spp. using ClustalW. Bt-MotS was classified as a MotP/MotS family (Figure 4). Bt-MotP and Bt-MotS were closely related to *B. alcalophilus* MotP (Ba-MotP) and MotS (Ba-MotS) (74% and 73% identity and 87% and 88% similarity, respectively). In addition, in the transmembrane region of the MotB and MotS subunits of *Bacillus* spp., the aspartic residue, which is a presumed coupling cation-binding site, is entirely conserved (Figure 5) [26]. It is known that valine residues are highly conserved in the H^+-coupled MotB subunit and that leucine residues are highly conserved in the

Na^+-coupled MotS subunit ten amino acids downstream of this aspartic acid. In Bt-MotS, the leucine residue was preserved at that position. These results indicated that Bt-MotP/Bt-MotS is a stator belonging to the MotP/MotS family.

3.4. *Swimming Assay of an E. coli Stator-Deficient Strain Expressing Bt-MotP/Bt-MotS*

It is unknown whether Bt-MotP/Bt-MotS uses Na^+ and K^+ as coupling ions. Therefore, the following experiment was carried out to confirm whether Bt-MotP/Bt-MotS uses these cations. An *E. coli* stator-deficient strain (ΔAB) expressing Bt-MotP/Bt-MotS (strain Bt-PS) was constructed. The swimming speeds of strain Bt-PS, strain ΔAB expressing Ba-MotP/Ba-MotS (strain Ba-PS) and strain ΔAB carrying pBAD24 in TES buffer solution (pH 7.0) were compared at several concentrations of KCl or NaCl (Figure ??). No motility was observed in the ΔAB strain carrying pBAD24 at any KCl or NaCl concentration. The Bt-PS and Ba-PS strains showed no motility in the absence of KCl or NaCl but showed KCl or NaCl concentration-dependent motility. These results suggested that Bt-MotP/Bt-MotS utilizes both Na^+ and K^+ as coupling cations.

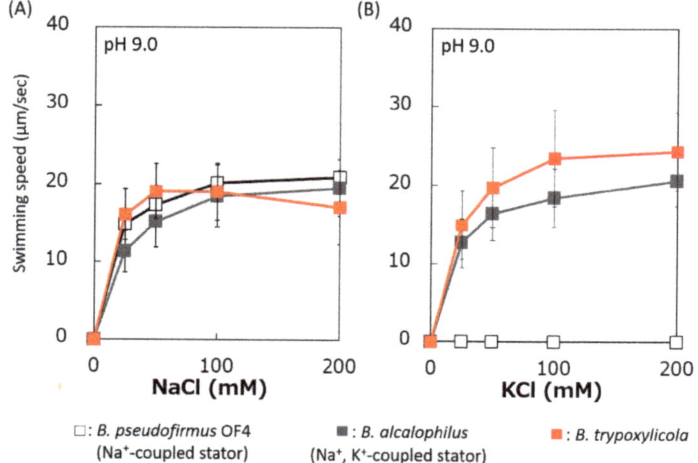

Figure 3. Effect of sodium and potassium ions on the swimming speed of *B. trypoxylicola*, *B. alcalophilus*, and *B. pseudofirmus* OF4. Swimming speeds of *B. trypoxylicola*, *B. alcalophilus*, and *B. pseudofirmus* OF4 cells were measured in 30 mM Tris-HCl containing 5 mM glucose and several NaCl (**A**) or KCl (**B**) concentrations at pH 9.0. The results represent the average swimming speed of 30 independent cells from three independent experiments. The error bars indicate the standard deviations.

3.5. *Growth of an E. coli K^+ uptake System-Deficient Strain Expressing Bt-MotP/Bt-MotS and Measurement of the Intracellular K^+ Concentration*

It was unknown whether the cells uptake K^+ via the stator complex Bt-MotP/Bt- MotS. Therefore, to confirm this, the following experiment was conducted. The major K^+ uptake system-deficient strain *E. coli* TK2420 cannot grow in medium containing ≤ 10 mM K^+; however, *E. coli* HB101 (wild-type) is capable of growing under this condition. Therefore, strain TK2420 expressing Bt-MotP/Bt-MotS (strain TK-BtPS) was constructed, and a growth experiment and intracellular K^+ concentration measurements were conducted under the conditions of 7 mM and 12 mM K^+ or less. Similarly, growth experiments and intracellular K^+ concentration measurements of strains HB101/pBAD24 (strain HB-pBAD), TK2420/pBAPS (strain TK-BaPS), and TK2420/pBAD24 (strain TK-pBAD) were also conducted (Figure 7 and Section 3.7). As a result, under the 7 mM K^+ growth condition, strain HB-pBAD showed good growth even though a growth lag was observed during the early stage of culture. Under similar conditions, two strains (TK-BaPS and TK-BtPS) grew, but growth was worse than that of strain HB-pBAD. On the other hand, TK-pBAD showed no growth. Under the 12 mM K^+ growth condition, all strains showed the same growth except strain TK-pBAD, which showed the slowest growth. Furthermore, under the 7 mM K^+ growth condition, strains HB-pBAD, TK-BaPS, and TK-BtPS clearly showed higher intracellular K^+ concentrations than strain TK-pBAD. Under the 12 mM K^+ growth condition, the intracellular K^+ concentration in all strains was more than 140 mM. These results suggest that Bt-MotP/Bt-MotS can take up K^+ into cells.

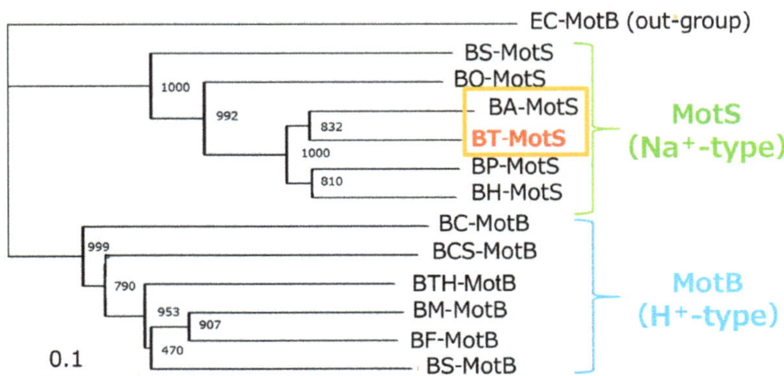

Figure 4. Phylogenetic tree of the stator subunits MotB and MotS of the flagellar motor from *Bacillus* spp. using the NJ method. A phylogenetic analysis of the flagellar motor stator of the genus *Bacillus* was performed. BT-MotS is shown in red. The Na^+- and K^+-coupled stator subunits, BT-MotS and BA-MotS are enclosed in ocher. The Na^+-driven stator is surrounded by yellow-green, and the H^+-driven stator is surrounded by light blue. The details are described in the Materials and Methods section. The scale bar indicates the number of amino acid substitutions per site. The numbers between the branches indicate the bootstrap values. Accession numbers for each protein are described in the Materials and Methods section. EC: *E. coli*, BS: *B. subtilis*, BO: *B. oceanisediminis*, BA: *B. alcalophilus*, BT: *B. trypoxylicola*, BP: *B. pseudofirmus*, BH: *B. halodurans*, BC: *B. clausii*, BCS: *B. cereus*, BTH: *B. thermoamylovorans*, BM: *B. megaterium*, and BF: *B. flexus*.

3.6. Swimming Assay of *E. coli* Stator-Deficient Strain Expressing Bt-MotP/Bt-MotS with/without a Flagellar Motor Inhibitor

Swimming inhibition assays of strains Bt-PS, *E. coli* W3110, and *B. pseudofirmus* were performed using CCCP, EIPA, and valinomycin. CCCP is a protonophore and an inhibitor of the H^+-coupled stator [41], EIPA is an amiloride analog and an inhibitor of the Na^+-coupled stator [15,25] and valinomycin is an ionophore of K^+ (Figure 9).

The swimming speed of strain Bt-PS decreased in response to increasing CCCP concentrations in only the presence of K^+, and swimming was completely stopped at 20 μM CCCP (Figure 9(A1,A2)). On the other hand, when Na^+ was present in the buffer, the Bt-PS strain was observed to swim. Since *E. coli* W3110 has an H^+-coupled motor, the swimming of *E. coli* W3110 was inhibited by elevated CCCP concentration under both KCl and NaCl conditions, and the swimming completely stopped at 25 μM CCCP. Since *B. pseudofirmus* has a Na^+-coupled motor, the swimming of *B. pseudofirmus* was not inhibited by CCCP in the presence of NaCl. The ion motive force (IMF) is composed of the cell membrane voltage (V_m) and the ion concentration gradient of inside and outside the cell. In general, *E. coli* cells maintain a lower Na^+ concentration and a higher K^+ concentration inside the cell compared to the outside [42,43]. K^+ concentration gradient (ΔpK) is smaller than Na^+ concentration gradient (ΔpNa). Therefore, the potassium motive force (potassium MF) has a higher dependence on the V_m than the sodium motive force (SMF). The inhibition of Bt-PS swimming in the presence of KCl may be due to the reduction of the V_m of the potassium MF by CCCP.

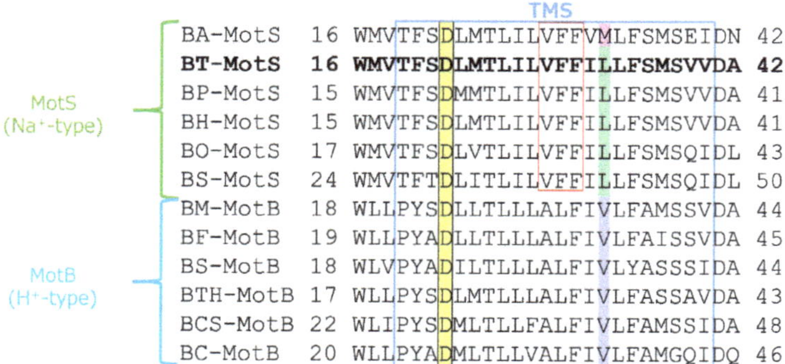

Figure 5. Multiple alignment of the regions containing the single transmembrane segment of MotB-type stator proteins from *Bacillus* spp. using ClustalW. Alignment of the flagellar motor stator of *Bacillus* was performed. The single transmembrane segment (TMS) is enclosed in blue. The sequence of BT-MotS is shown in bold. Aspartic acid residues at the putative coupling ion-binding site are highlighted yellow. A leucine residue is conserved in MotS and a valine residue is conserved in MotB and are highlighted green and purple, respectively. In BA-MotS, the methionine residue is highlighted pink. Na^+ and amiloride binding motif (VFF) are enclosed in red. Accession numbers for each protein are described in the Materials and Methods section. EC: *Escherichia coli*, BS: *B. subtilis*, BO: *B. oceanisediminis*, BA: *B. alcalophilus*, BT: *B. trypoxylicola*, BP: *B. pseudofirmus*, BH: *B. halodurans*, BC: *B. clausii*, BCS: *B. cereus*, BTH: *B. thermoamylovorans*, BM: *B. megaterium*, and BF: *B. flexus*.

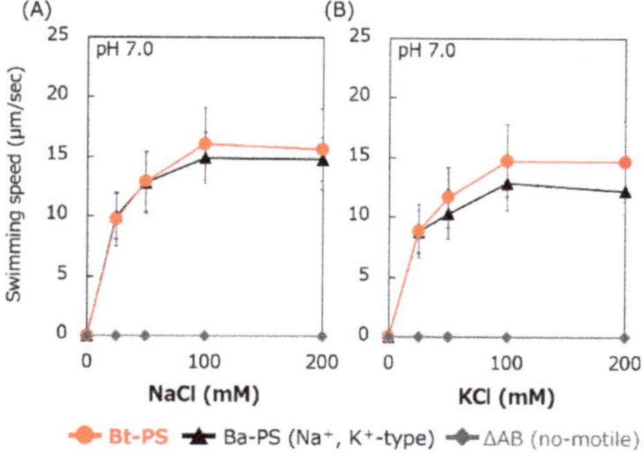

Figure 6. Effect of sodium and potassium ions on the swimming speed of *E. coli* mutant strains. The Bt-PS, Ba-PS, and ΔAB mutant strains were grown for 7 h at 30 °C in LB medium containing 0.1% arabinose with shaking. Cells were suspended in 1 mL of swimming buffer (pH 7.0) plus several NaCl concentrations (**A**) and several KCl concentrations (**B**) and were incubated at 30 °C for 10 min. Swimming buffer contained 30 mM TES, 5 mM glucose, and 0.1% arabinose, adjusted to pH 7.0 with *N*-methyl-D-glucamine. The results represent the average swimming speed of 30 independent cells from three independent experiments. The error bars indicate the standard deviations.

The swimming speed of the Bt-PS strain decreased in response to elevated EIPA concentrations under both KCl and NaCl conditions, and swimming completely stopped at 400 μM EIPA (Figure 9(B1,B2)). Since *E. coli* W3110 has an H^+-coupled motor, the swimming of *E. coli* W3110 was not inhibited by EIPA under both KCl and NaCl conditions. Since *B. pseudofirmus* has a Na^+-coupled motor, the swimming of *B. pseudofirmus* was inhibited by elevated EIPA concentrations in the presence of NaCl. The reason that the strain Bt-PS was sensitive to EIPA under both conditions is presumed to be that EIPA binds to the coupling ion transport pathway of the stator complex. Kuroda et al. reported that VFF sequence are present in many Na^+ coupled transport proteins and proposed that VFF sequence is a motif involved in Na^+ binding and amiloride binding [44]. A VFF sequence can be found in the transmembrane region of the Na^+-coupled MotS subunit (Figure 5). There are multiple reports that the EIPA and its analog phenamil inhibited the coupling ion transport pathway in the stator complex [17,45–48]. Previously, a similar inhibition experiment was carried out using Ba-MotP/Ba-MotS of *B. alcalophilus*, which can utilize both Na^+ and K^+ as coupling ions, and EIPA was inhibited under both K^+ and Na^+ conditions as in strain Bt-PS [22].

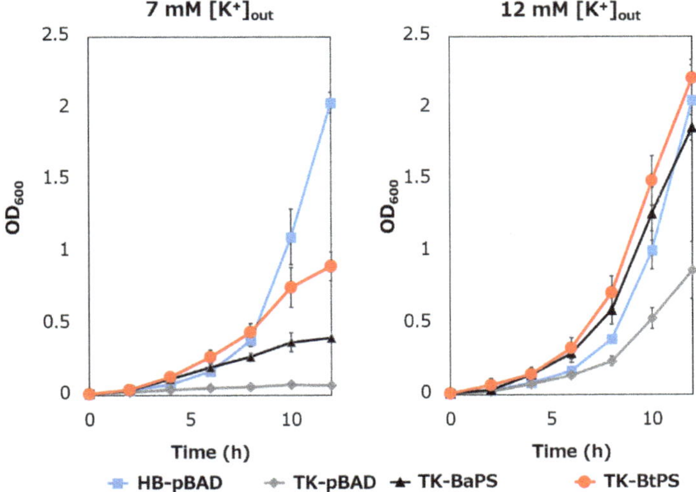

Figure 7. Effect of extracellular K^+ concentrations on the growth of *E. coli* and its K^+ uptake system-deleted mutant strains. The HB-pBAD, TK-pBAD, TK-BaPS, and TK-BtPS strains were grown for 12 h at 30 °C in a modified TK2420 medium (pH 7.0) contained 7 mM or 12 mM KCl with 0.2% arabinose with shaking. Growth was assessed hourly at OD_{600}.

The swimming speed of strain Bt-PS decreased in response to elevated valinomycin concentrations in only the presence of K^+, and swimming completely stopped at 50 μM valinomycin (Figure 9(C1,C2)). On the other hand, the swimming of Bt-PS was not inhibited by valinomycin in the presence of NaCl. Since *E. coli* W3110 has an H^+-coupled motor and *B. pseudofirmus* has a Na^+-coupled motor, *E. coli* W3110 and *B. pseudofirmus* swimming was not inhibited by valinomycin. The reason that Bt-PS swimming was inhibited by valinomycin in the presence of KCl was hypothesized to be the decrease in the potassium MF to rotate the flagella due to the valinomycin-induced loss of the ΔpK.

3.7. Functional Analysis of the Hybrid Stator with the Na^+-Coupled MotP/MotS Subunit Replaced with the Na^{+-} and K+-Coupled MotP/MotS Subunit

MotB-type (MotB, MotS, and PomB) subunits are thought to be particularly important for coupling cation selectivity of the stator, but details of the mechanism are still unknown. The data of Section 3.6 and Figure 9 suggested that the stator of *B. trypoxylicola* was both the Na^+- and K^+-coupled stators MotP and MotS. A leucine residue was the important amino acid for ion selectivity in the transmembrane region of the Bt-MotS subunit (Figure 5) [21]. Furthermore, the amino acid sequences in the transmembrane region between the Bt-MotS subunit and the Na^+-coupled MotS subunit of *B. halodurans* C-125 (Bh-MotS) were 100% identical (Figure 5). This suggests that the Bt-MotP subunit is important for the mechanism of K^+ selectivity of the *B. trypoxylicola* stator Bt-MotP/Bt-MotS. Therefore, we focused on the exchange of the Na^{+-} and K^+-coupled MotP and MotS subunits and the Na^+-coupled MotP and MotS subunits. The hybrid subunits were constructed using *B. trypoxylicola*-derived Na^+- and K^+-coupled Bt-MotP and Bt-MotS, *B. alcalophilus*-derived Na^{+-} and K^+-coupled Ba-MotP and Ba-MotS, and *B. pseudofirmus*-derived Na^+-coupled Bp-MotP and Bp-MotS. The constructed hybrid combinations are shown in Figure 10. The coupling ions of each hybrid stator shown in Figure 9 are based on the results in Figure 11.

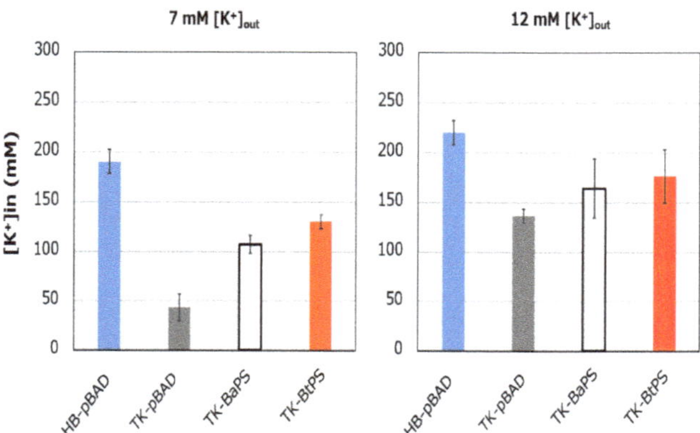

Figure 8. Effects of extracellular K^+ concentrations on the intracellular K^+ concentration of *E. coli* and its K^+ uptake system-deleted mutant strains. The HB-pBAD, TK-pBAD24, TK-BaPS, and TK-BtPS strains were grown for 12 h at 30 °C in modified TK2420 medium (pH 7.0) contained 7 mM or 12 mM KCl with 0.2% arabinose with shaking. K^+ concentrations per cell were measured as described in the Materials and Methods section.

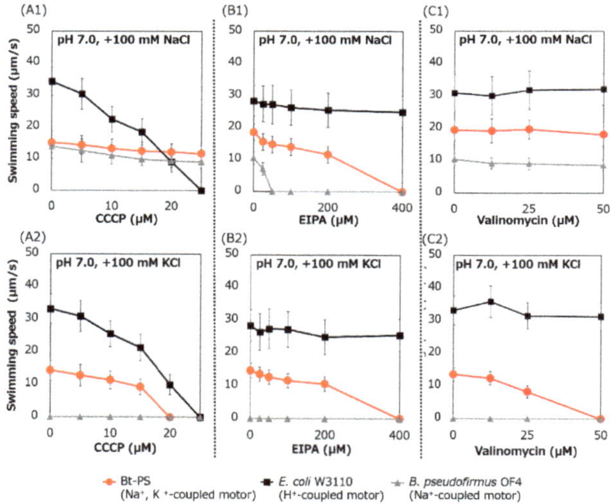

Figure 9. Swimming assay of BTPS (a stator-less *E. coli* mutant expressing Bt-MotPS) in the presence of flagellar motor inhibitor. The Bt-PS and *E. coli* W3110 strains were grown for 7 h at 30 °C in LB medium containing 0.1% arabinose with shaking. *B. pseudofirmus* OF4 was grown for 7 h at 30 °C in MYE medium (pH 10.5) with shaking. Cells were suspended in 1 mL of swimming buffer (pH 7.0) plus several carbonyl cyanide m-chlorophenyl hydrazine (CCCP) concentrations (**A1** and **A2**), plus several 5-(*N*-ethyl-*N*-isopropyl)-amiloride (EIPA) concentrations (**B1** and **B2**), plus several valinomycin concentrations (**C1** and **C2**) and were incubated at 30 °C for 10 min. Swimming buffer contained 30 mM TES, 5 mM glucose, 0.1% arabinose, 100 mM NaCl (**A1**, **B1** and **C1**) or 100 mM KCl (**A2**, **B2** and **C2**), adjusted to pH 7.0 with *N*-methyl-D-glucamine. The results represent the average swimming speed of 30 independent cells from three independent experiments. The error bars indicate the standard deviations.

To analyze the ion selectivity of the constructed hybrid stators, swimming analysis was performed using mutant strains expressing the hybrid stator constructed in the *B. subtilis* stator-deficient strain (Figure 11).

Strains OF4PS, PTPS, and PAPS showed Na^+-coupled swimming behavior. However, there was no K^+-dependent motility among them. These results suggest that the hybrid stators Bp-MotP/Bt-MotS and Bp-MotP/Ba-MotS are Na^+-driven stators that can use Na^+ as a coupling ion. These hybrid stators did not exhibit K^+ concentration-dependent motility while having Na^+- and K^+-coupled MotS (Bt-MotS from *B. trypoxylicola* or Ba-MotS from *B. alcalophilus*) subunits. Thus, the K^+ selectivity of the stator indicated that the MotP subunit is critical, but the MotS subunit is not. On the other hand, strains AAPS, which had *B. alcalophilus* MotP/MotS; and TTPS, which had a *B. trypoxylicola* MotP/MotS stator; and strains TPPS, APPS, TAPS, and ATPS with hybrid stators showed both Na^+- and K^+-coupled motility. From these results, the hybrid stators Bt-MotP/Bp-MotS, Ba-MotP/Bp-MotS, Bt-MotP/Ba-MotS, and Ba-MotP/Bt-MotS coupled with Na^+ and K^+ as their coupling cations. These strains commonly have Na^+- and K^+-coupled MotP subunits. Therefore, it was suggested that the MotP subunit that forms an ion channel together with the MotS subunit is more important for the K^+ selectivity of the flagellar motor stator than the MotS subunit itself.

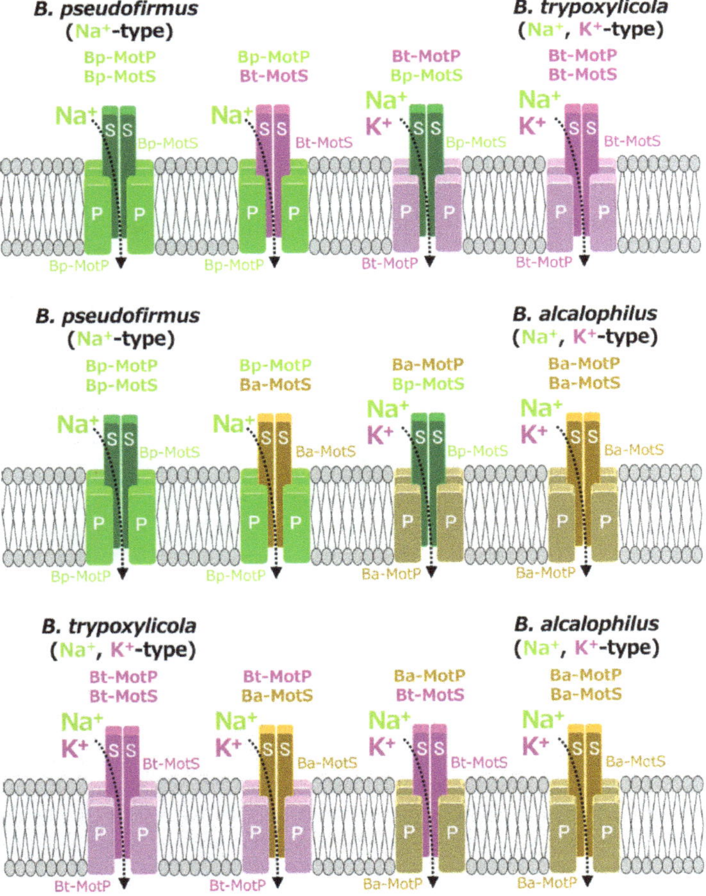

Figure 10. Schematic diagram of the stators of wild-type MotP/MotS and hybrid MotP/MotS. The coupling ions of each hybrid stator shown here are based on the results in Figure 11.

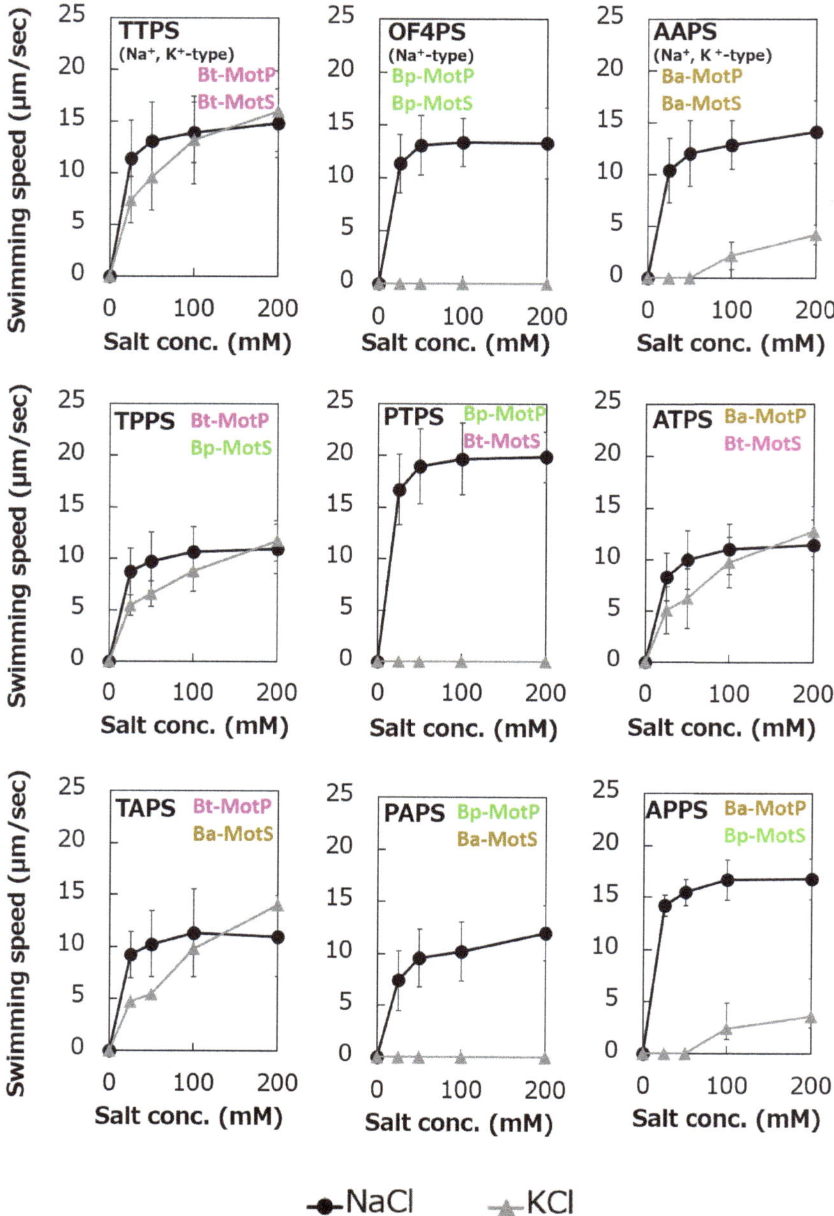

Figure 11. Swimming assays of hybrid stators between Na^+-type stators and Na^+ and K^+-type stators. *B. subtilis* strains were grown for 6 h at 37 °C in Spizizen I medium 1% xylose with shaking. Cells were suspended in 1 mL of swimming medium (pH 8.0) plus several NaCl concentrations plus several KCl concentrations and then incubated at 37 °C for 10 min. Swimming medium contained 0.04% tryptone, 0.02% yeast extract, and 5 mM glucose, adjusted to pH 8.0 with *N*-methyl-D-glucamine. The results represent the average swimming speed of 30 independent cells from three independent experiments. The error bars indicate the standard deviations.

4. Discussion

B. trypoxylicola, a species closely related to *B. alcalophilus*, requires K^+ for growth, suggesting that *B. trypoxylicola* has a homeostatic ability to utilize K^+ [22]. Generally, alkaliphilic bacteria isolated from soil require Na^+ for growth, whereas alkaliphilic bacteria isolated from human feces and the guts of insects require K^+ in addition to Na^+ for growth [3,49–51]. Gut portions of soil-feeding termites generally contain large amounts of potassium ions, and their pH is extremely alkaline [52,53]. Gut alkalinity helps solubilize and facilitate the uptake of soil organic matter [54]. Potassium-requiring alkaliphilic bacteria have been isolated from such environments [3,49,50]. *B. trypoxylicola* was isolated from the gut of the Japanese beetle [27]. This suggests that *B. trypoxylicola* has adaptively evolved into the K^+-rich gut environment for growth and motility. *B. trypoxylicola* was the second example of a flagellar motor that could use Na^+ and K^+ as coupling ions. Most of the MotP/MotS stators derived from alkaliphilic *Bacillus* spp. can function only with Na^+ [16,55]. The flagellar motor of *B. trypoxylicola* is the second example of a MotP/MotS stator that can use both Na^+ and K^+. Even though *B. trypoxylicola* exhibits no requirement of Na^+ for growth, the flagellar motor can utilize both Na^+ and K^+. This suggests that the ancestor of the MotP/MotS stator is originally a Na^+ type and that the MotP/MotS stator of *B. trypoxylicola* acquired the ability to use K^+ as a coupling ion in addition to Na^+ during the evolutionary process in a K^+-rich environment.

Previous studies have suggested that the transmembrane domain of the MotB-type (MotB, MotS, and PomB) subunit is particularly critical for the coupling ion selectivity of Na^+ and H^+ in the flagellar motor stator [21,25]. With respect to K^+ selectivity, MotS-M33 was reported to be important in the flagellar motor of *B. alcalophilus* [22]. However, a methionine residue is not conserved at a similar site in BT-MotS, and general MotS types conserve a leucine residue at this site. It is suggested that this site is not necessarily essential for K^+ selectivity. Furthermore, the amino acid sequence in the transmembrane segment of the Bt-MotS subunit was identical to that of the MotS subunit (Bh-MotS) of Na^+-driven MotP/MotS of *B. halodurans* C-125 (Figure 5). Therefore, it was concluded that the MotP subunit that forms an ion channel with MotS is important for the K^+ selectivity of the flagellar stator.

The swimming assay with various inhibitors against strain Bt-PS, strain Bt-PS showed sensitivity to CCCP, EIPA, and valinomycin in the presence of K^+. On the other hand, in the presence of Na^+, sensitivity was observed for only EIPA (Figure 9). Strain Bt-PS was sensitive to the H^+-coupled flagellar inhibitor CCCP in the presence of K^+. This indicates that the V_m of the potassium MF is dominant for rotation of the flagellar motor. The results for these inhibitors showed that the flagellar motor, which can utilize both K^+ and Na^+, contributes a great deal to the V_m when using the potassium MF. It was also shown that when using the SMF, the contribution of the ΔpNa is greater than that of the V_m. In addition, the results show for the first time that valinomycin is useful as an inhibitor of the K^+-coupled flagellar motor.

According to experiments using the hybrid stators MotA/MotS and MotP/MotB that replace the H^+-coupled stators MotA/MotB and Na^+-coupled stators MotP/MotS of *B. subtilis*, the MotB subunit defines H^+ selectivity, and the MotS subunit defines Na^+ selectivity [25]. In summary, when H^+ or Na^+ is selected as the coupling ion for ion selectivity of the flagellar motor stator of bacteria from the genus *Bacillus* and its closely related species, the transmembrane region of the MotB or MotS subunit is considered to be particularly important. On the other hand, this study suggested for the first time that the MotP subunit is important when selecting K^+ as a coupling ion. In the MotP subunit of *B. subtilis*, the structure of the pathway of the coupling ion is changed by the mutation of MotP-L170P of the third transmembrane region, and the motility of the MotS-D30E mutant is improved [21]. In this way, the MotP subunit may influence the ion selectivity of coupling ions. It is speculated that an ancestral type of Bt-MotP/Bt-MotS was originally considered the Na^+-coupled stator MotP/MotS, and in the process of adapting to a K^+-rich environment, the Bt-MotP subunit was mutated, and the structure of the coupling ion pathway evolved to use not only Na^+ but also K^+.

A recent study reported that *Aquifex aeolicus*, a hyperthermophilic bacterium thought to have branched off during the early stages of bacterial evolution, has a flagellar motor that utilized Na^+ as a coupling ion, and it has been proposed that the motor of the bacterial ancestor may have used Na^+ as a coupling ion [56]. It is believed that the initial flagellum motor adaptively evolved to an H^+-coupled stator and an Na^+-coupled stator, respectively, in accordance with a growing environment. A primitive stator is estimated to have evolved by a motor utilizing only H^+ or Na^+ by mutating the amino acid residues of the MotB-type subunit. After that, it is estimated that the Na^+-coupled stator evolved into a Na^+- and K^+-coupled stator to adapt to an environment (such as in the guts of living organisms) where K^+ exists abundantly. These processes are outlined in Figure 12.

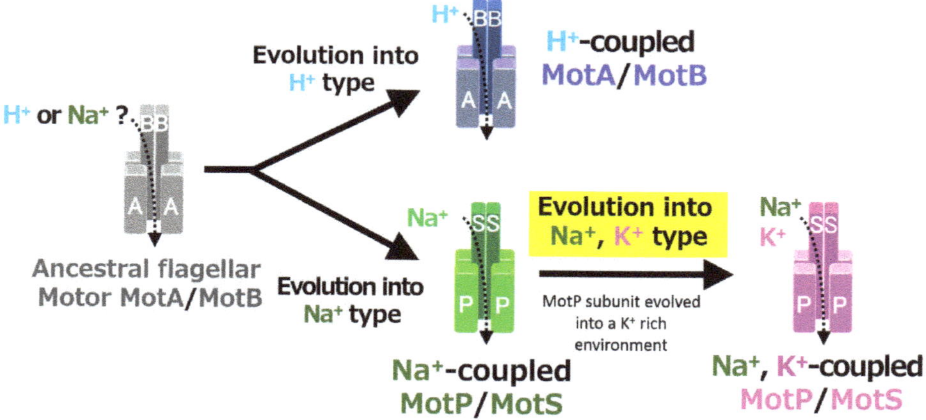

Figure 12. Evolutionary hypothesis of bacterial flagellar stator based on the results. It has not been clarified which ancestral flagellar motor stator utilized H^+ or Na^+ as the coupling ion of the flagellar motor. It is assumed that the ancestral stator evolved into H^+-coupled and Na^+-coupled stators. The MotB-type subunit is critical for the coupling ion selectivity of H^+ and Na^+. Furthermore, it is assumed that the Na^+-coupled stator evolved into the Na^+- and K^+-coupled stator. The MotP subunit is critical for the coupling ion selectivity of K^+. Therefore, *B. alcalophilus* and *B. trypoxylicola* are assumed to have evolved these properties to adapt to a potassium-rich environment.

In the future, the details of the K^+ selection mechanism of the flagellar motor stator should be clarified by focusing on the transmembrane regions of the MotP subunit.

5. Conclusions

Previously, the coupling ion selection of H^+ or Na^+ for flagellar motor stators of *Bacillus* spp. and its relatives suggested that the transmembrane region of the MotB or MotS subunit is particularly critical. This study suggests for the first time that the MotP subunit is critical when selecting K^+ as the coupling ion. The K^+- and Na^+-coupled MotP/MotS stator complexes from alkaliphilic *B. alcalophilus* and *B. trypoxylicola* were presumed to have evolved from Na^+-coupled MotP/MotS stator complexes during adaptation to a large potassium-rich environment. This work could provide a new perspective on the study of the ion selectivity of flagellar motor stators.

Supplementary Materials: The following are available online at http://www.mdpi.com/2218-273X/10/5/691/s1, Table S1: Primer list for this study.

Author Contributions: Conceptualization, M.I.; methodology, S.N. and M.I.; software, S.N.; validation, S.N. and M.I.; formal analysis, S.N.; investigation, M.I.; resources, M.I.; data curation, M.I.; writing—original draft preparation, S.N. and M.I.; writing—review and editing, M.I.; supervision, M.I.; project administration, M.I.; funding acquisition, M.I. All authors have read and agreed to the published version of the manuscript.

Funding: This research was funded by a Grant-in-Aid for Scientific Research on Innovative Areas of the Ministry of Education, Culture, Sports, Science and Technology of Japan, grant number JP24117005 (M.I.).

Acknowledgments: We would like to thank Arthur A. Guffanti for the critical discussions and reading of the manuscript.

Conflicts of Interest: The authors declare that they have no conflicts of interest. Ethical approval: this article does not describe any studies with human participants or animals performed by any of the authors.

References

1. Minamino, T.; Imada, K. The bacterial flagellar motor and its structural diversity. *Trends Microbiol.* **2015**, *23*, 267–274. [CrossRef] [PubMed]
2. Minamino, T.; Terahara, N.; Kojima, S.; Namba, K. Autonomous control mechanism of stator assembly in the bacterial flagellar motor in response to changes in the environment. *Mol. Microbiol.* **2018**, *109*, 723–734. [CrossRef] [PubMed]
3. Ito, M.; Takahashi, Y. Nonconventional cation-coupled flagellar motors derived from the alkaliphilic *Bacillus* and *Paenibacillus* species. *Extremophiles* **2017**, *21*, 3–14. [CrossRef] [PubMed]
4. Berg, H.C. The rotary motor of bacterial flagella. *Ann. Rev. Biochem.* **2003**, *72*, 19–54.
5. Sowa, Y.; Berry, R.M. Bacterial flagellar motor. *Q. Rev. Biophys.* **2008**, *41*, 103–132. [CrossRef] [PubMed]
6. Minamino, T.; Imada, K.; Namba, K. Molecular motors of the bacterial flagella. *Curr. Opin. Struct. Biol.* **2008**, *18*, 693–701. [CrossRef]
7. Terashima, H.; Kawamoto, A.; Morimoto, Y.V.; Imada, K.; Minamino, T. Structural differences in the bacterial flagellar motor among bacterial species. *Biophys. Physicobiol.* **2017**, *14*, 191–198. [CrossRef]
8. Kojima, S.; Blair, D.F. Solubilization and purification of the MotA/MotB complex of *Escherichia coli*. *Biochemistry* **2004**, *43*, 26–34.
9. Braun, T.F.; Al-Mawsawi, L.Q.; Kojima, S.; Blair, D.F. Arrangement of core membrane segments in the MotA/MotB proton-channel complex of *Escherichia coli*. *Biochemistry* **2004**, *43*, 35–45.
10. Zhou, J.; Lloyd, S.A.; Blair, D.F. Electrostatic interactions between rotor and stator in the bacterial flagellar motor. *Proc. Natl. Acad. Sci. USA* **1998**, *95*, 6436–6441.
11. Reid, S.W.; Leake, M.C.; Chandler, J.H.; Lo, C.J.; Armitage, J.P.; Berry, R.M. The maximum number of torque-generating units in the flagellar motor of *Escherichia coli* is at least 11. *Proc. Natl. Acad. Sci. USA* **2006**, *103*, 8066–8071. [CrossRef] [PubMed]
12. Manson, M.D.; Tedesco, P.; Berg, H.C.; Harold, F.M.; Van der Drift, C. A protonmotive force drives bacterial flagella. *Proc. Natl. Acad. Sci. USA* **1977**, *74*, 3060–3064. [PubMed]
13. Macnab, R.M. The bacterial flagellum: Reversible rotary propellor and type III export apparatus. *J. Bacteriol.* **1999**, *181*, 7149–7153. [PubMed]
14. Li, N.; Kojima, S.; Homma, M. Sodium-driven motor of the polar flagellum in marine bacteria *Vibrio*. *Genes Cells* **2011**, *16*, 985–999. [CrossRef] [PubMed]
15. Ito, M.; Hicks, D.B.; Henkin, T.M.; Guffanti, A.A.; Powers, B.D.; Zvi, L.; Uematsu, K.; Krulwich, T.A. MotPS is the stator-force generator for motility of alkaliphilic *Bacillus*, and its homologue is a second functional Mot in *Bacillus subtilis*. *Mol. Microbiol.* **2004**, *53*, 1035–1049. [CrossRef]
16. Fujinami, S.; Terahara, N.; Krulwich, T.A.; Ito, M. Motility and chemotaxis in alkaliphilic *Bacillus* species. *Future Microbiol.* **2009**, *4*, 1137–1149. [CrossRef]
17. Fujinami, S.; Terahara, N.; Lee, S.; Ito, M. Na^+ and flagella-dependent swimming of alkaliphilic *Bacillus pseudofirmus* OF4: A basis for poor motility at low pH and enhancement in viscous media in an "up-motile" variant. *Arch. Microbiol.* **2007**, *187*, 239–247. [CrossRef]
18. Kojima, S.; Blair, D.F. The bacterial flagellar motor: Structure and function of a complex molecular machine. *Int. Rev. Cytol.* **2004**, *233*, 93–134. [PubMed]
19. Paulick, A.; Delalez, N.J.; Brenzinger, S.; Steel, B.C.; Berry, R.M.; Armitage, J.P.; Thormann, K.M. Dual stator dynamics in the *Shewanella oneidensis* MR-1 flagellar motor. *Mol. Microbiol.* **2015**, *96*, 993–1001. [CrossRef]
20. Terahara, N.; Fujisawa, M.; Powers, B.; Henkin, T.M.; Krulwich, T.A.; Ito, M. An intergenic stem-loop mutation in the *Bacillus subtilis ccpA-motPS* operon increases *motPS* transcription and the MotPS contribution to motility. *J. Bacteriol.* **2006**, *188*, 2701–2705. [CrossRef]
21. Terahara, N.; Krulwich, T.A.; Ito, M. Mutations alter the sodium versus proton use of a *Bacillus clausii* flagellar motor and confer dual ion use on *Bacillus subtilis* motors. *Proc. Natl. Acad. Sci. USA* **2008**, *105*, 14359–14364. [CrossRef]

22. Terahara, N.; Sano, M.; Ito, M. A *Bacillus* flagellar motor that can use both Na^+ and K^+ as a coupling ion is converted by a single mutation to use only Na^+. *PLoS ONE* **2012**, *7*, e46248.
23. Imazawa, R.; Takahashi, Y.; Aoki, W.; Sano, M.; Ito, M. A novel type bacterial flagellar motor that can use divalent cations as a coupling ion. *Sci. Rep.* **2016**, *6*, 19773. [CrossRef]
24. Asai, Y.; Yakushi, T.; Kawagishi, I.; Homma, M. Ion-coupling determinants of Na^+-driven and H^+-driven flagellar motors. *J. Mol. Biol.* **2003**, *327*, 453–463.
25. Ito, M.; Terahara, N.; Fujinami, S.; Krulwich, T.A. Properties of motility in *Bacillus subtilis* powered by the H^+-coupled MotAB flagellar stator, Na^+-coupled MotPS or hybrid stators MotAS or MotPB. *J. Mol. Biol.* **2005**, *352*, 396–408. [PubMed]
26. Zhou, J.; Sharp, L.L.; Tang, H.L.; Lloyd, S.A.; Billings, S.; Braun, T.F.; Blair, D.F. Function of protonatable residues in the flagellar motor of *Escherichia coli*: A critical role for Asp 32 of MotB. *J. Bacteriol.* **1998**, *180*, 2729–2735. [PubMed]
27. Aizawa, T.; Urai, M.; Iwabuchi, N.; Nakajima, M.; Sunairi, M. *Bacillus trypoxylicola* sp. nov., xylanase-producing alkaliphilic bacteria isolated from the guts of Japanese horned beetle larvae (*Trypoxylus dichotomus septentrionalis*). *Int. J. Syst. Evol. Microbiol.* **2010**, *60*, 61–66. [CrossRef]
28. Attie, O.; Jayaprakash, A.; Shah, H.; Paulsen, I.T.; Morino, M.; Takahashi, Y.; Narumi, I.; Sachidanandam, R.; Satoh, K.; Ito, M.; et al. Draft Genome Sequence of *Bacillus alcalophilus* AV1934, a Classic Alkaliphile Isolated from Human Feces in 1934. *Genome Announc.* **2014**, *2*, e01175-14. [CrossRef]
29. Gronstad, A.; Jaroszewicz, E.; Ito, M.; Sturr, M.G.; Krulwich, T.A.; Kolsto, A.B. Physical map of alkaliphilic *Bacillus firmus* OF4 and detection of a large endogenous plasmid. *Extremophiles* **1998**, *2*, 447–453.
30. Ito, M.; Cooperberg, B.; Krulwich, T.A. Diverse genes of alkaliphilic *Bacillus firmus* OF4 that complement K^+-uptake-deficient *Escherichia coli* include an *ftsH* homologue. *Extremophiles* **1997**, *1*, 22–28.
31. Takahashi, Y.; Koyama, K.; Ito, M. Suppressor mutants from MotB-D24E and MotS-D30E in the flagellar stator complex of *Bacillus subtilis*. *J. Gen. Appl. Microbiol.* **2014**, *60*, 131–139.
32. Spizizen, J. Transformation of biochemically deficient strains of *Bacillus subtilis* by deoxyribonucleate. *Proc. Natl. Acad. Sci. USA* **1958**, *44*, 1072–1078.
33. Guzman, C.A.; Piatti, G.; Walker, M.J.; Guardati, M.C.; Pruzzo, C. A novel *Escherichia coli* expression-export vector containing alkaline phosphatase as an insertional inactivation screening system. *Gene* **1994**, *148*, 171–172. [CrossRef]
34. Grundy, F.J.; Henkin, T.M. The *rpsD* gene, encoding ribosomal protein S4, is autogenously regulated in *Bacillus subtilis*. *J. Bacteriol.* **1991**, *173*, 4595–4602. [PubMed]
35. *Promega Protocols and Applications Guide*, 3rd ed.; Promega: Madison, WI, USA, 1996; pp. 45–46.
36. Horton, R.M.; Cai, Z.L.; Ho, S.N.; Pease, L.R. Gene splicing by overlap extension: Tailor-made genes using the polymerase chain reaction. *Biotechniques* **1990**, *8*, 528–535. [PubMed]
37. Cohen-Bazire, G.; Sistrom, W.R.; Stanier, R.Y. Kinetic studies of pigment synthesis by non-sulfur purple bacteria. *J. Cell. Physiol.* **1957**, *49*, 25–68.
38. Epstein, W.; Buurman, E.; McLaggan, D.; Naprstek, J. Multiple mechanisms, roles and controls of K^+ transport in *Escherichia coli*. *Biochem. Soc. Trans.* **1993**, *21*, 1006–1010. [PubMed]
39. Bakker, E. Cell K^+ and K^+ transport systems in prokaryotes. In *Alkali Cation Transport Systems in Prokaryotes;* Bakker, E., Ed.; CRC Press: Boca Raton, FL, USA, 1993; pp. 205–224.
40. Ito, M.; Guffanti, A.A.; Zemsky, J.; Ivey, D.M.; Krulwich, T.A. Role of the *nhaC*-encoded Na^+/H^+ antiporter of alkaliphilic *Bacillus firmus* OF4. *J. Bacteriol.* **1997**, *179*, 3851–3857.
41. Atsumi, T.; McCarter, L.; Imae, Y. Polar and lateral flagellar motors of marine Vibrio are driven by different ion-motive forces. *Nature* **1992**, *355*, 182–184. [CrossRef]
42. Bakker, E.P. The role of alkali-cation transport in energy coupling of neutrophilic and acidophilic bacteria: An assesment of methods and concepts. *FEMS Microbiol. Rev.* **1990**, *75*, 319–334.
43. Booth, I.R. Regulation of cytoplasmic pH in bacteria. *Microbiol. Rev.* **1985**, *49*, 359–378. [PubMed]
44. Kuroda, T.; Shimamoto, T.; Mizushima, T.; Tsuchiya, T. Mutational analysis of amiloride sensitivity of the NhaA Na^+/H^+ antiporter from *Vibrio parahaemolyticus*. *J. Bacteriol.* **1997**, *179*, 7600–7602. [CrossRef] [PubMed]
45. Atsumi, T.; Sugiyama, S.; Cragoe, E.J., Jr.; Imae, Y. Specific inhibition of the Na^+-driven flagellar motors of alkalophilic *Bacillus* strains by the amiloride analog phenamil. *J. Bacteriol.* **1990**, *172*, 1634–1639. [PubMed]

46. Kojima, S.; Atsumi, T.; Muramoto, K.; Kudo, S.; Kawagishi, I.; Homma, M. *Vibrio alginolyticus* mutants resistant to phenamil, a specific inhibitor of the sodium-driven flagellar motor. *J. Mol. Biol.* **1997**, *265*, 310–318.
47. Kojima, S.; Asai, Y.; Atsumi, T.; Kawagishi, I.; Homma, M. Na^+-driven flagellar motor resistant to phenamil, an amiloride analog, caused by mutations in putative channel components. *J. Mol. Biol.* **1999**, *285*, 1537–1547. [PubMed]
48. Sugiyama, S.; Cragoe, E.J., Jr.; Imae, Y. Amiloride, a specific inhibitor for the Na^+-driven flagellar motors of alkalophilic *Bacillus*. *J. Biol. Chem.* **1988**, *263*, 8215–8219.
49. Thongaram, T.; Hongoh, Y.; Kosono, S.; Ohkuma, M.; Trakulnaleamsai, S.; Noparatnaraporn, N.; Kudo, T. Comparison of bacterial communities in the alkaline gut segment among various species of higher termites. *Extremophiles* **2005**, *9*, 229–238. [CrossRef] [PubMed]
50. Horikoshi, K. General Physiology of Alkaliphiles. In *Extremophiles Handbook*; Horikoshi, K., Ed.; Springer: Tokyo, Japan, 2011; pp. 99–118.
51. Vedder, A. *Bacillus alcalophilus* n. sp.; benevens enkele ervaringen met sterk alcalische voedingsbodems. *Anton. Leeuw.* **1934**, *1*, 143–147.
52. Bignell, D.E.; Oskarsson, H.; Anderson, J.M.; Ineson, P.; Wood, T.G. Structure, microbial associations and function of the so-called "mixed segment" of the gut in two soil-feeding termites, Procubitermes aburiensis and Cubitermes severus (Termitidae, Termitinae). *J. Zool.* **1983**, *201*, 445–480.
53. Brune, A.; Kuhl, M. pH profiles of the extremely alkaline hindguts of soil-feeding termites (Isoptera: Termitidae) determined with microelectrodes. *J. Insect Physiol.* **1996**, *42*, 1121–1127.
54. Thongaram, T.; Kosono, S.; Ohkuma, M.; Hongoh, Y.; Kitada, M.; Yoshinaka, T.; Trakulnaleamsai, S.; Noparatnaraporn, N.; Kudo, T. Gut of Higher Termites as a Niche for Alkaliphiles as Shown by Culture-Based and Culture-Independent Studies. *Microb. Environ.* **2003**, *18*, 152–159. [CrossRef]
55. Ito, M.; Fujinami, S.; Terahara, N. Bioenergetics: Cell motility and chemotaxis of extreme alkaliphiles. In *Extremophiles Handbook*; Horikoshi, K., Ed.; Springer: Tokyo, Japan, 2011; pp. 141–162.
56. Takekawa, N.; Nishiyama, M.; Kaneseki, T.; Kanai, T.; Atomi, H.; Kojima, S.; Homma, M. Sodium-driven energy conversion for flagellar rotation of the earliest divergent hyperthermophilic bacterium. *Sci. Rep.* **2015**, *5*, 12711. [CrossRef] [PubMed]

Article

Coupling Ion Specificity of the Flagellar Stator Proteins MotA1/MotB1 of *Paenibacillus* sp. TCA20

Sakura Onoe [1,†,‡], Myu Yoshida [1,†], Naoya Terahara [2] and Yoshiyuki Sowa [1,*]

1 Department of Frontier Bioscience, Research Center for Micro-Nano Technology, Hosei University, 3-7-2 Kajino-cho, Koganei, Tokyo 184-8584, Japan; sakura.onoe@protein.osaka-u.ac.jp (S.O.); miyu.yoshida.54@hosei.ac.jp (M.Y.)

2 Department of Physics, Chuo University, 1-13-27 Kasuga, Bunkyo-ku, Tokyo 112-8551, Japan; terahara.19r@g.chuo-u.ac.jp

* Correspondence: ysowa@hosei.ac.jp; Tel.: +81-42-387-7026

† These authors contributed equally to this work.

‡ Current address: Institute for Protein Research, Osaka University, 3-2 Yamadaoka, Suita, Osaka 565-0871, Japan.

Received: 28 March 2020; Accepted: 14 July 2020; Published: 20 July 2020

Abstract: The bacterial flagellar motor is a reversible rotary molecular nanomachine, which couples ion flux across the cytoplasmic membrane to torque generation. It comprises a rotor and multiple stator complexes, and each stator complex functions as an ion channel and determines the ion specificity of the motor. Although coupling ions for the motor rotation were presumed to be only monovalent cations, such as H^+ and Na^+, the stator complex MotA1/MotB1 of *Paenibacillus* sp. TCA20 (MotA1^{TCA}/MotB1^{TCA}) was reported to use divalent cations as coupling ions, such as Ca^{2+} and Mg^{2+}. In this study, we initially aimed to measure the motor torque generated by MotA1^{TCA}/MotB1^{TCA} under the control of divalent cation motive force; however, we identified that the coupling ion of MotA1^{TCA}MotB1^{TCA} is very likely to be a monovalent ion. We engineered a series of functional chimeric stator proteins between MotB1^{TCA} and *Escherichia coli* MotB. *E. coli* Δ*motAB* cells expressing MotA1^{TCA} and the chimeric MotB presented significant motility in the absence of divalent cations. Moreover, we confirmed that MotA1^{TCA}/MotB1^{TCA} in *Bacillus subtilis* Δ*motAB*Δ*motPS* cells generates torque without divalent cations. Based on two independent experimental results, we conclude that the MotA1^{TCA}/MotB1^{TCA} complex directly converts the energy released from monovalent cation flux to motor rotation.

Keywords: flagellar motor; coupling ion; divalent cation

---

1. Introduction

Most swimming bacteria can swim towards their favorable environments by rotating their helical flagella [1,2]. Each flagellum is rotated by a rotary molecular motor embedded in the bacterial cell envelope at its base [3]. The rotation of the bacterial flagellar motor is driven by the flux of the coupling ions across the cytoplasmic membrane. The motor comprises a rotor and multiple stator units. Each stator unit contains two types of membrane proteins called the Mot complex and functions to conduct the coupling ions via a channel in order to generate mechanical torque. Previous studies reported that various Mot complexes work in a wide range of bacterial species, such as MotA/MotB in H^+-driven motors of *Escherichia coli*, PomA/PomB in Na^+-driven motors of *Vibrio alginolyticus*, MotP/MotS in Na^+-driven motors of *Bacillus subtilis*, and MotP/MotS in Na^+, K^+, and Rb^+-driven motors of *Bacillus alcalophilus* [4–7]. All these Mot complexes share a common structure and function; a peptidoglycan (PG) binding domain to anchor the stator unit to the PG layer, a transmembrane (TM) domain to transduce the coupling ion(s) across the membrane, and a large cytoplasmic domain to interact with the rotor proteins [8]. Therefore, several functional chimeric motors comprise various combinations of rotor and stator proteins [9–13]. One such representative chimera motor used for functional analysis, is the Na^+-driven motor of *E. coli* that combines the rotor of the H^+-driven *E. coli* motor with the chimeric stator PomA/PotB [11,14–18]. The PomA/PotB stator contains PomA from *V. alginolyticus* and a fusion protein, which replaces the PG binding domain of PomB from *V. alginolyticus* with that of MotB from *E. coli* for efficient anchoring to the PG layer of *E. coli* cells [19].

Recently, $MotA1^{TCA}/MotB1^{TCA}$ in the flagellar motor of *Paenibacillus* sp. TCA20, which was isolated from a Ca^{2+} rich hot spring in Japan, was reported to use Mg^{2+} and Ca^{2+} as coupling ions for its rotation rather than the monovalent cations, which were utilized by all other flagellar motors [20]. The $MotA1^{TCA}/MotB1^{TCA}$ can interact with the rotor of the *B. subtilis* motor and then generate a torque [20]. This novel type of flagellar motor is presumed to shed light on the mechanism of energy conversion of the motor, because the charge number (z) contributes to the gained energy from the ion flux. In the bacterial flagellar motor, the free energy from a single ion passage across the cytoplasmic membrane comprises an electrical component (zeV_m) and a chemical component ($k_B T \ln (C_i/C_o)$), where e is the elementary charge, V_m is the transmembrane voltage, k_B is Boltzmann's constant, T is absolute temperature, and C_i and C_o are inside and outside concentrations of the coupling ion, respectively [3,21]. Therefore, the energy released from a single divalent ion moving down the electrochemical gradient is much larger than that of a single monovalent ion.

In the present study, we aimed to demonstrate the energy conversion efficiency of the electrochemical potential of divalent ions on the flagellar motor to understand the mechanism of converting chemical energy into mechanical rotational work by the ion passage. We initially replaced the PG binding domain of $MotA1^{TCA}/MotB1^{TCA}$ with that of *E. coli* MotB ($MotB^{EC}$), and analyzed the function of a series of chimera stator proteins in *E. coli* cells (Figure 1). Next, we measured the rotations of single motors driven by $MotA1^{TCA}/MotB1^{TCA}$ in *B. subtilis* cells (Figure 1). Surprisingly, our data revealed that $MotA1^{TCA}/MotB1^{TCA}$ couples monovalent ions, presumably protons, for its rotation, rather than divalent ions.

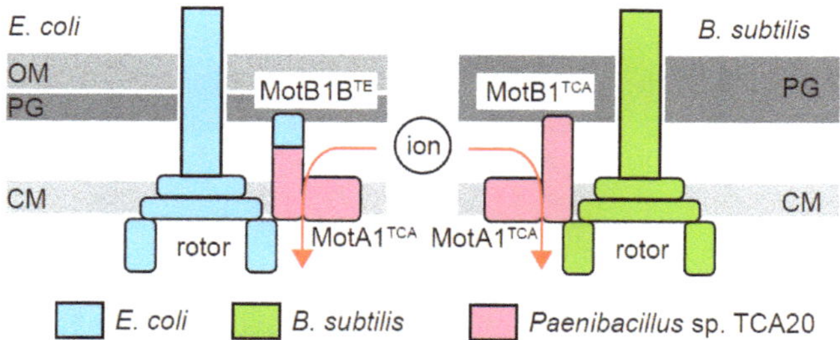

Figure 1. Schematic diagram of chimeric flagellar motors in *Escherichia coli* and *Bacillus subtilis*. In *E. coli*, the chimeric stator proteins, $MotA1^{TCA}/MotB1B^{TE}$, interact with an *E. coli* rotor. In *B. subtilis*, $MotA1^{TCA}/MotB1^{TCA}$ interacts with a *B. subtilis* rotor. The regions derived from *E. coli*, *B. subtilis*, and *Paenibacillus* sp. TCA20 are colored light blue, green, and magenta, respectively. OM: outer membrane, PG: peptidoglycan layer, CM: cytoplasmic membrane.

2. Materials and Methods

2.1. Strains, Plasmids, Growth Conditions, and Media

E. coli strains and plasmids used in this study are listed in Tables S1 and S2, respectively. For motility assay, plasmids were transformed into RP6665 to restore motility. For tethered cell assays, plasmids were transformed into JHC36 [15]. The *E. coli* cells were grown in T-broth (1% Bacto Tryptone, 0.5% sodium chloride) at 30 °C. Ampicillin was added at 50 µg/mL to preserve the plasmids. Inducer arabinose was added to the growth medium at 1 mM.

The *B. subtilis* strains used in this study are listed in Table S1. The *B. subtilis* cells were grown in Spizizen I medium plus 1 mM $MgCl_2$ and 0.6 mM IPTG at 37 °C [20]. Spizizen I medium (pH 8.0) contained 10% Spizizen salts, 0.5% glucose, 0.02% casamino acids, 0.1% yeast extract, 10 µg/mL tryptophan, and10 µg/mL lysine. Spizizen salts contained 85 mM K_2HPO_4, 40 mM KH_2PO_4, 15 mM $(NH_4)_2SO_4$, 6 mM sodium citrate, and 0.8 mM $MgSO_4$.

2.2. Construction of Plasmids Encoding Chimera Stator Proteins

The primers used for plasmid construction are listed in Table S3. To construct a series of chimera stator proteins between $MotB1^{TCA}$ and $MotB^{Ec}$, in vivo *E. coli* cloning (iVEC) was carried out following the recently published method [22]. To clone *motA1*^{TCA}*motB1*^{TCA} into pBAD24 by iVEC method, DNA fragments amplified by PCR using *Paenibacillus* sp. TCA20 genome as a template with primers 1200 and 1201 and using pBAD24 as a template with primers 1198 and 1204 were cotransformed to ME9783 strain, yielding pSHU157. To obtain pSHU161, PCR products from RP437 genome were used as a template with primers 1313 and 1316, and PCR products from pSHU157 were used as a template with primers 1317 and 1318 which were fused by iVEC method. pSHU162–pSHU167 were constructed with the same procedures used for pSHU161 via different primer combinations.

2.3. Strain Construction of B. subtilis

To construct the *B. subtilis* strain expressing $MotA1^{TCA}MotB1^{TCA}$ and sticky flagellar filaments, we followed the procedures reported as the rotation measurement of the *B. subtilis* flagellar motor [23] with minor modifications. Firstly, we removed two BamHI sites in *motA1*^{TCA}*motB1*^{TCA}. Silent mutations were introduced by site-directed mutagenesis by PCR using pSHU157 as a template with primers 1228 and 1229, and primers 1230 and 1231, yielding pSHU1347. Secondly, we added *motA1*^{TCA}*motB1*^{TCA} under an IPTG-inducible P_{grac} promoter. *motA1*^{TCA}*motB1*^{TCA} was amplified by PCR using pSHU1347 as a template with 1222 and 1223. The PCR product was digested with BamHI and SmaI and was then cloned into the BamHI and SmaI site of pHT01, yielding pSHU1348. Thirdly, we subcloned P_{grac}-*motA1*^{TCA}*motB1*^{TCA} into the pDR67 integration vector. P_{grac}-*motA1*^{TCA}*motB1*^{TCA} was amplified by PCR using pSHU1348 as a template with primers 1224 and 1225. The PCR product was digested with SmaI and SphI and then cloned into the SmaI and SphI sites of pDR-hagsticky, yielding pSHU1351. Eventually, P_{grac}-*motA1*^{TCA}*motB1*^{TCA} and P_{hag}-hagsticky were introduced to ΔABΔPSΔHag by selecting a chloramphenicol-resistant and amylase-negative phenotype, yielding SHU399.

2.4. Motility Assays

Motility of *E. coli* cells expressing the chimera stator proteins was tested by the motility plate and swimming assays. Semi-solid agar plates (0.25% Bacto Agar and 1 mM arabinose in T-broth) were inoculated from single colonies and incubated at 30 °C for 9 h. The cells were allowed to swim in 10 mM potassium phosphate buffer (pH 7.0) containing 1 mM EDTA or 1 mM $MgCl_2$. Cell suspensions with an optical density of around 0.8 at 600 nm were collected by centrifugation at 3000× *g* for 3 min, and washed thrice with the observation buffer. The speed of swimming cells was measured by tracking the cells using a custom-made program written in LabVIEW, after capturing their images at 60 fps by CMOS camera for 10 s.

2.5. Rotation Measurement by Tethered Cell Assays

To reduce the number of flagellar filaments extended from the cell body, the sticky flagellar filaments of the cultured cells were sheared by being passed through either a Pasteur pipette or a syringe with a 26G needle. Cells were harvested by centrifugation at 3000× *g* for 3 min and the pellet was resuspended in either phosphate buffer or HEPES-Tris buffer. This process was repeated thrice. Cells were incubated in a sample chamber for 15 min to attach on the coverslips via their sticky filament. After flushing the unbound cells with the same buffer, spinning cells, which attach on the coverslips via single filaments, were captured at 60 fps by CMOS camera for 10 s [24]. To observe the effect of the electrochemical potential of Mg^{2+} on the motor speed, the buffer in a sample chamber was exchanged by running the buffer containing appropriate concentrations of EDTA, $MgCl_2$, or CCCP (carbonyl cyanide m-chlorophenyl hydrazone). The rotational speed was determined by trajectory analysis of the cell body using a custom-made program written in LabVIEW.

3. Results

3.1. MotA1^{TCA} and Chimeras between MotB1^{TCA} and MotB^{Ec} Function as A Stator in E. coli

Most quantitative functional assays of the flagellar motor have been established in *E. coli* cells; however, MotA1^{TCA}/MotB1^{TCA} was reported to be nonfunctional as a stator in *E. coli* cells [20]. Therefore, we constructed a series of chimeras between MotB1^{TCA} and MotB^{Ec}, following the design of PomA/PotB in previous work (Figure 2a) [11]. MotB protein has a single TM domain with an ion-binding site and a large PG binding domain. All constructed chimeric proteins contain TM domain derived from MotB1^{TCA} and PG binding domain derived from MotB^{Ec}. The sites for swapping were chosen by comparing the sequence similarities between MotB1^{TCA} and MotB^{Ec}, and a series of chimeras were named as MotB1B^{TE1}–MotB1B^{TE7}.

To examine whether the constructed chimeric Mot complexes were able to rotate the flagellar rotor of *E. coli*, we transformed a series of plasmids encoding MotA1^{TCA} and MotB1B^{TE} to *E. coli* Δ*motAB* cells and checked their motility on semi-solid agar plates. We found that chimeras except MotA1^{TCA}/MotB1B^{TE6} and MotA1^{TCA}/MotB1B^{TE7} can restore the motility of *E. coli* Δ*motAB* cells (Figure 2b).

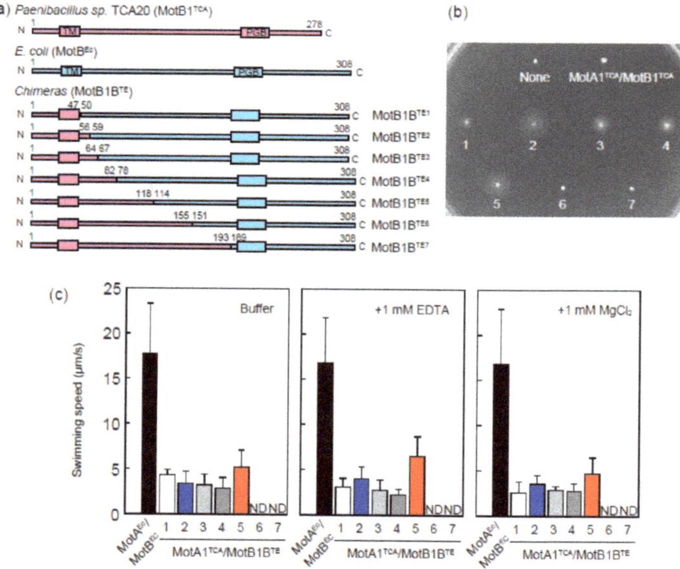

Figure 2. Characterization of chimeric stator in *E. coli*. (**a**) Design of a series of domain swap chimeras MotB1B^{TE}. The region derived from MotB1^{TCA} and MotB^{Ec} are colored light magenta and blue, respectively. TM and PGB represent the transmembrane and peptidoglycan binding domains, respectively. Numbers above each diagram indicate the amino acid sequence numbers. Depending on the swapping sites, chimera proteins MotB1B^{TE} are numbered from 1–7. (**b**) Motility of *E. coli* cells expressing chimera stator proteins on semi-solid agar plate. The numbers correspond to chimeras in (**a**). (**c**) Swimming speed of *E. coli* cells expressing chimera stator proteins in phosphate buffer (left) containing 1 mM EDTA (center) and 1 mM $MgCl_2$ (right). ND represents that no swimming cells were detected.

We measured the swimming speed of the *E. coli* Δ*motAB* cells expressing chimera stator proteins in phosphate buffer (Figure 2c, left panel). The swimming speed of *E. coli* Δ*motAB* cells expressing chimeras MotA1^{TCA}/MotB1B^{TE1}–MotB1B^{TE5} was ~4 µm/s, which is considerably slower than the cells expressing wild type MotA^{Ec}/MotB^{Ec}. A similar analysis was performed in the presence of 1 mM $MgCl_2$ (Figure 2c, right panel). An appropriate concentration was used for the swimming motility of *Paenibacillus* cells; however, no significant effect of Mg^{2+} was observed on the motility of *E. coli* Δ*motAB* cells expressing chimeras MotA1^{TCA}/MotB1B^{TE1}–MotB1B^{TE5}. Moreover, the motility was maintained despite Mg^{2+}-chelation by EDTA (Figure 2c, center panel). Therefore, presumably Mg^{2+} is not used as a coupling ion for motor rotation.

3.2. Chimeric Motor in E. coli Is Driven by Protons Rather than Divalent Ions

We further investigated the function of single motors driven by MotA1^{TCA}/MotB1B^{TE2} and MotA1^{TCA}/MotB1B^{TE5}, which efficiently support cell motility on semi-solid agar. We transformed the plasmids encoding MotA^{Ec}/MotB^{Ec}, MotA1^{TCA}/MotB1B^{TE2}, and MotA1^{TCA}/MotB1B^{TE5} into *E. coli* Δ*motAB* expressing sticky flagellar filament, namely EC, TE2, and TE5. Cells were tethered to a glass surface spontaneously via a single sticky filament, and their rotation was captured and analyzed. Figure 3a illustrates the dependence of Mg^{2+} concentrations on the motor speed of EC, TE2, and TE5, wherein the motor speeds of TE2 and TE5 were similar to that of EC at up to 5 mM $MgCl_2$, and were independent of the Mg^{2+} concentration. Figure 3b depicts the dependence of pH on the motor speed of EC, TE2, and TE5, wherein the motor speed of EC was constant in the range of pH 6–9, as previously reported [25]. The motors of TE2 and TE5 also rotated at an approximately constant speed despite the pH conditions. Since the buffer contains 10 mM EDTA-2K to chelate the divalent ions completely, the potential cations driving these motors are limited to be protons or potassium ions.

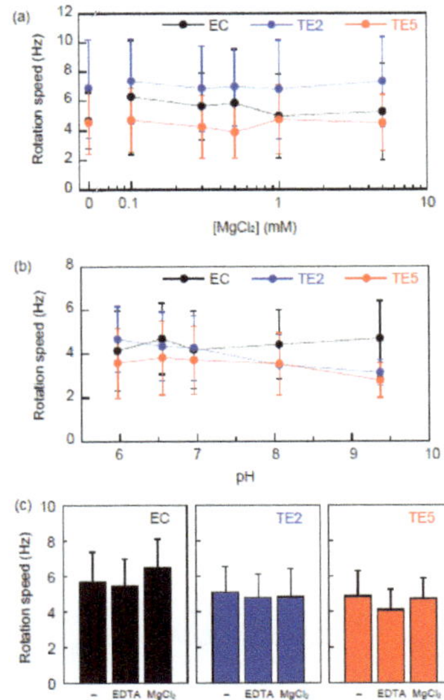

Figure 3. Rotational speed of single motors driven by chimera stators in *E. coli*. (**a**) Effect of Mg^{2+} concentrations on the motor speed. Buffer contains 10 mM potassium phosphate (pH 7.0) and indicated $MgCl_2$. In the absence of $MgCl_2$, 0.1 mM EDTA was added to the buffer. (**b**) Effect of pH on the motor speed. 10 mM potassium phosphate buffer containing 10 mM EDTA was adjusted to the desired pH by titration with HCl or KOH. (**c**) Motor speeds of EC (left), TE2 (center), and TE5 (right) in HEPES-Tris buffer. EDTA and $MgCl_2$ indicate the buffers containing 10 mM EDTA and 10 mM $MgCl_2$, respectively.

To test whether the motors of TE2 and TE5 require the potassium ions for rotation, we measured the motor rotation in 100 mM HEPES-Tris buffer (pH 7.0). The motors of EC, TE2, and TE5 rotated at a similar speed to that of the potassium phosphate buffer, even in the absence of potassium ion (Figure 3c). Addition of 10 mM EDTA or 10 mM $MgCl_2$ had no effect on these motor speeds. These results suggested that protons are very likely used as an energy source for motor rotation of TE2 and TE5 [21,26].

3.3. $MotA1^{TCA}$/$MotB1^{TCA}$ Couples Monovalent Cations in ΔmotABΔmotPS B. subtilis

$MotA1^{TCA}$/$MotB1^{TCA}$ was reported to interact with a rotor of *B. subtilis* and support the cell motility of *B. subtilis ΔmotABΔmotPS* cells. We found that AB1 cells (*ΔmotABΔmotPS* cells expressing $MotA1^{TCA}$/$MotB1^{TCA}$) swam very slowly (less than 2 μm/s) while wobbling, even in the buffer containing 10 mM $MgCl_2$, meaning that swimming analysis of AB1 cells would not be suitable for further detailed analysis. Therefore, we investigated the function of $MotA1^{TCA}$/$MotB1^{TCA}$ by a single motor assay in *B. subtilis* using the system of Hag-sticky filament, which was recently developed [23]. We initially constructed the strain expressing $MotA1^{TCA}$/$MotB1^{TCA}$ and Hag-sticky by following the procedures as previously reported [23], and named it as AB1-sticky. Moreover, we measured the rotational speed of WT-sticky, which expresses a wild-type flagellar motor and Hag-sticky as a control. The rotation speed of tethered AB1-sticky cells was about half that of WT-sticky cells in 10 mM potassium phosphate buffer (pH 8.0) containing 0.1 mM EDTA (Figure 4a). As in *E. coli* cells, the addition of 10 mM EDTA or 10 mM $MgCl_2$ had no effect on these motor speeds. Collapsing protonmotive force by the addition of 25 μM CCCP completely hindered the motor rotation of both WT-sticky and AB1-sticky. Furthermore, the rotational speeds of WT-sticky and AB1-sticky were constant within error in the range of pH 6–9 as well as in *E. coli* cells (Figure 4b). Therefore, we concluded that $MotA1^{TCA}$/$MotB1^{TCA}$ couples monovalent ions for its rotation.

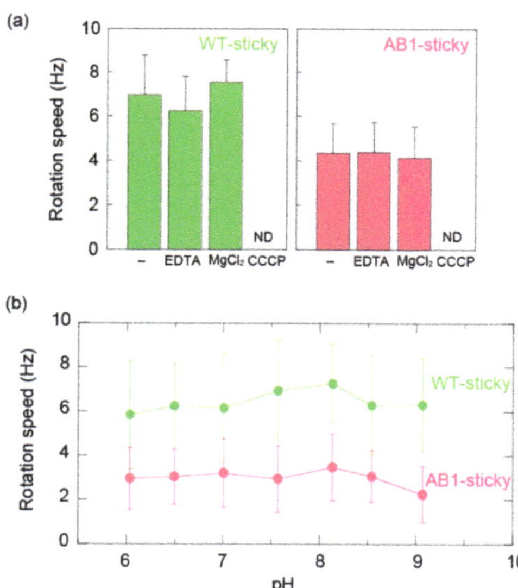

Figure 4. Rotational speed of single motors driven by $MotA1^{TCA}$/$MotB1^{TCA}$ in *B. subtilis*. (**a**) Motor speeds of WT-sticky (left) and AB1-sticky (right) in 10 mM potassium phosphate buffer containing 0.1 mM EDTA. EDTA, $MgCl_2$, and CCCP indicate buffers containing 10 mM EDTA, 10 mM $MgCl_2$, and 10 mM $MgCl_2$ and 25 μM CCCP, respectively. ND represents that no spinning tethered cells were detected. (**b**) Effect of pH on the motor speed. 10 mM potassium phosphate buffer containing 10 mM EDTA was adjusted to the desired pH by titration with HCl or KOH.

4. Discussion

Imazawa et al. reported that the flagellar motor of *Paenibacillus* sp. TCA20 is driven by the flux of divalent ions and MotA1^{TCA}/MotB1^{TCA}, which acts as a stator in the motor, couples the divalent ion flow to motor rotation [20]. This stator complex was also reported to interact with the rotors of *B. subtilis* and to use divalent ions for torque generation without requiring protonmotive force. In the present study, to investigate of energy conversion of MotA1^{TCA}/MotB1^{TCA}, we performed single motor assays in the *E. coli* and *B. subtilis* flagellar system, respectively. A stator complex of MotA1^{TCA} and chimera MotB1B$^{TE1–5}$ generated torque by interacting with a rotor of the *E. coli* flagellar motor. The chimeras with high MotB1^{TCA} occupancy, such as MotB1B^{TE6} and MotB1B^{TE7}, did not work in *E. coli*, presumably by not forming a structure to anchor to the PG layer of *E. coli* cells [19]. Our data indicated that the chimera stator complexes use monovalent ions, presumably protons, rather than divalent cations for torque generation, although they possess TM domain of MotA1^{TCA}/MotB1^{TCA}, which is a core of energy converter (Figure 3a,b). MotA1^{TCA}/MotB1^{TCA} functions as a stator in the stator-less *B. subtilis* cells, but does not require divalent cations for torque generation, which is not concordant to the previous report (Figure 4a). Some bacteria, such as *Bacillus clausii*, were reported to switch the ion specificity of their motor depending on the environmental pH [27]. Therefore, we checked the possibility of the ion specificity switching of chimeras in *E. coli* and MotA1^{TCA}/MotB1^{TCA} in *B. subtilis*; however, all stator complexes function in the absence of divalent cations in the range of pH 6–9. Collectively, the results from this study revealed that MotA1^{TCA}/MotB1^{TCA} does not directly couple the divalent ion to motor rotation, thus challenging the results of the previous report.

Although the reason for discrepancy between the results of this study and the previous reports remains unclear, one possible hypothesis that can be considered is the difference of measuring the flagellar rotation. Bacterial swimming is caused by the integration of complex processes, such as the flagellar motor rotation, motor switching related to chemotaxis, and bundle formation of flagella. [28,29]. If the divalent cations affect the chemotactic signaling and increase the switching frequency of the motor without changing the swimming speed, then the swimming speed may reduce or swimming may stop, leading to pseudodependence of divalent cations on motor function. Especially when analyzing slow swimming motions, the effects of fluctuations due to fluid flow and Brownian motion of cells must be carefully considered. In the present study, we measured the rotational speeds of single motors which is assumed to be the most reliable and robust assessment for the motor function itself. Alternatively, divalent cations might have a positive effect on the folding of stator complexes. As recently discovered in MotP/MotS of *B. subtilis*, binding of Na^+ to MotS induces the structural transition of its PG binding domain for anchoring to the PG layer [30]. If divalent cations activate the folding of the PG binding domain of MotB1^{TCA} to interact with a rotor, the fraction of stator complexes generating torque would increase its dependence on divalent cation concentrations. Here, MotA1^{TCA}/MotB1^{TCA} was overexpressed using a P_{lac} promotor without LacI; therefore, some portion of stator complexes might fold stochastically without requiring divalent cations and assemble in a motor to generate torque. This possibility would be tested by analyzing the effect of the expression level of MotA1^{TCA}/MotB1^{TCA} on the motor speed in the absence of divalent ions. Nevertheless, future studies are warranted.

Our experiments revealed that MotA1^{TCA}/MotB1^{TCA} in *B. subtilis* cells couples monovalent ions, presumably protons, to its rotation. To the best of our knowledge, the coupling ion specificity of the intact stator complex does not change depending on the expressed species; therefore, we speculate that the flagellar motor of *Paenibacillus* sp. TCA20 directly uses the proton flux for its energy source [11]. Nevertheless, the possibility that MotA1^{TCA}/MotB1^{TCA} uses divalent cations in extreme conditions, particularly the environment where *Paenibacillus* sp. TCA20 was isolated, cannot be ruled out completely. Considering this case, single motor assay would be suitable for revealing the motor properties.

Supplementary Materials: The following are available online at http://www.mdpi.com/2218-273X/10/7/1078/s1, Table S1: Strains used in this study, Table S2: Plasmids used in this study, Table S3: Primers used in this study.

Author Contributions: Conceptualization, N.T. and Y.S.; methodology, S.O., M.Y., N.T. and Y.S.; software, Y.S.; motility analysis, S.O., M.Y., N.T. and Y.S.; writing—original draft preparation, Y.S.; writing—review and editing, S.O., M.Y., N.T. and Y.S.; supervision, Y.S. All authors have read and agreed to the published version of the manuscript.

Funding: This work has been supported in part by MEXT/JSPS KAKENHI Grant Number JP15H01332, JP15K07034, JP18H02475, JP19H05404 and JP20K06564 to Y.S., Itoh Science Foundation to Y.S., and Takeda Science Foundation to Y.S.

Acknowledgments: We thank NBRP (NIG, Japan): *E. coli* for the gift of *E. coli* ME9783 strain. The authors thank Masahiro Ito, Yuka Takahashi, and Shun Naganawa for a gift of strains and their technical advices. The authors also thank Ikuro Kawagishi and Masatoshi Nishikawa for their fruitful discussions.

Conflicts of Interest: The authors declare no conflicts of interest.

References

1. Berg, H.C. The rotary motor of bacterial flagella. *Annu. Rev. Biochem.* **2003**, *72*, 19–54. [CrossRef] [PubMed]
2. Nakamura, S.; Minamino, T. Flagella-Driven Motility of Bacteria. *Biomolecules* **2019**, *9*, 279. [CrossRef] [PubMed]
3. Sowa, Y.; Berry, R.M. Bacterial flagellar motor. *Q. Rev. Biophys.* **2008**, *41*, 103–132. [CrossRef]
4. Yorimitsu, T.; Homma, M. Na^+-driven flagellar motor of *Vibrio*. *Biochim. Biophys. Acta* **2001**, *1505*, 82–93. [CrossRef]
5. Ito, M.; Hicks, D.B.; Henkin, T.M.; Guffanti, A.A.; Powers, B.D.; Zvi, L.; Uematsu, K.; Krulwich, T.A. MotPS is the stator-force generator for motility of alkaliphilic *Bacillus*, and its homologue is a second functional Mot in *Bacillus subtilis*. *Mol. Microbiol.* **2004**, *53*, 1035–1049. [CrossRef] [PubMed]
6. Terahara, N.; Sano, M.; Ito, M. A bacillus flagellar motor that can use both Na^+ and K^+ as a coupling ion is converted by a single mutation to use only Na^+. *PLoS ONE* **2012**, *7*, e46248. [CrossRef]
7. Thormann, K.M.; Paulick, A. Tuning the flagellar motor. *Microbiology* **2010**, *156*, 1275–1283. [CrossRef]
8. Minamino, T.; Terahara, N.; Kojima, S.; Namba, K. Autonomous control mechanism of stator assembly in the bacterial flagellar motor in response to changes in the environment. *Mol. Microbiol.* **2018**, *109*, 723–734. [CrossRef]
9. Asai, Y.; Kawagishi, I.; Sockett, R.E.; Homma, M. Hybrid motor with H^+-and Na^+-driven components can rotate *Vibrio* polar flagella by using sodium ions. *J. Bacteriol.* **1999**, *181*, 6332–6338. [CrossRef]
10. Asai, Y.; Kawagishi, I.; Sockett, R.E.; Homma, M. Coupling ion specificity of chimeras between H^+-and Na^+-driven motor proteins, MotB and PomB, in *Vibrio* polar flagella. *EMBO J.* **2000**, *19*, 3639–3648. [CrossRef]
11. Asai, Y.; Yakushi, T.; Kawagishi, I.; Homma, M. Ion-coupling determinants of Na^+-driven and H^+-driven flagellar motors. *J. Mol. Biol.* **2003**, *327*, 453–463. [CrossRef]
12. Takekawa, N.; Nishiyama, M.; Kaneseki, T.; Kanai, T.; Atomi, H.; Kojima, S.; Homma, M. Sodium-driven energy conversion for flagellar rotation of the earliest divergent hyperthermophilic bacterium. *Sci. Rep.* **2015**, *5*, 12711. [CrossRef]
13. Gosink, K.K.; Hase, C.C. Requirements for conversion of the Na^+-driven flagellar motor of *Vibrio cholerae* to the H^+-driven motor of *Escherichia coli*. *J. Bacteriol.* **2000**, *182*, 4234–4240. [CrossRef] [PubMed]
14. Sowa, Y.; Rowe, A.D.; Leake, M.C.; Yakushi, T.; Homma, M.; Ishijima, A.; Berry, R.M. Direct observation of steps in rotation of the bacterial flagellar motor. *Nature* **2005**, *437*, 916–919. [CrossRef] [PubMed]
15. Inoue, Y.; Lo, C.J.; Fukuoka, H.; Takahashi, H.; Sowa, Y.; Pilizota, T.; Wadhams, G.H.; Homma, M.; Berry, R.M.; Ishijima, A. Torque-speed relationships of Na^+-driven chimeric flagellar motors in *Escherichia coli*. *J Mol. Biol.* **2008**, *376*, 1251–1259. [CrossRef]
16. Lo, C.J.; Sowa, Y.; Pilizota, T.; Berry, R.M. Mechanism and kinetics of a sodium-driven bacterial flagellar motor. *Proc. Natl. Acad. Sci. USA* **2013**, *110*, E2544–E2551. [CrossRef]
17. Sowa, Y.; Homma, M.; Ishijima, A.; Berry, R.M. Hybrid-fuel bacterial flagellar motors in *Escherichia coli*. *Proc. Natl. Acad. Sci. USA* **2014**, *111*, 3436–3441. [CrossRef]
18. Ishida, T.; Ito, R.; Clark, J.; Matzke, N.J.; Sowa, Y.; Baker, M.A.B. Sodium-powered stators of the bacterial flagellar motor can generate torque in the presence of phenamil with mutations near the peptidoglycan-binding region. *Mol. Microbiol.* **2019**, *111*, 1689–1699. [CrossRef]

19. Kojima, S.; Imada, K.; Sakuma, M.; Sudo, Y.; Kojima, C.; Minamino, T.; Homma, M.; Namba, K. Stator assembly and activation mechanism of the flagellar motor by the periplasmic region of MotB. *Mol. Microbiol.* **2009**, *73*, 710–718. [CrossRef] [PubMed]
20. Imazawa, R.; Takahashi, Y.; Aoki, W.; Sano, M.; Ito, M. A novel type bacterial flagellar motor that can use divalent cations as a coupling ion. *Sci. Rep.* **2016**, *6*, 19773. [CrossRef]
21. Lo, C.J.; Leake, M.C.; Pilizota, T.; Berry, R.M. Nonequivalence of membrane voltage and ion-gradient as driving forces for the bacterial flagellar motor at low load. *Biophys. J.* **2007**, *93*, 294–302. [CrossRef]
22. Nozaki, S.; Niki, H. Exonuclease III (XthA) Enforces In Vivo DNA Cloning of *Escherichia coli* To Create Cohesive Ends. *J. Bacteriol.* **2019**, *201*, e00660-18. [CrossRef] [PubMed]
23. Terahara, N.; Noguchi, Y.; Nakamura, S.; Kami-Ike, N.; Ito, M.; Namba, K.; Minamino, T. Load-and polysaccharide-dependent activation of the Na^+-type MotPS stator in the *Bacillus subtilis* flagellar motor. *Sci. Rep.* **2017**, *7*, 46081. [CrossRef] [PubMed]
24. Kasai, T.; Sowa, Y. Measurements of the Rotation of the Flagellar Motor by Bead Assay. *Methods Mol. Biol.* **2017**, *1593*, 185–192.
25. Chen, X.; Berg, H.C. Solvent-isotope and pH effects on flagellar rotation in *Escherichia coli*. *Biophys. J.* **2000**, *78*, 2280–2284. [CrossRef]
26. Gabel, C.V.; Berg, H.C. The speed of the flagellar rotary motor of *Escherichia coli* varies linearly with protonmotive force. *Proc. Natl. Acad. Sci. USA* **2003**, *100*, 8748–8751. [CrossRef]
27. Terahara, N.; Krulwich, T.A.; Ito, M. Mutations alter the sodium versus proton use of a *Bacillus clausii* flagellar motor and confer dual ion use on *Bacillus subtilis* motors. *Proc. Natl. Acad. Sci. USA* **2008**, *105*, 14359–14364. [CrossRef]
28. Turner, L.; Ryu, W.S.; Berg, H.C. Real-time imaging of fluorescent flagellar filaments. *J. Bacteriol.* **2000**, *182*, 2793–2801. [CrossRef]
29. Wadhams, G.H.; Armitage, J.P. Making sense of it all: Bacterial chemotaxis. *Nat. Rev. Mol. Cell Biol.* **2004**, *5*, 1024–1037. [CrossRef]
30. Terahara, N.; Kodera, N.; Uchihashi, T.; Ando, T.; Namba, K.; Minamino, T. Na^+-induced structural transition of MotPS for stator assembly of the *Bacillus* flagellar motor. *Sci. Adv.* **2017**, *3*, eaao4119. [CrossRef]

MDPI

Article

Fluctuations in Intracellular CheY-P Concentration Coordinate Reversals of Flagellar Motors in *E. coli*

Yong-Suk Che [1], Takashi Sagawa [2,†], Yuichi Inoue [2,†], Hiroto Takahashi [2], Tatsuki Hamamoto [1,‡], Akihiko Ishijima [1] and Hajime Fukuoka [1,*]

1 Graduate School of Frontier Biosciences, Osaka University, 1-3 Yamadaoka, Suita, Osaka 565-0871, Japan; cheyong@fbs.osaka-u.ac.jp (Y.-S.C.); tatsuki.hamamoto@oist.jp (T.H.); ishijima@fbs.osaka-u.ac.jp (A.I.)

2 Institute of Multidisciplinary Research for Advanced Materials, Tohoku University, Aoba-ku, Sendai 980-8577, Japan; t.sagawa@sigma-koki.com (T.S.); y.inoue@sigma-koki.com (Y.I.); hiroto.takahashi.b7@tohoku.ac.jp (H.T.)

* Correspondence: f-hajime@fbs.osaka-u.ac.jp; Tel.: +81-6-6879-4429

† Present address: Sigmakoki Co., Ltd., Tokyo Head Office, 1-19-9 Midori, Sumida-ku, Tokyo 130-0021, Japan.

‡ Present address: Okinawa Institute of Science and Technology Graduate University, 1919-1 Tancha, Onna-son, Kunigami-gun, Okinawa 904-0495, Japan.

Received: 15 October 2020; Accepted: 9 November 2020; Published: 12 November 2020

Abstract: Signal transduction utilizing membrane-spanning receptors and cytoplasmic regulator proteins is a fundamental process for all living organisms, but quantitative studies of the behavior of signaling proteins, such as their diffusion within a cell, are limited. In this study, we show that fluctuations in the concentration of the signaling molecule, phosphorylated CheY, constitute the basis of chemotaxis signaling. To analyze the propagation of the CheY-P signal quantitatively, we measured the coordination of directional switching between flagellar motors on the same cell. We analyzed the time lags of the switching of two motors in both CCW-to-CW and CW-to-CCW switching ($\Delta\tau_{CCW\text{-}CW}$ and $\Delta\tau_{CW\text{-}CCW}$). In wild-type cells, both time lags increased as a function of the relative distance of two motors from the polar receptor array. The apparent diffusion coefficient estimated for $\Delta\tau$ values was ~9 μm^2/s. The distance-dependency of $\Delta\tau_{CW\text{-}CCW}$ disappeared upon loss of polar localization of the CheY-P phosphatase, CheZ. The distance-dependency of the response time for an instantaneously applied serine attractant signal also disappeared with the loss of polar localization of CheZ. These results were modeled by calculating the diffusion of CheY and CheY-P in cells in which phosphorylation and dephosphorylation occur in different subcellular regions. We conclude that diffusion of signaling molecules and their production and destruction through spontaneous activity of the receptor array generates fluctuations in CheY-P concentration over timescales of several hundred milliseconds. Signal fluctuation coordinates rotation among flagella and regulates steady-state run-and-tumble swimming of cells to facilitate efficient responses to environmental chemical signals.

Keywords: chemotaxis; signal transduction; diffusion; response regulator; CheY; flagellar motor; *E. coli*

---

1. Introduction

Diffusion of signaling proteins within a cell and their interaction with the molecules responsible for the input and output of the signal play an essential role in all biological signal transduction systems [1–3]. In both prokaryotes and eukaryotes, cells use transmembrane receptors to sense environmental stimuli and to generate intracellular signals in the form of messenger molecules. These intracellular messenger molecules diffuse through the cytoplasm to their targets, where they regulate cellular functions such as gene expression and locomotion. In an extracellular example, neurotransmitters released from the presynaptic membrane diffuse across the synaptic cleft to receptors embedded in the postsynaptic membrane to initiate action potentials. Therefore, diffusion of signaling molecules plays an important role in both intracellular and extracellular sensory transduction. However, quantitative studies of behavior in signaling systems including diffusion of signaling molecules are limited because the input-output relationship in signaling pathways is difficult to measure directly with high temporal and spatial resolution. Precise knowledge of the in vivo kinetics and localization of the reactions that generate and degrade signals, and of the diffusion parameters for signaling proteins, are essential to characterize sensory input-output pathways fully.

Chemotaxis enables *Escherichia coli* cells swimming in a liquid environment to track and navigate chemical gradients with high precision [4,5]. *E. coli* uses chemoreceptor proteins, located primarily at the cell pole(s), to detect specific chemicals and to monitor pH and temperature. The receptors employ a His-Asp two-component phosphorelay to transmit sensory messages to the flagellar motors [4]. *E. coli* cells swim by rotating their left-handed helical flagellar filaments: counter-clockwise (CCW) rotation produces forward swimming; clockwise (CW) rotation triggers random turning movements, called tumbles. The chemoreceptors assemble in large signaling arrays connected by two cytoplasmic proteins: CheA, a histidine autokinase, and CheW, a scaffolding protein that couples CheA activity to receptor control. Receptor arrays modulate the autophosphorylation activity of CheA to control the flux of phosphoryl groups from CheA to the response regulator CheY. Phospho-CheY (CheY-P) is the intracellular messenger that binds to a flagellar motor to induce CW rotation [6–10]. CheY-P molecules appear to reach the flagellar motors through intracellular diffusion [1]. CheZ, a dedicated CheY-P phosphatase, degrades the CW signal, but it is located mainly at the polar receptor arrays through interaction with a variant form of CheA [11,12]. Thus, CheY-P is both generated and degraded at receptor arrays, but CheY-P molecules that escape from CheZ in the array encounter additional, although less concentrated, CheZ molecules as they diffuse through the cytoplasm.

Previously, we demonstrated that under steady-state conditions with no external stimuli, two flagellar motors on the same *E. coli* cell coordinately switch their rotational direction [13]. The switching time lag ($\Delta\tau_{correlation}$) of two coordinated motors depended on their relative distance from the receptor array at the cell pole. In addition, we showed that binding and dissociation of CheY-P molecules at the motors triggered CW and CCW rotation episodes, respectively [6]. To explain these results, we proposed that increases and decreases in CheY-P concentration directly and coordinately regulate the rotational direction of motors under steady-state conditions. The $\Delta\tau_{correlation}$ represents the delay in the propagation of CheY-P concentration changes to motors that are positioned at different distances from the receptor array.

In the present study, to clarify the mechanism of the intracellular signaling of *E. coli* under steady-state, we asked how motor coordination changes when CheZ is uniformly distributed throughout the cell rather than localized at the polar receptor array. Our working hypothesis predicts that the propagation of increases and decreases in the CheY-P concentration depends on the distance from the polar receptor array, owing to the diffusion of CheY and CheY-P in the cytoplasm, and the rates of phosphorylation and dephosphorylation of CheY by polarly localized CheA and CheZ. To test this model, we first demonstrated that, in a wild-type cell, the $\Delta\tau$ for coordinated switching between two motors (either CCW-to-CW or CW-to-CCW) depends on the relative distance between the two motors and their distance from a polar receptor array. In contrast, there was no distance-dependency for $\Delta\tau$ of CW-to-CCW switching when CheZ was distributed uniformly throughout the cytoplasm. Moreover, using photorelease of serine from a caged compound, we found that, when CheZ is uniformly distributed, there was no distance-dependency in the response to an instantaneously applied attractant signal. These results are consistent with the predictions of a simulation based on the localization of the CheZ phosphatase and the diffusion of CheY and CheY-P molecules in the cytoplasm. We conclude that the increases and decreases in CheY-P concentration are generated by spontaneous bursts of receptor array activity and that the rotational direction of multiple flagellar motors is coordinated under steady-state conditions. This signal fluctuation depends on the diffusion coefficient of the signaling molecule and on the signal-generating and signal-destroying reactions in the polar receptor array. We propose that the steady-state fluctuations in CheY-P concentration thus serve to regulate run-and-tumble swimming of the cell through coordinated switching of the flagellar motors. Thus, an *E. coli* cell would maintain the chemotaxis system on high alert to respond efficiently to environmental chemical signals rather than conserve energy by keeping the system in a resting mode.

2. Materials and Methods

2.1. *E. coli Strains, Plasmids, and Cell Growth*

The strains and plasmids are listed in Table S1. All strains were derived from the K12 strain RP437, which is wild-type for chemotaxis [14]. The replacement of the wild-type cheA gene with cheA(M98L) to make $CheA_S^-$ cell [15], the replacement of the wild-type fliC gene with the fliC-sticky gene [16], and the replacement of the wild-type cheZ gene with cheZ(F98S) [11] were carried out using the λ red recombinase and tetracycline sensitivity selection method [17,18]. LB broth (1% bactotryptone (BD, Sparks, MD, USA), 0.5% yeast extract (BD, Sparks, MD, USA), 0.5% NaCl (Wako, Osaka, Japan)) was used for culture growth, transformations, and plasmid isolation. Tryptone broth (TB) (1% bactotryptone, 0.5% NaCl) was used to grow cells for measurements of motor rotation. Growth conditions were described in Supplemental Methods. For all measurement, the cells were suspended in 10NaMB (10 mM potassium phosphate buffer (Wako, Osaka, Japan), pH 7.0; 0.1 mM EDTA-2K (Wako, Osaka, Japan), pH 7.0; 10 mM NaCl, 75 mM KCl (Wako, Osaka, Japan)).

2.2. *Measurement of Rotation of Multiple Flagellar Motors*

Cells were prepared and measured by a method similar to that described in our previous report [13] (see also Supplementary Methods). For measurement of rotation of the motor, a polystyrene bead, diameter (φ) 0.5 μm (Polysciences; Warrington, PA, USA), was attached to the sticky flagellar filaments [16]. The sticky filaments made by this mutant FliC readily adsorbed polystyrene beads without modification of beads. The phase-contrast images of beads through the objective lens (UPlanFl 40× NA 0.75 Ph2; Olympus, Tokyo, Japan) were recorded with a high-speed charge-coupled device (CCD) camera (IPX-VGA210LMCN; Imperx, Boca Raton, FL, USA) at 1250 or 1255 frames/s. This high-speed CCD camera was controlled by the measurement software Real Time Video Nanometry (RTVN), which we developed using LabVIEW 2009 (National Instruments, Austin, TX, USA) [19]. Images of bead were fitted by a two-dimensional Gaussian function for every sampling frame, and the position of a bead was expressed as the peak position of X and Y coordinates of a fitted Gaussian curve. The bead position was approximated by an ellipse function every 500 frames, and the bead position was corrected to approximate a perfect circle centered on the origin. The rotation angle was calculated for every two sampling frames, and a time trace of rotational velocity and rotational direction were estimated by repeating this process every video frame.

To observe the location of GFP-CheW, a blue laser beam (wavelength, 488 nm) (Sapphire 488-20-SV; Coherent, Hercules, CA, USA) was focused on the back focal plane of the objective lens. After recording the bead rotation with the high-speed CCD camera, the fluorescence image of GFP-CheW and the phase-contrast images of the bead and cell were recorded at 120 frames/s with a second CCD camera (DMK23G618; The Imaging Source, Bremen, Germany). The distance between the fluorescent focus at the cell pole derived from the GFP-CheW fluorescence and the rotational center of the rotating bead was measured and defined as the distance from the receptor array to motor.

2.3. *Measurement of the Response Time to the Instantaneously Applied Photoreleased Serine*

Cell preparation, the microscopic system, and the conditions for photoreleasing serine from caged serine were the same as in our previous study [19] (Figure S3; see also Supplementary Methods). Caged serine surrounding an *E. coli* cell in a microscopic field was photolyzed by irradiation with a violet laser beam (wavelength, 405 nm) (KBL-90C-A; Kimmon Koha, Tokyo, Japan, or OBIS 405-50 LX; Coherent,

Hercules, CA) for 80 ms. The violet laser beam was uniformly applied to the irradiated area (diameter, 32 μm), and the energy density of the laser beam in the irradiated area was 370 $mW \cdot mm^{-2}$. The rate constant for the photolysis of caged serine in the irradiated area was 0.16 s^{-1}. When RTVN detected the switching of the flagellar motor from the CCW to CW, it opens a mechanical shutter (UHS1 ZM 2; Uniblitz, San Diego, CA, USA) positioned in front of the laser beam for 80 ms. The distance between the polar receptor array and the flagellar motor was quantified by the polar localization of GFP-CheW and the rotational center of the rotating bead as described above.

2.4. Correlation Analysis

To analyze the correlation in the switching between flagellar motors, the rotational velocity was classified into three states by the following procedure. The time-trace of the rotational velocity was filtered by the Chug-Kennedy filtering algorithm (C-K filter) [20], using an analytical window of 100 data-points and a weight of 10. Rotational velocities of more than +20 Hz, between +20 Hz and −20 Hz, and less than −20 Hz were assigned as CCW rotation (+1), pause (0), and CW rotation (−1), respectively. The correlation analysis was performed by applying Equation (1) to the time traces of the rotational directions between two flagellar motors:

$$Z(\tau) = \frac{\frac{1}{N}\sum_{t=1}^{N}\left[x(t)\cdot y(t+\tau) - \overline{x(t)}\cdot\overline{y(t)}\right]}{\sqrt{\frac{1}{N}\sum_{t=1}^{N}\left[x(t)-\overline{x(t)}\right]^2}\cdot\sqrt{\frac{1}{N}\sum_{t=1}^{N}\left[y(t)-\overline{y(t)}\right]^2}} \quad (1)$$

where Z is the function used for the correlation analysis, t is time, τ is the time difference, N is the total number of sampling points, and $x(t)$ and $y(t)$ are the time traces of the rotational directions of two motors, respectively. In cells producing GFP-CheW, this function was applied to the motor closer to the fluorescent focus at one of the cell poles. We analyzed motors on the cells less than 3 μm long. All correlations Z (τ) were calculated ($-1 \leq Z \leq 1$) by Equation (1) using the traces for 1 or 2 min.

2.5. Estimation of $\Delta\tau_{correlation}$, $\Delta\tau_{CCW\text{-}CW}$, and $\Delta\tau_{CW\text{-}CCW}$

To analyze the time difference of switching between two motors from correlation analysis, the near 0-sec peak determined by correlation analysis was fitted by a Gaussian function, and the peak time of the fitted curve was defined as $\Delta\tau_{correlation}$. $\Delta\tau_{CCW\text{-}CW}$ and $\Delta\tau_{CW\text{-}CCW}$ were analyzed as follows. In cells for which coordinated switching between two motors was detected, all the coordinated switching events were extracted from a trace, and the time difference of switching ($\Delta\tau$) between 2 different motors was calculated for every extracted switching event. The time difference was calculated for both CCW-to-CW switching ($\Delta\tau_{CCW\text{-}CW}$) and CW-to-CCW switching ($\Delta\tau_{CW\text{-}CCW}$), respectively. Their average values were plotted against [$M2^2$-$M1^2$], where M1 and M2 are the distances from the polar chemoreceptor array to the proximal and distal motors, respectively. The apparent diffusion coefficients for CCW-to-CW and CW-to-CCW switching signals were estimated from the slope of the approximation line to the plots for $\Delta\tau_{CCW\text{-}CW}$ and $\Delta\tau_{CW\text{-}CCW}$, respectively. To estimate the diffusion coefficient for CW-to-CCW switching, only plots having a plus value were used.

2.6. Simulation of the Change in the Intracellular Concentration of CheY-P and the Estimation of Response Time for Photoreleased Serine

To investigate the effect of polar localization of CheZ on changes in the CheY-P concentration, we performed a particle-based simulation. In the simulation, a 2 μm-long and 0.8 μm-wide rectangle was assumed, and 9000 CheY particles were initially placed randomly within this rectangle. These particles could diffuse freely in two dimensions independently of each other. The particles were reflected by each wall. The step size of a particle every time-interval was calculated from the following equations:

$$dx = \sqrt{2 \times D \times \Delta t} \quad (2)$$

$$dy = \sqrt{2 \times D \times \Delta t} \quad (3)$$

where dx and dy are the step sizes of the particle in each time increment, D is the diffusion coefficient, and Δt is the time-interval of the calculation. In the calculation, 0.002 ms was used as Δt, and 11.7 $\mu m^2/s$ [13] was used as the diffusion coefficient of the CheY and CheY-P molecules.

When the receptor array is in an inactive state, CheA is not phosphorylated. On the other hand, when the receptor array is in an active state, CheA is auto-phosphorylated at the constant rate k_1 in the reaction scheme shown as follows:

$$A \xrightarrow{k_1} AP \quad (4)$$

where A is non-phosphorylated CheA and AP is auto-phosphorylated CheA (CheA-P). We used 30 s^{-1} (k_1) as a value close to the past report [2]. The number of CheA-P was counted every Δt. Phosphorylation of CheY and dephosphorylation of CheY-P occur in the reaction scheme as follows:

$$Y + AP \xrightarrow{k_2} YP + A \quad (5)$$

$$YP + Z \xrightarrow{k_3} Y + Z \quad (6)$$

where YP is CheY-P and Z is CheZ. k_2 and k_3 are the rate constants for CheY phosphorylation and CheY-P dephosphorylation, respectively. In all simulations, k_2 was 1.0×10^8 M^{-1} s^{-1} as previously reported [2]. The value of k_3 was changed according to the calculation (see each figure and figure legend). In all simulations, 4500 CheA molecules and 3200 CheZ molecules were used to approximate physiological conditions [21]. In the simulation for the cell in the presence of CheZ localization (the $CheA_S^+$ cell), all 4500 CheA molecules were placed at the left edge of the rectangle (within 20 nm from the edge). Probability of phosphor-transfer to each CheY particle (P_{YP}) in this area was calculated as follows:

$$P_{YP} = k_2 \cdot [AP] \cdot \Delta t \quad (7)$$

where $[AP]$ is the concentration of CheA-P estimated from the number of CheA-P by assuming a 0.02 μm-long, 0.8-μm wide, and 0.8 μm high volume located at the left edge of rectangle. In total, 2600 CheZ molecules were placed at the left edge of the rectangle, corresponding to their binding to 2600 $CheA_S$ molecules at cell pole (21), while 600 CheZ molecules were placed at the bulk-cytoplasmic area of the rectangle. The probability of dephosphorylation of each CheY-P particle at the left edge ($P_{dephos\text{-}Y@pole}$) and cytoplasm ($P_{dephos\text{-}Y@cyto}$) were calculated as follows:

$$P_{dephos-Y@pole} = k_3 \cdot [Z]_{pole} \cdot \Delta t \quad (8)$$

$$P_{dephos-Y@cyto} = k_3 \cdot [Z]_{cyto} \cdot \Delta t \quad (9)$$

where $[Z]_{pole}$ is the concentration of CheZ at the cell pole estimated from the number of CheZ by assuming a 0.02 μm-long, 0.8-μm wide, and 0.8 μm high volume is located at the left edge of the rectangle. $[Z]_{cyto}$ is the concentration of CheZ at the cytoplasm estimated from the number of CheZ by assuming a 1.98 μm-long, 0.8-μm wide, and 0.8 μm high volume is located at the cytoplasmic area of the rectangle. Thus, CheY was phosphorylated at the left edge of the rectangle, and CheY-P was rapidly dephosphorylated at the left edge of the rectangle and dephosphorylated more slowly in the remainder of the rectangle.

In the simulation for the cell in the absence of CheZ localization (the $CheA_S^-$ cell), all CheA molecules were again located at the left edge of this rectangle, but all 3200 CheZ molecules were placed at the cytoplasmic area of the rectangle. Thus, all CheY molecules were again phosphorylated within 20 nm from the left edge of the rectangle, but CheY-P molecules were dephosphorylated throughout the whole rectangle. The P_{YP} and the $P_{dephos\text{-}Y@cyto}$ was estimated for each CheY or CheY-P particle and for every Δt by the same method as described above. In both simulations, assuming $CheA_S^+$ and $CheA_S^-$ cells, CheY molecules were phosphorylated from 0 to 1 s, and CheY-P was dephosphorylated constantly from 0 to 2 s. The motors were positioned at 0.05, 0.2, 0.4, 0.6, 0.8, 1.0, 1.2, 1.4, 1.6, and 1.8 μm from the left edge of the rectangle. A 0.2 μm-long and 0.8 μm-wide rectangle spanning the larger rectangle was centered on each motor, and the number of CheY-P molecules within this area was counted. The number of molecules was converted to the concentration by assuming a 0.2 μm-long, 0.8-μm wide, and 0.8 μm high volume. The simulations were performed at least 10 times for each value of k_3 with both the $CheA_S^+$ cell and the $CheA_S^-$ cell. The response time for CW to CCW rotation to instantaneously applied attractant stimulus was estimated from the average of the traces at each value of k_3 as the time required for the CheY-P concentration to fall below the threshold concentration of 3.2 μM for CW-to-CCW switching after the CheA activity was inhibited [22].

3. Results

3.1. Coordination of Flagellar Motors under Steady-State Conditions

We monitored the rotation of two motors on the same cell with small beads attached to their flagellar filaments, as illustrated in Figure 1A (see also Materials and Methods). The cells expressed GFP-CheW to determine the position of their chemoreceptor array relative to the two motors. The rotational motion of each bead attached to a flagellar stub was followed with a high-speed CCD camera to obtain a time-trace of the rotational velocity and directional switching of the two motors [13]. Two motors on a single wild-type cell coordinately switched their rotational directions, both from CCW to CW and from CW to CCW (Figure 1B). A correlation analysis for the time traces of motor 1 and motor 2 (see Materials and Methods) showed a major peak near 0 s, indicating a sub-second time delay between switching of the two flagellar motors (Figure 1C). These results are consistent with our previous report [13].

3.2. Coordination between Flagellar Motors in the Absence of Polar Localization of CheZ

To investigate whether localization of CheZ phosphatase activity affects the signaling process, we compared the coordination of motor switching in cells that produce both full-length CheA ($CheA_L$) and the N-terminally truncated form of CheA known as CheA short ($CheA_S$) with cells that produce $CheA_L$ only (Figure 2A). In wild-type cells, a large fraction of CheZ localizes at the chemoreceptor array through a reversible interaction with $CheA_S$ [11,12], an alternate translation product of the *cheA* gene initiated at codon ATG-98 [23,24]. The $CheA_S$ protein is not essential for overall chemotaxis [25], and its expression can be eliminated by mutations that alter *cheA* codon 98. In this work, we used mutant strains encoding CheA-M98L to eliminate $CheA_S$ synthesis [15] (Table S1). In these strains, CheZ should be uniformly distributed in the cytoplasm, with none being concentrated at the receptor array [11]. Accordingly, CheY-P dephosphorylation should occur throughout the cytoplasm instead of predominantly at the cell poles [12]. We confirmed the expected localization patterns of CheZ by imaging CheZ-GFP in $CheA_S^+$ and $CheA_S^-$ cells (Figure 2A). Approximately 85% of the $CheA_S^+$ cells exhibited polar localization of CheZ-GFP. In contrast, almost all of the $CheA_S^-$ cells exhibited a uniform cytoplasmic distribution of CheZ-GFP.

Like the motors on $CheA_S^+$ cells (Figure 1B), motors on $CheA_S^-$ cells coordinately switched their rotational directions, both from CCW to CW and from CW to CCW (Figure 2B). A correlation analysis of the rotational directions of the proximal motor 1 and the distal motor 2 showed a major peak near 0 s, indicating a sub-second time delay between switching of the two flagellar motors (Figure 2C). This sub-second coordination prevailed in 84/86 cells (98%) that we observed (Figure 2D, 45/45 cells measured at 1250 fps and Figure S1, 39/41 cells measured at 1255 fps) and was still apparent in the averaged correlation profile (Figure 2D and Figure S1, red lines). These results indicate that multiple flagellar motors coordinately switch their rotational direction in both $CheA_S^+$ and $CheA_S^-$ cells under steady-state conditions in the absence of chemoeffector stimuli. Therefore, polar localization of CheZ is not essential for steady-state coordination of motor switching.

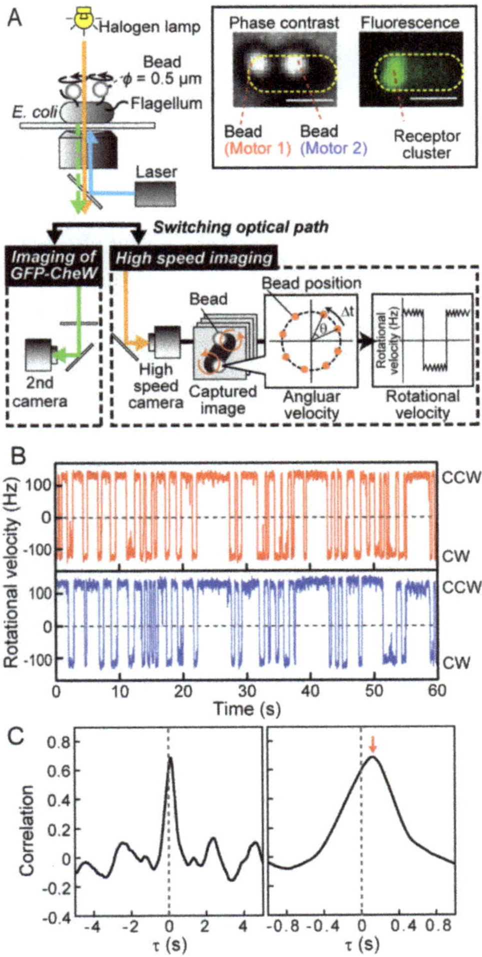

Figure 1. Coordination of steady-state switching of flagellar motors on a wild-type ($CheA_S^+$) cell. (**A**) Schematic diagram of the measurement system. The cell was stuck to a coverslip, and polystyrene beads (φ = 0.5 μm) were attached to the sticky flagellar stubs to calculate the angular velocity of the motor from the position of each bead. The phase-contrast image of each bead was recorded with a high-speed charge-coupled device (CCD) camera (1250 or 1255 frames/s). A time trace of rotational velocity and rotational direction of the motor was calculated from the bead position in each sampling frame. The fluorescence image was recorded with a second CCD camera by switching optical path. Phase-contrast images of the cells and beads (left inset) and fluorescence imaging of the cells (right inset) are also shown. Bar, 1 μm. The yellow-dotted ellipses indicate the cell bodies, and the positions of motors 1 and 2 are approximated from the position of the beads to which they are attached. The position of the polar receptor array was determined by the localization of GFP-fusion of CheW (GFP-CheW). (**B**) The time traces of the rotational velocities of motors 1 (red: proximal to the receptor array) and 2 (blue; distal from receptor array) on a wild-type ($CheA_S^+$) cell. The plus and minus values represent counterclockwise (CCW) and clockwise (CW) rotations, respectively. (**C**) Cross-correlation profile between the time traces of motors 1 and 2, which are depicted in (**B**) (left), with a magnified version of it (right). The analysis was based on proximal motor 1. Correlations were calculated using Equation (1), as shown in the Materials and Methods. The red arrow indicates the time of peak correlation ($\Delta\tau_{correlation}$ = +0.121 s).

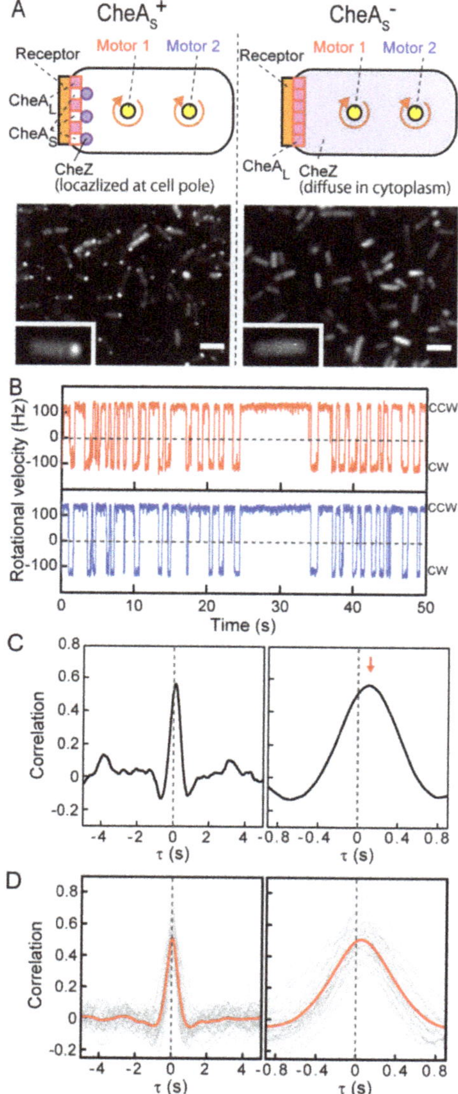

Figure 2. Coordination of steady-state switching of flagellar motors on a CheA$_S^-$ cell. (**A**) Localization of CheZ-EGFP in a CheA$_S^+$ cell (left) and a CheA$_S^-$ cell (right). A magnified image of each strain is shown at the bottom. Bar = 3 μm. (**B**) The time traces of the rotational directions of the proximal motor 1 (red) and the distal motor 2 (blue) on a CheA$_S^-$ cell. (**C**) A cross-correlation profile of the time traces of motors 1 and 2, which are depicted in (**B**) (left) and magnified (right). The analysis was based on proximal motor 1. The red arrow indicates the time of peak correlation ($\Delta\tau_{correlation}$ = +0.110 s). (**D**) Gray lines indicate a correlation analysis for CheA$_S^-$ cells measured at 1250 fps. The traces of 45 cells that showed coordination of switching are shown. The analysis was performed on the cells with monopolar or bipolar localization of GFP-CheW. The average of the $\Delta\tau_{correlation}$ was 68 ± 73 ms (mean ± SD). In a monopolar cell, the correlation was calculated based on the motor closer to receptor array. In a bipolar cell, the correlation was calculated based on the motor closer to more brighter receptor array. The red line shows the average traces of the correlation analyses from 45 cells.

3.3. Distance-Dependent Time Lags in Steady-State Motor Switching

Although motor switching events were nearly synchronous, close inspection of the rotation traces in a $CheA_S^+$ cell revealed consistent time lags between switching of the two motors (Figure 3A; corresponding to the longer time traces in Figure 1B). The motor closer to the polar chemoreceptor array (proximal motor) always switched before the motor farther from the array (distal motor). This held true for both CCW-to-CW and CW-to-CCW switching (Figure 3A, light- and dark-green hatches, black arrows). Magnified rotational time traces for a $CheA_S^-$ cell also revealed time lags between motor switching events. For CCW-to-CW switching, the motor closer to the chemoreceptor array (Figure 3B, red trace) preceded the motor that was farther from the array (Figure 3B, blue trace), as in $CheA_S^+$ cells. However, in contrast to the $CheA_S^+$ case, CW-to-CCW switching of the motor closer to the receptor array did not always precede the switching of the more-distant motor in $CheA_S^-$ cells. For example, the proximal motor (motor 1) preceded the distal motor (motor 2) in one switching event (Figure 3B, dark-green hatches and black arrows), but motor 2 preceded motor 1 in another switching event (Figure 3B, dark-green hatches and magenta arrows). In 55% of CW-to-CCW switching, motor 1 delayed to motor 2 in this cell.

To quantify these switching behaviors, we analyzed the relationship between the distance of each motor from the polar receptor array and the time difference in the onset of switching, which was estimated from the peak time of the correlation profile ($\Delta\tau_{correlation}$). The $\Delta\tau_{correlation}$ values were plotted against [$M2^2$-$M1^2$], where M1 and M2 are the distances from the polar chemoreceptor array to the proximal and distal motors, respectively. In both $CheA_S^+$ and $CheA_S^-$ cells, the aggregate $\Delta\tau$ values scaled with [$M2^2$-$M1^2$] (Figure 3C,D, left).

Next, we analyzed the relation between $\Delta\tau$ and [$M2^2$-$M1^2$] for CCW-to-CW switching and for CW-to-CCW switching (Figure 3A,B, light- and dark-green hatched area; see Materials and Methods). For CCW-to-CW switching, the $\Delta\tau_{CCW\text{-}CW}$ values for both $CheA_S^+$ and $CheA_S^-$ cells scaled with [$M2^2$-$M1^2$] (Figure 3C,D, middle). For CW-to-CCW switching in $CheA_S^+$ cells, the $\Delta\tau_{CW\text{-}CCW}$ values also scaled with [$M2^2$-$M1^2$] (Figure 3C, right, closed, and opened purple). However, $CheA_S^-$ cells exhibited both positive and negative $\Delta\tau_{CW\text{-}CCW}$ values evenly distributed around 0 s; there was no correlation with the corresponding [$M2^2$-$M1^2$] values (Figure 3D, right).

To confirm that the different CW-to-CCW switching behavior of the $CheA_S^-$ cells arose from a uniform cytoplasmic distribution of CheZ, we repeated the switching experiments with $CheA_S^+$ cells that had a mutant form of CheZ. The CheZ-F98S protein has CheY-P phosphatase activity, but cannot bind to $CheA_S$ [11]. The behavior of these cells was comparable to that of $CheA_S^-$ cells (Figure S2).

These results indicate that, in $CheA_S^+$ cells, the intracellular signals that induce both CCW-to-CW and CW-to-CCW switching events emanate at the polar receptor arrays and propagate to the flagellar motors by diffusion, accounting for the distance-dependent time delay in motor switching responses. We estimated the apparent diffusion coefficient for CCW-to-CW signals and for CW-to-CCW signals in $CheA_S^+$ cells from the slopes of linear fits to the $\Delta\tau$ versus distance data (Figure 3C, middle and right). The values (9.3 ± 2.7 and 9.6 ± 0.6 $\mu m^2/s$ (mean ± SD), respectively) were in agreement with estimates obtained in previous studies [1,13,19].

We note that not all $CheA_S^+$ cells exhibited distance-dependency of $\Delta\tau$ values in CW-to-CCW switching (Figure 3C, right). Because loss of polar localization of CheZ in $CheA_S^-$ cells abolishes the distance-dependency of the switching times, we thought that the anomalous $CheA_S^+$ cells might have a more uniform cytoplasmic distribution of CheZ. Indeed, ~15% of $CheA_S^+$ cells did not show polar localization of CheZ (Figure 2A). Moreover, plasmid-encoded expression of $CheA_S$ at elevated levels in $CheA_S^+$ cells restored the distance-dependency of $\Delta\tau$ values for CW-to-CCW switching in all cells that were examined (Figure 3C, right, gray).

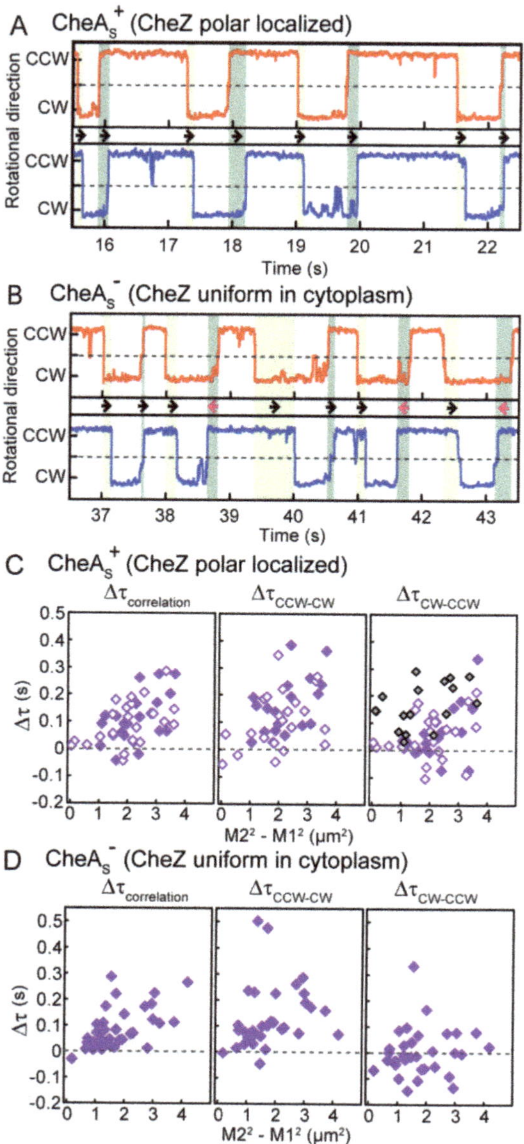

Figure 3. Time lags between switching of the two motors as a function of the distance from the polar chemoreceptor array. (**A**) The time traces of the rotational directions of the proximal motor 1 (red) and the distal motor 2 (blue) of a $CheA_S^+$ cell, shown in Figure 1B, over a short time period. Light-green and dark-green hatched areas indicate the time lags of switching between the two motors from CCW-to-CW and CW-to-CCW, respectively. Forward arrows indicate that switching of motor 1 preceded that of motor 2. $\Delta\tau_{CCW\text{-}CW}$, $\Delta\tau_{CW\text{-}CCW}$, and $M2^2$-$M1^2$ values of this cell are 0.16 s, 0.03 s, and 1.21 μm^2, respectively. (**B**) The time traces of the rotational directions of the proximal motor 1 (red) and the distal motor 2 (blue) of a $CheA_S^-$ cell, shown in Figure 2B, over a short time period. Colored hatches are as in (**A**). Forward arrows indicate that switching of motor 1 preceded that of motor 2, and reverse arrows

indicate that the switching of motor 1 was delayed relative to switching of motor 2. $\Delta\tau_{CCW\text{-}CW}$, $\Delta\tau_{CW\text{-}CCW}$, and M2^2-M1^2 values of this cell are 0.231 s, −0.026 s, and 1.26 μm^2, respectively. (**C**) The relationship between the $\Delta\tau$ value and [M2^2-M1^2] for CheA$_S$$^+$ cells. Closed purple squares show the plots measured in this study (n = 22 cells). Open purple squares show the plots estimated in our previous study [13] (n = 22 cells in $\Delta\tau_{correlation}$, and n = 19 cells in $\Delta\tau_{CCW\text{-}CW}$ and $\Delta\tau_{CW\text{-}CCW}$). The data were reused from reference [13] with permission. Gray squares show the relationship between the $\Delta\tau$ value and [M2^2-M1^2] in CheA$_S$$^+$ cells in which CheA$_S$ was overproduced (n = 17 cells). Spearman's rank correlation coefficient for $\Delta\tau_{correlation}$, $\Delta\tau_{CCW\text{-}CW}$ and $\Delta\tau_{CW\text{-}CCW}$ were 0.49, 0.53, and 0.36, respectively. For the estimation of the coefficient for $\Delta\tau_{CW\text{-}CCW}$, the data of the cell that CheAs is overexpressed was excluded. (**D**) The relationship between the $\Delta\tau$ value and [M2^2-M1^2] for CheA$_S$$^-$ cells (n = 33 cells). Spearman's rank correlation coefficient for $\Delta\tau_{correlation}$, $\Delta\tau_{CCW\text{-}CW}$, and $\Delta\tau_{CW\text{-}CCW}$ were 0.61, 0.42, and -0.05, respectively. (**A–D**) To evaluate the propagation of CheY-P precisely, cells with a monopolar localization of GFP-CheW were chosen for analysis. The correlation analyses were made based on the motor closer to the polar chemoreceptor array.

3.4. The Effect of Polar Localization of CheZ on Distance-Dependent Motor Response Times to an Instantaneously Applied Chemoattractant

To determine whether the coordinated CW-to-CCW switching in steady-state is induced by a decrease in CheY-P concentration, we measured the response of CheA$_S$$^-$ cells to an instantaneously applied serine signal, which was photoreleased from caged serine (Figure S3). The area surrounding the targeted cell (φ = 32 μm) was irradiated with a violet laser (φ = 405 nm, 370 mW mm^{-2}) to release free serine. In this experiment, the cellular response was measured in the presence of 1 mM caged serine. The concentration of photoreleased serine reached 10.6 μM during 80 ms of laser irradiation, and the serine concentration decreased at sub-seconds, because of diffusion, after laser irradiation was terminated [19].

When the Tsr chemoreceptor binds serine, CheY-P production is inhibited by the suppression of CheA autophosphorylation, and the CheY-P concentration is decreased by CheZ. As CheY-P concentration decreases, the fraction of CheY-P molecules bound to FliM decrease, and the rotational direction of the motor reverses from CW to CCW. To investigate whether the CW-to-CCW switching is induced by a decrease in CheY-P concentration, the shutter shielding the laser beam from the cells was opened in response to CCW-to-CW switching of a targeted motor (serine was photoreleased following CCW-to-CW switching) (Figure 4A, red line). We then measured the response time, which was defined as the duration of CW flagellar rotation immediately after laser irradiation (Figure 4A). The response time includes five successive events: (1) the time required for serine molecules to diffuse to and become bound by Tsr; (2) the time required for inactivation of CheA activity after serine binds to Tsr; (3) the time required for the CheY-P concentration to be decreased by the phosphatase activity of both array-localized CheZ (mostly), bulk-cytoplasmic CheZ (rarely), and diffusion of CheY-P in CheA$_S$$^+$ cells, or the time required for CheY-P concentration to be decreased by bulk-cytoplasmic CheZ and diffusion of CheY-P in CheA$_S$$^-$ cells; (4) the time required for CheY-P to dissociate from the motor; (5) the time required for switching from CW-to-CCW rotation within the CheY-P-depleted motor.

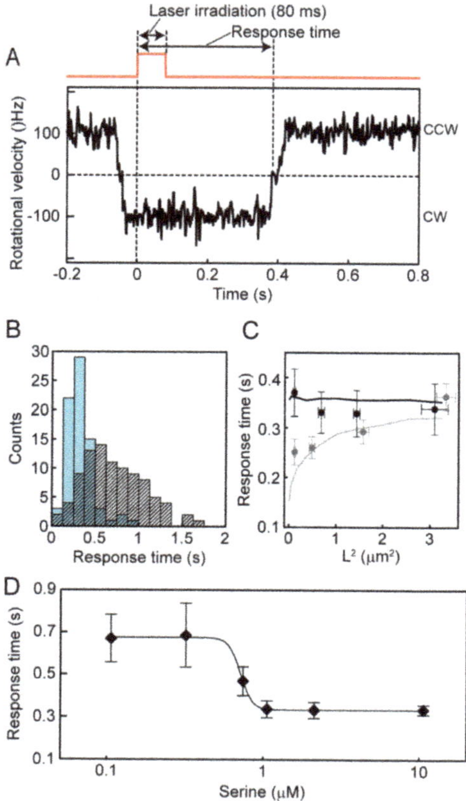

Figure 4. Measurement of the response to an instantaneously applied attractant stimulus of photoreleased serine. (**A**) Typical response of a $CheA_S^-$ cell after laser irradiation in the presence of 10.6 μM released serine (1000 μM of caged serine was contained in motility buffer). The laser shutter was open at 0 s for 80 ms (red line) immediately following a CCW-to-CW switching event. The interval between the initiation of laser irradiation and the first CW-to-CCW switch was defined as the response time. (**B**) Histogram of the response time of $CheA_S^-$ cells that exhibit monopolar and bipolar localization of GFP-CheW. Blue bars show the histogram of the response times obtained from 76 cells exposed to 10.6 μM released-serine. Gray hatched bars delineate the histogram of the CW duration obtained from 96 trials for 76 cells in the presence of 1000 μM caged serine without laser irradiation. (**C**) The relationship between the response time and the square of the distance from the polar receptor array to motor (L^2). Black plots show the relationship for $CheA_S^-$ cells that show monopolar localization of GFP-CheW. Error bars show the standard error of the mean. The response times in the presence of 10.6 μM released-serine are shown (n = 45 cells). The black line is the estimation of the response time from the simulation shown in Figure 5C when the rate constant for dephosphorylation of CheY-P by CheZ (k_3) is 1.6×10^6 M^{-1} s^{-1}. The experimentally measured response time of $CheA_S^-$ cells was well fitted by this value of k_3. Gray plots show the relationship for $CheA_S^+$ cells that show monopolar localization of GFP-CheW (the data were reused from reference [19] with permission). The gray line is the estimation of the response time from the simulation shown in Figure 5B when k_3 is 4.0×10^6 M^{-1} s^{-1}. The experimentally measured response time of $CheA_S^+$ cells was well fitted by this value of k_3. (**D**) Relationship between the response time of $CheA_S^-$ cells and the concentration of released-serine. The response times for 0.11 (n = 9 cells), 0.32 (n = 9 cells), 0.74 (n = 9 cells), 1.1 (n = 11 cells), 2.1 (n = 12 cells), and 10.6 (n = 44 cells) μM released serine are shown. The black line shows the fitted curve using the Hill equation. Error bars show the standard error of the mean.

A typical result is shown in Figure 4A. The shutter was opened at 0 s for 80 ms, and the direction of flagellar rotation switched to CCW 396 ms later (Figure 4A). In the presence of photoreleased serine, the average response time of $CheA_S^-$ cell was 329 ± 162 ms (Figure 4B, blue). In the presence of caged serine without laser irradiation, the average CW duration immediately following TTL (Transistor-transistor-logic) from the A/D (Analog/Digital) converter was 793 ± 514 ms (Figure 4B, black; see Materials and Methods). These results were consistent with the response of $CheA_S^+$ cells reported previously [19,26]. Thus, the duration of CW rotation immediately following laser irradiation was also significantly shortened by photoreleased serine in $CheA_S^-$ cells. Therefore, the response time is a reliable parameter to assess the response time of $CheA_S^-$ cells to serine.

Next, we compared the relation between the square of the distance from the polar receptor array to each motor (L^2) and the response times of $CheA_S^+$ and $CheA_S^-$ cells. In our previous work, we showed that the response time of $CheA_S^+$ cells depended on L^2 (Figure 4C, gray plots) [19]. On the other hand, the response time of $CheA_S^-$ cells was independent of L^2 (Figure 4C, black plots). These results indicate that, in $CheA_S^+$ cells, the polar localization of CheZ delays the decrease in CheY-P concentration depending on the distance from the receptor array, whereas in $CheA_S^-$ cells, the CheY-P concentration decreases at the same rate throughout the cell because CheZ is distributed uniformly throughout the cytoplasm of the cell.

To investigate the effect of polar localization of CheZ on the sensitivity of the cellular response to serine, the relationship between the response time and the concentration of photoreleased serine was compared in the presence and absence of polar localization of CheZ. In $CheA_S^-$ cells, the average CW duration after photorelease of serine was ~670 ms when the released-serine concentration was <0.32 μM, whereas a response time of ~330 ms was observed when the released-serine concentration was >1.0 μM (Figure 4D). The $K_{1/2}$ estimated from the Hill equation was 0.72 ± 0.01 μM (mean ± SE), which is 4.3-fold higher than that of $CheA_S^+$ cells (0.17 ± 0.11 μM) [19]. This result is consistent with a previous report of the higher phosphatase activity of CheZ in the presence of $CheA_S$ in vitro [27]. Therefore, polar localization of CheZ enhances the sensitivity for serine. A similar tendency was reported for cells expressing CheZ-F98S, which does not localize to the cell pole, by detecting ensemble FRET from the cell population [12]. This higher sensitivity to serine would be due to the higher phosphatase activity of CheZ assembled into receptor array in $CheA_S^+$ cells, as described in the next section. A higher sensitivity for serine would be useful for detecting and migrating within shallow gradients at low serine concentrations.

3.5. Simulation of the Change in CheY-P Concentration and Response Time in the Presence and Absence of Polar Localization of CheZ

To discuss the distance-dependency of time lags of switching between two motors ($\Delta\tau_{CCW\text{-}CW}$ and $\Delta\tau_{CW\text{-}CCW}$) in $CheA_S^+$ cells and the lack of distance-dependency of $\Delta\tau_{CW\text{-}CCW}$ in $CheA_S^-$ cells, we modeled the change in CheY-P concentration by simulating the diffusion of CheY and CheY-P molecules and the area of phosphorylation of CheY and of dephosphorylation of CheY-P, respectively. In a simulated $CheA_S^+$ cell, CheY and CheY-P molecules diffuse two-dimensionally in a rectangle (Figure 5A, top). Each CheY molecule is phosphorylated within a narrow area at the left side of the rectangle (the cell pole), and each CheY-P molecule is mostly dephosphorylated within the same area (see Materials and Methods). In this simulation, both the increase and decrease in CheY-P concentration showed delays dependent on distance from the cell pole, where the phosphorylation of CheY and the dephosphorylation of CheY-P occur (Figure 5B and Figure S4C). In the simulated $CheA_S^-$ cell, CheY is phosphorylated at the cell pole, whereas CheY-P is dephosphorylated in the bulk-cytoplasmic area at the same rate regardless of distance from the polar receptor array (Figure 5A, bottom; see Materials and Methods). In this situation, the increase in CheY-P concentration showed a delay dependent on the distance from the cell pole, similar to the $CheA_S^+$ case. However, CheY-P concentration decreased at the same rate independent of distance from the cell pole (Figure 5C and Figure S4D). Therefore, the distance-dependency of $\Delta\tau_{CCW\text{-}CW}$ and $\Delta\tau_{CW\text{-}CCW}$ in coordinated switching in wild-type cells can be explained by fluctuations in CheY-P concentration, taking into consideration the area of phosphorylation of CheY and dephosphorylation of CheY-P and the diffusion of CheY and CheY-P molecules.

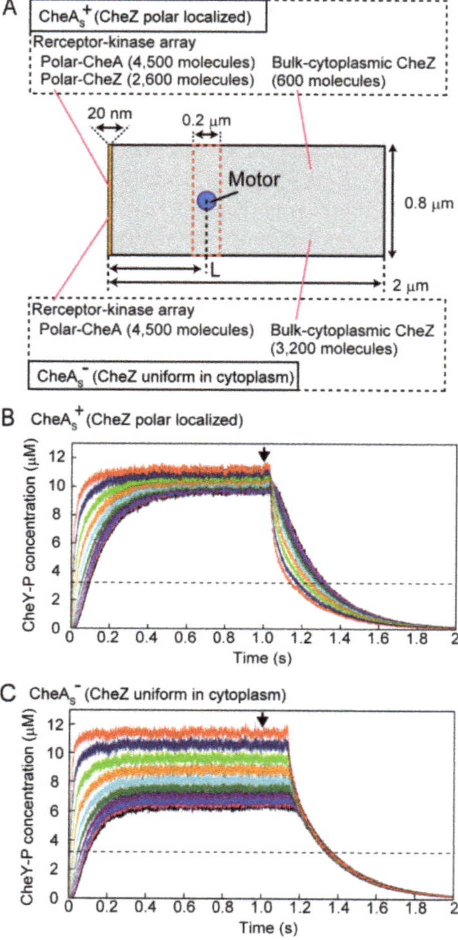

Figure 5. Simulation for the change in CheY-P concentration in the presence and absence of polar localization of CheZ. (**A**) A scheme depicting the method used to estimate the CheY-P concentration around a flagellar motor. Distance between the polar receptor-kinase array and motor shown as L. The number of CheY-P molecules within a 0.2-μm-long and 0.8-μm-wide area surrounding a motor (red dotted area) was counted and was converted to the concentration (see Materials and Methods). (**B**) Simulation for a $CheA_S^+$ cell when the rate constant for dephosphorylation of CheY-P (k_3) of polar localized CheZ was 4.0×10^6 M^{-1} s^{-1}. Distances between the cell pole and the motor are, in μm, 0.05 (red), 0.2 (dark blue), 0.4 (green), 0.6 (orange), 0.8 (cyan), 1.0 (green), 1.2 (violet), 1.4 (blue), 1.6 (magenta), and 1.8 (black), respectively. At each position, the average trace from 13 simulations is shown. (**C**) Simulation for a $CheA_S^-$ cell when the k_3 value for cytoplasmic CheZ is 1.6×10^6 M^{-1} s^{-1}. Distances from the cell pole and colors are the same as for the $CheA_S^+$ cell. At each position, the average trace from 10 simulations is shown. The results of the simulation using other k_3 values are shown in Figure S4. In both simulations, the activity of CheA was turned on at 0 s, and a plateau level of CheY-P was reached within 1 s. CheA activity was turned off at 1 s (downward arrow).

To address the distance-dependency of serine response times in CheA$_S$$^+$ cells and the lack of distance-dependency in CheA$_S$$^-$ cells, the response time was estimated in this simulation as the time required to decrease the CheY-P concentration below 3.2 μM after the inactivation of CheA activity (Figure 5B,C, arrows and dotted lines); the CW bias of the flagellar motor is known to change drastically around this concentration of CheY-P [22]. In the CheA$_S$$^+$ simulation, response times matched well with the distance-dependency of response times that we had measured experimentally (Figure 4C, gray line, and Figure S4E, blue line). The CheA$_S$$^-$ simulation explained the lack of distance-dependency in the experimentally measured response times (Figure 4C black line, and Figure S4E red line). These results indicate that the distance-dependency of the response time in CheA$_S$$^+$ cells and its lack in CheA$_S$$^-$ cells are correlated with the diffusion of CheY and CheY-P molecules and the area of phosphorylation of CheY and of dephosphorylation of CheY-P, respectively. Thus, in wild-type (CheA$_S$$^+$) cells, the polar localization of CheZ delays the decrease in CheY-P concentration depending on the distance from the receptor array. In contrast, in CheA$_S$$^-$ cells, the CheY-P concentration decreases at the same rate throughout the cell because CheZ is distributed uniformly throughout the cytoplasm.

By comparing the experimentally measured cellular response time to photoreleased serine for a motor near the receptor array (~0.1 μm2 from cell pole), the response time of CheA$_S$$^-$ cells was about 100 ms longer than that of CheA$_S$$^+$ cells (Figure 4C). To explain this difference in response time using the simulation, the rate constant of the phosphatase activity of CheZ in the receptor array (see Materials and Methods) must be set ~2.5-fold higher than that of CheZ in the cytoplasm (Figure 4C, black and gray lines). The higher activity of polar-localized CheZ is consistent with a previous report of the phosphatase activity of CheZ in the presence and absence of CheA$_S$ in vitro [27]. Therefore, the assembly of CheZ within the receptor array also enhances its phosphatase activity in living *E. coli* cells.

4. Discussion

4.1. Mechanism for Coordination of Switching among Flagellar Motors on an E. coli Cell

To clarify the mechanism of intracellular signaling during chemotaxis in *E. coli*, we previously measured the coordination of rotational switching of two different flagellar motors on the same cell [13]. That study showed that two flagellar motors on the same cell coordinately switch their rotational direction in the absence of external stimuli. The switching time lag ($\Delta\tau_{\text{correlation}}$) of two coordinated motors depended on their relative distance from the receptor array at the cell pole. A mutant cell lacking the CheZ protein did not exhibit coordinated switching. Coordinated switching was also inhibited by the expression of a constitutively active mutant form of CheY, which mimics the CW rotation-stimulating function of wild-type CheY-P, regardless of its phosphorylation state [13,28]. These results suggested that fluctuations in CheY-P concentration regulate the coordination of switching between flagellar motors.

In the present study, we asked how motor coordination changes when CheZ is uniformly distributed throughout the cell rather than localized at the receptor array. We found that time lags in CCW-CW switching ($\Delta\tau_{\text{CCW-CW}}$) increased as a function of the distance of the two motors from the pole containing the receptor array, both in the presence and absence of polar localization of CheZ (Figure 3). On the other hand, the distance-dependency of time lags in CW-CCW switching ($\Delta\tau_{\text{CW-CCW}}$) disappeared with the absence of polar localization of CheZ (in CheA$_S$$^-$ cells), whereas it was retained in the presence of polar localization of CheZ (in CheA$_S$$^+$ cells). Therefore, we propose that, in CheA$_S$$^+$ cells, dephosphorylation of CheY-P by polarly localized CheZ allows the CheY-P concentration to drop more rapidly at motors close to the receptor array. In CheA$_S$$^-$ cells with uniformly distributed CheZ, the CheY-P concentration decreases at the same rate at any distance from the receptor array. These results are consistent with the distance-dependency and distance-independency of response times to photoreleased serine in CheA$_S$$^+$ and CheA$_S$$^-$ cells, respectively (Figure 4C). CheA localizes at the cell pole in both CheA$_S$$^+$ and CheA$_S$$^-$ cells, so phosphorylation of CheY always occurs at a cell pole. Therefore, the CheY-P concentration increases more rapidly at motors close to the receptor array in both $CheA_S^+$ and $CheA_S^-$ cells, and in both cell types there is a delay in CCW-CW switching for motors distal from the receptor array relative to motors proximal to the array (Figure 6).

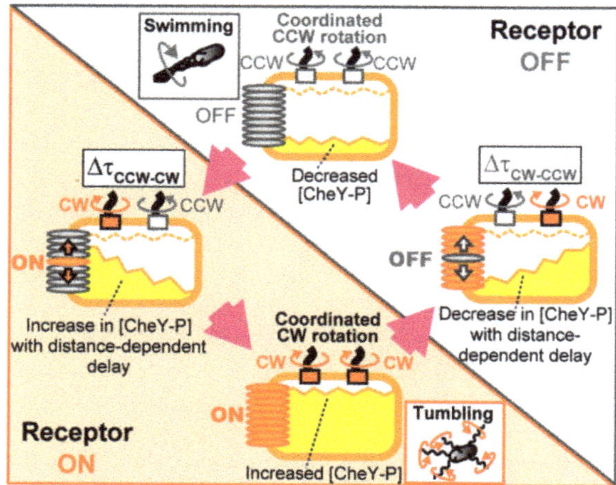

Figure 6. A model for intracellular signaling in an *E. coli* cell under steady-state conditions with no external stimuli. The orange-hatched and white-hatched triangles indicate times when the receptor array is active or inactive, respectively. When the receptor units (the receptor/CheW/CheA complexes) are in an inactive state (depicted as gray ellipses located at cell pole), flagellar motors on the same cell rotate CCW. When one or small numbers of the receptor units are activated (orange ellipse depicted in left cell), its activation is propagated through the receptor array and/or sub-arrays (multiple interconnected segments in the array) because of the cooperativity among the receptor units. The CheY-P concentration increases at the cell pole through the activity of CheA, and the increase propagates through the cytoplasm by diffusion with a delay dependent on the distance from the receptor array (yellow area within a cell). Therefore, two flagellar motors coordinately switch rotational direction from CCW to CW with a delay (left and bottom cells). On the other hand, when one or several of the receptor units is inactivated (gray ellipse depicted in right cell), its inactivation is propagated through the receptor array and/or sub-arrays because of the cooperativity among the receptor units. The CheY-P concentration is decreased at the cell pole through the activity of CheZ with a delay dependent on the distance from receptor array due to the diffusion of CheY-P molecules. Therefore, two flagellar motors coordinately reverse their rotational direction from CW to CCW with a delay (right and top cell). In the steady-state, the spontaneous blinking in activity of receptor array causes the fluctuation in CheY-P concentration that coordinates the switching of rotational direction among the flagellar motors.

Recent theoretical studies have proposed that intrinsic motor-to-motor coupling caused by hydrodynamic interactions between motors is responsible for coordinated motor switching [29]. This possibility was explored in our previous investigation, but we did not see coordination of motors on different cells that were very close to one another [13]. We only saw coordination between motors on the same cell, indicating that hydrodynamic interaction between motors is not the main contributor to coordinated switching.

We simulated switching coordination between motors, assuming a steep sigmoidal relation of CW bias and switching frequency to CheY-P concentration, as shown by Cluzel et al. [22]. Thus, CW bias and switching frequency reflect CheY-P concentration (see Supplementary Methods). Switching coordination was seen in simulations when fluctuations in CheY-P concentration were taken into account (Figure S5A–C). In contrast, motors stochastically switched their rotational direction at a constant CheY-P concentration (Figure S5D–F). Another theoretical model from the Namba-Shibata group reproduced the strong coordination of rotational switching between two motors caused by fluctuations of the CheY-P concentration arising from spontaneous fluctuations in the kinase activity of the receptor array [30]. We conclude that fluctuation in the CheY-P concentration is the main contributor to the coordination of switching between motors on a cell.

4.2. Coordinated Motor Switching via Diffusive Signal Propagation from Receptor Arrays

As shown in the present study, both $\Delta\tau_{CCW\text{-}CW}$ and $\Delta\tau_{CW\text{-}CCW}$ increase as a function of the distance of the two motors from the cellular pole containing the receptor array in $CheA_S^+$ cells. The apparent diffusion coefficients estimated from these data were 9.3 and 9.6 $\mu m^2/s$, respectively, in agreement with estimates for CheY-P obtained in previous studies [1,13,19]. The distance-dependency of $\Delta\tau_{CCW\text{-}CW}$ in $CheA_S^+$ cells was explained well by a simulation that took into consideration the phosphorylation of CheY by polarly localized CheA, the dephosphorylation of CheY-P by polarly localized CheZ, and the diffusion of CheY and CheY-P molecules (Figure 5 and Figure S4). The distance-dependency of $\Delta\tau_{CCW\text{-}CW}$ in $CheA_S^-$ cells with uniformly distributed CheZ was also explained by considering the phosphorylation of CheY by polarly localized CheA, the dephosphorylation of CheY-P by bulk-cytoplasmic CheZ, and the diffusion of CheY and CheY-P molecules. However, the distance-dependency of $\Delta\tau_{CW\text{-}CCW}$ in $CheA_S^-$ cells was lost in this simulation because the uniformly distributed CheZ causes CheY-P levels to fall at the same rate throughout the cell. Therefore, fluctuation depends on the diffusion of CheY and CheY-P molecules and signal-producing and signal-destroying reactions in the receptor array (Figure 6).

To produce fluctuations in CheY-P concentration over timescales of several hundred milliseconds, a receptor array, which is composed of more than 10,000 protein molecules, would work as one single or several large signaling units, which is possible because of the high cooperativity derived from the network architecture [31]. Multiple interconnected segments in the receptor array (sub-arrays), which work independently of each other, and/or the entire receptor array, would spontaneously blink between active and inactive states in the absence of chemoeffectors. A theoretical model to explain the spontaneous blinking of array activity was proposed by the Namba-Shibata group [30]. Their model incorporated two states (active and inactive) of each receptor unit and cooperativity among the units constituting the array. Both CheB and CheY are phosphorylated by CheA when the receptor array is activated. In this situation, CheB-P demethylates the receptor to bring the receptor array into an inactive state. On the other hand, when the receptor array is inactive, methylation of the receptors promoted by the constitutive activity of CheR brings the array to an active state. Therefore, changes in the methylation level produced by the activity of CheB-P and CheR could affect the spontaneous blinking in the activity of the receptor array, as mentioned by Shimizu et al. [32]. The fluctuation in CheY-P concentration measured in a single cell through FRET between CheY-YFP and CheZ-CFP has also been reported for cells possessing or lacking CheR and CheB [33,34]. The sampling rate in the FRET experiments (1–0.2 Hz) is very different from our high-speed imaging (~1250 Hz), so further experiments are required to verify that fluctuations in CheR and CheB activity lead to fluctuations in CheY-P production.

4.3. Behavioral and Evolutionary Implications of the Blinking Array Model

Despite the stochastic nature of the run-and-tumble swimming pattern of *E. coli* cells, it was recently reported that a swimming cell expressing wild-type CheY coordinates rotational switching between motors. This coordination of switching in swimming cells was not observed in the presence of a constitutively active mutant form of CheY [35]. These results indicate that the coordination of motor switching occurs in a swimming cell, and the run-and-tumble behavior is caused not only through stochastic switching of the flagellar motors but also coordination of motor switching regulated by fluctuations in CheY-P concentration (Figure 6). Turner et al. reported that the distribution of directional changes from run to run during a tumble is narrow and biased in the forward direction when a smaller number of flagella are unwound from the flagellar bundle, whereas the distribution is wide when a larger number of flagella are unwound from the bundle [36]. The coordination of switching among flagellar motors leads to the unwinding of a larger number of flagella from the bundle during a tumble; therefore, the cell undergoes a bigger change in the direction of swimming. Fluctuations in CheY-P concentration under steady-state conditions could be a strategy that enables *E. coli* to explore its environment more widely by regulating run-and-tumble swimming through the coordinated switching of motors.

Bacillus subtilis, a rod-shaped Gram-positive bacterium evolutionarily very distant from *E. coli*, is also peritrichously flagellated, has a polar receptor array, and carries out chemotaxis by biasing the run-and-tumble swimming pattern [37]. However, the chemotaxis system of *B. subtilis* is quite different from that of *E. coli*; CheY-P induces CCW flagellar rotation, and thus promotes runs, and chemoattractants cause receptors to stimulate the activity of CheA in phosphorylating CheY. CheY-P is dephosphorylated at the cytoplasmic face of the flagellar basal body by the FlaY protein instead of by CheZ. Therefore, the flagellar motor acts as a sink to deactivate the run signal. Are the flagellar motors of *B. subtilis* coordinated? If so, how do they achieve coordinated switching? It will be interesting to model the dynamics of CheY-P in *B. subtilis* to see whether they also lead to the coordination of multiple flagella. *B. subtilis* might be using a very different paradigm from the one that is used by *E. coli*.

A swimming bacterium hydrolyzes ATP to produce CheY-P under steady-state conditions. The cell must remain in a state of run-and-tumble readiness to respond quickly when chemoeffectors are encountered. There must be a substantial advantage to maintain the chemotaxis system on high alert rather than to decrease energy consumption by keeping it in a resting mode. Here, we report how this system is organized in *E. coli*. The challenge remains to determine whether and how other bacteria may have evolved different CheY-dependent mechanisms to coordinate locomotor behavior.

Supplementary Materials: The following are available online at http://www.mdpi.com/2218-273X/10/11/1544/s1, Supplementary Methods, Table S1: Bacterial strains and plasmids, Figure S1: The switching coordination between two motors in $CheA_S^-$ cell measured at 1255 fps, Figure S2: The switching coordination between two motors in a mutant cell, which has the substitution for F98S in CheZ (CheZ(F98S) cell), Figure S3: Schematic diagram of the measurement system to measure the cellular response time to serine signal photoreleased from caged serine, Figure S4: Simulation for the change in CheY-P concentration in the presence and absence of the polar localization of CheZ, Figure S5: Simulation for the coordination of the directional switching between motors [38,39].

Author Contributions: All measurements and data analysis were carried out by Y.-S.C. and H.F. Strains were designed and constructed by H.F. and H.T. Computational programs for data analysis were developed by T.S., Y.I., T.H., A.I., and H.F. The work was planned by H.F. and the first draft of the paper was written by H.F. All authors have read and agreed to the published version of the manuscript.

Funding: This work was supported by Grants-in-Aid for Scientific Research from MEXT/JSPS KAKENHI, 19H05797 (to A.I. and H.F.), 23115004, 16H00803, and 16H04777 (to A.I.), and 26440073 (to H.F.).

Acknowledgments: We thank John S. Parkinson and Michael D. Manson for their critical reading, providing some useful comments, and grammatical correction of this manuscript. We thank Kazushi Kinbara and Takahiro Muraoka for the caged serine.

Conflicts of Interest: The authors declare that they have no conflict of interest.

References

1. Segall, J.E.; Ishihara, A.; Berg, H.C. Chemotactic Signaling in Filamentous Cells of *Escherichia coli*. *J. Bacteriol.* **1985**, *161*, 51–59. [CrossRef] [PubMed]
2. Lipkow, K. Changing Cellular Location of CheZ Predicted by Molecular Simulations. *PLoS Comput. Biol.* **2006**, *2*, e39. [CrossRef] [PubMed]
3. Kholodenko, B.N. Cell-Signalling Dynamics in Time and Space. *Nat. Rev.* **2006**, *7*, 165–176. [CrossRef] [PubMed]
4. Wadhams, G.H.; Armitage, J.P. Making Sense of It All: Bacterial Chemotaxis. *Nat. Rev.* **2004**, *5*, 1024–1037. [CrossRef]
5. Macnab, R. Flagela and Motility. In *Eschericia coli and Salmonella*; Neidhardt, F.C., Ed.; American Society for Microbiology: Washington, DC, USA, 1996; pp. 123–145.
6. Fukuoka, H.; Sagawa, T.; Inoue, Y.; Takahashi, H.; Ishijima, A. Direct Imaging of Intracellular Signaling Components That Regulate Bacterial Chemotaxis. *Sci. Signal.* **2014**, *7*, ra32. [CrossRef]

7. Stewart, R.C. Kinetic Characterization of Phosphotransfer between CheA and CheY in the Bacterial Chemotaxis Signal Transduction Pathway. *Biochemistry* **1997**, *36*, 2030–2040. [CrossRef]
8. Sourjik, V.; Berg, H.C. Binding of the *Escherichia coli* Response Regulator Chey to Its Target Measured In Vivo by Fluorescence Resonance Energy Transfer. *Proc. Natl. Acad. Sci. USA* **2002**, *99*, 12669–12674. [CrossRef]
9. Bren, A.; Eisenbach, M. The N Terminus of the Flagellar Switch Protein, Flim, Is the Binding Domain for the Chemotactic Response Regulator, CheY. *J. Mol. Biol.* **1998**, *278*, 507–514. [CrossRef]
10. Welch, M.; Oosawa, K.; Aizawa, S.; Eisenbach, M. Phosphorylation-Dependent Binding of a Signal Molecule to the Flagellar Switch of Bacteria. *Proc. Natl. Acad. Sci. USA* **1993**, *90*, 8787–8791. [CrossRef]
11. Cantwell, B.J.; Draheim, R.R.; Weart, R.B.; Nguyen, C.; Stewart, R.C.; Manson, M.D. CheZ Phosphatase Localizes to Chemoreceptor Patches via CheA-Short. *J. Bacteriol.* **2003**, *185*, 2354–2361. [CrossRef]
12. Vaknin, A.; Berg, H.C. Single-Cell FRET Imaging of Phosphatase Activity in the *Escherichia coli* Chemotaxis System. *Proc. Natl. Acad. Sci. USA* **2004**, *101*, 17072–17077. [CrossRef] [PubMed]
13. Terasawa, S.; Fukuoka, H.; Inoue, Y.; Sagawa, T.; Takahashi, H.; Ishijima, A. Coordinated Reversal of Flagellar Motors on a Single *Escherichia coli* Cell. *Biophys. J.* **2011**, *100*, 2193–2200. [CrossRef] [PubMed]
14. Parkinson, J.S.; Houts, S.E. Isolation and Behavior of *Escherichia coli* Deletion Mutants Lacking Chemotaxis Functions. *J. Bacteriol.* **1982**, *151*, 106–113. [CrossRef] [PubMed]
15. Skidmore, J.M.; Ellefson, D.D.; McNamara, B.P.; Couto, M.M.P.; Wolfe, A.J.; Maddock, J.R. Polar Clustering of the Chemoreceptor Complex in *Escherichia coli* Occurs in the Absence of Complete CheA Function. *J. Bacteriol.* **2000**, *182*, 967–973. [CrossRef] [PubMed]
16. Ryu, W.S.; Berry, R.M.; Berg, H.C. Torque-Generating Units of the Flagellar Motor of *Escherichia coli* Have a High Duty Ratio. *Nature* **2000**, *403*, 444–447. [CrossRef]
17. Datsenko, K.A.; Wanner, B.L. One-Step Inactivation of Chromosomal Genes in *Escherichia coli* K-12 Using PCR Products. *Proc. Natl. Acad. Sci. USA* **2000**, *97*, 6640–6645. [CrossRef]
18. Maloy, S.R.; Nunn, W.D. Selection for Loss of Tetracycline Resistance by *Escherichia coli*. *J. Bacteriol.* **1981**, *145*, 1110–1111. [CrossRef]
19. Sagawa, T.; Kikuchi, Y.; Inoue, Y.; Takahashi, H.; Muraoka, T.; Kinbara, K.; Ishijima, A.; Fukuoka, H. Single-Cell *E. coli* Response to an Instantaneously Applied Chemotactic Signal. *Biophys. J.* **2014**, *107*, 730–739. [CrossRef]
20. Chung, S.; Kennedy, R.A. Forward-Backward Non-Linear Filtering Technique for Extracting Small Biological Signals from Noise. *J. Neurosci. Methods* **1991**, *40*, 71–86. [CrossRef]
21. Li, M.; Hazelbauer, G.L. Cellular Stoichiometry of the Components of the Chemotaxis Signaling Complex. *J. Bacteriol.* **2004**, *186*, 3687–3694. [CrossRef]
22. Cluzel, P. An Ultrasensitive Bacterial Motor Revealed by Monitoring Signaling Proteins in Single Cells. *Science* **2000**, *287*, 1652–1655. [CrossRef] [PubMed]
23. Smith, R.A.; Parkinson, J.S. Overlapping Genes at the CheA Locus of Escherichia coli. *Proc. Natl. Acad. Sci. USA* **1980**, *77*, 5370–5374. [CrossRef] [PubMed]
24. Kofoid, E.C.; Parkinson, J.S. Tandem Translation Starts in the CheA Locus of *Escherichia coli*. *J. Bacteriol.* **1991**, *173*, 2116–2119. [CrossRef] [PubMed]
25. Sanatinia, H.; Kofoid, E.C.; Morrison, T.B.; Parkinson, J.S. The Smaller of Two Overlapping CheA Gene Products Is Not Essential for Chemotaxis in *Escherichia coli*. *J. Bacteriol.* **1995**, *177*, 2713–2720. [CrossRef] [PubMed]
26. Segall, J.E.; Manson, M.D.; Berg, H.C. Signal Processing Times in Bacterial Chemotaxis. *Nature* **1982**, *296*, 855–857. [CrossRef]
27. Wang, H.; Matsumura, P. Characterization of the CheAS/CheZ Complex: A Specific Interaction Resulting in Enhanced Dephosphorylating Activity on CheY-Phosphate. *Mol. Microbiol.* **1996**, *19*, 695–703. [CrossRef]
28. Bourret, R.B.; Hess, J.F.; Simon, M.I. Conserved Aspartate Residues and Phosphorylation in Signal Transduction by the Chemotaxis Protein CheY. *Proc. Natl. Acad. Sci. USA* **1990**, *87*, 41–45. [CrossRef]
29. Hu, B.; Tu, Y. Coordinated Switching of Bacterial Flagellar Motors: Evidence for Direct Motor-Motor Coupling? *Phys. Rev. Lett.* **2013**, *110*, 158703. [CrossRef]
30. Namba, T.; Shibata, T. Propagation of Regulatory Fluctuations Induces Coordinated Switching of Flagellar Motors in Chemotaxis Signaling Pathway of Single Bacteria. *J. Theor. Biol.* **2018**, *454*, 367–375. [CrossRef]
31. Bray, D.; Levin, M.D.; Morton-Firth, C.J. Receptor Clustering as a Cellular Mechanism to Control Sensitivity. *Nature* **1998**, *393*, 85–88. [CrossRef]

32. Shimizu, T.S.; Aksenov, S.; Bray, D. A Spatially Extended Stochastic Model of the Bacterial Chemotaxis Signalling Pathway. *J. Mol. Biol.* **2003**, *329*, 291–309. [CrossRef]
33. Keegstra, J.M.; Kamino, K.; Anquez, F.; Lazova, M.D.; Emonet, T.; Shimizu, T.S. Phenotypic Diversity and Temporal Variability in a Bacterial Signaling Network Revealed by Single-Cell FRET. *eLife* **2017**, *6*, e27455. [CrossRef] [PubMed]
34. Colin, R.; Rosazza, C.; Vaknin, A.; Sourjik, V. Multiple Sources of Slow Activity Fluctuations in a Bacterial Chemosensory Network. *eLife* **2017**, *6*. [CrossRef] [PubMed]
35. Mears, P.J.; Koirala, S.; Rao, C.V.; Golding, I.; Chemla, Y.R. *Escherichia coli* Swimming Is Robust Against Variations in Flagellar Number. *eLife* **2014**, *3*, e01916. [CrossRef]
36. Turner, L.; Ryu, W.S.; Berg, H.C. Real-Time Imaging of Fluorescent Flagellar Filaments. *J. Bacteriol.* **2000**, *182*, 2793–2801. [CrossRef]
37. Szurmant, H.; Ordal, G.W. Diversity in Chemotaxis Mechanisms among the Bacteria and Archaea. *Microbiol. Mol. Biol. Rev.* **2004**, *68*, 301–319. [CrossRef]
38. Morales, V.M.; Bäckman, A.; Bagdasarian, M. A Series of Wide-Host-Range Low-Copy-Number Vectors That Allow Direct Screening for Recombinants. *Gene* **1991**, *97*, 39–47. [CrossRef]
39. Guzman, L.M.; Belin, D.; Carson, M.J.; Beckwith, J. Tight Regulation, Modulation, and High-Level Expression by Vectors Containing the Arabinose PBAD Promoter. *J. Bacteriol.* **1995**, *177*, 4121–4130. [CrossRef]

Publisher's Note: MDPI stays neutral with regard to jurisdictional claims in published maps and institutional affiliations.

Article

A Factor Produced by *Kaistia* sp. 32K Accelerated the Motility of *Methylobacterium* sp. ME121

Yoshiaki Usui [1], Yuu Wakabayashi [1], Tetsu Shimizu [2], Yuhei O. Tahara [3,4], Makoto Miyata [3,4], Akira Nakamura [2] and Masahiro Ito [1,5,*]

1. Graduate School of Life Sciences, Toyo University, Oura-gun, Gunma 374-0193, Japan; ysak4415@gmail.com (Y.U.); canonoftheend@gmail.com (Y.W.)
2. Faculty of Life and Environmental Sciences, and Microbiology Research Center for Sustainability (MiCS), University of Tsukuba, Tsukuba, Ibaraki 305-8572, Japan; sakuratettyan@hotmail.com (T.S.); nakamura.akira.fm@u.tsukuba.ac.jp (A.N.)
3. Department of Biology, Graduate School of Science, Osaka City University, Osaka 558-8585, Japan; taharayuhei@gmail.com (Y.O.T.); miyata@sci.osaka-cu.ac.jp (M.M.)
4. The OCU Advanced Research Institute for Natural Science and Technology (OCARINA), Osaka City University, Osaka 558-8585, Japan
5. Bio-Nano Electronics Research Centre, Toyo University, Kawagoe, Saitama 350-8585, Japan

* Correspondence: masahiro.ito@toyo.jp; Tel.: +81-273-82-9202

Received: 9 March 2020; Accepted: 14 April 2020; Published: 16 April 2020

Abstract: Motile *Methylobacterium* sp. ME121 and non-motile *Kaistia* sp. 32K were isolated from the same soil sample. Interestingly, ME121 was significantly more motile in the coculture of ME121 and 32K than in the monoculture of ME121. This advanced motility of ME121 was also observed in the 32K culture supernatant. A swimming acceleration factor, which we named the K factor, was identified in the 32K culture supernatant, purified, characterized as an extracellular polysaccharide (5–10 kDa), and precipitated with 70% ethanol. These results suggest the possibility that the K factor was directly or indirectly sensed by the flagellar stator, accelerating the flagellar rotation of ME121. To the best of our knowledge, no reports describing an acceleration in motility due to coculture with two or more types of bacteria have been published. We propose a mechanism by which the increase in rotational force of the ME121 flagellar motor is caused by the introduction of the additional stator into the motor by the K factor.

Keywords: symbiosis; coculture; motility; *Methylobacterium*; *Kaistia*

1. Introduction

Microorganisms often establish symbiotic relationships with other species. Because bacteria share their habitat with other microorganisms, increasing studies on cocultivation by intentionally mixing and culturing different bacteria have been conducted. For example, Olson et al. reported that the interspecific interaction between *Candida albicans* and *Candida glabrata* increased biofilm formation and virulence-related gene expression in a composition-dependent manner [1]. Onaka et al. reported that coculturing actinomycetes and bacteria produces antibiotics that are not produced under monoculture [2]. Thus, coculture studies are expected to reveal new bacterial properties not observed during monoculture. To date, cocultivation studies have mainly focused on growth, probiotics, and metabolic products [3–5].

Methylobacterium sp. ME121 and *Kaistia* sp. 32K were isolated from the same soil sample during a search for bacteria capable of assimilating L-glucose [6,7]. ME121 has a unipolar flagellum and is motile (Supplementary Figure S1A, Movies S1 and S2), whereas 32K has no flagellum and is thus not motile (Supplementary Figure S1B). We accidentally discovered that the swimming speeds of

ME121 grown in a ME121-32K coculture showed a significantly accelerated motility compared with that in an ME121 monoculture. Nakamura et al. reported a deceleration in bacterial swimming speed upon the cocultivation of the lactose-fermenting bacteria *Lactococcus lactis* subsp. lactis and *Salmonella enterica serovar Typhimurium* [5]. The motility of *Salmonella* was either decreased or lost due to acidic substances produced by the lactic acid bacteria. Another study demonstrated the enhancement of bacterial motility in an *Escherichia coli* monoculture due to the production of an attractant [8]. However, no report has been published describing an acceleration in motility by a coculture with two or more types of bacteria.

This study was initiated based on the observation that the swimming speed of ME121 increased in a mixed culture with 32K. Therefore, we attempted to elucidate the mechanism underlying this advanced motility by investigating the properties of the substances derived from the 32K culture supernatant.

2. Materials and Methods

2.1. Bacterial Strains and Growth Media

Methylobacterium sp. ME121 and *Kaistia* sp. 32K were used in this study.

For ME121, a Met medium (10.0 g of peptone, 2.0 g of yeast extract, 1.0 g of $MgSO_4$, and 5 mL of methanol per liter) was used for the preculture, whereas for 32K, an LM medium (10.0 g of tryptone, 5.0 g of yeast extract, and 1.0 g of D-mannitol per liter) was used for the preculture. Methanol was filter-sterilized with Millex-LG filters (Merck KGaA, Darmstadt, Germany, pore size: 0.2 μm).

A D-glucose synthetic medium (1.07 g of NH_4Cl, 0.81 g of $MgCl_2$, 0.75 g of KCl, 1.74 g of KH_2PO_4, 1.36 g of K_2HPO_4, 2 mL of Hutner's trace elements, and 0.90 g of D-glucose per liter) was used for the monoculture and the coculture. Hutner's trace elements were prepared by dissolving 22.0 g of $ZnSO_4{\cdot}7H_2O$, 11.4 g of H_3BO_3, 5.06 g of $MnCl_2{\cdot}7H_2O$, 1.16 g of $CoCl_2{\cdot}5H_2O$, 1.57 g of $(NH_4)_6Mo_7O_{24}{\cdot}4H_2O$, and 1.57 g of $FeSO_4{\cdot}7H_2O$ in 500 mL of sterile water. EDTA·2Na (50 g) was dissolved while warming 300 mL of Milli-Q water. The pH was adjusted in the 6.5–6.8 range with KOH after the addition of each component, and its final volume was adjusted to 1 L. The solution was stored at 4 °C for approximately two weeks until its color changed from light green to purple red. It was then sterilized by filtration and used for the D-glucose synthetic medium. *E. coli* W3110 was used as the control for the tethered cell assay of ME121.

2.2. Monoculture and Coculture Conditions

The cells of ME121 and 32K were cultured in 10 mL of the Met and LM media (28 °C, 300 rpm, 48 h), respectively, washed with saline, and suspended in 2 mL of the D-glucose synthetic medium.

For the monoculture, the cells were inoculated into Φ24-mm test tubes containing 10 mL of the D-glucose synthetic medium to ensure an initial optical density (OD_{600}) of 0.08, and then they were cultured (28 °C, 300 rpm). For the coculture, 5 mL of each bacterial suspension (OD_{600} = 0.08) were mixed in the same test tube and cultured (28 °C, 300 rpm).

A tethered cell assay was conducted for bacterial flagellar rotation analysis, in which a single colony of *E. coli* W3110 was cultured in 2 mL of an LB medium (30 °C, 200 rpm, 14 h). *E. coli* W3110 was precultured in 2 mL of the LB medium (OD_{600} = 0.01; 30 °C, 200 rpm, 7 h).

2.3. Combined Cultures Established in a Beppu Flask

In this paper, "coculture" refers to mixed cultures of two types of bacteria, whereas "dialysis culture" refers to separate cultures of two types of bacteria established in a Beppu flask [9] (Nihon Pall Corporation, Tokyo, Japan) with two tanks partitioned by a membrane filter (pore size: 0.2 μm), as shown in Figure 1. In this "dialysis coculture" the cultures are established in a Beppu flask with one tank inoculated with the pure culture of ME121 and the other tank inoculated with the mixed culture of ME121 along with 32K, as shown in Figure 1.

Culture type	The left tank	The right tank
A. monoculture	ME121	—
B. dialysis culture	ME121	32K
C. coculture	—	ME121 & 32K
D. dialysis coculture	ME121	ME121 & 32K

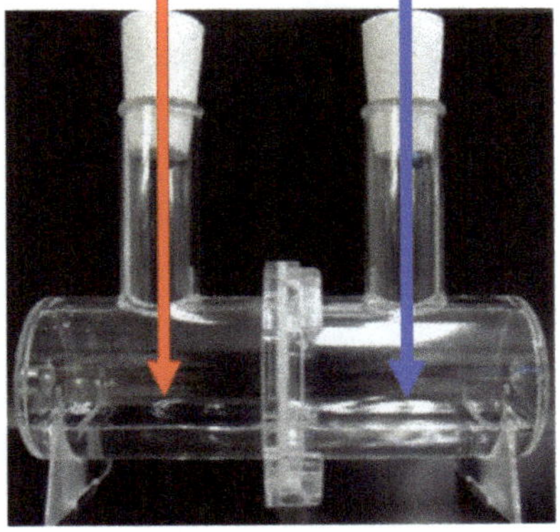

Figure 1. Establishing combined cultures in Beppu flask. The combination of inocula in the culture tanks separated by a membrane filter (Supor 200 hydrophilic polyether sulfone; pore size: 0.2 μm; diameter: 44 mm) was (A) monoculture, (B) dialysis culture, (C) coculture and (D) dialysis coculture. The optical density (OD_{600}) measurement on the left tank in the coculture (C) was not performed.

Briefly, 10 mL of the synthetic D-glucose medium or cells suspended in the same medium were inoculated into both culture tanks (OD_{600} = 0.08) and cultured (28 °C, 300 rpm). For the coculture of ME121 and 32K, 5 mL of each bacterial suspension (OD_{600} = 0.08) were inoculated in the D-glucose synthetic medium and placed in the same culture tanks. The OD_{600} of the left culture tank in Figure 1 was measured every 24 h.

2.4. ME121 Culture in 32K Culture Supernatant

To evaluate the growth and swimming speed of ME121 cultured in the 32K culture supernatant, the cells of ME121 cultured in 10 mL of the Met medium (28 °C, 300 rpm, 48 h) were washed with saline and suspended in 2 mL each of the D-glucose synthetic medium and the 32K culture supernatant. Each was suspended into a test tube containing 10 mL of the D-glucose synthetic medium and the 32K culture supernatant (OD_{600} = 0.08), then cultured (28 °C, 300 rpm, 72 h). The OD_{600} of each culture was measured every 24 h.

2.5. Motility Assay

Bacterial motility was observed under a dark-field microscope (Leica DMRE; Leica geosystem, Tokyo Japan) while maintaining the culture solution at 28 or 32 °C on a microscope stage (Type: MP-2000, Leica) [10]. The swimming speed of ME121 was almost the same when the plate temperature was 28 or 32 °C. We recorded a video of the observed movements using a digital color camera (Leica DF310 FX). The speed of each swimming cell was calculated using 2D movement measurement capture 2D-PTV software (Digimo, Tokyo, Japan) and the captured movie. Three independent experiments were conducted, and at least 60 bacterial cells were measured. Statistical analysis was performed by a Microsoft Excel *t*-test.

2.6. Swimming Speed of ME121 in the 32K Culture Supernatant

A single colony of ME121 was cultured in 10 mL of the Met medium (28 °C, 300 rpm, 48 h), and 50 μL of the preculture broth was inoculated into a test tube containing 10 mL of the Met medium and then cultured (28 °C, 300 rpm, 24 h). The culture broth (1 mL) was centrifuged (room temperature, 9100× *g*, 5 min). The cells of ME121 were resuspended in 1 mL of the swimming assay medium.

Five types of swimming assay media were tested: (i) the Met medium, (ii) the synthetic D-glucose medium, (iii) the synthetic Met medium, (iv) the carbon-free synthetic medium, and (v) the 32K culture supernatant. The synthetic Met medium contained 5 mL of methanol instead of 0.9 g of D-glucose per liter, as was in the D-glucose medium. The carbon-free synthetic medium was used to remove D-glucose from the D-glucose medium.

Immediately after suspension, the microbial cells were kept at 28 or 32 °C on a glass heater; the motility of the bacteria was observed with a dark-field microscope, and their appearance was video-recorded. Three independent experiments were conducted with at least 100 bacterial cells.

2.7. Various Treatments of the 32K Culture Supernatant

2.7.1. Heat Treatment

The 32K culture supernatant was heated at 121 °C for 40 min using an autoclave, after which it was returned to room temperature.

2.7.2. Lipid Removal Treatment

A 10 mL aliquot of the 32K culture supernatant was added to 20 mL of a chloroform/ethanol mixture (2:1) and mixed thoroughly. After separating the aqueous layer, ethanol was removed with a rotary evaporator (N-1100, Eyela, Tokyo, Japan). Distillation was performed in an eggplant flask placed in a 37 °C water bath for 20 min. Sterile water was added to the aqueous layer until a final volume of 10 mL.

2.7.3. Protein Removal Treatment with an Enzyme

The amount of protein contained in the 32K culture supernatant was calculated using the Lowry method. Proteinase K (Merck Millipore) was added to the 32K culture supernatant to a final concentration of 0.090 Anson units/mL and incubated at 37 °C for 1 h. After incubation, the enzyme was kept at 75 °C for 10 min for inactivation, and the 32K culture supernatant was returned to room temperature.

2.7.4. Ethanol Precipitation

The 32K culture supernatant (50 mL) was dispensed in an eggplant flask and frozen at −30 °C. The frozen 32K culture supernatant was lyophilized overnight with a freeze-drier (VD-250R; Taitec Co., Ltd., Japan) and then dissolved in 5 mL of sterilized water. The concentrated 32K culture supernatant (5 mL) was desalted by dialysis (4 °C, 24 h) with a Spectra/Por 6 instrument (Spectrum Laboratories; diameter: 11.5 mm; membrane material: standard regenerated cellulose membrane [standard RC membrane], molecular weight: 3500 Da cutoff). Using 1 L of Milli-Q water, the external solution was changed three times. The desalted 32K culture supernatant (approximately 6 mL) was lyophilized overnight and then dissolved in 6 mL of sterilized water. To the suspension, 14 mL of 99.5% ethanol was added, and the mixture was allowed to stand overnight at −30 °C. The sample was centrifuged (4 °C, 13,000× *g*, 1 h); the supernatant (nonpolar fraction) was separated from the precipitate (polar fraction), and the former was discarded. The polar fraction was dissolved in 10 mL of sterilized water, and the ethanol was removed with a rotary evaporator. The whole polar fraction (approximately 10 mL) was frozen at −30 °C, lyophilized overnight, and dissolved in 50 mL of sterilized water.

2.7.5. Dialysis

The 32K culture supernatant (5 mL) was inoculated at 4 °C for 24 h using the Spectra/Por®6 instrument (Spectrum Laboratories, Rancho Dominguez, CA, USA; diameter: 11.5 mm; membrane material: standard RC membrane; molecular weight: 3500 Da cutoff). For the external solution, a carbon-free medium (1 L) was used and dialyzed by exchanging the external solution three times.

2.7.6. Ultrafiltration

The 32K culture supernatant (5 mL) was added to a centrifugal filtration filter and centrifuged (4 °C, 2200× *g*, 90 min; RLX-105, Tomy Seiko). This solution was used for the motility assay. The following membranes were used as centrifugal filtration filters: an Amicon Ultra-15 centrifugal filter unit (Merck Millipore; membrane material: ultra-cell regenerated cellulose membrane; nominal molecular weight limit: 10,000 Da) and the VIVA SPIN 15R (Sartorius; membrane material: Hydrosart; nominal molecular weight limit: 5000 Da).

2.8. Swimming Speed of ME121 at Various pH Values

ME121 was cultured in 10 mL of the Met medium. After harvesting, the cells were resuspended in the D-glucose synthetic medium (pH 5.0, 5.5, or 6.0), after which the swimming speed was recorded as described previously. The pH was adjusted with 6 N HCl. Three independent experiments were conducted with at least 100 bacterial cells.

2.9. Preparation of the 32K Culture Supernatant

32K was cultured in 100 mL of the LM medium (28 °C, 200 rpm, 48 h) and centrifuged (4 °C, 9100× *g*, 5 min). The cells were resuspended in 50 mL of saline and centrifuged (4 °C, 9100× *g*, 5 min). The pellet was inoculated in a 5 L jar fermenter (BMS-05; Able Co., Ltd., Tokyo, Japan) containing 2.9 L of the D-glucose synthetic medium to achieve an initial OD_{600} = 0.08, and then it was cultured at 28 °C using a stirring blade rotating at 750 rpm and an aeration rate of 3 L/min. After the 32K growth reached the stationary phase, the culture supernatant was harvested by centrifugation (4 °C, 14,000× *g*, 30 min). The culture supernatant (2.5 L) was sterilized using a Nalgene Rapid-Flow Polyethersulfone (PES) membrane filter unit (pore diameter: 0.2 μm; Thermo Fisher Scientific).

2.10. Refining of the Motility-Accelerating Factor

Bacterial motility was simultaneously observed with growth measurements using a dark-field microscope at 32 °C on a glass heater every 12 h until ME121 lost its motility. Three independent experiments were conducted with at least 60 bacterial cells.

2.11. Preparation of the K Factor

The 32K culture supernatant (50 mL) was dialyzed overnight against the Milli-Q water (molecular weight: 3500 Da cutoff, 4 °C). The desalted culture supernatant was lyophilized overnight in a freeze-dryer (EYELA FDU-2200). After dissolving the dried sample in 6 mL of sterilized water, ethanol was added to a final concentration of 70%, and the mixture was allowed to stand at −30 °C overnight. Next, the precipitate and the supernatant were separated by centrifugation (4 °C, 13,000× *g*, 1 h; Tomy Seiko MX-305, Tokyo). Ethanol was removed in the supernatant using rotary evaporation (37 °C, 10 min; EYELA, Type N-1210B). The precipitated fraction was resuspended in 10 mL of sterilized water, and ethanol was removed as described above. The suspension was again lyophilized overnight, and the lyophilized powder was resuspended in a 50 mM acetic acid/NaOH buffer solution (pH 5.0). The solution was loaded on a DE-52 DEAE-cellulose column (2.5 × 50 cm; GE Healthcare Japan, Hino, Japan) that had been equilibrated with a 50 mM acetic acid/NaOH buffer (pH 5.0) [11]. The column was washed with 500 mL of the buffer at a flow rate of 110 mL/h. The column was eluted at the same rate with 250 mL of a buffer containing 0.2 M NaCl, followed by a linear gradient elution from 0.2

to 0.6 M NaCl in the buffer (700 mL) at a flow rate of 60 mL/h. Fractions (12 mL) containing neutral sugars were pooled, dialyzed against deionized water, and concentrated in a rotary evaporator (43 °C). Neutral sugars were determined using the anthrone reagent method with glucose as a reference [12].

2.12. Analysis of the Monosaccharide Composition of the K Factor

Trifluoroacetic acid (100 μL; 4 M) was added to the K factor (lyophilized product; 1.28 mg), and the mixture was incubated at 100 °C for 3 h. The hydrolysate was dried, solidified, dissolved in 100 μL of ultrapure water, and centrifuged (4 °C, 10,000 g, 10 min). The supernatant (50 μL) was recovered. Then, 50 μL of the supernatant was diluted 10-fold with ultrapure water, *N*-acetylated using acetic anhydride, and subjected to fluorescence labeling with the *p*-aminobenzoic acid ethyl ester (ABEE) reagent [13]. Thereafter, the monosaccharides labeled with fluorescence were recovered from the water layer by chloroform extraction and used for analysis. The analytical conditions were as follows: boric acid buffer/acetonitrile; flow rate, 0.5 mL/min; detection, fluorescence (Ex: 305 nm; Em: 360 nm); BioAssist EZ (Tosoh, Tokyo, Japan); and column, PN-PAK C18 (3.0 × 75 mm).

2.13. Flagellar Motor Rotation

The ME121 preculture broth (50 μL) was inoculated into a test tube containing 10 mL of the Met medium and was cultured (28 °C, 300 rpm, 24 h). This culture broth (1 mL) was subjected to shearing 20 times using a 1 mL syringe and injection needle. The ME121 culture medium (40 μL) was poured between a glass slide (S 1225; Matsunami Glass Industry Co., Ltd., Osaka, Japan) and a glass cover slip (Thickness No. 1; Matsunami Glass Industry Co., Ltd.), so that the side of the cover slip faced downward. It was allowed to stand at 28 °C for 20 min, and the flagella were then adsorbed onto the side of the glass cover slip. The D-glucose synthetic medium (40 μL) was poured between the glass cover slip and the glass slide to wash away the unadsorbed bacterial cells. This was performed twice. Subsequently, cell rotation with flagellar rotation was observed using a dark-field microscope and was video recorded. After recording, the solution was exchanged into the 32K culture supernatant in the same manner. This operation was also performed twice, and cell rotation was video-recorded as described above. The rotation speed per second was calculated using the ImageJ version 1.50i software (National Institutes of Health). At least three independent experiments were conducted, and the rotational speed of at least 100 cells was measured. The same operation was repeated for *E. coli* W3110, which served as the control.

2.14. Fluorescent Staining of the Flagella

The fluorescent staining of flagella was performed according to the method of Kinoshita et al. with slight modifications [14]. Briefly, the ME121 culture in the Met medium (1 mL) was transferred into a 1.5 mL tube and centrifuged (24 °C, 9100× *g*, 3 min). The pellet was resuspended in 1 mL of a phosphate buffer (0.81 g of $MgCl_2$, 1.36 g of KH_2PO_4, and 1.36 g of K_2HPO_4 per liter), centrifuged (24 °C, 9100× *g*, 3 min), and resuspended in 0.1 mL of the phosphate buffer. Cy *N*-Hydroxysuccinimide (NHS) ester monoreactive dye (GE Healthcare) was added to the bacterial suspension. Subsequently, dyeing was carried out at room temperature in the dark for 30 min to 1 h. Next, the suspension was centrifuged (24 °C, 9100× *g*, 3 min), and the pellet was suspended in 1 mL of the phosphate buffer to remove the excess fluorescent reagent. Thereafter, the suspension was centrifuged (24 °C, 9100× *g*, 3 min), and the pellet was resuspended in 50 μL of the D-glucose synthetic medium or the 32K culture supernatant. While maintaining the temperature at 32 °C with a glass heater, the swimming behavior of the bacteria was observed with a dark-field microscope. To each cell suspension, 50 μL of a 2% methylcellulose solution was added and mixed. The flagellar structure of the stained cells was observed under a phase-contrast fluorescence microscope. Cell images were acquired using the Leica DF310 FX camera. The pitch of each flagellar filament was analyzed using the ImageJ version 1.50i software. Three independent experiments were conducted with at least 100 bacterial cells.

2.15. *Quick-Freeze, Deep-Etch Electron Microscopy*

ME121 and 32K cell suspensions in the logarithmic growth phase were collected by centrifugation (room temperature, 8000× *g*, 5 min) and suspended in a buffer consisting of 10 mM HEPES (pH 7.6), 150 mM NaCl, and 1 mM $MgCl_2$ to achieve a 20-fold higher cell density. The cell suspensions were mixed with a slurry that included mica flakes, placed on a piece of rabbit lung, and frozen with a CryoPress (Valiant Instruments, St. Louis, MO, USA) that was cooled by liquid helium [15]. The slurry was used to retain an appropriate amount of water before freezing. The specimens were fractured and etched for 15 min at −104 °C in a JFDV freeze-etching device (JEOL Ltd., Akishima, Japan) [16]. The exposed cells were rotary shadowed by platinum at an angle of 20 degrees to a 2 nm thickness and backed with carbon. The replicas were floated off on full-strength hydrofluoric acid, rinsed in water, cleaned with a commercial bleach, rinsed again in water, and picked up onto copper grids as described [17,18]. They were observed under a JEM-1010 transmission electron microscope (JEOL, Tokyo, Japan) at 80 kV equipped with a FastScan-F214 (T) Charge Coupled Device (CCD) camera (TVIPS, Gauting, Germany).

3. Results

3.1. *Growth and Swimming Speed of ME121 in Several Culture Conditions*

This study was initiated by the accidental discovery that a coculture with the nonmotile bacterium 32K accelerated the swimming speed of the motile bacterium ME121. Therefore, we focused on the swimming acceleration product (herein termed as K factor) in the 32K culture supernatant, and we investigated its properties.

When ME121 and 32K were separately cultured in the synthetic D-glucose medium, the growth of ME121 was not as good as when cocultured (Figure 2). When Beppu flasks were used, the swimming speed of ME121 in the logarithmic growth phase was the fastest among other growth phases under any other culture conditions (Figure 3).

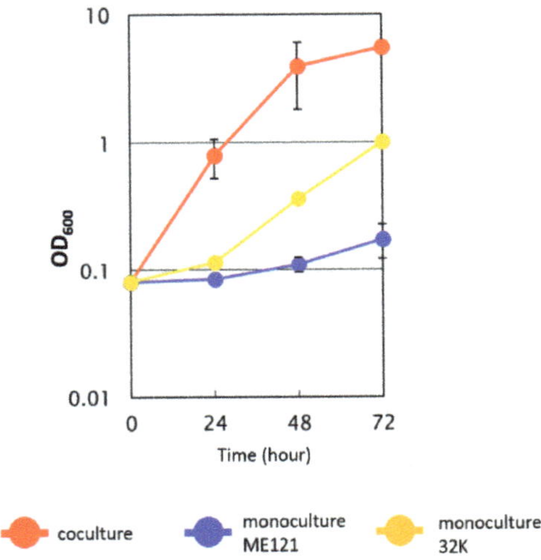

Figure 2. Growth curve of the monocultures of ME121 and of 32K and of coculture with ME121 and 32K in synthetic D-glucose medium in a test tube. Error bars indicate standard error.

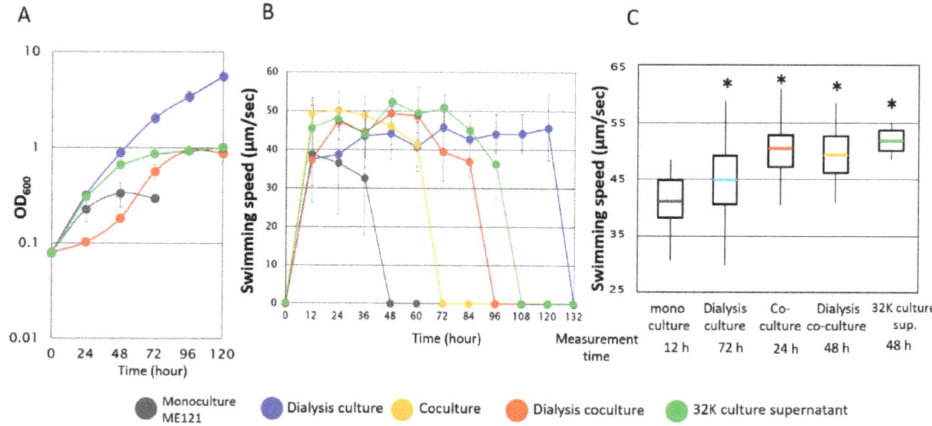

Figure 3. Growth and swimming speed of ME121 under several culture conditions in the synthetic D-glucose medium using a Beppu flask. (**A**) Growth curve, (**B**) swimming duration, and (**C**) swimming speed at exponential growth. In (**A**), the OD_{600} measurement on the left side of a Beppu flask during the coculture was not carried out. In (**C**), half of the values are within the box, and thick lines in the middle indicate average values. The line extending vertically indicates the remaining values, and the ends of each line indicate the maximum and the minimum values. * Significant difference from the D-glucose medium ($p < 0.001$). Statistical analysis was performed with a Microsoft Excel t-test. Values are expressed as the mean of three independent experiments, and the swimming speed of at least 60 cells was measured. Error bars indicate standard error.

In the monoculture, ME121 reached the stationary phase in 48 h, while in contrast, growth was observed in the dialysis culture and the dialysis coculture beyond 48 h (Figure 3A). Significant growth in various cocultures, as well as prolonged swimming period and an accelerated swimming speed of ME121 (Figure 3 and Supplementary Movies S1 and S2) compared with the monoculture was observed. No motility was observed when ME121 reached the stationary phase in each culture condition. When the fastest swimming speeds during the logarithmic growth phase of ME121 were compared, significant differences in swimming speeds were observed between the monoculture and other cultures (Figure 3C). In the dialysis culture, the coculture, and the dialysis coculture, the swimming speeds of ME121 were faster than that in the monoculture.

To elucidate the mechanism of the accelerated swimming speed of ME121, we investigated whether the 32K supernatant accelerated the swimming speed of ME121. We found that swimming speed peaked in the logarithmic growth phase (Figure 3). Significant ME121 growth and motility duration were observed when the 32K culture supernatant was used instead of the synthetic D-glucose medium (Figure 3A,B). A significant acceleration in the swimming speed was also observed (Figure 3C and Supplementary Movies S3 and S4).

Suspensions in the Met medium, the synthetic D-glucose medium, the synthetic Met medium, or the Carbone (C)-free synthetic medium showed no increase in ME121 swimming speed; only the 32K culture supernatant significantly increased ME121's swimming speed (Figure 4). Therefore, the K factor is not a metabolite of ME121. An enhancement in ME121 growth was also observed in the 32K culture supernatant. From this finding, we speculated that the K factor accelerated the growth of ME121, in addition to accelerating its swimming speed.

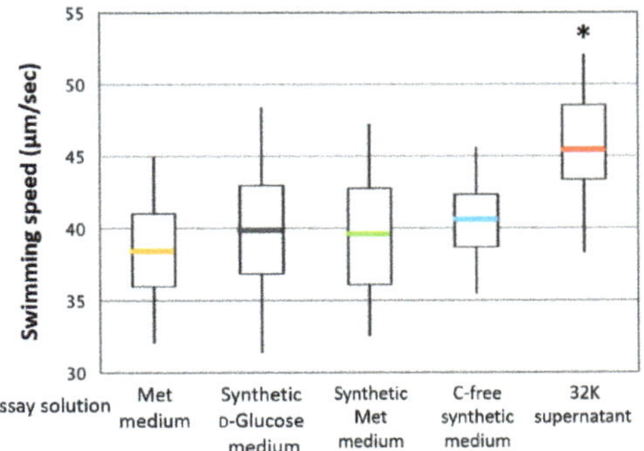

Figure 4. Swimming speed of ME121 in the 32K culture supernatant and other media. The distribution of the results of the motility test of the ME121 using the 32K culture supernatant was compared. The swimming speed of 100 cells was measured under each condition. The explanation of the box-and-whisker plot is shown in the legend of Figure 3. * Significant difference from the synthetic D-glucose medium ($p < 0.001$). Value are expressed as the mean of three independent experiments. For each growth, the synthetic D-glucose medium was used.

3.2. Analysis of the Monosaccharide Composition in the Ethanol-Precipitated Fraction of the 32K Culture Supernatant

By analyzing the swimming speed of ME121 in the 32K culture supernatant solution that had been heated and treated with enzymes, the characteristics of the K factor of ME121 were estimated (Figure 5).

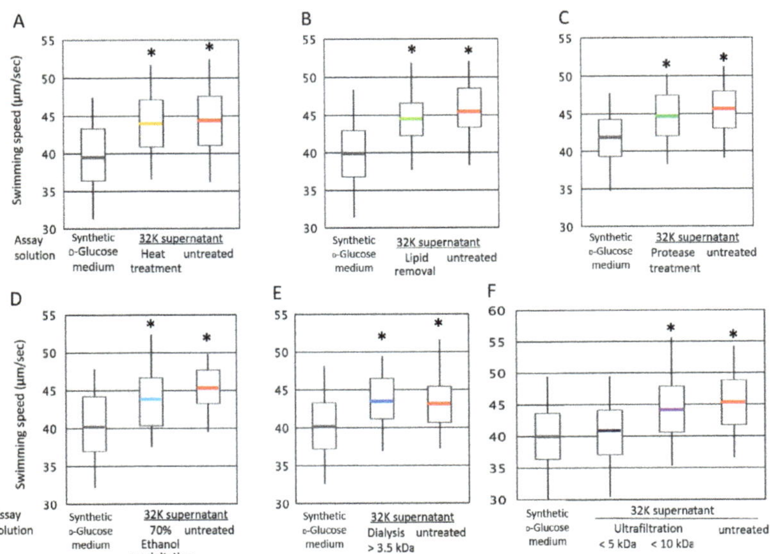

Figure 5. Swimming speed of ME121 in various treatments of the 32K culture supernatant: (**A**) heat, (**B**) lipid removal, (**C**) protein removal by enzyme, (**D**) ethanol precipitation, (**E**) dialysis, and (**F**) ultrafiltration. The variance of the results of the ME 121 motility test was compared. The explanation of the box-and-whisker plot is shown in the legend of Figure 3. * Significant difference from the synthetic D-glucose medium ($p < 0.001$). Values are expressed as the mean of three independent experiments, and the swimming speed of at least 100 cells was measured.

Heat treatment did not significantly change the swimming speed of ME121 (Figure 5A). This observation was similarly observed in treatments involving lipid and protein removal (Figure 5B,C). Therefore, the K factor is heat-stable and non-volatile, but it is neither a lipid or a substance affected by protease. Upon the ethanol treatment, a polar substance was precipitated. However, the aqueous solution of the precipitated polar substance had no significant effect on the ME121 swimming speed (Figure 5D). This suggested that the K factor is a polar substance precipitated in 70% ethanol.

The molecular weight of the K factor was estimated by dialysis and ultrafiltration. No significance differences in ME121 swimming speed were observed upon dialysis with a molecular weight cutoff of 3.5 kDa (Figure 5E). The molecular weight fraction of the 32K culture supernatant containing a substance of 10 kDa or less exerted the same effect on ME121 swimming speed as the 32K culture supernatant. However, when ME121 was suspended in the molecular weight fraction of the 32K culture supernatant that contained substances of 5 kDa or more, the swimming speed was almost the same as when ME121 was suspended in the synthetic D-glucose medium (Figure 5F). This result suggested that the molecular weight of the K factor is approximately 5–10 kDa.

3.3. Preparation of the K Factor

As a result of the anion-exchange chromatography fractionation of the ethanol precipitation from the 32K culture supernatant using a DEAE cellulose column, substances containing neutral sugars with 0.23–0.32 M NaCl were eluted (Figure 6).

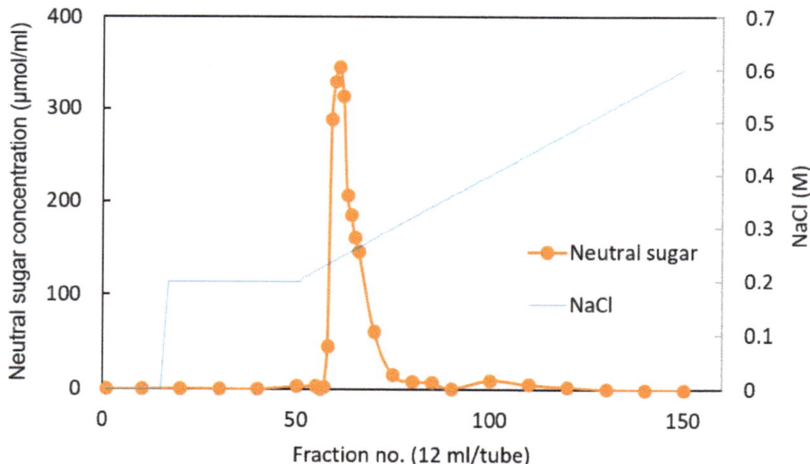

Figure 6. Purification of the ethanol precipitation of the 32K strain culture supernatant through a Diethylaminoethyl (DEAE)-cellulose column. Neutral sugars were quantified using the anthrone sulfate method. Fraction no. 58 to 80 were collected, desalted, freeze-dried, and subjected to the composition analysis indicated. Each fraction was collected in 12 mL portions. The concentration of NaCl was determined from the refractive index of the fractions.

Fractions 58 to 80, containing the highest neutral sugar concentration, were collected, desalted, lyophilized, and subjected to monosaccharide composition analysis.

3.4. Analysis of Monosaccharide Composition in the Ethanol-Precipitated Fraction (K Factor) of the 32K Culture Supernatant

The fluorescent pre-labeling method was used to analyze the monosaccharide composition (Table 1, Supplementary Figure S2). The neutral sugars glucose and galactose were present at a ratio of approximately 1:1 and accounted for approximately 55% of the total K factor. Uronic acids and amino sugars were not detected. Therefore, the K factor was presumed to be a kind of extracellular polysaccharide (EPS) composed of neutral sugars.

Table 1. Monosaccharide composition of the ethanol-precipitated fraction of the 32K culture supernatant.

No.	Component Name	pmol	Per g of Sample	
			μmol	mg
1	Glucuronic acid	ND	ND	ND
2	Galacturonic acid	ND	ND	ND
3	Galactose	139	1.46×10^3	263
4	Mannose	3.7	39	7.1
5	Glucose	145	1.53×10^3	275
6	Arabinose	3.1	32	4.8
7	Ribose	ND	ND	ND
8	*N-acetyl-mannosamine*	ND	ND	ND
9	Xylose	1.2	13	1.9
10	*N-acetyl-glucosamine*	ND	ND	ND
11	Fucose	ND	ND	ND
12	Rhamnose	1.8	19	3.0
13	*N-acetyl-galactosamine*	ND	ND	ND

3.5. Elucidation of the ME121 Motility-Accelerating Mechanism of the K Factor

To elucidate the ME121 motility-accelerating mechanism, the measurement of the rotation of the flagellar motor was performed using a tethered cell assay, along with measurement of the pitch of the flagellar fibers of ME121 (Figure 7). When ME121 was exposed to the 32K culture supernatant, the rotational speed of the flagellar motor increased by approximately 25% (Figure 7A). In W3110, the rotational speed of the flagellar motor was not affected by the 32K culture supernatant (Figure 7B). This suggested that the acceleration was caused by an increase in the rotational power of the flagellar motor and that the increase in the rotational force of the flagellar motor was specific to ME121. Increased motility was also observed immediately after suspending ME121 in the 32K culture supernatant.

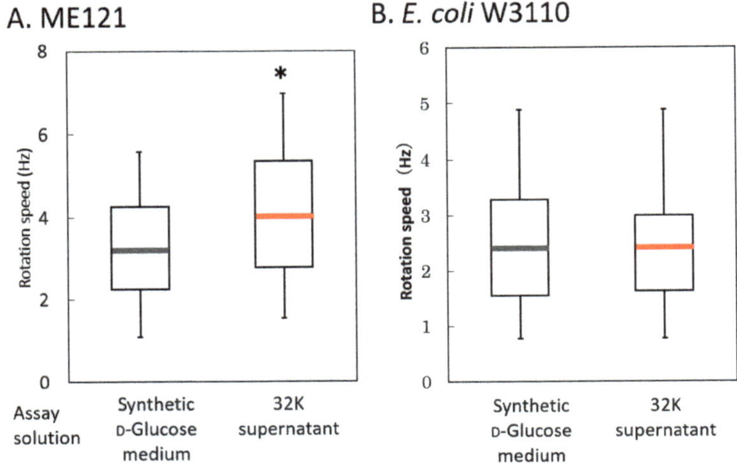

Figure 7. Rotation measurement of the flagellar motor using a tethered cell assay in ME121 (**A**) and W3110 (**B**). The variance of the results of the rotational measurement experiment of the flagellar motor of ME121 was compared with W3110. The vertical axis shows rotation speed (Hz). The explanation of the box-and-whisker plot is shown in the legend of Figure 3. * Significant difference from the synthetic D-glucose medium ($p < 0.001$). Values are expressed as the mean of three independent experiments, and the swimming speed of at least 100 cells was measured.

Next, the pitch of the flagellar fibers was determined in the presence or absence of the 32K culture supernatant. Specifically, fluorescent staining was used to investigate whether this pitch affected the swimming speed of ME121. The addition of the 32K culture supernatant also significantly accelerated the swimming speed of the stained flagellum ME121 (Supplementary Figure S3), indicating that fluorescent staining had no effect on the motility acceleration of ME121. The pitch of the flagellar fibers of ME121 was also measured (Figure 8). When the 32K culture supernatant was added, the pitch of the flagellar fibers of ME121 was shortened by approximately 10%.

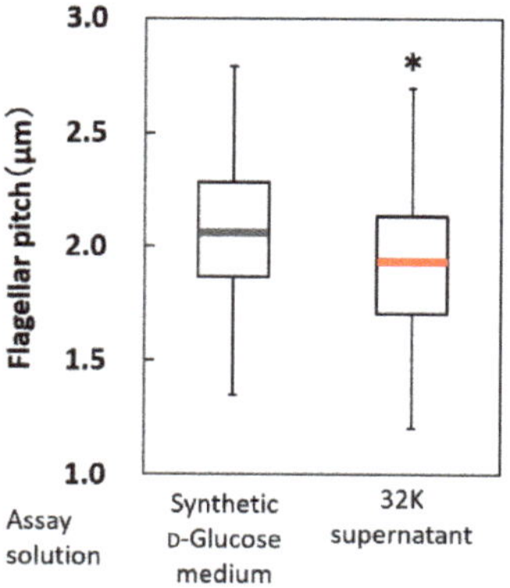

Figure 8. Flagellar pitch of strain ME121 with and without the 32K culture supernatant treatment. The variance of the measurement results of the pitch length of flagellar fiber of ME121 was compared with/without the 32K culture supernatant treatment. The explanation of the box-and-whisker plot is shown in the legend of Figure 3. * Significant difference from the synthetic D-glucose medium ($p < 0.001$). Values are expressed as the mean of three independent experiments, and the swimming speed of at least 100 cells was measured.

The pH values of the synthetic D-glucose medium and the 32K culture supernatant were 6.74 and 5.56, respectively. To investigate the pH influence, the pH of the synthetic D-glucose medium was adjusted to match that of the 32K culture supernatant, after which the swimming speed of ME121 was analyzed (Figure 9). Even after the pH adjustment, the swimming speed of ME121 did not accelerate, suggesting that environmental pH was not involved in the accelerated motility of ME121.

Lastly, the elevated viscosity of the swimming environment did not increase the ME121 motility (Supplementary Figure S4). Thus, the acceleration of the ME121 swimming speed was not related to changes in environmental viscosity.

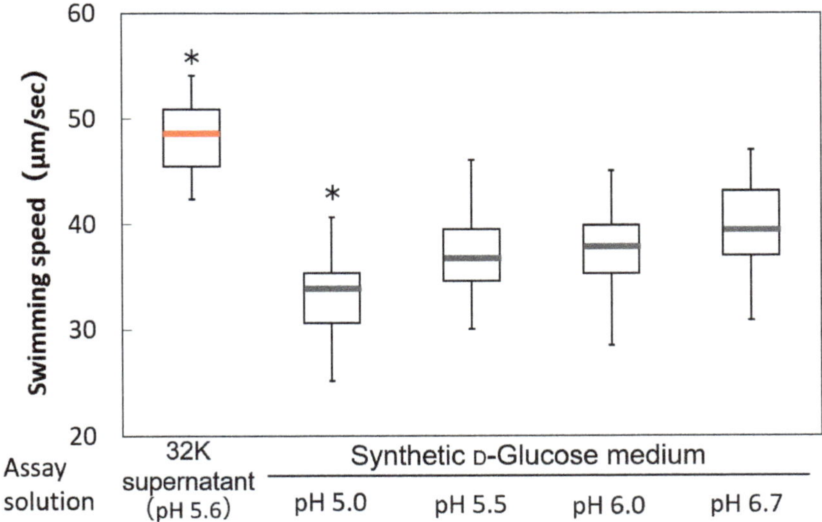

Figure 9. Swimming speed of ME121 using the 32K culture supernatant and the synthetic D-glucose medium at several pH values. The variance of the swimming speed measurement results of ME121 was compared using the 32K culture supernatant and the synthetic D-glucose medium with several pH values. The experimental method is similar to that outlined in Figure 8. The explanation of the box-and-whisker plot is shown in the legend of Figure 3. * Significant difference from the synthetic medium (pH 6.7; $p < 0.001$). Values are expressed as the mean of three independent experiments, and the swimming speed of at least 100 cells was measured.

4. Discussion

4.1. 32K-Derived Products Stimulated the ME121 Growth and Swimming Speed

When ME121 and 32K were separately cultured in the synthetic D-glucose medium, growth was not as good as in the coculture. When using Beppu flasks, a significant improvement in growth, a prolonged swimming period, and an accelerated swimming speed were observed in all other culture conditions, compared with the ME121 monoculture. The growth of ME121 was initially slow in the dialysis culture, but its final OD_{600} was higher than that in the ME121 monoculture, and swimming duration was also prolonged for up to 84 h. In each culture condition, no swimming was observed once the stationary growth phase was reached. Plant-related *Methylobacterium* species and *Vibrio alginolyticus* release polar flagella after prolonged stationary culturing [19,20]. In *E. coli*, when nutrients are depleted, YcgR, a cyclic dimer guanosine monophosphate (c-di-GMP) binding protein, is activated, and YcgR and the flagellar switch complex proteins FliG and FliM interact to stop rotation without detaching the flagella [21,22]. Studies on the cessation of swimming of *Methylobacterium* species during prolonged stationary culturing and the mode of flagella are still poor and require for further investigation.

The swimming speed of ME121 was faster in the dialysis culture, the coculture, and the dialysis coculture than in the monoculture. In the dialysis coculture, the coculture in the right tank assimilated glucose in the medium and might have delayed the growth of the ME121 monoculture in the left tank. Therefore, the final OD_{600} reached by supplying the 32K-derived products was higher than that of the monoculture. It was concluded that these phenomena were due to 32K-derived products that stimulated the growth and swimming speed in the dialysis culture, the coculture, and the dialysis coculture conditions of ME121.

4.2. The Growth and Motility of ME121 Using the 32K Culture Supernatant

When the 32K culture supernatant was used as the medium, the growth of ME121 notably improved, and the swimming period was prolonged, indicating that the K factor was contained in the 32K culture supernatant. The increased swimming speed of ME121 was also observed immediately after suspension in the 32K culture supernatant. Therefore, the K factor was not metabolized by ME121. We expect that the 32K culture supernatant contains the K factor and growth factors promoting ME121. The growth factors may have been the same as the K factor.

4.3. The Motility of ME121 Using Treated Solutions (Heating, Dialysis, Enzyme Treatment, etc.) of the 32K Culture Supernatant

The K factor is a polar substance with a molecular weight of approximately 5–10 kDa. It is neither a lipid nor a protein cleaved by proteases. Therefore, we considered that the K factor is a sugar-containing complex or a sugar chain-modified protein.

4.4. Analysis of the Monosaccharide Composition of the Ethanol-Precipitated Fraction of the 32K Culture Supernatant

Because the neutral sugars glucose and galactose were present in a ratio of approximately 1:1 and accounted for approximately 55% of the K factor, the K factor was considered to be a kind of EPS composed of neutral sugars (Table 1). Demir and Salman reported that accelerated bacterial motility by attractants caused an increase in flagellar motor torque [23]. Bacteria are also known to move in the direction of attractants, such as amino acids and sugars [24]. Therefore, we report an unprecedented case of EPS-derived macromolecules driving the acceleration of bacterial swimming speed.

4.5. A Hypothesis on the Mechanism of Motion Acceleration of the Flagellar Motor of ME121 by the K Factor

The rotation speed of the ME121 flagellar motor was increased by approximately 25% when the 32K culture supernatant was added. However, in the case of W3110, an acceleration of the flagellar motor rotation speed of *E. coli* was not observed. This may be suggestive of the ME121 flagellar motor-specificity of the K factor. Because the acceleration of the rotation speed was immediately observed once the 32K culture supernatant was added, it is likely that the acceleration of ME121 swimming speed was due to an increase in the rotational force of the flagellar motor.

Furthermore, the addition of the 32K culture supernatant shortened the flagellar fiber pitch of ME121 by approximately 10%. Changes in flagellar pitch occur when environmental pH changes [25]. The pH of the synthetic D-glucose medium and the 32K culture supernatant were 6.74 and 5.56, respectively, suggesting that the short pitch of the flagellar filaments of ME121 may be affected by pH. However, no acceleration of the ME121 swimming speed was observed, indicating that differences in environmental pH are not involved in the acceleration of the ME121 motility.

From the abovementioned results, we proposed a mechanism underlying motility acceleration. The flagellar motor is considered to have a structure consisting of a stator and a rotor that is more dynamic than previously believed [26–28]. Upon activation, the stator is incorporated into the flagellar motor and generates a rotational driving force to rotate. *Bacillus subtilis* has two flagellar motor stators, MotAB and MotPS [29,30]. Though MotAB is more often used, *B. subtilis* uses MotPS, instead of MotAB, by recognizing the viscosity of the environment in the hydrophilic region of MotPS [31]. In our study, however, ME121 did not exhibit an improved motility even when the viscosity of the environment increased. Thus, the improvement in the ME121 swimming speed is not related to changes in viscosity.

ME121 has a MotA/MotB stator (Accession no. ME121_1986 and ME121_1987). The attachment and detachment of the stator from the motor are dynamic and dependent on the load on the motor [32]. The maximum speed of the motor increases as additional stators are recruited to the motor. This is one of the proposed mechanisms by which the K factor increases ME121 swimming speed. In other words, the increase in the rotational force of the ME121 flagellum motor is caused by the introduction of the additional stator into the motor by the K factor. In other species, such as *Vibrio* spp., additional components (MotX and MotY) are known to increase bacterial swimming speed [33]. However, no such additional components in the flagellar motor of *Methylobacterium* spp. are known. No proteins homologous to MotX and MotY were identified in the ME121 genome sequence. We plan to elucidate this mechanism in the future by generating a ME121 mutant whose swimming speed is not increased by the K factor.

By recognizing the motility accelerating factor, ME121 was inferred to generate a stronger rotational driven force by incorporating more stators into the flagellar motor than usual. The coculture of ME121 and 32K improved the growth of both strains. 32K secreted the K factor by metabolizing nutrients in the medium. It was considered that the ME121 swimming speed was thereby accelerated. From the draft genome sequence of ME121, 33 methyl-accepting chemotaxis sensor (MCP) proteins involved in chemotaxis have been identified [6]. However, studies on chemotaxis and MCP in *Methylobacterium* spp. remain limited. It has been known that even nonmotile bacteria, such as 32K, have passive motility (colony spreading) [34].

In nature, bacteria often exist as agglomerations known as biofilms. Biofilm formation and bacterial motility are strongly related [35]. Generally, a motility defective mutant is related to poor biofilm formation [30]. Biofilms are formed by pathogenic bacteria on medical materials, such as intravascular indwelling catheters, artificial heart valves, and artificial joints, which cause infections [34]. In biofilm formation in cocultures using the opportunistic pathogens *Pseudomonas aeruginosa* and *Staphylococcus aureus*, the predominance of *P. aeruginosa* in the biofilm indicates that diguanylate cyclase is involved in matrix polysaccharide biosynthesis and its control from the early stage to the mature stage [36]. The modulation of the second messenger c-di-GMP levels is linked to bacterial swimming and biofilm formation [37].

We think that how ME121 senses the K factor, i.e., the mechanism of the accelerated motility, is associated with biofilm formation. Interestingly, when biofilm formation was studied using phylogenetically isolated *Methylobacterium*, biofilm formation in cocultures is enhanced compared with monocultures [38]. Therefore, biofilm formation in the coculture of ME121 and 32K strains should be further investigated. It would also be interesting to investigate whether the increase in the motility of *Methylobacterium* spp. is observed in the coculture of *Methylobacterium* spp. and *Kaistia* spp. isolated from different environments.

5. Conclusions

In this study, we suggested that the acceleration of ME121 motility is caused by the metabolites of 32K supernatant and not by the contact stimulation between cells. Our findings suggested that the K factor is an extracellular saccharide of 5–10 kDa produced by 32K and contains neutral sugars. We further inferred that motility was accelerated by the enhancement of the motor torque of the ME121 flagellar motor. We will promote the further elucidation of the swimming acceleration mechanism by the K factor.

Supplementary Materials: The following are available online at http://www.mdpi.com/2218-273X/10/4/618/s1, Figure S1: Quick-freeze deep-etch replica TEM imaging of a *Methylobacterium* sp. ME121 cell (A) and a *Kaistia* sp. 32K cell (B); Figure S2: Chromatogram of the monosaccharide composition in the ethanol precipitate fraction of the 32K culture supernatant; Figure S3: Swimming speed of ME121 in the 32K culture supernatant after fluorescence staining; Figure S4: Swimming speed of strain ME121 in a swimming assay buffer containing several concentrations of Ficoll 400; Video S1: (A) ME121 swimming in the monoculture of strain ME121 and (B) ME121 swimming in coculture of both strains ME121 and 32K strains; Video S2: (A) ME121 swimming in the synthetic D-glucose medium and (B) ME121 swimming in 32K culture supernatant.

Author Contributions: A.N. and M.I. designed the research; Y.U., Y.W., T.S., Y.O.T. and M.M. conducted the research; A.N. and M.I. analyzed the data; and Y.U. and M.I. wrote the paper. All authors have read and agreed to the published version of the manuscript.

Funding: This research was funded by a Grant-in-Aid for Scientific Research on Innovative Areas of the Ministry of Education, Culture, Sports, Science and Technology of Japan, grant number 24117005 (M.I.). The electron microscopy was supported by the Grant-in-Aid for Scientific Research on Innovative Areas "Harmonized Supramolecular Motility Machinery and Its Diversity", grant number JP24117001 (M.M.).

Acknowledgments: We would like to thank Arthur A. Guffanti for the critical discussions and reading of the manuscript.

Conflicts of Interest: The authors declare that they have no conflicts of interest. Ethical approval: This article does not describe any studies with human participants or animals performed by any of the authors.

References

1. Olson, M.L.; Jayaraman, A.; Kao, K.C. Relative Abundances of *Candida albicans* and *Candida glabrata* in In Vitro Coculture Biofilms Impact Biofilm Structure and Formation. *Appl. Environ. Microbiol.* **2018**, *84*, e02769-17. [CrossRef]
2. Onaka, H.; Mori, Y.; Igarashi, Y.; Furumai, T. Mycolic acid-containing bacteria induce natural-product biosynthesis in *Streptomyces* species. *Appl. Environ. Microbiol.* **2011**, *77*, 400–406. [CrossRef]
3. Kumar, S.; Treloar, B.P.; Teh, K.H.; McKenzie, C.M.; Henderson, G.; Attwood, G.T.; Waters, S.M.; Patchett, M.L.; Janssen, P.H. *Sharpea* and *Kandleria* are lactic acid producing rumen bacteria that do not change their fermentation products when co-cultured with a methanogen. *Anaerobe* **2018**. [CrossRef] [PubMed]
4. Barnett, A.M.; Roy, N.C.; Cookson, A.L.; McNabb, W.C. Metabolism of Caprine Milk Carbohydrates by Probiotic Bacteria and Caco-2:HT29(-)MTX Epithelial Co-Cultures and Their Impact on Intestinal Barrier Integrity. *Nutrients* **2018**, *10*, 949. [CrossRef] [PubMed]
5. Nakamura, S.; Morimoto, Y.V.; Kudo, S. A lactose fermentation product produced by *Lactococcus lactis* subsp. lactis, acetate, inhibits the motility of flagellated pathogenic bacteria. *Microbiology* **2015**, *161*, 701–707. [CrossRef]
6. Fujinami, S.; Takeda-Yano, K.; Onodera, T.; Satoh, K.; Shimizu, T.; Wakabayashi, Y.; Narumi, I.; Nakamura, A.; Ito, M. Draft Genome Sequence of *Methylobacterium* sp. ME121, Isolated from Soil as a Mixed Single Colony with *Kaistia* sp. 32K. *Genome Announc.* **2015**, *3*. [CrossRef]
7. Shimizu, T.; Takaya, N.; Nakamura, A. An L-glucose catabolic pathway in *Paracoccus* species 43P. *J. Biol. Chem.* **2012**, *287*, 40448–40456. [CrossRef] [PubMed]
8. Karmakar, R.; Naaz, F.; Tirumkudulu, M.S.; Venkatesh, K.V. *Escherichia coli* modulates its motor speed on sensing an attractant. *Arch. Microbiol.* **2016**, *198*, 827–833. [CrossRef] [PubMed]
9. Ohno, M.; Okano, I.; Watsuji, T.; Kakinuma, T.; Ueda, K.; Beppu, T. Establishing the independent culture of a strictly symbiotic bacterium *Symbiobacterium thermophilum* from its supporting *Bacillus* strain. *Biosci. Biotechnol. Biochem.* **1999**, *63*, 1083–1090. [CrossRef] [PubMed]
10. Takahashi, Y.; Koyama, K.; Ito, M. Suppressor mutants from MotB-D24E and MotS-D30E in the flagellar stator complex of *Bacillus subtilis*. *J. Gen. Appl. Microbiol.* **2014**, *60*, 131–139. [CrossRef]
11. Aono, R.; Ito, M.; Horokoshi, K. Occurrence of teichuronopeptide in cell walls of group 2 alkaliphilic *Bacillus* spp. *J. Gen. Microbiol.* **1993**, *139*, 2739–2744. [CrossRef]
12. Spiro, R.G. Analysis of sugars found in glycoproteins. *Method. Enzymol.* **1966**, *8*, 3–26. [CrossRef]
13. Wang, W.T.; LeDonne, N.C., Jr.; Ackerman, B.; Sweeley, C.C. Structural characterization of oligosaccharides by high-performance liquid chromatography, fast-atom bombardment-mass spectrometry, and exoglycosidase digestion. *Anal. Biochem.* **1984**, *141*, 366–381. [CrossRef]
14. Kinosita, Y.; Kikuchi, Y.; Mikami, N.; Nakane, D.; Nishizaka, T. Unforeseen swimming and gliding mode of an insect gut symbiont, Burkholderia sp. RPE64, with wrapping of the flagella around its cell body. *ISME J.* **2018**, *12*, 838–848. [CrossRef]
15. Heuser, J.E. Procedure for freeze-drying molecules adsorbed to mica flakes. *J. Mol. Biol.* **1983**, *169*, 155–195. [CrossRef]
16. Numata, T.; Murakami, T.; Kawashima, F.; Morone, N.; Heuser, J.E.; Takano, Y.; Ohkubo, K.; Fukuzumi, S.; Mori, Y.; Imahori, H. Utilization of photoinduced charge-separated state of donor-acceptor-linked molecules for regulation of cell membrane potential and ion transport. *J. Am. Chem. Soc.* **2012**, *134*, 6092–6095. [CrossRef]
17. Kunoh, T.; Suzuki, T.; Shiraishi, T.; Kunoh, H.; Takada, J. Treatment of leptothrix cells with ultrapure water poses a threat to their viability. *Biology (Basel)* **2015**, *4*, 50–66. [CrossRef]
18. Tulum, I.; Tahara, Y.O.; Miyata, M. Peptidoglycan layer and disruption processes in *Bacillus subtilis* cells visualized using quick-freeze, deep-etch electron microscopy. *Microscopy (Oxf.)* **2019**, *68*, 441–449. [CrossRef]

19. Doerges, L.; Kutschera, U. Assembly and loss of the polar flagellum in plant-associated methylobacteria. *Naturwissenschaften* **2014**, *101*, 339–346. [CrossRef]
20. Zhuang, X.Y.; Guo, S.; Li, Z.; Zhao, Z.; Kojima, S.; Homma, M.; Wang, P.; Lo, C.J.; Bai, F. Live cell fluorescence imaging reveals dynamic production and loss of bacterial flagella. *Mol. Microbiol.* **2020**. [CrossRef]
21. Paul, K.; Nieto, V.; Carlquist, W.C.; Blair, D.F.; Harshey, R.M. The c-di-GMP binding protein YcgR controls flagellar motor direction and speed to affect chemotaxis by a "backstop brake" mechanism. *Mol. Cell.* **2010**, *38*, 128–139. [CrossRef] [PubMed]
22. Boehm, A.; Kaiser, M.; Li, H.; Spangler, C.; Kasper, C.A.; Ackermann, M.; Kaever, V.; Sourjik, V.; Roth, V.; Jenal, U. Second messenger-mediated adjustment of bacterial swimming velocity. *Cell* **2010**, *141*, 107–116. [CrossRef] [PubMed]
23. Demir, M.; Salman, H. Bacterial thermotaxis by speed modulation. *Biophys. J.* **2012**, *103*, 1683–1690. [CrossRef] [PubMed]
24. Bi, S.; Sourjik, V. Stimulus sensing and signal processing in bacterial chemotaxis. *Curr. Opin. Microbiol.* **2018**, *45*, 22–29. [CrossRef]
25. Darnton, N.C.; Berg, H.C. Force-extension measurements on bacterial flagella: Triggering polymorphic transformations. *Biophys. J.* **2007**, *92*, 2230–2236. [CrossRef]
26. Miyata, M.; Robinson, R.C.; Uyeda, T.Q.P.; Fukumori, Y.; Fukushima, S.I.; Haruta, S.; Homma, M.; Inaba, K.; Ito, M.; Kaito, C.; et al. Tree of motility—A proposed history of motility systems in the tree of life. *Genes Cells* **2020**, *25*, 6–21. [CrossRef]
27. Fukuoka, H.; Wada, T.; Kojima, S.; Ishijima, A.; Homma, M. Sodium-dependent dynamic assembly of membrane complexes in sodium-driven flagellar motors. *Mol. Microbiol.* **2009**, *71*, 825–835. [CrossRef]
28. Leake, M.C.; Chandler, J.H.; Wadhams, G.H.; Bai, F.; Berry, R.M.; Armitage, J.P. Stoichiometry and turnover in single, functioning membrane protein complexes. *Nature* **2006**, *443*, 355–358. [CrossRef]
29. Terahara, N.; Fujisawa, M.; Powers, B.; Henkin, T.M.; Krulwich, T.A.; Ito, M. An intergenic stem-loop mutation in the *Bacillus subtilis ccpA-motPS* operon increases *motPS* transcription and the MotPS contribution to motility. *J. Bacteriol.* **2006**, *188*, 2701–2705. [CrossRef]
30. Ito, M.; Hicks, D.B.; Henkin, T.M.; Guffanti, A.A.; Powers, B.D.; Zvi, L.; Uematsu, K.; Krulwich, T.A. MotPS is the stator-force generator for motility of alkaliphilic *Bacillus*, and its homologue is a second functional Mot in *Bacillus subtilis*. *Mol. Microbiol.* **2004**, *53*, 1035–1049. [CrossRef]
31. Ito, M.; Terahara, N.; Fujinami, S.; Krulwich, T.A. Properties of motility in *Bacillus subtilis* powered by the H^+-coupled MotAB flagellar stator, Na^+-coupled MotPS or hybrid stators MotAS or MotPB. *J. Mol. Biol.* **2005**, *352*, 396–408. [CrossRef]
32. Nirody, J.A.; Berry, R.M.; Oster, G. The Limiting Speed of the Bacterial Flagellar Motor. *Biophys. J.* **2016**, *111*, 557–564. [CrossRef]
33. Takekawa, N.; Kojima, S.; Homma, M. Mutational analysis and overproduction effects of MotX, an essential component for motor function of Na^+-driven polar flagella of *Vibrio*. *J. Biochem.* **2017**, *161*, 159–166. [CrossRef]
34. Kaito, C.; Sekimizu, K. Colony spreading in *Staphylococcus aureus*. *J. Bacteriol.* **2007**, *189*, 2553–2557. [CrossRef]
35. Dufrene, Y.F.; Persat, A. Mechanomicrobiology: How bacteria sense and respond to forces. *Nat. Rev. Microbiol.* **2020**, *18*, 227–240. [CrossRef]
36. Chew, S.C.; Yam, J.K.H.; Matysik, A.; Seng, Z.J.; Klebensberger, J.; Givskov, M.; Doyle, P.; Rice, S.A.; Yang, L.; Kjelleberg, S. Matrix Polysaccharides and SiaD Diguanylate Cyclase Alter Community Structure and Competitiveness of *Pseudomonas aeruginosa* during Dual-Species Biofilm Development with *Staphylococcus aureus*. *mBio* **2018**, *9*. [CrossRef]
37. Opoku-Temeng, C.; Sintim, H.O. Targeting c-di-GMP Signaling, Biofilm Formation, and Bacterial Motility with Small Molecules. *Methods Mol. Biol.* **2017**, *1657*, 419–430. [CrossRef]
38. Xu, F.F.; Morohoshi, T.; Wang, W.Z.; Yamaguchi, Y.; Liang, Y.; Ikeda, T. Evaluation of intraspecies interactions in biofilm formation by *Methylobacterium* species isolated from pink-pigmented household biofilms. *Microbe Environ.* **2014**, *29*, 388–392. [CrossRef]

Article

A Novel Lysophosphatidic Acid Acyltransferase of *Escherichia coli* Produces Membrane Phospholipids with a *cis*-vaccenoyl Group and Is Related to Flagellar Formation

Yosuke Toyotake [1,2], Masayoshi Nishiyama [1], Fumiaki Yokoyama [1], Takuya Ogawa [1], Jun Kawamoto [1] and Tatsuo Kurihara [1,*]

1 Institute for Chemical Research, Kyoto University, Uji, Kyoto 611-0011, Japan; toyotake@fc.ritsumei.ac.jp (Y.T.); mnishiyama@phys.kindai.ac.jp (M.N.); yokoyama@mbc.kuicr.kyoto-u.ac.jp (F.Y.); ogawa.tky@mbc.kuicr.kyoto-u.ac.jp (T.O.); jun_k@mbc.kuicr.kyoto-u.ac.jp (J.K.)

2 Department of Biotechnology, College of Life Sciences, Ritsumeikan University, 1-1-1 Noji-higashi, Kusatsu, Shiga 525-8577, Japan

* Correspondence: kurihara@scl.kyoto-u.ac.jp or kurihara@mbc.kuicr.kyoto-u.ac.jp; Tel.: +81-774-38-4710

Received: 26 March 2020; Accepted: 7 May 2020; Published: 11 May 2020

Abstract: Lysophosphatidic acid acyltransferase (LPAAT) introduces fatty acyl groups into the *sn*-2 position of membrane phospholipids (PLs). Various bacteria produce multiple LPAATs, whereas it is believed that *Escherichia coli* produces only one essential LPAAT homolog, PlsC—the deletion of which is lethal. However, we found that *E. coli* possesses another LPAAT homolog named YihG. Here, we show that overexpression of YihG in *E. coli* carrying a temperature-sensitive mutation in *plsC* allowed its growth at non-permissive temperatures. Analysis of the fatty acyl composition of PLs from the *yihG*-deletion mutant (Δ*yihG*) revealed that endogenous YihG introduces the *cis*-vaccenoyl group into the *sn*-2 position of PLs. Loss of YihG did not affect cell growth or morphology, but Δ*yihG* cells swam well in liquid medium in contrast to wild-type cells. Immunoblot analysis showed that FliC was highly expressed in Δ*yihG* cells, and this phenotype was suppressed by expression of recombinant YihG in Δ*yihG* cells. Transmission electron microscopy confirmed that the flagellar structure was observed only in Δ*yihG* cells. These results suggest that YihG has specific functions related to flagellar formation through modulation of the fatty acyl composition of membrane PLs.

Keywords: lysophosphatidic acid acyltransferase; membrane phospholipid diversity; swimming motility; flagellar formation

1. Introduction

Phospholipids (PLs) are the primary component of biological membranes. They consist of a phosphate-containing head group and two fatty acyl groups. The structural diversity of these fatty acyl groups affects the physical state of biological membranes such as fluidity, permeability, rigidity, and thickness [1–4]. Bacteria maintain the ideal physical state of their membrane in response to environmental changes by modulating the fatty acyl groups of membrane PLs. For example, as environmental temperature decreases, bacteria generally introduce lower-melting-point fatty acyl groups such as unsaturated fatty acyl groups and branched-chain fatty acyl groups into membrane PLs [5–13]. However, it is still not fully understood how bacteria regulate the fatty acyl composition of membrane PLs in response to environmental changes.

During the de novo synthesis of PLs [14–17], fatty acyl groups are incorporated into the *sn*-1 and *sn*-2 position by glycerol-3-phosphate acyltransferases and lysophosphatidic acid acyltransferases

(LPAATs), respectively [18–23]. In some bacteria, such as *Neisseria meningitidis*, *Pseudomonas fluorescens*, *Pseudomonas aeruginosa*, and *Rhodobacter capsulatus*, multiple LPAAT homologs have been identified and characterized [24–27]. These LPAATs potentially play different roles in vivo to contribute to the generation of diversity in membrane PLs.

Shewanella livingstonensis Ac10 is a psychrotrophic bacterium used as a model of cold-adapted organisms [9,28–31]. This bacterium has five LPAAT homologs (SlPlsC1 to SlPlsC5) [32]. We previously reported that SlPlsC1 plays a major role in the synthesis of PLs containing polyunsaturated fatty acyl groups [32,33], while SlPlsC4 is mainly responsible for the synthesis of PLs containing branched-chain fatty acyl groups (i13:0 and i15:0) [34]. Some marine bacteria such as *Alteromonas mediterranea* and *Colwellia psychrerythraea* have a putative SlPlsC4 ortholog. These bacteria also have an SlPlsC1 ortholog, suggesting that the multiple LPAAT homologs introduce specific fatty acyl groups into membrane PLs for their adaptation to the marine environment, as has been shown in *S. livingstonensis* Ac10. Likewise, it is conceivable that uncharacterized LPAAT homologs also exist in other bacterial species to generate membrane lipid diversity to allow environmental adaptation.

It has long been believed that *Escherichia coli* has only one essential LPAAT homolog, named PlsC—the deletion of which is lethal [35]. However, we found that *E. coli* possesses an SlPlsC4 ortholog named YihG. YihG was originally thought to be a second poly(A) polymerase [36], but this claim has subsequently been denied [37]. Even though YihG can be considered as an inner membrane protein belonging to a 1-acyl-*sn*-glycerol-3-phosphate *O*-acyltransferase family based on its conserved catalytic motif [38,39], the sequence identity between YihG and *E. coli* PlsC is 17.9%, and thus YihG has not been recognized as a functional LPAAT homolog. Sutton and co-workers previously reported that overproduction of YihG suppresses the hyperinitiation of DNA replication and resulting growth defect in *E. coli*, presumably by regulating the level of ATP-binding DnaA (DnaA-ATP), an activator for the initiation of DNA replication [40]. However, no one has yet clarified whether YihG has LPAAT activity or investigated how YihG affects the membrane lipid composition. YihG is conserved in some γ-proteobacteria such as *P. aeruginosa*, *Salmonella typhimurium*, and *Vibrio cholerae*. Thus, we hypothesized that an SlPlsC4 ortholog has a physiological function in various γ-proteobacteria including some enteric and pathogenic bacteria.

In this study, we demonstrated that YihG is a functional LPAAT homolog by complementation assay using the *E. coli* strain JC201, a temperature-sensitive *plsC* mutant. We found that YihG has a different substrate specificity from PlsC, and that endogenous YihG contributes to the synthesis of PLs containing a *cis*-vaccenoyl group at the *sn*-2 position. The lack of YihG caused enhanced flagellar formation and swimming motility in the liquid medium. Thus, *E. coli* YihG appears to regulate bacterial swimming motility through modulation of the composition of fatty acyl groups in membrane PLs.

2. Materials and Methods

2.1. Bacterial Strains, Plasmids, and Growth Conditions

The bacterial strains and plasmids used in this study are listed in Table S1. *E. coli* K-12 strain BW25113 and its *yihG*-knockout mutant (Δ*yihG*) were obtained from the National BioResource Project (NIG, Mishima, Japan). *E. coli* JC201 strain, which carries a temperature-sensitive mutation in *plsC*, was used in the complementation assay to evaluate LPAAT activity [35]. The *E. coli* cells were cultivated in lysogeny broth [LB; 1% (*w*/*v*) Bacto Tryptone, 0.5% (*w*/*v*) Bacto Yeast extract, and 1% (*w*/*v*) NaCl], tryptone broth [TB; 1% (*w*/*v*) Bacto Tryptone and 1% (*w*/*v*) NaCl], and M9-based minimal medium [48 mM Na_2HPO_4, 22 mM KH_2PO_4, 19 mM NH_4Cl, 8.6 mM NaCl, 0.1 mM $CaCl_2$, 1 mM $MgSO_4$, and 0.4% (*w*/*v*) glucose]. Growth was monitored by measuring OD_{600} with a UV–visible spectrophotometer (UV-2450, Shimadzu, Kyoto, Japan). Antibiotics were used, when required, at the following concentrations: kanamycin (30 μg/mL) and chloramphenicol (34 μg/mL).

2.2. Construction of YihG- and PlsC-Expression Plasmids

PCR primers used in this study are listed in Table S2. The plasmid named pBAD-Cm^R was created from pBAD28 [41] by digestion with ScaI and self-ligation to remove the ampicillin resistance gene. DNA fragments coding for the C-terminal hexa-histidine-tagged YihG (YihG-His$_6$) and PlsC (PlsC-His$_6$) were obtained by PCR using the BW25113 genome as a template. The resulting DNA fragments were individually introduced into the SalI-HindIII site in pBAD-Cm^R using an In-Fusion Advantage PCR cloning kit (TaKaRa Bio, Otsu, Japan) following the manufacturer's instructions. The resulting plasmids were designated pBAD/*yihG-his*$_6$ and pBAD/*plsC-his*$_6$, respectively.

2.3. JC201 Complementation Assay

JC201 cells harboring pBAD-Cm^R or pBAD/*yihG-his*$_6$ were cultured at 30 °C in LB until the OD$_{600}$ reached 1.2 to 1.4. The cell cultures were normalized to an OD$_{600}$ of 1.0 and diluted to 10^{-2}, 10^{-3}, 10^{-4}, 10^{-5}, and 10^{-6} in fresh LB. Three microliters of the serial dilutions was spotted onto 1.5% (*w*/*v*) agar LB plates containing 0 to 2% (*w*/*v*) L-arabinose. The plates were incubated at 30 and 42 °C until colonies were formed.

JC201 cells harboring pBAD-Cm^R, pBAD/*yihG-his*$_6$, or pBAD/*plsC-his*$_6$ were cultured at 30 °C in LB until an OD$_{600}$ reached 1.0 to 1.8. They were diluted to an OD$_{600}$ of 0.01 in LB containing 0.5% (*w*/*v*) and 1% (*w*/*v*) L-arabinose and grown at 37 °C for 9 h to monitor their growth rates. To analyze membrane lipids as described below, cells expressing YihG and PlsC with 1% (*w*/*v*) L-arabinose were grown and harvested by centrifugation at room temperature when their OD$_{600}$ reached 0.6 to 0.8.

2.4. Total PL Extraction and Analysis by Electrospray Ionization Tandem Mass Spectrometry (ESI–MS/MS)

BW25113 cells harboring pBAD-Cm^R and Δ*yihG* cells harboring pBAD-Cm^R or pBAD/*yihG-his*$_6$ were grown in TB containing 0.2% (*w*/*v*) L-arabinose at 37 °C to an OD$_{600}$ of 0.4 to 0.6 and harvested by centrifugation at room temperature. The cells were flash frozen in liquid nitrogen and stored at −80 °C until use. The frozen cells were lyophilized, and total PLs were extracted using the Bligh and Dyer method [42]. JC201 cells harvested as described above were not lyophilized and were instead directly used for PL extraction. The extracted PLs were analyzed by ESI–MS/MS [a triple-quadrupole Sciex API 3000™ LC/MS/MS System (Applied Biosystems, Foster City, CA, USA)], as described previously [34].

2.5. Analysis of the sn-1 and sn-2 Fatty Acyl Groups of PLs

The PLs prepared as described above were hydrolyzed with Phospholipase A2 (PLA2, P6534, Sigma, St. Louis, MO). The resulting *sn*-2 fatty acyl groups were extracted by the Dole's method [34,43,44] and analyzed by gas chromatography–mass spectrometry [GC–MS, Clarus 680 gas chromatograph interfaced with Clarus SQ 8C mass spectrometer (Perkin Elmer, Wellesley, MA) equipped with an Agilent J&W GC column DB-1 (Agilent Technologies Inc., Santa Clara, CA, USA)], as described previously [45]. Lysophospholipids (LPLs) were analyzed using the total lipid extracts from an aliquot of the PLA2 reaction product by ESI–MS as described above.

2.6. Motility Assay in Soft Agar Plates

The BW25113 and Δ*yihG* cells were grown in LB at 37 °C to an OD$_{600}$ of 0.8 to 1.2. Two microliters of each cell culture was spotted on TB 0.2% (*w*/*v*) agar plates. For Δ*yihG* cells harboring the plasmids, TB 0.2% (*w*/*v*) agar plates containing 0.02% (*w*/*v*) or 0.2% (*w*/*v*) L-arabinose were used. The plates were incubated at 37 °C for an appropriate time, as noted in the Results section.

2.7. Microscopic Observation of Swimming Cells

The BW25113 and Δ*yihG* cells were grown at 37 °C in TB to an OD$_{600}$ of 0.6 to 1.3. The cultivated cells were diluted with fresh TB medium. Motility of the cells was observed at room temperature under a microscope (Ti–E, Nikon, Tokyo, Japan) [46]. The phase-contrast images of cells near the cover slip were recorded at video rate. All assays were repeated with four different cultures. The fraction of the swimming cells was obtained by dividing the number of cells that swam with a speed of >2 μm/s by that of all cells in the focal plane. The speed of each of the swimming cells was analyzed using a custom-made plugin (Version 0.7.1) of Image J [47].

2.8. Flagellin Preparation and Analysis by Western Blot Analysis

The BW25113 and Δ*yihG* cells were grown at 37 °C in TB to an OD_{600} of 0.5 to 0.7. Δ*yihG* cells harboring the plasmids were grown in TB containing 0.2% (*w*/*v*) L-arabinose. The culture was transferred into a tube and vigorously shaken to shear off cell-associated flagella. The sample was centrifuged to spin down the cells, and the supernatant was passed through a filter with a pore size of 0.45 μm. The filtrated supernatant was concentrated with an Amicon Ultra device (30,000 MWCO), and the flagella in the supernatant were precipitated with trichloroacetic acid. The precipitates were subjected to SDS-PAGE and Western blot analysis using the rabbit anti-flagellin antibody [48]. The signal was detected using a peroxidase-conjugated anti-rabbit IgG antibody (Sigma, St. Louis, MO, USA) and Chemi-Lumi One Ultra (Nacalai Tesque, Kyoto, Japan).

2.9. Transmission Electron Microscope (TEM) Observation of the Flagellar Structures

TEM observation of the flagellar structures was performed according to the method described by Furuno et al. [49]. The BW25113 and Δ*yihG* cells were grown at 37 °C on TB 1.5% (*w*/*v*) agar plates until colonies were formed. Some colonies were scraped with a toothpick and gently suspended in 2.5 μL of water to prevent detachment of the flagellar from the cells. Two microliters of the suspension was adsorbed onto a hydrophilized carbon-coated copper grid and then treated twice with 2% (*w*/*v*) phosphotungstic acid solution. The TEM images were obtained with a JEM-1400 transmission electron microscope (JEOL, Ltd., Tokyo, Japan) at an accelerating voltage of 120 kV. Images were acquired using a charge-coupled device camera (a built-in camera in the JEM-1400).

3. Results

3.1. Identification of YihG in E. coli as an SlPlsC4 Ortholog

Analysis of the *E. coli* genome using the BLAST program with the SlPlsC4 amino acid sequence (accession number, BBD74888) as a query revealed that *E. coli* YihG (accession number, AIN34165), a putative membrane acyltransferase, is an SlPlsC4 ortholog. The pairwise sequence alignment using the EMBOSS Needle global alignment tool (https://www.ebi.ac.uk/Tools/psa/emboss_needle/) showed that the amino acid sequence of YihG shares 39.1% identity with that of SlPlsC4. YihG contains highly conserved acyltransferase motifs I–III but does not contain a motif IV like SlPlsC4 (Figure S1) [32,39]. These results suggested that YihG has a similar enzymatic activity to SlPlsC4.

3.2. Overexpression of YihG in an E. coli plsC Mutant Allows its Growth at Non-Permissive Temperatures

To examine whether YihG is a functional LPAAT homolog, we performed in vivo complementation assays using *E. coli* strain JC201, carrying a temperature-sensitive mutation in *plsC*. This strain grows normally at 30 °C, but not at higher temperatures [35]. The pBAD derivatives, expressing YihG or PlsC upon induction by L-arabinose, were introduced into JC201 cells, and the transformants were tested for their ability to grow at non-permissive temperatures. In plate assays, cells expressing the recombinant YihG grew well at 42 °C in the presence of 2% L-arabinose but showed no or marginal growth in the presence of 0 to 1% L-arabinose (Figure 1A and Figure S2). No complementation was observed in cells harboring the empty vector under the same conditions (Figure 1A and Figure S2). Thus, it was demonstrated that YihG, like PlsC, can act as an LPAAT in vivo.

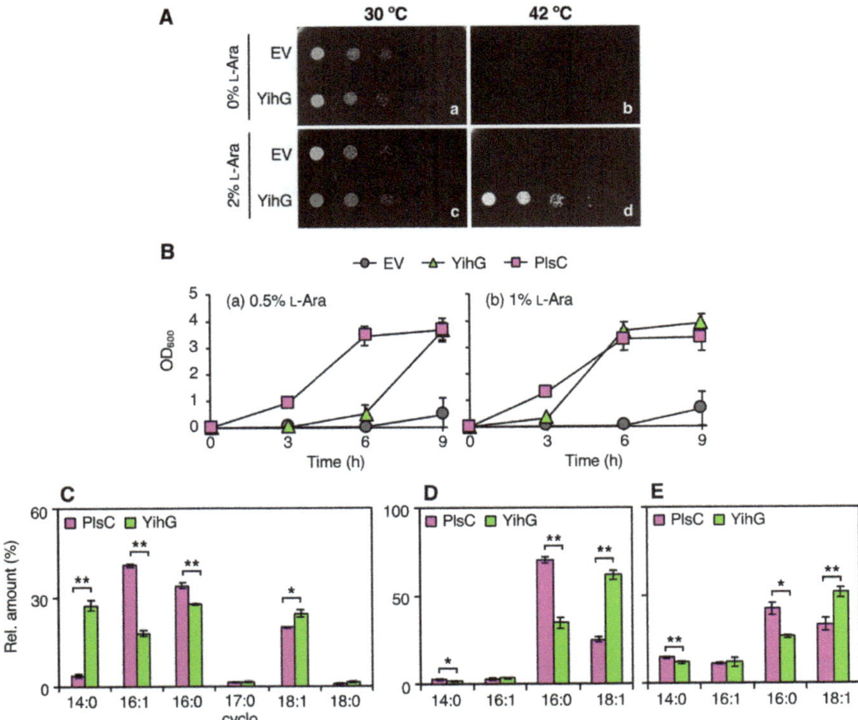

Figure 1. Overexpression of YihG and PlsC in JC201 cells. (**A**) Serially diluted JC201 cells harboring the pBAD-Cm^R empty vector (EV) or pBAD/*yihG-his$_6$* (YihG) were grown on LB plates containing 0% or 2% L-arabinose at 30 °C (a, c) and 42 °C (b, d) for 12–14 h. (**B**) JC201 cells harboring the pBAD-Cm^R empty vector (EV, gray), pBAD/*yihG-his$_6$* (YihG, light green), or pBAD/*plsC-his$_6$* (PlsC, magenta) were grown in LB containing 0.5% (a) or 1% (b) L-arabinose at 37 °C. Each data point is the average of three biological replicates ± SD. (**C–E**) The fatty acyl composition of phospholipids (PLs) from JC201 cells overexpressing YihG and PlsC. Total PL extracts were hydrolyzed by PLA2, and the resulting fatty acids were analyzed by GC–MS (**C**), whereas the resulting lysophosphatidylethanolamines (LPEs) and lysophosphatidylglycerols (LPGs) were analyzed by ESI–MS (D and E, respectively). The graphs show the relative amounts of the fatty acids and lysophospholipids (LPLs) from JC201 cells harboring pBAD/*plsC-his$_6$* (PlsC, magenta) or pBAD/*yihG-his$_6$* (YihG, light green). The cyclopropane derivative of 16:1 is indicated as 17:0cyclo (**C**). Data are shown as the mean ± SD (n = 3). Statistical analysis was performed using Welch's t-test: *, $p < 0.05$; **, $p < 0.01$.

To investigate the physiological contribution of YihG to cell growth, we compared the growth rates of JC201 cells overexpressing YihG with JC201 cells overexpressing PlsC (Figure 1B). JC201 cells harboring the empty vector hardly grew at 37 °C. However, cells overexpressing PlsC showed vigorous growth at 37 °C in the presence of both 0.5% and 1% L-arabinose. The growth rate of cells overexpressing YihG at 37 °C was much slower than that of cells overexpressing PlsC in the presence of 0.5% L-arabinose, but these strains grew similarly in the presence of 1% L-arabinose. These results indicated that YihG, expressed following L-arabinose induction, suppresses the growth defect of an *E. coli plsC* mutant at non-permissive temperatures in a quantity-dependent manner.

3.3. In vivo Substrate Specificity of YihG is Different from that of PlsC

To compare the in vivo substrate specificities of YihG and PlsC, we analyzed the fatty acyl composition of PLs from JC201 cells overexpressing YihG or PlsC grown at 37 °C in the presence of 1% L-arabinose. The *sn*-2 ester bonds of PLs were hydrolyzed by PLA2, and the resulting free fatty acids were analyzed by GC–MS, whereas the resulting lysophosphatidylethanolamines (LPEs) and lysophosphatidylglycerols (LPGs) were analyzed by ESI–MS. As shown in Figure 1C, the palmitoleoyl group (16:1) and the palmitoyl group (16:0) were more abundant in cells overexpressing PlsC, whereas the myristoyl group (14:0) and the *cis*-vaccenoyl group (18:1) were more abundant in cells overexpressing YihG, indicating that YihG preferably introduces 14:0 and 18:1 into the *sn*-2 position of PLs compared with PlsC. As shown in Figure 1D and E, 16:0-LPE and 16:0-LPG were the most abundant LPLs in cells overexpressing PlsC, whereas 18:1-LPE and 18:1-LPG were the most abundant LPLs in cells overexpressing YihG, indicating that YihG preferably introduces fatty acyl groups into the PLs containing 18:1 at the *sn*-1 position compared with PlsC. Thus, we concluded that YihG and PlsC have distinct in vivo substrate specificities.

3.4. Deletion of Endogenous YihG Affects Membrane PL Composition

To investigate the role of endogenous YihG in PL biosynthesis in *E. coli* cells, we analyzed the PL composition of Δ*yihG* cells grown at 37 °C in TB. The PL extracts were subjected to ESI–MS/MS analysis, and the compositions of PE and PG molecular species were determined by their ion peak intensities (Figure 2A,B). The fatty acyl chains in each PL species are summarized in Tables S3 and S4. In Δ*yihG* cells, the levels of 34:2-PE, 34:2-PG, 36:2-PE, and 36:2-PG containing 18:1 decreased compared with those in wild-type cells. The amounts of these PLs were increased when the YihG-expression plasmid was introduced into Δ*yihG* cells. These results suggest that YihG is involved in the biosynthesis of PLs containing 18:1. The amounts of 34:1-PE and 34:1-PG containing 18:1 also decreased in Δ*yihG* cells compared with those in wild-type cells. However, these PLs further decreased when the YihG-expression plasmid was introduced into Δ*yihG* cells. A possible reason for this will be described in the Discussion.

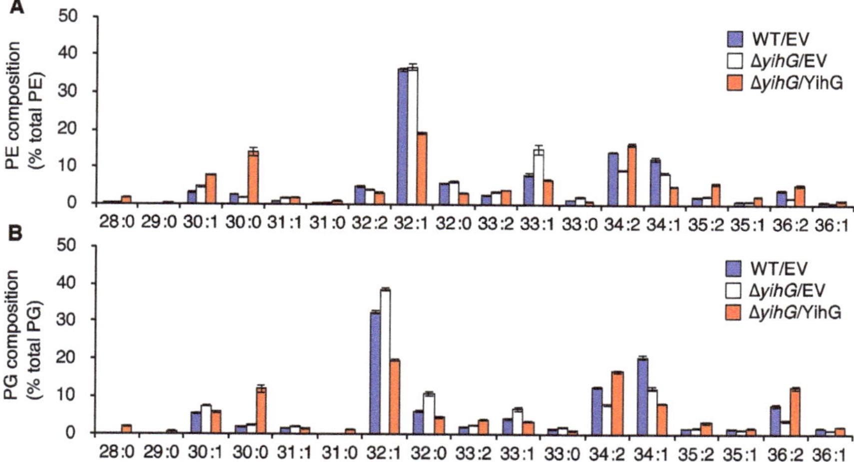

Figure 2. Effects of *yihG* disruption on PL composition. Total PLs extracted from cells grown at 37 °C to an OD_{600} of 0.6–0.8 were analyzed by ESI-MS/MS. The graphs show the composition of PE (**A**) and PG (**B**) derived from wild-type cells harboring the pBAD-Cm^R empty vector (WT/EV, blue) and Δ*yihG* cells harboring the pBAD-Cm^R empty vector (Δ*yihG*/EV, white) or pBAD/*yihG-his*$_6$ (Δ*yihG*/YihG, red). Data are shown as the mean ± SD ($n = 3$).

To confirm that endogenous YihG introduces 18:1 into the *sn*-2 position of PLs, we analyzed the *sn*-2 fatty acyl group composition of PLs from Δ*yihG* cells. The *sn*-2 ester bonds of PLs were hydrolyzed by PLA2. The resulting free fatty acids were subjected to GC–MS analysis, and the composition of the fatty acids was determined by their peak intensities (Figure 3). When compared with wild-type cells, the amount of 18:1 linked to the *sn*-2 position significantly decreased in Δ*yihG* cells. This result indicated that YihG introduces 18:1 into the *sn*-2 position of PLs. Consistently, the amount of 18:1 significantly increased when the YihG-expression plasmid was introduced into Δ*yihG* cells. The amount of 14:0 drastically increased when the YihG-expression plasmid was introduced into Δ*yihG* cells. The reason for this could be explained by the substrate preference of YihG for 14:0 (Figure 1C).

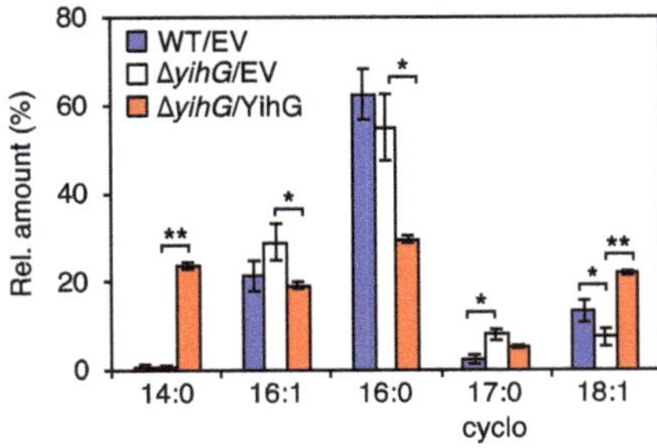

Figure 3. Effects of *yihG* disruption on the fatty acyl group at the *sn*-2 position of PLs. Total PL extracts were hydrolyzed by PLA2, and the resulting fatty acids were analyzed by GC–MS. The graph shows the relative amounts of the fatty acids from wild-type cells harboring the pBAD-Cm^R empty vector (WT/EV, blue) and Δ*yihG* cells harboring the pBAD-Cm^R empty vector (Δ*yihG*/EV, white) or pBAD/*yihG-his$_6$* (Δ*yihG*/YihG, red). Data are shown as the mean ± SD ($n = 3$). The cyclopropane derivative of 16:1 is indicated as 17:0cyclo. Statistical analysis was performed using Welch's *t*-test: *, $p < 0.05$; **, $p < 0.01$.

3.5. The Deletion of Endogenous YihG Causes Enhanced Swimming Motility

To investigate the physiological role of YihG in *E. coli* cells, we characterized the growth phenotype of Δ*yihG* cells. Δ*yihG* cells were cultured in LB and minimal medium at various temperatures. We found that the lack of YihG has no or only marginal effects on the growth of *E. coli* cells under these conditions (Figure S3).

We subsequently monitored bacterial motility by spotting the cell cultures on TB soft agar plates. *E. coli* BW25113 is known as a motility-impaired strain due to the low transcription level of motility-related genes compared with other motile *E. coli* strains [50]. However, interestingly, Δ*yihG* cells showed a much larger swimming halo compared with wild-type cells after 12 h incubation at 37 °C (Figure 4A), although the growth rate of Δ*yihG* cells was very similar to that of wild-type cells in TB at 37 °C (Figure S4). Motility of Δ*yihG* cells harboring the YihG-expression plasmid was also assayed. These cells were incubated on plates containing 0.02% or 0.2% L-arabinose for 15 h (Figure 4B). The induction of YihG with 0.02% L-arabinose did not affect the motility of Δ*yihG* cells. However, cells overexpressing YihG in the presence of 0.2% L-arabinose did not show the enhanced motility compared with cells harboring the empty vector. These results suggested that YihG is related to the bacterial motility.

We further checked the motility phenotype of Δ*yihG* cells in solution (Figure 4C,D, and Video S1). The cells were grown at 37 °C in TB to an OD_{600} of 0.6 to 0.8 and observed under the microscope at room temperature. Wild-type cells did not show any directional swimming motion but diffused freely in solution. In contrast, more than half of Δ*yihG* cells swam smoothly in solution, and only a small proportion of cells showed a jiggling motion. The fraction of the swimming cells was 59 ± 4% (mean ± SD, four assays), and their speed was 15 ± 5 μm/s (mean ± SD, 59 cells). These results indicated that the motility machinery, the flagellum, functions well in Δ*yihG* cells compared to wild-type cells.

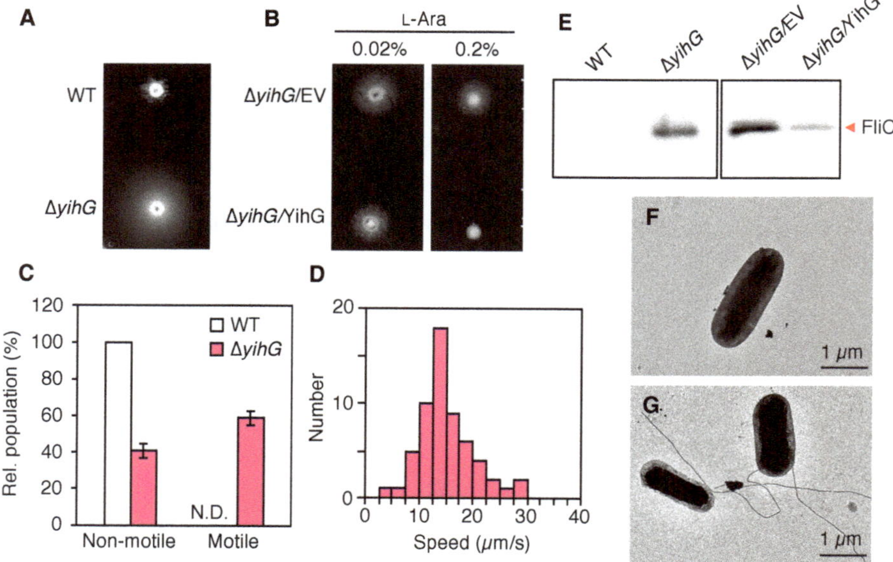

Figure 4. Characterization of the swimming motility of the *E. coli* cells. (**A**) Motilities of wild-type and Δ*yihG* cells on the 0.2% agar plate. The plate was incubated at 37 °C for 12 h. (**B**) Motilities of Δ*yihG* cells harboring the pBAD-Cm^R empty vector (Δ*yihG*/EV) or pBAD/*yihG-his*$_6$ (Δ*yihG*/YihG) on the 0.2% agar plates containing 0.02% or 0.2% L-arabinose. The plates were incubated at 37 °C for 15 h. (**C**) Motility of the *E. coli* cells in solution. Wild-type (white) and Δ*yihG* (pink) cells were grown at 37 °C to an OD_{600} of 0.6–0.8 and observed under the microscope. Relative populations of motile and non-motile cells were calculated. Data are shown as the average of four biological replicates ± SD. N.D., not detected. (**D**) Histogram of the swimming speed of motile Δ*yihG* cells. (**E**) Effects of *yihG* disruption on the flagellar formation of *E. coli* cells. Flagellin was separated from wild-type and Δ*yihG* cells grown at 37 °C to an OD_{600} of 0.6–0.8 and analyzed by Western blot analysis. Flagellin was also separated from Δ*yihG* cells harboring the pBAD-Cm^R empty vector (Δ*yihG*/EV) or pBAD/*yihG-his*$_6$ (Δ*yihG*/YihG) grown at 37 °C to an OD_{600} of 0.6-0.8 and analyzed by Western blot analysis. (**F**,**G**) TEM observation of the flagellar structures of the *E. coli* cells. Wild-type (**F**) and Δ*yihG* (**G**) cells were grown at 37 °C on TB agar plates and observed under the TEM.

E. coli cells swim by rotating flagellar filaments [51–54], and the speed of swimming cells is dependent on the number of flagellar filaments [55]. Therefore, we analyzed the flagellar expression of wild-type and Δ*yihG* cells. *E. coli* flagellin (FliC) was isolated from the cells and analyzed by Western blot analysis. As shown in Figure 4E, FliC was highly expressed in Δ*yihG* cells compared to wild-type cells. In addition, this phenotype was suppressed by introduction of the YihG-expression plasmid into the mutant and expression of YihG by addition of 0.2% L-arabinose. Finally, the flagellar formation was confirmed by electron microscopy. Wild-type cells showed no evidence of flagellar filaments (Figure 4F). In contrast, approximately half of Δ*yihG* cells had one or two flagellar filaments (Figure 4G). Taking these data together, we concluded that loss of YihG promotes the formation of the functional flagella, leading to the enhanced swimming motility.

4. Discussion

To assess the in vivo function of an SlPlsC4 ortholog of *E. coli* named YihG, we conducted a complementation assay using *E. coli* JC201 cells. Overproduction of YihG suppressed the temperature-sensitive phenotype of *E. coli* JC201 carrying a mutated *plsC* (Figure 1A,B and Figure S2), demonstrating that YihG has LPAAT activity. GC–MS analysis of *sn*-2 fatty acyl groups and ESI–MS analysis of 1-acyl LPLs (Figure 1C–E) revealed that YihG facilitates the synthesis of PLs containing 14:0 and 18:1 at the *sn*-2 position and 18:1 at the *sn*-1 position, whereas PlsC facilitates the synthesis of PLs containing 16:1 and 16:0 at the *sn*-2 position and 16:0 at the *sn*-1 position. Thus, the in vivo substrate specificities of YihG and PlsC are clearly different from each other. To our knowledge, this is the first report showing that *E. coli* expresses two functional LPAAT homologs with different substrate specificities. It is notable that deletion of *plsC* is lethal, and that endogenous YihG cannot suppress the growth defect of JC201 cells at non-permissive temperatures. This may be due to low expression of the endogenous YihG. In fact, an increase in the L-arabinose concentration led to a corresponding increase in the growth rate of JC201 cells harboring arabinose-controlled YihG-expression plasmid (Figure 1B).

Next, to unveil the physiological role of endogenous YihG in *E. coli*, we analyzed the effects of *yihG* disruption on PL biosynthesis in *E. coli* BW25113. ESI–MS/MS analysis of PLs (Figure 2) and GC–MS analysis of the *sn*-2 fatty acyl groups (Figure 3) revealed that endogenous YihG plays a major role in the synthesis of PLs containing 18:1 at the *sn*-2 position. However, 18:1 at the *sn*-2 position of PLs still remained in Δ*yihG* cells (levels were approximately a half of that in wild-type cells). Thus, endogenous PlsC also contributes to the synthesis of PLs containing 18:1 at the *sn*-2 position. In fact, PL species containing 18:1 (34:2-PE, 34:2-PG, 34:1-PE, 34:1-PG, 36:2-PE, and 36:2-PG) were still synthesized in Δ*yihG* cells, even though some of these PLs possibly contain 18:1 at their *sn*-1 positions. The amount of some PL species containing 18:1 (34:1-PE and 34:1-PG) further decreased when the YihG-expression plasmid was introduced into Δ*yihG* cells (Figure 2). These results may be due to overexpression of *yihG* under the control of the P_{BAD} promoter [41]. Overproduced YihG is supposed to facilitate incorporation of 14:0 on the acyl carrier protein of fatty acid synthase into 16:0-LPA to produce 30:0-PE and 30:0-PG (Figures 2 and 3) and cause deficiency of 16:0-LPA for the synthesis of other PL species, such as 34:1-PE and 34:1-PG. Consistent with this speculation, the abundance of 32:1-PE and 32:1-PG containing 16:0 also drastically decreased when YihG was overexpressed in Δ*yihG* cells (Figure 2).

Various bacteria, including *N. meningitidis*, *P. fluorescens*, *P. aeruginosa*, *R. capsulatus*, and *S. livingstonensis* Ac10, produce multiple LPAAT homologs [24–27,32]. However, there is no report describing the occurrence of an LPAAT homolog that preferentially produces PLs containing 18:1 in bacterial cells under physiological conditions. Thus, YihG is a new type of LPAAT homolog that plays a major role in the synthesis of PLs containing 18:1 at the *sn*-2 position under physiological conditions. Although YihG was identified as an ortholog of SlPlsC4 derived from *S. livingstonensis* Ac10, the in vivo substrate specificities of YihG and SlPlsC4 are different from each other. YihG prefers 14:0 and 18:1 as acyl donor substrates (Figure 1C), whereas SlPlsC4 prefers i13:0, 14:0, and i15:0 as acyl donor substrates [34]. This is possibly explained by the differences in the size and shape of their hydrophobic tunnels which accommodate the acyl chains of the acyl-acyl carrier protein or coenzyme A [22]. To understand the molecular basis of substrate specificity of YihG and SlPlsC4, biochemical and structural analyses should be conducted in the future.

In bacteria, PLs that contain low-melting-point fatty acids, such as monounsaturated fatty acids, contribute to the maintenance of membrane fluidity for their optimal growth at low temperatures. In fact, a decrease in monounsaturated fatty acids in the membrane leads to growth defects in some bacteria at low temperatures [7]. However, in the case of *E. coli*, the growth rate of Δ*yihG* cells at 18 °C was barely distinguishable from that of wild-type cells (Figure S3). Thus, PLs containing 18:1 produced by YihG are not essential for the growth of *E. coli* at low temperatures. Δ*yihG* cells still contain a large amount of 16:1 and 18:1, constituting 36.5% of the total fatty acids at the *sn*-2 position of PLs (Figure 3), and this may contribute to the maintenance of the membrane fluidity for the growth of this bacterium.

Interestingly, the deletion of YihG enhanced swimming motility and caused abnormal flagellar production (Figure 4). The expression of motility-related genes is tightly regulated in *E. coli*, and these genes are arranged in hierarchical order into three classes (I, II, and III) [51]. At the top of the hierarchy is the class I operon containing the *flhDC* genes called the master operon. Motile *E. coli* strains like MG1655 and W3110 contain insertion sequence elements IS*1* or IS*5* in the regulatory region of the *flhDC* promoter, and this leads to dramatic activation of the master operon that is otherwise silenced [56,57]. In contrast, poorly motile *E. coli* strains like BW25113 lack such an insertion element in the corresponding region [57,58]. According to this fact, the following reasons may account for the abnormal flagellar production by *E. coli* BW25113 in the absence of *yihG*. First, *flhDC* expression may be derepressed in Δ*yihG* cells due to envelope stress induced by changing membrane PL composition [59], given that a certain bacterial signal transduction system responds to the perturbations in membrane lipid properties [60–63]. Second, YihG may affect the cellular ratio of DnaA-ATP/DnaA-ADP [40], which not only regulates DNA replication [64,65], but also affects *flhDC* expression [66,67]. An ATP/ADP switch of DnaA defines its DNA-binding activity [68], and this conversion is in part promoted by highly unsaturated membrane PLs [69,70]. Thus, membrane PLs generated by YihG may regulate the cellular DnaA-ATP level, coordinating the DNA replication and the *flhDC* expression. YihG is also conserved in some γ-proteobacteria such as *S. typhimurium* and *V. cholerae*, and flagella play an important role in the adhesion and invasion of these pathogenic cells [71]. We propose that YihG may affect the virulence of these bacteria by regulating flagellar formation according to environmental changes. The details of the mechanism by which YihG affects the flagellar formation should be clarified in the future.

5. Conclusions

In this study, we characterized the in vivo function of YihG, a novel *E. coli* LPAAT homolog, and further investigated its physiological roles. Analysis of the fatty acyl composition of PLs from JC201 cells overexpressing YihG and PlsC revealed that YihG facilitates the synthesis of PLs containing 14:0 and 18:1 at the *sn*-2 position and 18:1 at the *sn*-1 position, whereas PlsC facilitates the synthesis of PLs containing 16:1 and 16:0 at the *sn*-2 position and 16:0 at the *sn*-1 position, thus demonstrating that *E. coli* has two functional LPAAT homologs with different substrate specificities. Analysis of the fatty acyl composition of PLs from Δ*yihG* cells revealed that endogenous YihG introduces 18:1 into the *sn*-2 position of PLs. Phenotypic analysis revealed that the lack of YihG causes high expression of FliC and enhances swimming motility but does not affect cell growth or morphology. In addition, the flagellar structure was observed only in Δ*yihG* cells. These results suggested that PlsC is responsible for the synthesis of the majority of membrane PLs, whereas YihG has more specific functions related to flagellar formation via modulation of the fatty acyl composition of membrane PLs.

Supplementary Materials: The following are available online at http://www.mdpi.com/2218-273X/10/5/745/s1, Figure S1: Multiple sequence alignment of YihG, SlPlsC4, and PlsC; Figure S2: Overexpression of YihG in JC201 cells; Figure S3: Growth characteristics of Δ*yihG* cells in LB and minimal medium; Figure S4: Growth characteristics of Δ*yihG* cells in TB; Table S1: Bacterial strains and plasmids used in this study; Table S2: Sequences of primers used in this study; Table S3: Fatty acyl groups of PEs from wild-type and Δ*yihG* cells harboring the plasmids found in Figure 2A; Table S4: Fatty acyl groups of PGs from wild-type and Δ*yihG* cells harboring the plasmids found in Figure 2B. Video S1: Observation of swimming Δ*yihG* cells in liquid medium.

Author Contributions: Conceptualization, Y.T. and T.K.; biochemical and molecular biological analysis, Y.T.; swimming analysis, Y.T. and M.N.; TEM observation, F.Y.; writing—original draft preparation, Y.T. and T.K.; writing—review and editing, Y.T., M.N., F.Y., T.O., J.K., and T.K.; supervision, M.N. and T.K.; project administration, T.K.; funding acquisition, M.N., T.O., and T.K. All authors have read and agreed to the published version of the manuscript.

Funding: This work was supported by JSPS KAKENHI (Grant Number JP16K04908 and JP17KT0026 to M.N., JP17K15259 to T.O. and JP18H02127 to T.K.).

Acknowledgments: TEM observation was performed in collaboration with the Analysis and Development System for Advanced Materials (ADAM) in the Research Institute for Sustainable Humanosphere, Kyoto University. We thank Michio Homma, Associate Seiji Kojima, and Kimika Maki (Nagoya University) for their discussions and technical support.

Conflicts of Interest: The authors have no conflict of interest to declare.

References

1. Cornell, B.A.; Separovic, F. Membrane thickness and acyl chain length. *Biochim. Biophys. Acta Biomembr.* **1983**, *733*, 189–193. [CrossRef]
2. Lewis, B.A.; Engelman, D.M. Lipid bilayer thickness varies linearly with acyl chain length in fluid phosphatidylcholine vesicles. *J. Mol. Biol.* **1983**, *166*, 211–217. [CrossRef]
3. Mykytczuk, N.C.S.; Trevors, J.T.; Leduc, L.G.; Ferroni, G.D. Fluorescence polarization in studies of bacterial cytoplasmic membrane fluidity under environmental stress. *Prog. Biophys. Mol. Biol.* **2007**, *95*, 60–82. [CrossRef]
4. Mostofian, B.; Zhuang, T.; Cheng, X.; Nickels, J.D. Branched-chain fatty acid content modulates structure, fluidity, and phase in model microbial cell membranes. *J. Phys. Chem. B* **2019**, *123*, 5814–5821. [CrossRef] [PubMed]
5. Nagamachi, E.; Shibuya, S.; Hirai, Y.; Matsushita, O.; Tomochika, K.; Kanemasa, Y. Adaptational changes of fatty acid composition and the physical state of membrane lipids following the change of growth temperature in *Yersinia enterocolitica*. *Microbiol. Immunol.* **1991**, *35*, 1085–1093. [CrossRef] [PubMed]
6. Annous, B.A.; Becker, L.A.; Bayles, D.O.; Labeda, D.P.; Wilkinson, B.J. Critical role of anteiso-C15:0 fatty acid in the growth of *Listeria monocytogenes* at low temperatures. *Appl. Environ. Microbiol.* **1997**, *63*, 3887–3894. [CrossRef]
7. Allen, E.E.; Facciotti, D.; Bartlett, D.H. Monounsaturated but not polyunsaturated fatty acids are required for growth of the deep-sea bacterium *Photobacterium profundum* SS9 at high pressure and low temperature. *Appl. Environ. Microbiol.* **1999**, *65*, 1710–1720. [CrossRef] [PubMed]
8. Knoblauch, C.; Sahm, K.; Jorgensen, B.B. Psychrophilic sulfate-reducing bacteria isolated from permanently cold Arctic marine sediments: Description of *Desulfofrigus oceanense* gen. nov., sp. nov., *Desulfofrigus fragile* sp. nov., *Desulfofaba gelida* gen. nov., sp. nov., *Desulfotalea psychrophila* gen. nov., sp. nov. and *Desulfotalea arctica* sp. nov. *Int. J. Syst. Bacteriol.* **1999**, *49*, 1631–1643.
9. Kawamoto, J.; Kurihara, T.; Yamamoto, K.; Nagayasu, M.; Tani, Y.; Mihara, H.; Hosokawa, M.; Baba, T.; Sato, S.B.; Esaki, N. Eicosapentaenoic acid plays a beneficial role in membrane organization and cell division of a cold-adapted bacterium, *Shewanella livingstonensis* Ac10. *J. Bacteriol.* **2009**, *191*, 632–640. [CrossRef]
10. Wang, F.; Xiao, X.; Ou, H.Y.; Gai, Y.; Wang, F. Role and regulation of fatty acid biosynthesis in the response of *Shewanella piezotolerans* WP3 to different temperatures and pressures. *J. Bacteriol.* **2009**, *191*, 2574–2584. [CrossRef]
11. Poerschmann, J.; Spijkerman, E.; Langer, U. Fatty acid patterns in *Chlamydomonas* sp. as a marker for nutritional regimes and temperature under extremely acidic conditions. *Microb. Ecol.* **2004**, *48*, 78–89. [CrossRef] [PubMed]
12. Yumoto, I.; Hirota, K.; Iwata, H.; Akutsu, M.; Kusumoto, K.; Morita, N.; Ezura, Y.; Okuyama, H.; Matsuyama, H. Temperature and nutrient availability control growth rate and fatty acid composition of facultatively psychrophilic *Cobetia marina* strain L-2. *Arch. Microbiol.* **2004**, *181*, 345–351. [CrossRef] [PubMed]
13. Drouin, P.; Prévost, D.; Antoun, H. Physiological adaptation to low temperatures of strains of *Rhizobium leguminosarum* bv. viciae associated with *Lathyrus* spp. *FEMS Microbiol. Ecol.* **2000**, *32*, 111–120. [PubMed]
14. Kennedy, E.P.; Weiss, S.B. The function of cytidine coenzymes in the biosynthesis of phospholipides. *J. Biol. Chem.* **1956**, *222*, 193–214. [PubMed]
15. Zhang, Y.M.; Rock, C.O. Thematic review series: Glycerolipids. Acyltransferases in bacterial glycerophospholipid synthesis. *J. Lipid Res.* **2008**, *49*, 1867–1874. [CrossRef] [PubMed]
16. Yao, J.; Rock, C.O. Phosphatidic acid synthesis in bacteria. *Biochim. Biophys. Acta Mol. Cell Biol. Lipids* **2013**, *1831*, 495–502. [CrossRef]
17. Parsons, J.B.; Rock, C.O. Bacterial lipids: Metabolism and membrane homeostasis. *Prog. Lipid Res.* **2013**, *52*, 249–276. [CrossRef]
18. Lightner, V.A.; Larson, T.J.; Tailleur, P.; Kantor, G.D.; Raetz, C.R.H.; Bell, R.M.; Modrich, P. Membrane phospholipid synthesis in *Escherichia coli*. Cloning of a structural gene (*plsB*) of the *sn*-glycerol-3-phosphate acyltransferase. *J. Biol. Chem.* **1980**, *255*, 9413–9420.
19. Green, P.R.; Merrill, A.H.; Bell, R.M. Membrane phospholipid synthesis in *Escherichia coli*. Purification, reconstitution, and characterization of *sn*-glycerol-3-phosphate acyltransferase. *J. Biol. Chem.* **1981**, *256*, 11151–11159.

20. Lu, Y.J.; Zhang, Y.M.; Grimes, K.D.; Qi, J.; Lee, R.E.; Rock, C.O. Acyl-phosphates initiate membrane phospholipid synthesis in Gram-positive pathogens. *Mol. Cell* **2006**, *23*, 765–772. [CrossRef]
21. Coleman, J. Characterization of the *Escherichia coli* gene for 1-acyl-*sn*-glycerol-3-phosphate acyltransferase (*plsC*). *MGG Mol. Gen. Genet.* **1992**, *232*, 295–303. [CrossRef]
22. Robertson, R.M.; Yao, J.; Gajewski, S.; Kumar, G.; Martin, E.W.; Rock, C.O.; White, S.W. A two-helix motif positions the lysophosphatidic acid acyltransferase active site for catalysis within the membrane bilayer. *Nat. Struct. Mol. Biol.* **2017**, *24*, 666–671. [CrossRef]
23. Li, Z.; Tang, Y.; Wu, Y.; Zhao, S.; Bao, J.; Luo, Y.; Li, D. Structural insights into the committed step of bacterial phospholipid biosynthesis. *Nat. Commun.* **2017**, *8*, 1691. [CrossRef]
24. Shih, G.C.; Kahler, C.M.; Swartley, J.S.; Rahman, M.M.; Coleman, J.; Carlson, R.W.; Stephens, D.S. Multiple lysophosphatidic acid acyltransferases in *Neisseria meningitidis*. *Mol. Microbiol.* **1999**, *32*, 942–952. [CrossRef]
25. Cullinane, M.; Baysse, C.; Morrissey, J.P.; O'Gara, F. Identification of two lysophosphatidic acid acyltransferase genes with overlapping function in *Pseudomonas fluorescens*. *Microbiology* **2005**, *151*, 3071–3080. [CrossRef]
26. Baysse, C.; Cullinane, M.; Dénervaud, V.; Burrowes, E.; Dow, J.M.; Morrissey, J.P.; Tam, L.; Trevors, J.T.; O'Gara, F. Modulation of quorum sensing in *Pseudomonas aeruginosa* through alteration of membrane properties. *Microbiology* **2005**, *151*, 2529–2542. [CrossRef]
27. Aygun-Sunar, S.; Bilaloglu, R.; Goldfine, H.; Daldal, F. Rhodobacter capsulatus OlsA is a bifunctional enzyme active in both ornithine lipid and phosphatidic acid biosynthesis. *J. Bacteriol.* **2007**, *189*, 8564–8574. [CrossRef]
28. Sato, S.; Kawamoto, J.; Sato, S.B.; Watanabe, B.; Hiratake, J.; Esaki, N.; Kurihara, T. Occurrence of a bacterial membrane microdomain at the cell division site enriched in phospholipids with polyunsaturated hydrocarbon chains. *J. Biol. Chem.* **2012**, *287*, 24113–24121. [CrossRef]
29. Yokoyama, F.; Kawamoto, J.; Imai, T.; Kurihara, T. Characterization of extracellular membrane vesicles of an Antarctic bacterium, *Shewanella livingstonensis* Ac10, and their enhanced production by alteration of phospholipid composition. *Extremophiles* **2017**, *21*, 723–731. [CrossRef]
30. Tokunaga, T.; Watanabe, B.; Sato, S.; Kawamoto, J.; Kurihara, T. Synthesis and functional assessment of a novel fatty acid probe, ω-ethynyl eicosapentaenoic acid analog, to analyze the *in vivo* behavior of eicosapentaenoic acid. *Bioconjug. Chem.* **2017**, *28*, 2077–2085. [CrossRef]
31. Kawai, S.; Kawamoto, J.; Ogawa, T.; Kurihara, T. Development of a regulatable low-temperature protein expression system using the psychrotrophic bacterium, *Shewanella livingstonensis* Ac10, as the host. *Biosci. Biotechnol. Biochem.* **2019**, *83*, 2153–2162. [CrossRef]
32. Cho, H.N.; Kasai, W.; Kawamoto, J.; Esaki, N.; Kurihara, T. Characterization of 1-acyl-*sn*-glycerol-3-phosphate acyltransferase from a polyunsaturated fatty acid-producing bacterium, *Shewanella livingstonensis* Ac10. *Trace Nutr. Res.* **2012**, *29*, 92–99.
33. Ogawa, T.; Tanaka, A.; Kawamoto, J.; Kurihara, T. Purification and characterization of 1-acyl-*sn*-glycerol-3-phosphate acyltransferase with a substrate preference for polyunsaturated fatty acyl donors from the eicosapentaenoic acid-producing bacterium *Shewanella livingstonensis* Ac10. *J. Biochem.* **2018**, *164*, 33–39. [CrossRef]
34. Toyotake, Y.; Cho, H.N.; Kawamoto, J.; Kurihara, T. A novel 1-acyl-*sn*-glycerol-3-phosphate *O*-acyltransferase homolog for the synthesis of membrane phospholipids with a branched-chain fatty acyl group in *Shewanella livingstonensis* Ac10. *Biochem. Biophys. Res. Commun.* **2018**, *500*, 704–709. [CrossRef]
35. Coleman, J. Characterization of *Escherichia coli* cells deficient in 1-acyl-*sn*-glycerol-3-phosphate acyltransferase activity. *J. Biol. Chem.* **1990**, *265*, 17215–17221.
36. Cao, G.J.; Pogliano, J.; Sarkar, N. Identification of the coding region for a second poly(A) polymerase in *Escherichia coli*. *Proc. Natl. Acad. Sci. USA* **1996**, *93*, 11580–11585. [CrossRef]
37. Mohanty, B.K.; Kushner, S.R. Residual polyadenylation in poly(A) polymerase I (*pcnB*) mutants of *Escherichia coli* does not result from the activity encoded by the *f310* gene. *Mol. Microbiol.* **1999**, *34*, 1109–1119. [CrossRef]
38. Heath, R.J.; Rock, C.O. A conserved histidine is essential for glycerolipid acyltransferase catalysis. *J. Bacteriol.* **1998**, *180*, 1425–1430. [CrossRef]
39. Yamashita, A.; Nakanishi, H.; Suzuki, H.; Kamata, R.; Tanaka, K.; Waku, K.; Sugiura, T. Topology of acyltransferase motifs and substrate specificity and accessibility in 1-acyl-*sn*-glycero-3-phosphate acyltransferase 1. *Biochim. Biophys. Acta Mol. Cell Biol. Lipids* **2007**, *1771*, 1202–1215. [CrossRef]

40. Babu, V.M.P.; Itsko, M.; Baxter, J.C.; Schaaper, R.M.; Sutton, M.D. Insufficient levels of the *nrdAB*-encoded ribonucleotide reductase underlie the severe growth defect of the Δ*hda E. coli* strain. *Mol. Microbiol.* **2017**, *104*, 377–399. [CrossRef]
41. Guzman, L.M.; Belin, D.; Carson, M.J.; Beckwith, J. Tight regulation, modulation, and high-level expression by vectors containing the arabinose P_{BAD} promoter. *J. Bacteriol.* **1995**, *177*, 4121–4130. [CrossRef]
42. Bligh, E.G.; Dyer, W.J. A rapid method of total lipid extraction and purification. *Can. J. Biochem. Physiol.* **1959**, *37*, 911–917. [CrossRef]
43. Dole, V.P. A relation between non-esterified fatty acids in plasma and the metabolism of glucose. *J. Clin. Investig.* **1956**, *35*, 150–154. [CrossRef]
44. Dole, V.P.; Meinertz, H. Microdetermination of long-chain fatty acids in plasma and tissues. *J. Biol. Chem.* **1960**, *235*, 2595–2599.
45. Ito, T.; Gong, C.; Kawamoto, J.; Kurihara, T. Development of a versatile method for targeted gene deletion and insertion by using the *pyrF* gene in the psychrotrophic bacterium, *Shewanella livingstonensis* Ac10. *J. Biosci. Bioeng.* **2016**, *122*, 645–651. [CrossRef]
46. Nishiyama, M. High-pressure microscopy for tracking dynamic properties of molecular machines. *Biophys. Chem.* **2017**, *231*, 71–78. [CrossRef]
47. Nishiyama, M.; Arai, Y. Tracking the movement of a single prokaryotic cell in extreme environmental conditions. In *The Bacterial Flagellum. Methods in Molecular Biology*; Minamino, T., Namba, K., Eds.; Humana Press: New York, NY, USA, 2017; Volume 1593, pp. 175–184.
48. Nishiyama, M.; Kojima, S. Bacterial motility measured by a miniature chamber for high-pressure microscopy. *Int. J. Mol. Sci.* **2012**, *13*, 9225–9239. [CrossRef]
49. Furuno, M.; Atsumi, T.; Yamada, T.; Kojima, S.; Nishioka, N.; Kawagishi, I.; Homma, M. Characterization of polar-flagellar-length mutants in *Vibrio alginolyticus*. *Microbiology* **1997**, *143*, 1615–1621. [CrossRef]
50. Wood, T.K.; González Barrios, A.F.; Herzberg, M.; Lee, J. Motility influences biofilm architecture in *Escherichia coli*. *Appl. Microbiol. Biotechnol.* **2006**, *72*, 361–367. [CrossRef]
51. Berg, H.C. The rotary motor of bacterial flagella. *Annu. Rev. Biochem.* **2003**, *72*, 19–54. [CrossRef]
52. Kojima, S. Dynamism and regulation of the stator, the energy conversion complex of the bacterial flagellar motor. *Curr. Opin. Microbiol.* **2015**, *28*, 66–71. [CrossRef]
53. Nakamura, S.; Minamino, T. Flagella-driven motility of bacteria. *Biomolecules* **2019**, *9*, 279. [CrossRef]
54. Miyata, M.; Robinson, R.C.; Uyeda, T.Q.P.; Fukumori, Y.; Fukushima, S.; Haruta, S.; Homma, M.; Inaba, K.; Ito, M.; Kaito, C.; et al. Tree of motility—A proposed history of motility systems in the tree of life. *Genes Cells* **2020**, *25*, 6–21. [CrossRef]
55. Sakai, T.; Inoue, Y.; Terahara, N.; Namba, K.; Minamino, T. A triangular loop of domain D1 of FlgE is essential for hook assembly but not for the mechanical function. *Biochem. Biophys. Res. Commun.* **2018**, *495*, 1789–1794. [CrossRef]
56. Barker, C.S.; Prüß, B.M.; Matsumura, P. Increased motility of *Escherichia coli* by insertion sequence element integration into the regulatory region of the *flhD* operon. *J. Bacteriol.* **2004**, *186*, 7529–7537. [CrossRef]
57. Tamar, E.; Koler, M.; Vaknin, A. The role of motility and chemotaxis in the bacterial colonization of protected surfaces. *Sci. Rep.* **2016**, *6*, 19616. [CrossRef]
58. Wang, X.; Wood, T.K. IS5 inserts upstream of the master motility operon *flhDC* in a quasi-Lamarckian way. *ISME J.* **2011**, *5*, 1517–1525. [CrossRef]
59. Inoue, K.; Matsuzaki, H.; Matsumoto, K.; Shibuya, I. Unbalanced membrane phospholipid compositions affect transcriptional expression of certain regulatory genes in *Escherichia coli*. *J. Bacteriol.* **1997**, *179*, 2872–2878. [CrossRef]
60. Lai, H.C.; Soo, P.C.; Wei, J.R.; Yi, W.C.; Liaw, S.J.; Horng, Y.T.; Lin, S.M.; Ho, S.W.; Swift, S.; Williams, P. The RssAB two-component signal transduction system in *Serratia marcescens* regulates swarming motility and cell envelope architecture in response to exogenous saturated fatty acids. *J. Bacteriol.* **2005**, *187*, 3407–3414. [CrossRef]
61. Amin, D.N.; Hazelbauer, G.L. Influence of membrane lipid composition on a transmembrane bacterial chemoreceptor. *J. Biol. Chem.* **2012**, *287*, 41697–41705. [CrossRef]
62. Keller, R.; Ariöz, C.; Hansmeier, N.; Stenberg-Bruzell, F.; Burstedt, M.; Vikström, D.; Kelly, A.; Wieslander, Å.; Daley, D.O.; Hunke, S. The *Escherichia coli* Envelope Stress Sensor CpxA responds to changes in lipid bilayer properties. *Biochemistry* **2015**, *54*, 3670–3676. [CrossRef]

63. Blanka, A.; Düvel, J.; Dötsch, A.; Klinkert, B.; Abraham, W.R.; Kaever, V.; Ritter, C.; Narberhaus, F.; Häussler, S. Constitutive production of c-di-GMP is associated with mutations in a variant of *Pseudomonas aeruginosa* with altered membrane composition. *Sci. Signal.* **2015**, *8*, ra36. [CrossRef]
64. Fuller, R.S.; Kornberg, A. Purified dnaA protein in initiation of replication at the *Escherichia coli* chromosomal origin of replication. *Proc. Natl. Acad. Sci. USA* **1983**, *80*, 5817–5821. [CrossRef]
65. Fuller, R.S.; Funnell, B.E.; Kornberg, A. The dnaA protein complex with the *E. coli* chromosomal replication origin (*oriC*) and other DNA sites. *Cell* **1984**, *38*, 889–900. [CrossRef]
66. Mizushima, T.; Tomura, A.; Shinpuku, T.; Miki, T.; Sekimizu, K. Loss of flagellation in *dnaA* mutants of *Escherichia coli*. *J. Bacteriol.* **1994**, *176*, 5544–5546. [CrossRef]
67. Mizushima, T.; Koyanagi, R.; Katayama, T.; Miki, T.; Sekimizu, K. Decrease in expression of the master operon of flagellin synthesis in a *dnaA46* mutant of *Escherichia coli*. *Biol. Pharm. Bull.* **1997**, *20*, 327–331. [CrossRef]
68. Speck, C.; Weigel, C.; Messer, W. ATP- and ADP-DnaA protein, a molecular switch in gene regulation. *EMBO J.* **1999**, *18*, 6169–6176. [CrossRef]
69. Yung, B.Y.M.; Kornberg, A. Membrane attachment activates dnaA protein, the initiation protein of chromosome replication in *Escherichia coli*. *Proc. Natl. Acad. Sci. USA* **1988**, *85*, 7202–7205. [CrossRef]
70. Castuma, C.E.; Crooke, E.; Kornberg, A. Fluid membranes with acidic domains activate DnaA, the initiator protein of replication in *Escherichia coli*. *J. Biol. Chem.* **1993**, *268*, 24665–24668.
71. Haiko, J.; Westerlund-Wikström, B. The role of the bacterial flagellum in adhesion and virulence. *Biology* **2013**, *2*, 1242–1267. [CrossRef]

MDPI
St. Alban-Anlage 66
4052 Basel
Switzerland
Tel. +41 61 683 77 34
Fax +41 61 302 89 18
www.mdpi.com

Biomolecules Editorial Office
E-mail: biomolecules@mdpi.com
www.mdpi.com/journal/biomolecules

www.ingramcontent.com/pod-product-compliance
Lightning Source LLC
LaVergne TN
LVHW071551170726
843515LV00008B/2262
* 9 7 8 3 0 3 6 5 1 3 3 8 6 *